Soil Health Restoration and Management

Ram Swaroop Meena
Editor

Soil Health Restoration and Management

Springer

Editor
Ram Swaroop Meena
Department of Agronomy
Institute of Agricultural Sciences (BHU)
Varanasi, Uttar Pradesh, India

ISBN 978-981-13-8569-8 ISBN 978-981-13-8570-4 (eBook)
https://doi.org/10.1007/978-981-13-8570-4

This Springer imprint is published by the registered company Springer Nature Singapore Pte Ltd.
The registered company address is: 152 Beach Road, #21-01/04 Gateway East, Singapore 189721, Singapore

Contents

About the Editor

Ram Swaroop Meena was born in a farmer family in VOP, Harsana, Tehsil, Laxmangarh, Alwar District, Rajasthan, India. Dr. Meena had his schooling in the same village and graduated in Agriculture in 2003 from the Sri Karan Narendra Agriculture University, Jobner, Jaipur (Rajasthan). He has obtained his Master's and Doctorate in Agronomy from the Swami Keshwanand Rajasthan Agricultural University, Bikaner (Rajasthan), securing first division in all the classes with triple NET, Junior Research Fellowship (JRF), and Senior Research Fellowship (SRF) from the Indian Council of Agricultural Research, and RGNF Award from the University Grants Commission, Government of India (UGC, GOI). He has been awarded Raman Research Fellowship by the Ministry of Human Resource Development (MHRD), GOI. He has completed his postdoctoral research on soil carbon sequestration under Prof. Rattan Lal, distinguished scientist, and Director, Carbon Management and Sequestration Center (CMASC), Ohio State University, USA. He is working on soil sustainability, crop productivity, and resources use efficiency under current climatic era. He has supervised 17 postgraduate and 4 Ph.D. students, and he has 9 years of research and teaching experience at the undergraduate/postgraduate/Ph.D. level. He is working on the three externally funded running projects including the Department of Science and Technology (DST), GOI, and involved in many academic and administrative activities going on at the institute/university level. He has published more than 100 research and review papers in peer-reviewed reputed journals and contributed in the edited books with 25 book chapters at national and international levels. He has published 4 books on the national level and another 6 on the international level. He has worked as an expert in the National Council of Educational Research and Training (NCERT), MHRD, GOI, to develop the two books for school education at XI and XII standards. He has been awarded several awards, namely, Young Scientist, Young Faculty, Global Research, Excellence in Research, Honorable Faculty Award, etc. He is a member of 9 reputed national and international societies and is working as a General Secretary, Editor, and Member of the editorial board in 12 national and international peer-reviewed reputed journals. He has attended several national and international conferences in the country and abroad. He is contributing to the agricultural extension activities on farmers' level as associate coordinator in training, meetings, workshops, and farmers' fair.

Carbon Footprint in Eroded Soils and Its Impact on Soil Health

1

Mehraj U. Din Dar, Shakeel Ahmad Bhat, Ram Swaroop Meena, and Aamir Ishaq Shah

Abstract

Climate change, soil degradation, and losses in the biodiversity have led the soil to become one of the most vulnerable resources on the earth. The tremendous scientific advancement made until now has made possible through the protection, monitoring, and surveillance of soil resources at national and global levels. However protection and management of soil resources still have to face the complex challenges, which prevent the effective planning of policies in the sector and implementation, that vary generally from place to place. Though, there is still not sufficient support for the protection and sustainable management of the soil resources in the world. The soils contain appreciable amount of terrestrial carbon (c), which plays an essential role in its balance at global level through the regulation of dynamic, biogeochemical processes and exchange of greenhouse gases (GHGs) with the atmosphere. Soil organic carbon (SOC) stocks are estimated to be 1500 ± 230 Gt C in the first meter of land, almost twice the atmospheric 828 Gt C as carbon dioxide (CO_2). Burning of fossil fuels, the use of the earth and change of land cover (which includes agriculture), is the largest anthropogenic source of C in the atmosphere and within the agriculture systems, the land have a global source of

M. U. D. Dar (✉)
Department of Soil and Water Engineering, Punjab Agricultural University, Ludhiana, Punjab, India
e-mail: mehrdar-coaeswe@pau.edu

S. A. Bhat
College of Agricultural Engineering and Technology, Sher i Kashmir University of Agricultural Sciences and Technology of Kashmir, Srinagar, Jammu and Kashmir, India

R. S. Meena
Department of Agronomy, Institute of Agricultural Sciences (BHU), Varanasi, Uttar Pradesh, India

A. I. Shah
Department of Hydrology, Indian Institute of Technology(IIT), Roorkee, Uttrarakhand, India

R. S. Meena (ed.), *Soil Health Restoration and Management*,
https://doi.org/10.1007/978-981-13-8570-4_1

GHGs. These processes and emissions are highly influenced by the use of land pattern, land use change, plant cover, and soil management. The SOC stocks in the upper layers of the soil (800 Gt C in 0–40) cm) are particularly sensitive and receptive to such changes in land use and management, which provides the chance to influence the amount of CO_2 in the atmosphere. This can be attained by keeping existing soil C stocking soils with high soil organic C content or by soil C sequestration. The aim of this chapter is to produce the obtainable information of C stocks in various types of soils and agroclimatic zones. Soil erosion/soil degradation and main management operations and strategies influence C sequestration, and vast-scale policy interventions are needed in Indian environment.

Keywords
Carbon pool · Carbon footprints · Climate change dynamics · Soil degradation · Soil erosion · Soil organic carbon

Abbreviations

CDIAC	Carbon Dioxide Information Analysis Center
FAI	Fertilizer Association of India
FAO	Food and Agriculture Organization
GHGs	Greenhouse Gases
ICRISAT	International Crops Research Institute for the Semi-Arid Tropics
LIC	Lithogenic Inorganic Carbon
Mha	Million Hectares
MRT	Mean Residence Time
NBSS LUP	National Bureau of Soil Survey and Land Use Planning
OC	Organic Carbon
POC	Particulate Organic Matter
RMPs	Recommended Management Practices
SAT	Semiarid Tropics
SOC	Soil Organic Carbon
SOM	Soil Organic Matter

1.1 Introduction

India remains primarily an agricultural economy. With the total geographical area of 328.7 million hectares (Mha), 141 Mha area is purely cultivated for agriculture, from which only 63 Mha or 44% is net irrigated area producing over 56% of total food grains (FAI 2011). While it generates only 44% of total food, pluvial agriculture is crucial as they contribute to the production of cereals (90%), legumes (87%), and oilseeds (74%). These products, made under rainfed agricultural ecosystems, are important for ensuring food and nutrition security for the growing population. Due to the rainfall, temperature regimes, original materials, vegetation, and topography, the country has a great diversity of land use patterns. India is equipped with a broad range of soil types, viz., inceptisol (95.8 Mha), entisols (80.1 Mha), alfisols (79.7 Mha), vertisols (26.3 Mha), aridisols (14.6 Mha), mollisols (8, 0 Mha), ultisols (0.8Mha), oxisols (0.3 Mha), and various soils (23.1 Mha). Soil deterioration is the main limitation that threatens the food security of the country. Of the 328.7 Mha of the total area, 120.72 Mha (36.70%) is degraded to a certain extent. Among the main degradation processes, 73.27 Mha (60.27%) is influenced by water erosion, 12.40 Mha (10.30%) of wind erosion, 17.45 Mha (14.50%) by chemical degradation, and 1.07 Mha (0.9%) due to physical degradation (ICAR 2010). Another 18.2 million hectares (5.5%) area is unsuitable for agriculture due to the limitations of the polar ice caps, saline solution, barren mountains, and stony physiography (Sehgal and Abrol 1994). Due to the long-term use of extractive agricultural practices in mountain areas, most of the soil degrades to a greater degree and is distinguished by a low concentration of organic matter in the soil (SOM), structure and inclination of unfavorable soil, low nutrient reserves, and marginal productivity (Sharma et al. 1994; Ashoka et al. 2017). The scarcity of land and the low soil per capita require the farming of marginal lands, albeit with high ecological, economic, and environmental costs. Semiarid tropics (SAT) include regions in which precipitation overshoots potential evapotranspiration for only 2 ± 4.5 months (SAT dry) or 4.5 ± 7 months (SAT wet) with an average temperature in each case of N18 °C (Troll 1965). Semiarid or arid tropics cover almost 6.5 million km^2 of land in 55 countries around the world and are inhabited by more than 2000 million people (ICRISAT 2010). About 60% of these arid lands are found in developing countries. Approximately 170 million hectares of the SAT area are located in India. The average cereal yields in these areas of the developing countries are 1.5 Mg/ha or about half of those found in irrigated areas (3.1 Mg/ha). The uncertainty of precipitation, degraded land, low application of inputs, biotic stresses, inadequate crop management, lack of a targeted awareness program, and adequate political platform are the major reasons of this little productivity (Rockstrom et al. 2010; Buragohain et al. 2017). Most soils are heavily deprived in SOC reserves. The concentration of SOC is a measure of soil quality and yield. Therefore, restoring the quality of deteriorated soils requires a higher concentration of SOC in the root area. The term "soil C sequestration" involves the elimination of atmospheric C dioxide from plants via photosynthesis and biomass transfer C to the soil as compost. The plan is to augment SOC density, enhance distribution depth, and stabilize SOC encapsulate within stable microaggregates for C to be prevented from microbial processes and have a long dwell time signifying (MRT). In this manner, the adoption of recommended management practices (RMPs) in agroecosystems and those that causes a positive C balance on the soil is an important strategy for SOC/terrestrial seizure. Therefore, the changes in land use can be an essential tool for the seizure of the SOC, considering that the misuse of land and the mismanagement of land have caused the deprivation of SOC with the consequent emission of C and other greenhouse gases (GHG) in the atmosphere. This could substantially offset the emissions of fossil fuels (Kaupppi et al. 2001) and also enhance soil quality and create ecosystem services. SOC's ability to dissipate atmospheric carbon dioxide (CO_2) can be highly improved when deteriorated land and ecosystems are reinstated and small agricultural land becomes a land repair use or replanted with perennial vegetation. The addition of SOC in the subsoil through the creation of plants with a deep root system or an illuviation through pedogenesis can raise its MRT. The application of ecological practices to the management of natural resources (e.g., the nutrient cycle, C budget soil mix favorable for macro invertebrates, and greater soil biodiversity) can be an essential factor for the improvement of soil quality and capturing of SOC percent (Lavelle 2000). SOC concentration on the top layer generally raises as a solid bio input (Graham et al. 2002; Dadhich and Meena 2014), even though the specific empirical relationship depends on the soil moisture content and temperature regimes. The nutrient availability, which includes (NPKS), consistency, and climate besides the amount of input and the quality of biomass may also be important for finding out the SOC group. Major of the research conducted so far on SOC seizure in agroecosystem lands is limited to cold and temperate regions. There is a small information obtainable on this topic in the tropical and subtropical regions of India (Velayutham et al. 2000; Lal 2004; Srinivasarao et al. 2011, 2012a, b, c, d, e, f, 2013a, b; Dadhich et al. 2015). Using the extensive database of the National Bureau of Soil Survey and Land Use Planning (NBSS & LUP), Bhattacharyya et al. (2009) estimated that the soil of India contains only 9.55 and 24.04 Gt organic C (SOC) out of about 13.69 and 46.50 Gt total C is in the first 0.3 and 1 m of soil, respectively. Sreenivas et al. (2016) estimated the SOC and inorganic soil C, and the total soil C size is 22.72 ± 0.93 and 12.83 ± 0.2, respectively, and 1.35 and 35.55 ± 1.87 Gt, in the first 1 m in India. Similarly, using modeling approaches, Falloon et al. (2007) determined C for Indian soils is in the range of 6.5–8.5 Gt, while Banger et al. (2015) estimated the stock between 20.5 and 23.4 Gt. Therefore, the Indian share to the global SOC set is in the range of 20–25 Gt for the first 1 m. The C seizure rate for India would be around 23–28 per thousand compared to the global need of 4 per thousand. A closer look at the regional distribution of stocks of SOC predominant agro observation in India has shown that semiarid region (116.4 mha) and subregion (105.0 mha) contribute nearly about 56% of the total supply of SOC (Bhattacharyya et al. 2008; Datta et al. 2017a). Forests (9.38 Gt), monoculture lands, and double crops (8.81 Gt) contribute 80% to the total reserve of SOC (Sreenivas et al. 2016). Incidentally, Sreenivas et al. (2016) also reported the maximum average density of SOC in land under plantation is 253 t ha^{-1}, then by forests (139.9 t ha^{-1}), and agricultural land (58.5–67.4 t ha^{-1}). Given the decrease in forest areas and the accessibility of 147 Mha of deteriorated land (Maji et al. 2010), requiring instantly rehabilitation measures can make a needly attempt to raise plantations on land to improve the availability of SOC in the country. Long-term experiments (25–40 years) with balanced fertilizer (NPK) and 05–10 Mg ha^{-1} of organic waste in cultivated lands showed an increase in the content of only 10–20% SOC initial soil value (Mandal 2011; Pathak et al. 2011; Datta et al. 2017b). Assuring C accumulation rate of only 0.13–0.27 t C ha^{-1} $year^{-1}$ (equivalently, 0043–0089 Gt $year^{-1}$) is based on different rice culture systems (Mandal et al. 2007, 2008). Lal (2004) also estimated the C seizure rates in India vary from 0.024 to 0.036Gt/ year. Recently Banger et al. (2015) reported C seizure rates in a range of 0027–0045 Gt year^{-1}. Therefore, these catch rates of C 0024–0089 Gt year^{-1} and about 25 Gt of SOC stock may be possible to contribute about 1–4 per thousand SOC lost prevalently through emission C. Long-term experiments have shown that it does not cause exhaustion and maintaining the SOC level requires at least 0.31–5.16 t ha^{-1} year^{-1} to be added to the soil via crop residues in different agrochemicals. Ecological areas of the country depend mainly on the dryness index locations (Mandal 2011; Dhakal et al. 2015). The promotion of pulses (for the single accumulation of SOC property), deviating a part of the fertilizer contribution and efficient use of available plant residues (679 Mt. per year) and of solid urban waste (64.8 Mt. year) with green manure and appropriate cultivation systems (based on rice), can contribute to improving or at least slowing downward trends in the SOC warehouse in India soil. Land covered landscapes on earth are geomorphologically dynamic and within the continuous actions of erosion and deposition. In recent years, researchers have made advancement in understanding how the geomorphological processes influence C storage in soil (C) (Doetterl et al. 2015; Kumar et al. 2017a). Overall soil erosion (mainly water flow and spray) has been estimated to redistribute 1.6 Gt of organic SOC annually (Müller-Nedebock and Chaplot 2015), and much of the eroded C becomes to deposit locally. Yet, less is familiar about how minerals in these sediments interact and affect eroded SOC redistribution. Although the goal and degree of stabilization of the eroded SOC during sediment transport changes greatly (Kirkels et al. 2014; McCorkle et al. 2016; Dhakal et al. 2016), some researches have shown that sedimentation of eroded soils can accelerate organic C mineral protection (OC) (Doetterl et al. 2015; Wang et al. 2014). Behind the erosive SOC control is time, which exerts a strong influence of the first order on both biological and chemical activity (Amundson 2001; Giardina et al. 2014; Kumar et al. 2018). The climatic and geomorphological conditions, the collaborative entries of C from plants in soils, and the microbial decomposition (Luizão et al. 2004; Luo et al. 2017; Yoo et al. 2006) the same stated for alteration and erosion processes (Dixon et al. 2016; Riebe et al. 2004; Zapata-Rios et al. 2015) which control soil texture (Attal et al. 2015; Riebe et al. 2015) and thicknesses (Follain et al. 2006; Scarpone et al. 2016). Times and mechanisms of physical soil erosion change not only with topography but also with climate (Montgomery and Dietrich 2002), and the classification of substantial particles occurs in some types of erosion mechanisms (Doetterl et al. 2015; Heimsath et al. 2002; Shi et al. 2012). For example, soil erosion of decomposed and transported macroaggregated water media redistributes particles preferably enriched with OC from upstream positions, thus increasing SOC content in deposition areas (Kinnell 2001; Oakes et al. 2012; Yadav et al. 2018a). Yet, it is known that the transport and loss of particulate organic C (POC) from water erosion are significantly influenced by time (Müller-Nedebock and Chaplot 2015). These are examples that highlight a critical knowledge barrier in the climate which can transmit its signal on erosion of SOC redistribution landscape by regulating the mechanisms of prevalent erosion. The importance of combining the biological perspectives against contrasting geomorphic climate as the main driver of SOC dynamics is underlined by

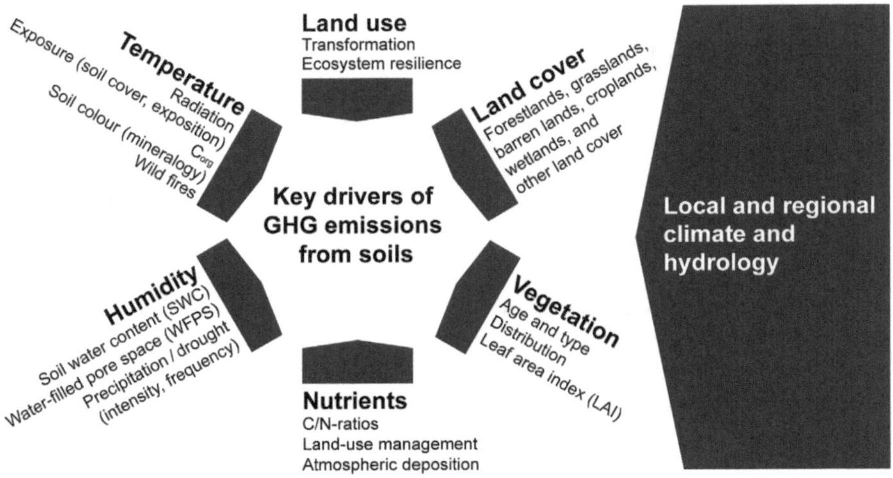

Fig. 1.1 Key drivers of GHG emissions from soils. (Oertel et al. 2016)

increasing appreciation of mineralogical controls on C cycling in the soil (Kleber et al. 2015; Lawrence et al. 2015), and recently it was shown that SOC is subject to interactive effects between climate and geochemical factors (e.g., soil mineralogy) and not only climatic factors (Doetterl et al. 2015; Kramer and Chadwick. 2016). The content of the clay and iron oxide (FeO) in soils, which often impairs soil C substitution (Doetterl et al. 2015; Lawrence et al. 2015; Masiello et al. 2004), is affected by the regional climate (Dahlgren et al. 1997;Long et al. 2016; Kumar et al. 2017b) and varies with the locations of the scale topographic landscape (Berhe et al. 2014; Doetterl et al. 2015; Yadav et al. 2017a). Finally, the topography may be associated with dramatic microclimate disparity, which may influence soil properties and plant biomass (Luizão et al. 2004; Yoo et al. 2006; Meena and Meena 2017), and the stabilization of organic matter (Berhe et al. 2014; Chaplot et al. 2001) between photosynthesis and ecosystem respiration. Figure 1.1 shows the different factors of greenhouse gas emissions from the soil.

1.2 Status of Land Degradation in India

Land degradation is not properly addressed; still raising awareness is essential, as land management decisions can guide for more sustainable and adaptive farming system. Total geographical area of India is 328.7 Mha from which 304.9 Mha includes the reporting area with an area of 264.5 Mha used for agriculture, forestry, grazing, and other biomass production. The acuteness and extent of land degradation in India have been formerly evaluated by many organizations (Table 1.1). As reported by the National Bureau of Land Survey and Land Use Planning (NBSS&LUP 2004), about 146.8 Mha area is degraded. The water erosion is the main degradation issue in India, with consequent soil loss and soil deformation.

Table 1.1 Extent of land degradation in India, as assessed by different organizations

Organizations	Assessment year	References	Degraded areas (Mha)
National Commission on Agriculture	1976	NCA (1976)	148.1
Ministry of Agriculture-soil and water conservation division	1978	MoA (1978)	175.0
Department of Environment	1980	Vohra (1980)	95.0
National Wasteland Development Board	1985	NWDB (1985)	123.0
Society for Promotion of Wastelands Development	1984	Bhumbla and Khare (1984)	129.6
National Remote Sensing Agency	1985	NRSA (2000)	53.3
Ministry of Agriculture	1985	MoA (1985)	173.6
Ministry of Agriculture	1994	MoA (1994)	107.4
NBSS&LUP	1994	NBSS&LUP (1994)	187.7
NBSS&LUP	2004	NBSS&LUP (2005)	146.8

According to first approximation, the examination of obtainable soil loss data, average soil erosion rate was nearly about 16.4 tons ha^{-1} $year^{-1}$, with a total annual soil loss of about 5.3 billion tons of soil from the whole country. Almost 29% of the total land eroded is permanently deposited in the sea, while 61% is moved from one place to another, and the rest 10% is deposited in the water basins.

Most of the area cultivated in our country has been cultivated for hundreds of years and had achieved its topmost impoverishment many years ago. In this regard, it should be recalled that the combined lack of nitrogen is the limiting factor of India (Report of the Royal Commission on Agriculture in India) (Royal Commission on Agriculture 1928). The Green Revolution led to technological progress that led to the employment of short-lived high-productivity varieties which assisted in intensifying land use in a year by augmenting the area fed by irrigation and much increased in the utilization of chemicals, e.g., pesticides and fertilizers. Agricultural production in our country has raised from 50 to 250 Mt. more in the previous five decades. However, it has other effects, including the environmental pollution and loss of plant biodiversity. Extensive soil degradation resulting from unsuitable agricultural practices has a direct and negative influence on food security and income of farmers. Primarily, the degradation is resulted from erosion, with consequent loss of soil due to the action of wind and water, or refining, with consequent salinization. Maheswarappa et al. (2011) noted that (i) the C-sustainability index was high in 1960 and was indicative of the minimal use of input before the start of the Green Revolution and (ii) subsequently, the C-sustainability index reduced due to more revenue based on C, where there is a linear relationship between inputs and outputs of C activity and agricultural practices can cause soil deterioration in various ways depending on land use, cultivated and management practices adopted. Some of the most common causes of soil degradation include cultivation in fragile deserts and marginal soils, without conservation measures, clear-cutting deforestation, and agricultural impoverishment of soil nutrients due to poor agricultural practices, overgrazing of pastures, urban expansion, commercial development, and soil, including the disposal of industrial waste in arable land. Intensive farming practices, in particular wheat (*Triticum aestivum* L.) and rice (*Oryza sativa* L.) in India, are practically extracting soil nutrients. Consumption ratio and unbalanced 6.2: 4: 1 (N: P: K) 1990–1991 were extended to 7: 2.7: 1 in 2000–2001 and 5: 2: 1 in 2009–2010 compared to a goal ratio of 4: 2: 1. As wheat increases with time, the number of defective elements in Indian soil raised from one (N) in 1950 to nine (N, P, K, S, B, Cu, Fe, Mn, and Zn) in 2005–2006; although the utilization of fertilizers has augmented several times, overall consumption is still less in majority of the countries. The diffuse Zn deficit, followed by S, Fe, Cu, Mn, and B, is common throughout India. Every year nearly about 20 Mt. of the three main nutrients are eliminated by growing cultures (Tendon 1992; Meena et al. 2016a), but the corresponding addition through inorganic fertilizers and organic fertilizer does not reach this crop. Another estimate suggests that over the previous 50 years, the interval between extraction and nutrient additions were 8–10 Mt. N + P_2O_5 + K_2O per year (Tandon 2004). Moreover, the loss of nutrients via soil erosion is also the reason for the depletion of soil fertility, which represents an annual loss of eight million tons of plant nutrients of 5.3 billion tons of soil loss (Prasad et al. 2000; Meena et al. 2018a). NBSS and LUP data (Sehgal and Abrol 1994) showed that almost 3.7 Mha experience nutrient loss and/or SOM (soil organic matter) exhaustion. Burning crop residuum for heating, cooking, or simply eliminating them is a widespread issue in India and contributes to the loss of SOM. As per Ministry of the New and Renewable energy (MNRE 2009), ~ 500 Mt. of crop residues and ~ 185 Mt. generated is burned each year. Cultivation production of waste is higher in Uttar Pradesh (60 Mt) followed by Punjab (51 Mt) and Maharashtra (46 Mt). Among the different crops, cereals generate 352Mt residue followed by textile plants (66 Mt), oil (29 Mt), legumes (13 Mt), and sugarcane (*Saccharum officinarum*) (12 Mt). The rice ~ (34%) and wheat ~ (22%) are the dominant cereals that contribute to the generation of crop residues (NAAS 2012). The development of irrigation channels (such as the Indira Gandhi Nahar Project) has been allied with common problems of ponding and salinity in the regions of the Indo-Gangetic Plains (IGP). In arid, semiarid, and subhumid areas, vast areas have become sterile due to the formation of saline-sodium soils owing to poor irrigation management. The fracture of the soil due to lack of irrigation management leads to the bypass water flow and the consequent leaching of the nitrate (Barman et al. 2013; Verma et al. 2015b).

1.3 Soil Organic and Inorganic Carbon Stocks in India

1.3.1 Soil Organic Carbon

Several external factors contribute to the change in soil SOC levels. They are climate, hydrology, parent material, inherent soil fertility, vegetation patterns, biological activity, and land use (Jenny 1941). The chief renewable source to enhance the soil with organic C is the accumulation of phyto-mass. Yet, the expansion of farmland to meet global food requirements implies a cost of reducing C stocks (West et al. 2010; Yadav et al. 2017c). Normally, the processes that operate in the organic matter cycle appear to be alike in dry or hot areas and in temperate zones where C levels are generally higher (Batjes and Sombroek 1997; Syers 1997). The deposition of C is favored by the low temperatures, the acidic environment, and the anaerobic conditions, which do not favor the decomposition of the organic substance. Inventory of the storage of organic C in soil and its spatial distribution, as well as temporal dynamics, is essential for understanding the C cycle and climate change, as well as for developing future land use policies. As per Lal et al. (1998), an augment in the SOC content of 0.01% may substantially reduce the negative consequences of the annual rise in atmospheric CO_2 concentration.

1.3.2 Inorganic Soil Carbon

The SIC set can be grouped into inorganic lithogenic C (LIC) and pedogenic inorganic C (PIC). In arid and semiarid regions, covering one third of the earth's surface, SIC groupings and dynamics are important because the rate of accumulation of SIC is generally greater than in other biomes (Lal 2006). Of the two SIC groups, the PIC group plays a crucial role in the global C cycle. Up to now, studies on conservation dynamics and SIC have focused mainly on local or regional assessments (Monger and Matrinez-Rios 2000; Rasmussen 2006a, b; Singh et al. 2007). National or global SC estimates have been poorly documented with respect to the SOC group estimates (Lal et al. 1998; Guo et al. 2006; Meena et al. 2016b). Research has also demonstrated the experience of significant changes in the SIC series of different soil types linked to climate change (Adams and post 1999; Landi et al. 2003), land use (Mikhailova and Post 2006; Papiernik et al. 2007; Sartori et al. 2007; Yadav et al. 2017b) and deposition of atmospheric nutrients (Goddard et al. 2007).

1.4 Carbon Footprints in Degraded Soils

Soil degradation, reduction of soil quality by reducing biomass productivity and the ability to moderate the environment, has serious adverse effects on the whole SOC. In other terms, the low level of SOC in Indian soils is partially due to the serious issue of soil degradation. Measurement of soil degradation with different processes changes widely (Biswas et al. 1991). GLASOD report, (FAO 1994) shows that the affected area is 32.8 Mha from water erosion, nearly 10.8 Mha from wind erosion, 29.4 Mha from fertility decrease, 3.1 Mha from salinization, and 4.1 Mha from flooding. The total area damaged by various degradation processes is 45 Mha. Another problem is desertification of 102 Mha of 162 or 63% of the dryland area that is prone to some degree of desertification (Table 1.2). The main cause of decreases in the SOC group of degraded soils is the decrease in productivity of biomass and low amounts of plant residues returned to soil and roots. A common example of the low level of SOC is in Haryana soils affected by salt, Andhra Pradesh, and

Table 1.2 Estimates of soil degradation in India. (FAO 1994; Yadav 1996; Karale et al. 1991)

Process	GLASOD (Mha)	Others (Mha)
Water erosion	32.8	74–177
Wind erosion	10.8	13–39
Soil fertility decline	29.4	26
Waterlogging	3.1	7–9
Salinization	4.1	7–26
Lowering of water table	0.2	Na
Total area	45.1	175

West Bengal. Also in the superficial layer from 0 to 10 cm, the SOC set can be lesser than 5 g kg^{-1} (Singh and Bandyopadhyay 1996). Anthropogenic soil erosion seriously and rapidly impairs the SOC suite. The SOC pool is preferably removed by surface and wind runoff as it concentrates close to the surface of the soil and has a low density ranges from 1.2 to 1.5 mg/m^3 compared to 2.5–2.7 mg/m^3 for the SOC pool. As a result, it eroded sediment enriched with the SOC group compared to the field soil with an enrichment ratio between 1.5 and 5.0 (Lal et al. 1999). The SOC loss due to erosion and contamination can be higher even on the gentle slopes from 0.5% to 3.0% (Banerjee et al. 1991). Erosion of soil occurs in four steps. It involves detachment, decomposition, transport, and storage of soil particles. The detachment and decomposition in the soil are caused by the soil off or interruption of the aggregates by the impact of the drop, the cutting action of water or wind blowing, and the collision between the particles. The decomposition of the aggregates exposes SOC until now enclosed and physically preserved to microbial processes. Though the chance of SOC moved together with eroded sediment is controlled by a chain of complex and interactive processes, fragment of it is digested and leads to the release of CO_2 and CH_4 under aerobic conditions in anaerobic environments. Lal (1995) hypothesized that 20% of SOCs moved by erosion are mineralized. Erosion of soil in India, soil displaced and redistributed over the landscape and transported to the aquatic ecosystems and deposition sites, is reported around 3 Pg of sediment soil per year of the total loss; 1.2 Pg (40%) of soil erosion happens at a rate of 10–20 Mg ha^{-1} $year^{-1}$, 0.5 Pg (16%) at 20–40 Mg ha^{-1} $year^{-1}$, and an extra 5 Pg (16%) to 40–80 Mg ha^{-1} $year^{-1}$. Assuming the concentration of SOC is 8–12 g kg^{-1}(1%) in the eroded sediments, the total Cd that is placed by the erosion process is about 29.8 Tg C/a. Suppose 20% of the eroded SOC is mineralized, the emission of C induced by erosion in India is measured at 6 Tg C/a. This compares with the emission of 15 Tg C/year induced by erosion in the United States (Lal et al. 1998) and 1.1 Pg C/a in the world (Lal 1995). Moreover, the adoption of effective conservation measures which reduce erosion can lead to decrease in C emissions from ecosystems prone to erosion.

1.5 Carbon Pool Estimation

To calculate spatially explicit maps of soil C density, different computational approaches, such as interpolation of point data (Agboadoh 2011), multiple regression, regression trees (Martin et al. 2011), artificial neural networks (Alvarez et al. 2011), and digital landscape modeling or mapping (Thompson et al. 2006; Minasny et al. 2013; Verma et al. 2015a). Recently Sreenivas et al. (2014) used data mining random forests (RF) with a large number of metrics derived from remote sensing to map the density of SOCs for the states of Andhra Pradesh and Karnataka (India) with a spatial resolution of 1 km. Globally, the estimated group size of SOC in the depth of 100 cm is 1200–1600 Pg (Lal 2004; Batjes 1996), while the size of the SIC was estimated at 695 and 1738 Pg (Batjes 1996; Eswaran et al. 1993; Sombroe et al. 1993) for a similar depth. While SOC reserves stored predominantly in humid temperate regions, most SIC stocks are stored in arid and semiarid regions (Diaz-Hernandez et al. 2003; Yadav et al. 2018b). The estimated size of the SOC group in India from different authors varies between 21 and 27 Pg (Dadhwal and Nayak 1993). In India, the density of SOC for different types of forests and the group SOC for forest ecosystems have also estimated at 4.13 Pg (Chhabra et al. 2003; Meena et al. 2015a) and 5.25 Pg (Velmurugan et al. 2014) at a top 50 cm floor depth. The Indian SIC group was estimated at 22.46 Pg from Bhattacharyya et al. (2008). Furthermore, several site-specific regional studies in arid and semiarid regions have revealed various soil C densities in these areas. These reports were made using soil samples taken for a prolonged period and analyzed under different laboratory testing procedures that could result in inconsistencies in the estimate. In the world, there are efforts to standardize analytical procedures and nomenclature of soil properties and methods for spatial prediction with little data (Arrouays et al. 2014) (http://www.globalsoilmap.net). With a high and very heterogeneous spatial distribution of C densities, the reliable estimate of the soil C pool depends on its precise spatial evaluation, as well as on the suitable sampling of point densities with an appropriate validation mechanism. This study was taken as part of the ISRO Geosphere-Biosphere Program (IGBP) to establish maps of space density of organic C and inorganic soil explicitly and estimate the size of these groups using remote sensing and spatial modeling mining approach. The approach of Sreenivas et al. (2014) applies both SOC and SIC with geospatial modeling approach to a resolution of 250 m. In their previous research, Sreenivas et al. (2014) demonstrated the applicability of modeling RF to predict the spatially highest SOC density which is only 30 cm. Current research gives a spatially explicit distribution of SOC and SIC density across India. The extended scope of this spatial modeling approach will be to develop a predictive model for assessing the impact of climate change and land use on C density and reserves (Fig. 1.2).

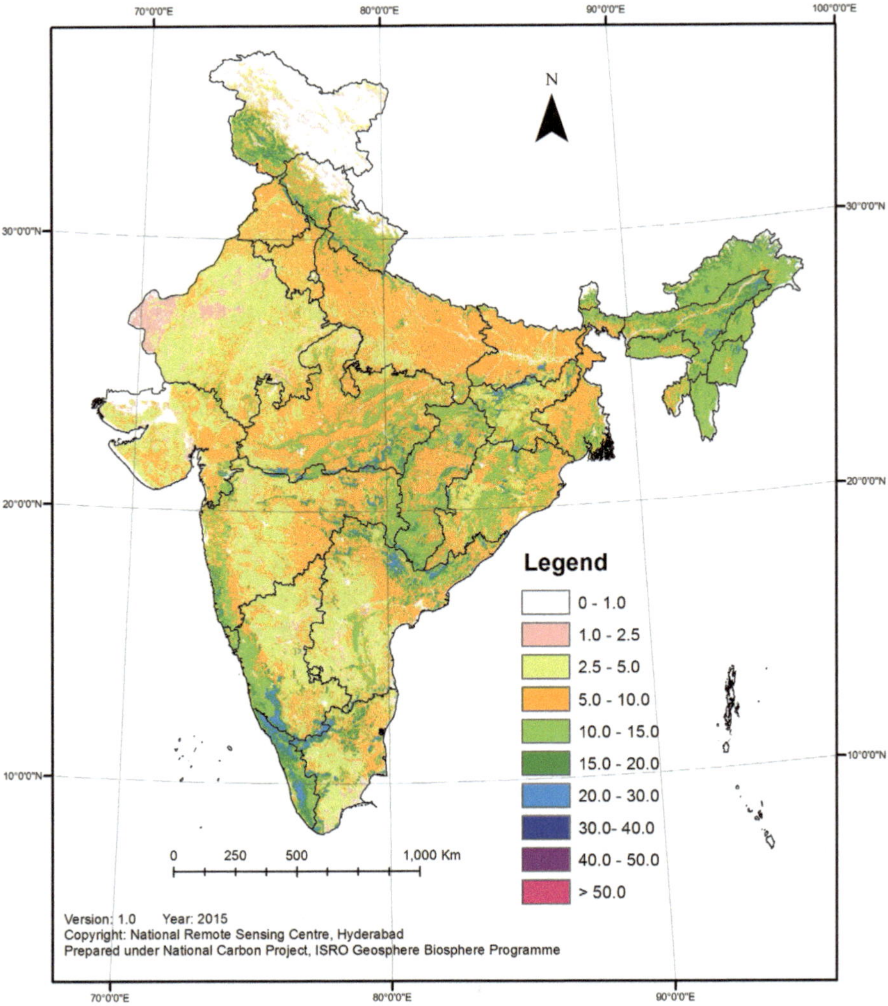

Fig. 1.2 Soil organic carbon density map of India. (Source: NRSC 2015)

1.6 Soil Erosion Rates and Carbon Dynamics Estimation

Soil erosion is one of the major threats to European soils. Soil erosion deteriorates ecosystems and can reduce crops, alarming food security and farmers' income. Soil water erosion can transport agricultural wastes into watercourses, and even the wind-eroded soil can carry soil particles and contamination of other sites. Due to erosion, the ground can be at loss of C (Berhe et al. 2007; Lal 2004). Safeguarding European soil from erosion is therefore utmost important in the European Commission's thematic soil protection strategy. In compliance with the requirements of good agronomic and environmental conditions (GAEC), member states are obliged to prevent erosion and to preserve organic matter of the soil with minimum maintenance of land cover and minimum land management conditions that respect specific for the site (Borrelli et al. 2016; Meena et al. 2017a). The susceptibility of a soil to water erosion is calculated by the erosive rain potential, the inclination of the soil surface, the situation of the soil in the basin, and the vegetation cover on the soil surface (FAO 2015). Normal rates of soil erosion due to water have been estimated for representative agroecological conditions. The cultivation of mountainous land under common agriculture and horticulture, without additional cover soil in temperate climates, is prone to erosion rates ranging from 10 to 20 tons ha^{-1} $year^{-1}$, while average rates are usually less than10 tons/ha/year. Grasslands and pastures in the tropical and subtropical mountain areas can experience erosion rates same as that of tropical farmland. Because of the absence of field boundaries, which usually act as sediment and runoff barriers and favor infiltration, these pastures can also be vulnerable to the formation of ravines. It may not have much influence on the soil, but it can make the soil unreachable and therefore unusable. Pastures in temperate zones are characterized by erosion rates usually much lower and usually less than 1 ton/year. These pastures are used little intensively and manage well than (sub) tropical pastures (FAO 2015). Erosion of soil improves soil deterioration with a long-term decrease in soil productivity and soil capacity to perform ecosystem functions. The latest models show that nearly 20% of the surface of the 27 is subject to loss of soil above 10 tons ha^{-1} $year^{-1}$. The total geographical area affected by extreme erosion (both wind and water erosion) is 132 million hectare in Europe, 50 Mha in America,20 Mha in Oceania, 407 Mha in Asia, 93 Mha in South America, 267 Mha in Africa, and 78 Mha in North America (Lal 2004), central, and eastern. Still there are several areas with high erosion rates such as the Himalayan-Tibetan ecosystem in Southern Asia, the Loess in China, subhumid and semiarid sub-Saharan Africa regions, the Andean region, the highlands of Central America, Haiti, and the Caribbean region (Lal 2004; Scherr and Yadav 1996). As it is founded that organic matter in the soil is proportional to soil quality (Bastida et al. 2006), the reduction of soil C is usually an indication of soil degradation (Lal 2004). The effects of erosion on soil deterioration, productivity, and food safety are commonly known; still, their effects on dynamics of C and greenhouse gas emissions are not properly understood (Lal 2004). The soil erosion is a process that involves detachment, rupture, transport, and deposition of sediments. During these steps, the organic soil C is affected. Moreover, soil erosion affects the dynamics of organic soil C due to its impact on (1) the alteration of soil clusters; (2) removal of soil organic C from the ground by runoff or sandstorms; (3) mineralization of organic matter in situ; (4) mineralization of the displaced organic soil C, redistribution of organic soil C on the landscape, and organic matter transport in the soil, rivers, and storms; (5) reaggregation of the soil via the creation of organic mineral complexes; and (6) heavy sediment enriched in C in landfills, reservoirs, flood plains, and burials ocean floor (Lal 2004; Meena et al. 2017b). As a consequence, the C (and other elements) cycle is severely affected by erosion of soil, which affects the inflow and outflow of the soil system, storage, distribution within the soil matrix, and time of residence in the soil (Berhe et al. 2014). The erosion of soil distributes about 75 Gt of upper layer of the soil, and the total C content removed by soil erosion is about 4–6 Pg/year (Berhe et al. 2007). It is interesting to note that portion of the eliminated C can easily be mineralized, which represents an erosion-induced emission of 0.8 to 1.2 Pg C/year (Lal 2004). However, the effect of erosion on the dynamics of C depends to a large extent on the type and position of the survey taken into account. For example, Berhe et al. (2008) noted that the soil redistribution C with soil erosion and its deposition determined a noticeable rise in the C slope reduction. Furthermore, these researchers have found that the C erosion was restored by the new C photosynthesis and C organic rate decomposition in the depositional configuration. It was slower than the eroded position, hinting that soil erosion could be a sinking process of the C (Berhe et al. 2008).

1.7 Types of Soil Erosion and Soil Organic Carbon Depletion

The soils contain a significant amount of C. They are the third largest C store in the world after the ocean and geological ponds. An augment in the soil organic C content (SOC) causes a greater structure and stability of the aggregates, which means better conditions for the flora and fauna. Furthermore, humus is the most significant nutrient conservation factor of the soil. Soil erosion is the major soil degradation factors that play an important role in the redistribution of the SOC. The erosion of the sheet eliminates the highest soil layer with the high SOC content. The added protected SOC becomes vulnerable due to the aggregate influence of decay resulting from raindrop erosion (Kerényi 1991). SOC is strongly correlated with the clay content of soil (Fuchs et al. 2010). Colloid components usually migrate together as a consequence, and the content of SOC generally increases sediments in the clay content; however, depending on the environmental conditions, it may not be necessary for all cases (Stavi and Lal 2011). Wang et al. (2010) reports a significant SOC enrichment with a limited increase in clay content. Erosion processes alter SOC stocks of the land unit during transporting rich sediments in SOC to a unit of agricultural land, oxidize SOC reserves, and develop C dioxide (CO_2) atmosphere anhydride and cause SOC loss to surface outflow. Therefore, soil erosion, its transport, and storage processes redistribute SOC landscape, improve oxidation, and make a source of SOC and a sink. Still, the SOC redistributed to the underlying soils is not SOC seized, if it produced outside the boundaries of the known ground unit. To make active C sequestration in soil (C) sink, plants in a land unit take CO_2 from the atmosphere and store it in the humus fraction or SOC unit within agricultural land. Geological or natural erosion, a significant terrestrial process, is modeled on the Earth's surface and developed some of the major fertile soils (alluvial and loess) from the initial time. Nevertheless, the acceleration of the process from human activities (e.g., deforestation, burning of the removal/biomass, drainage, plowing, and change of land use in the changing of natural ecosystems to agricultural ecosystems of management) has negatively impacted the SOC stock. Water resources have had an impact and have had a negative impact on the net primary productivity of the earth as a source of SOC reserves and the environment (Lal et al. 1998; Meena et al.

2018b). Prior to determining effects of water and wind erosion and the associated transport and storage processes in SOC reserves, researchers have established a pellucid standard for reloading SOC stocks. This process is known as the seizure of SOC (Olson et al. 2014a, b; Sundermeier et al. 2005; Lal et al. 1998; Mann 1986). The statement of C sequestration in the soil requires that an agricultural land unit covered with borders be identified to monitor seizure, storage, and loss of C in the soil. There are mainly three groups of land units (field of study), generally used by researchers to measure changes in facts and the storage abduction of SOC: (1) soil eroding, (2) deposition, and (3) mixed landscape (erosion combined and deposition). The SOC of these groups of land units is influenced differently by the processes of water, wind, transport, and deposition erosion. The basic effects of the soil erosion process are physical, chemical, and biological changes of the SOC reserves. Soil processes and soil erosion mechanisms negatively influence seizure amounts and SOC rate and endanger soil productivity (Lal 2004) and augment emission of greenhouse gases (GHG) (MEA 2005). Land units, particularly sloped and eroded ones, in intensive soil processing systems, can lose a significant amount of SOC and sediments seized by wind and water erosion. Reported: "The rating action of wind or water removes a large percentage of clay and soil humus and leaves behind sand, gravel and less productive soil. Most of the fertile soil is associated with clay and humus." Land use change is an important factor for the SOC stock, especially in slope prone to soil erosion because of changes in land use or conventional local forest and erosive farming systems of terraced crops (Olson et al. 2012, 2013a, b). The role of the erosion process in the change of the C reserves in the soil occurs through a preferential transport of the organic light component and the change of the biological and physical properties of the soil, which are the drivers of soil sequestration C. Therefore, it is necessary to understand the impacts of soil erosion related to agricultural use, especially in intensive agricultural production systems, where the erosion process becomes a determinant of the sources of C and the loss as CO_2 instead of a significant one. Contributory retention of C in the soil within land unit (Lance et al. 1986), Olson et al. (2013b) and Young et al. (2014) reported that the agricultural parcels almost level (<1%) of the highlands were also affected by erosion, transport, and deposition of sediments rich in SOC. It is often presumed that tracks with almost flat slopes (<1%) were not significantly influenced by erosion. Soil processing can increase the loss of calcium C in soil in conventional agricultural systems. Intensive processing plays an important role in increasing the loss of C in the soil through the oxidation of organic matter, the destruction of soil clusters, and the decrease in the rate of infiltration of water causing significant water erosion and surface runoff of the rich in C. Olson et al. (2013b) used a flying ash procedure and estimated the SOC concentration of Muscatune and Sable soil for an almost leveled agricultural area (<1%) at Monmouth, Illinois. The results of the field trails (Salemme and Olson 2014; Olson et al. 2014c, 2013a, 2011; Kumar et al. 2012; Gennadiyev et al. 2010; Meena et al. 2017c) have been reported, and after about 150 years of agricultural practices and accelerated erosion, the lands have less SOC reserves of the similar soil than before in a native forest or grassland. Some researchers (Salemme and Olson 2014; Olson et al. 2014c, 2013a, 2011; Kumar et al. 2012;

Gennadiyev et al. 2010; Meena et al. 2015e) estimated that the change of prairie to agricultural land for 90–140 years yielded in a loss of 20–51% of the SOC stock, and changing of forest to agricultural land for 100–150 years brings about SOC stock loss of 10–52%. Thus, it is necessary to understand the influence of soil erosion in the manner of agricultural use, particularly in intensive agriculture systems, where soil erosion process becomes a driver in creating C sources and loss as CO_2 than an important contributor to soil C retention within a land unit (Lance et al. 1986). Figure 1.3 shows the fate of organic carbon taken away by different erosion processes.

The organic C of soil (SOC) is an essential parameter of natural ecosystems. A key role is played in the maintaining of food production and biomass via its positive effect on the nutrient availability, water retention, and biodiversity (Lal 2004; Verma et al. 2015). SOC also decreases the risk of soil compaction, crusting on surface erosion of soil (Meena et al. 2015b). It is reported that 60 Gt of C each year is exchanged between the surface and the earth's atmosphere (Folger 2009). Due to large quantities of SOC accumulated in global soil, about 2500 Gt C at a depth of 20 cm (Robert 2002), degradation of soil can accelerate the rotation of SOC and augment the power of CO_2 emissions in terrestrial ecosystems (Lal 2004). The erosion of the SOC is one of the main mechanisms of soil degradation (Lal 2004); however, not much is known about its effect on soil C erosion. The soil water cycle

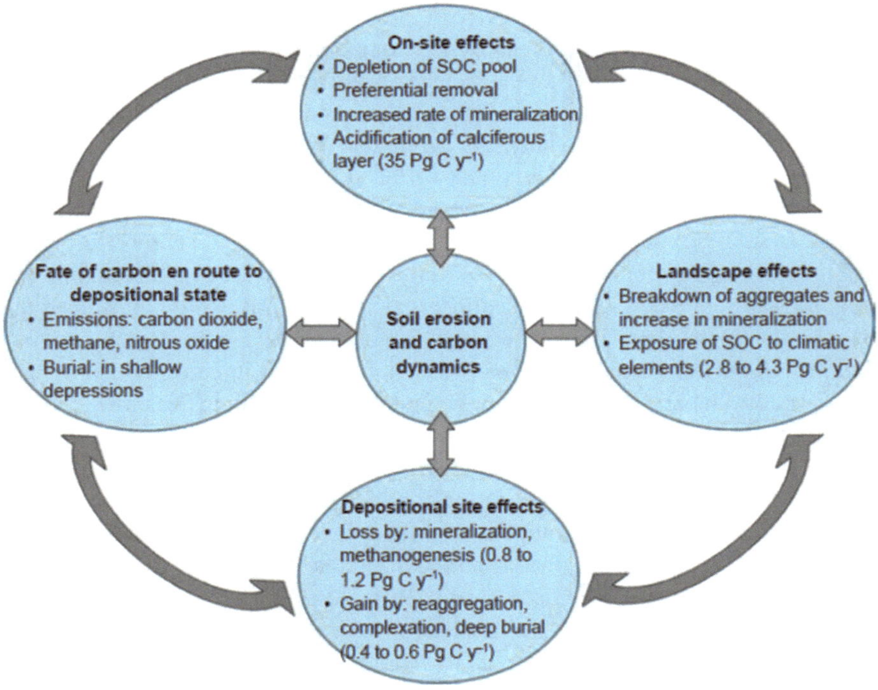

Fig. 1.3 Fate of SOC transport by erosional processes. Modified and adapted from Lal (2004)

is the process whereby the soil material (whether organic or inorganic) from its starting position is removed by both the action of drop energy and outflow. Water erosion influences SOC of (1) complete transport and disposal of inert soil aggregates (Goebel et al. 2005) and (2) preferential removal of SOC, resulting from the decomposition of aggregate soil, either by the impact of rain or runoff (erosion sheet) (Lal 2004). The intrinsic quality of the soil determines its ability to be influenced by water erosion, with rich clay soils and aggregates SOC showing more stable soil and sandy soils with a low SOC content.

1.8 Climate Change and Carbon Dynamics in Eroded Soils

One of the most important environmental problems is the way in which change in climate will affect the general dynamics of C on the planet. Soil, as one of the major organic compartments of C in the planet's soil, can seriously be affected by climate change, but how and how much is unknown. For example, soil respiration due to microbial CO_2 and breathed by the root is the second largest flow of terrestrial C (Schimel 1995; Meena et al. 2015c) and is influenced by warming (Bond-Lamberty and Thomson 2010). The northern bogs are rich in SOM and comprises of one third of the total organic C of the planet's soil. If the decomposition of these soils or organic C is increased by factors derived from climate change, there will be a visible effect on the global climate. For example, Soil CH_4 emissions in high-latitude regions may rise due to heating, permafrost defrosting and CO_2 fertilization (Koven et al. 2011; Meena et al. 2014). Dorrepaal et al. (2007) noted that 1 °C warming increased the total pore ecosystem CO_2 release rates by around 60% and this effect was maintained for 8 years. Approximately 69% of the stimulated CO_2 comes from C in the permafrost peat. Thus, climate warming can accelerate organic decrease of soil C in the permafrost with long-term positive global climate feedback. However, impact C heating cycle depends on the temperature set in which the heating takes place in the central Swiss Alps (Ferrari et al. 2016), while, on the contrary, Shen et al. (2009) suggested that land in arid areas would probably sequester CO_2 with a future raise in precipitation and, however, release C with decreases in precipitation, in the future, as also discussed (Albaladejo et al. 2013). Episodic availability of water erosion clearly affects the cycle of elements in arid and semiarid ecosystems (Austin et al. 2004; Gebauer and Ehleringer 2000; Varma et al. 2017b). High temperature and humidity inputs erratic generation model pulsed biological activities, influencing the rotation of C and N (Collins et al. 2008), so that organic material tends to accumulate in periods of drought in which the growth of plants and microbes is limited. In addition, drought affects the quality and composition of humic acids in semiarid soils, but these effects can be improved by organic soil improvers. Liu et al. (2009) have suggested that the availability of soil water is more significant than the temperature of microbes and microbial respiration regulation of biomass in a temperate semiarid steppe.

1.9 Management Strategies for Carbon Storage and Soil Building

Conversion of land from a native or naturally conventional agricultural ecosystem can result in a 30% loss of the original SOC stock (no longer remains within the land unit) (Lal et al. 1998; Lal et al. 1999; Meena et al. 2015d). The other 70% SOC is exposed to wind erosion or water and transported as rich sediments SOC. It can be stored and redeposited on terrestrial upslope soils or deep-sea deposition. If the transported SOC is deposited in an adjacent land holding unit, it may augment the SOC stock of that unit (Olson et al. 2014a, b; Meena and Yadav 2014). This process is a redistribution of the SOC and seized and stored. The total stock SOC of the depositional unit is a combination of SOC sequestered per SOC deposition unit and previously stored adjacent units transported by the erosion plants. The change caused by erosion actions SOC generates an unbalanced sink C and a source that leads to a significant loss of stocks C as indicated above via different routes (mineralization, leaching, outflow, etc.). These results are accelerated by modern management practices and with agriculture such as soil erosion. It has been documented that inclined soil use to vary soil between crops and act as a vector plowing storage of atmospheric CO_2 in the soil during the regeneration of natural fallow (CO_2 stored by bad voluntary grasses and other plants and in some cases without plowing when fallow), ultimately, transfers through the process of erosion of the water from the steepest slopes to plains, where it accumulates (Chaplot et al. 2009). The variation in the distribution of C in the ground within a ground unit cannot be explained as C sequestration of the ground due to the absence of the unit role of the ground unit in the capture and storage of atmospheric CO_2 (Olson et al. 2014a, b; Varma et al. 2017a).

1.10 Strategies for Building the Soil C

As already mentioned, soil processing practices generally lead to organic loss of C in the soil with formation and emission of CO_2 in the atmosphere. Conservation tillage, including minimal tillage and tillage, to retain soil C and decrease the amount of fossil fuel for agriculture should be practised. Kern and Johnson (1993) estimated that the amount of C in the soil that would be lost or seized and the amount of fossil fuel necessary for agriculture accounted for three different conservation scenarios, between 1990 and 2020, were 27% (scenario 1), 57% (scenario 2), and 76% (scenario 3). In the case of scenario 1, the level of conservative processing remained constant at the current level of 1990 of 27% for 30 years; for scenarios 2 and 3, it was simulated that force started in 27% in 1990 and increased linearly to 57% and 76%, respectively, in the first 20 years of analysis and therefore remains constant in the remaining 10 years. Today the low organic C (and N) in some of the planet's soils limits the functionality of the soil and its ecosystem services. Suitable for soil regeneration, reduced plowing or reforestation practices are essential to promote C sequestration in the soil. The mitigation of climate change through the sequestration of C depends on the creation of a stable plant cover. Changes in land use, such as due to afforestation and management of rapidly growing tree species, can affect regional C sequestration by incorporating C dioxide (CO_2) into biomass of plants and soil (Jandl et al. 2007). Furthermore, the creation of mixed forests can avoid high rates of mineralization of the organic substance (Jandl et al. 2007). The organic C content in soil can reduce not only the nature of agricultural land but also the abandonment of agricultural land under different climates (Bastida et al. 2006; Chen et al. 2016; Ram and Meena 2014). Hence the reforestation of abandoned farmland suggested as an appropriate strategy to increase SOC content and improve the abduction of C (Lal 2006). In Burkina Faso, the potential of the biofuel crop *Jatropha curcas L.* for the sequestration of C has been demonstrated in the long term (Baumert et al. 2016). A meta-analysis carried by Deng and Shangguan (2017) evaluated the dynamics of C and N in soil after afforestation and the factors that influenced SOM in China. Temperature, precipitation, land use, soil depth, tree species, and forest age were the chief factors that affected the SOC content after afforestation. Moreover, Hu et al. (2016) have showed that the root instead of the litter of *Populus simonii* Carr. controls the sequestration of C in the soil and concluded that deep trees with large root biomass could be used to promote C capture in the soil. A meta-analysis on C accumulation in agricultural land after afforestation displayed that the main factors influencing SOC reserves after profitability were previous land use, clay content, preimplantation disorder, and type of planted trees (Laganière et al. 2010; Meena and Yadav 2015). It was also noted that (1) the impact of reforestation on C populations was more pronounced on cultivated land than on pasture or on gaseous areas, (2) broad-leaved species have a greater ability to increase SOC than coniferous species, and (3) soils with a high clay content (> 33%) have a greater ability to accumulate SOC compared to land with low clay content. The afforestation of pastures and agricultural systems usually increase C capture in soils, but the extent, timing, and direction of organic soil dynamics depend on site conditions, management practices, and previous use of earth (Hernández et al. 2016; Sihag et al. 2015). The positive effects of afforestation in SOC sequestration are sometimes contradictory due to the multiple factors that control the dynamics of C (Laganière et al. 2010). For example, Hernández et al. (2016) did not detect significant changes in SOC stocks after 8 years of afforestation of native pastures with eucalyptus or pine in a temperate region of Uruguay. Figure 1.4 shows different strategies of soil carbon buildup in semiarid areas.

1.11 Conclusions

Erosion and management of soil agricultural land units influencing SOC stock unit decline agricultural land with associated changes in the abduction, storage, and loss of SOC. It is essential to recognize that the processes of wind and water erosion, the transport, and the deposition of sediments rich in SOC within a landscape unit contribute to the redistribution of SOC reserves, especially within the boundaries of the units of agricultural land, bodies of water over those land units, or in the atmosphere, which includes the dynamic interactions between soil, plant, and

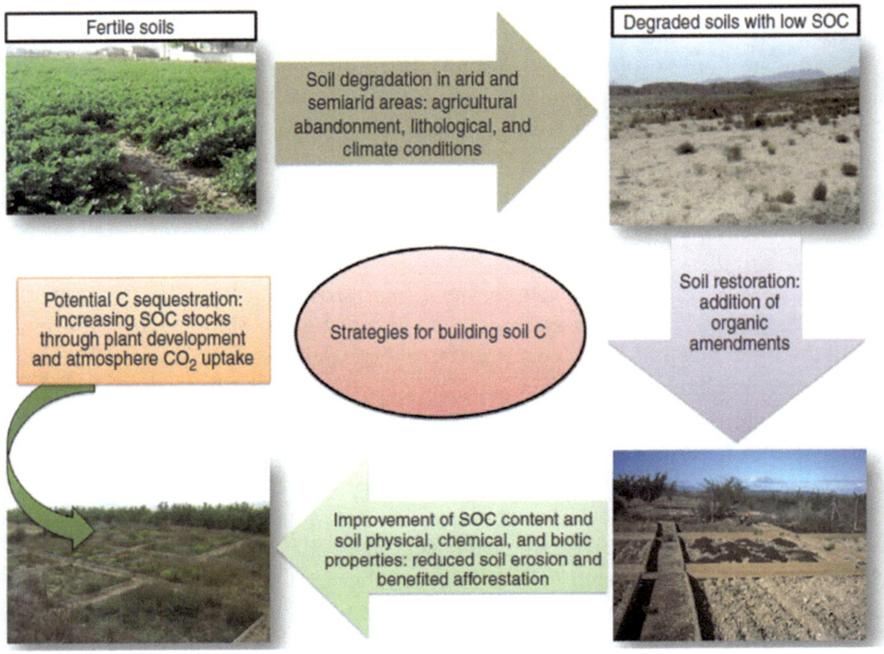

Fig. 1.4 Strategies for building SOC in arid and semiarid areas through the application of organic amendments

atmospheric CO_2 within the designated unit. The lack of such dynamic conductors to the conclusion that erosion is a destructive process alters soil C (organic and inorganic) and causes changes. The loss of a significant amount of relatively stable SOC retained in the soil system for millennia and negatively affects net primary productivity and efficiency in the use of inputs. The selection of the agricultural land unit and its location to study and determine the SOC stock can influence the results and their interpretations. A unit of abraded earth subinformará stock of SOC, a unit of deposition agricultural land, is overestimating the stock of SOC, while a mixed agricultural land unit (combined and erosion deposit) may have a result of the distribution of several SOCs and uncertain due to losses, due to decomposition, leaching, and runoff. Past and current knowledge of soil erosion can be of great importance in suggesting where and how the erosion of the future could be a problem.

Future Prospective

Future erosion will depend on future erosion rates of water and wind, both restrained by climate change and change in land use. Water erosion rates are likely to respond to the increase in precipitation in a nonlinear manner, with disproportionately higher increases in the rainy years. However, there are still gaps in knowledge on this topic and on scenarios that influence C content in soil. Therefore, it is a matter of debate about how erosion will respond to climate change factors and how this combination will affect C soil reserves and ecosystem services. These control factors will ultimately influence the response of the activity and composition of the plant and microbial community and, consequently, the CO_2 emissions of soil to the atmosphere. The promotion of C sequestration in arid and semiarid regions may imply reforestation strategies. To this end, an improvement of the soil conditions is necessary through the application of organic changes. However, the C budget of modified and reforested soils should be studied according to the expected impacts of climate change.

References

Adams JM, Post WM (1999) A preliminary estimate of changing calcrete C storage on land since the last glacial maximum. Glob Planet Chang 20:243–256

Agboadoh DMY (2011) Estimation and mapping of soil organic C stocks in croplands of the Bechem Forest District Ghana. MS thesis submitted to ITC

Albaladejo J, Ortiz R, García-Franco N, Ruiz-Navarro A, Almagro M, García-Pintado J, Martínez-Mena M (2013) Land use and climate change impacts on soil organic C stocks in semi-arid Spain. J Soil Sediments 13:265–277

Alvarez R, Steinbach HS, Bono A (2011) An artificial neural network approach for predicting soil C budget in agroecosystem. Soil Sci Soc Am J 75:965–975

Amundson R (2001) The C budget in soils. Annu Rev Earth Planet Sci 29:535–562

Arrouays D, Grundy MG, Hartemink AE, Hempel JW, Heuvelink GB, Hong SY, Lagacherie P, Lelyk G, McBratney AB, McKenzie NJ, Mendonca-Santos ML (2014) Chapter three-Global Soil Map: toward a fine-resolution global grid of soil properties. Adv Agron 125:93–134

Ashoka P, Meena RS, Kumar S, Yadav GS, Layek J (2017) Green nanotechnology is a key for eco-friendly agriculture. J Clean Prod 142:4440–4441

Attal M, Mudd SM, Hurst MD, Weinman B, Yoo K, Naylor M (2015) Impact of change in erosion rate and landscape steepness on hillslope and fluvial sediments grain size in the Feather River basin (sierra Nevada, California). Earth Surf Dyn 3:201–222

Austin AT, Yahdjian L, Stark JM, Belnap J, Porporato A, Norton U, Ravetta DA, Schaeffer SM (2004) Water pulses and biogeochemical cycles in arid and semiarid ecosystems. Oecologia 141:221–235

Banerjee SK, Chinnaman S, Jha MN (1991) Forest soils of north and Northeast Himalayas and constraints limiting their productivity'. In Biswas TD et al (eds) Soil-related constraints in crop production, Indian Society of Soil Science Bulletin 15, New Delhi, India, pp 164–176

Banger K, Tian H, Tao B, Lu C, Ren W, Yang J (2015) Magnitude, spatiotemporal patterns, and controls for soil organic C stocks in India during 1901–2010. Soil Sci Soc Am J 79:864–875

Barman D, Sangar C, Mandal P, Bhattacharjee R, Nandita R (2013) Land degradation: its control, Management and environmental benefits of Management in Reference to agriculture and aquaculture. Environ Ecol 31:1095–1103

Bastida F, Moreno JL, Hernández T, García C (2006) Microbiological degradation index of soils in a semiarid climate. Soil Biol Biochem 38:3463–3473

Batjes NH (1996) Total C and nitrogen in soils of world. Eur J Soil Sci 47:151–163

Batjes NH, Sombroek WG (1997) Possibilities for C sequestration in tropical and subtropical soils. Glob Chang Biol 3:161–173

Baumert S, Khamzina A, Vlek PLG (2016) Soil organic C sequestration in *Jatropha curcas* systems in Burkina Faso. Land Degrad Dev 27:1813–1819

Berhe A, Harte A, Harden J, Torn WMS (2007) The significance of erosion-induced terrestrial C sink. Bioscience 57:3374–3346

Berhe AA, Harden JW, Torn MS, Harte J (2008) Linking soil organic matter dynamics and erosion-induced terrestrial C sequestration at different landform positions. J Geophys Res 113:G04039

Berhe A, Arnold A, Stacy C, Lever E, McCorkle R, Araya ESN (2014) Soil erosion control son biogeochemical cycling of C and nitrogen. Nat Educ Knowl 5(2)

Bhattacharyya T, Pal DK, Chandran P, Ray SK, Mandal C, Telpande B (2008) Soil C storage capacity as a tool to prioritize areas for C sequestration. Curr Sci 95:482–484

Bhattacharyya R, Prakash V, Kundu S, Srivastava AK, Gupta HS (2009) Soil properties and their relationships with crop productivity after 30 years of different fertilization in the Indian Himalayas. Arch Agron Soil Sci 55(6):641–661

Bhumbla DR, Khare A (1984) Estimate of wastelands in India. Society for promotion of wastelands development. Allied, New Delhi, p 18

Biswas TD, Narayanasamy G, Goswami NN, Sekhon GS, Sastry TG (1991) Soil- related constraints in crop production, Indian Society of Soil Science Bulletin vol. 15, New Delhi, India, 176pp

Bond-Lamberty B, Thomson A (2010) Temperature-associated increases in the global soil respiration record. Nature 564:579–582

Borrelli P, Paustian K, Panagos P, Jones A, Schütt B, Lugato E (2016) Effect of good agricultural and environmental conditions on erosion and soil organic C balance: a national case study. Land Use Policy 50:408–421

Buragohain S, Sharma B, Nath JD, Gogaoi N, Meena RS, Lal R (2017) Impact of ten years of biofertilizer use on soil quality and rice yield on an inceptisol in Assam, India. Soil Res. https://doi.org/10.1071/SR17001

Chaplot V, Bernoux M, Walter C, Curmi P, Herpin U (2001) Soil C storage prediction in temperate hydromorphic soils using a morphologic index and digital elevation model. Soil Sci 166:48–60

Chaplot VP, Podwojewewski K, Phachomphon, Valentin C (2009) Soil erosion impact on soil organic C spatial variability on steep tropical slopes. Soil Sci Soc Am J 73(3):769–779

Chen C, Zhang J, Lu M, Qin C, Chen Y, Yang L, Huang Q, Wang J, Zhengu S, Shen Q (2016) Microbial communities of an arable soil treated for 8 years with organic and inorganic fertilizers. Biol Fertil Soils 52:455–467

Chhabra A, Palria S, Dadhwal VK (2003) Soil organic C pool in Indian forests. For Ecol Manag 173:187–199

Collins SL, Sinsabaugh RL, Crenshaw C, Green L, Porras-Alfaro A, Stursova M, Zeglin LH (2008) Pulse dynamics and microbial processes in arid land ecosystems. J Ecol 96:413–420

Dadhich RK, Meena RS (2014) Performance of Indian mustard (*Brassica juncea* L.) in response to foliar spray of thiourea and thioglycollic acid under different irrigation levels. Indian J Ecol 41(2):376–378

Dadhich RK, Meena RS, Reager ML, Kansotia BC (2015) Response of bio-regulators to yield and quality of Indian mustard (*Brassica juncea* L. Czernj. and Cosson) under different irrigation environments. J Appl Nat Sci 7(1):52–57

Dadhwal VK, Nayak SRI (1993) A preliminary estimate of biogeochemical cycle of C for India. Sci Cult 59:9–13

Dahlgren RA, Boettinger JL, Huntington GL, Amundson RG (1997) Soil development along an elevational transect in the western Sierra Nevada, California. Geoderma 78:207–236

Datta R, Baraniya D, Wang YF, Kelkar A, Moulick A, Meena RS, Yadav GS, Ceccherini MT, Formanek P (2017a) Multi-function role as nutrient and scavenger off reeradical in soil. Sustain MDPI 9:402. https://doi.org/10.3390/su9081402

Datta R, Kelkar A, Baraniya D, Molaei A, Moulick A, Meena RS, Formanek P (2017b) Enzymatic degradation of lignin in soil: a review. Sustain MDPI 9:1163. https://doi.org/10.3390/su9071163, 1–18

Deng L, Shangguan ZP (2017) Afforestation drives soil C and nitrogen changes in China. Land Degrad Dev 28:151–165

Dhakal Y, Meena RS, De N, Verma SK, Singh A (2015) Growth, yield and nutrient content of mungbean (*Vigna radiata* L.) in response to INM in eastern Uttar Pradesh, India. Bangladesh J Bot 44(3):479–482

Dhakal Y, Meena RS, Kumar S (2016) Effect of INM on nodulation, yield, quality and available nutrient status in soil after harvest of green gram. Legum Res 39(4):590–594
Díaz-hernández JL, Fernández EB, González JL (2003) Organic and inorganic C in soils pf semi-arid regions: a case study from the Guadix–Baza basin (Southeast Spain). Geoderma 114:65–80
Dixon JL, Chadwick OA, Vitousek PM (2016) Climate-driven thresholds for chemical weathering in postglacial soils of New Zealand. J Geophys Res Earth Surf 121:1619–1634
Doetterl S, Stevens A, Six J, Merckx R, Van Oost K, Casanova Pinto M, Casanova-Katny A, Muñoz C, Boudin M, Zagal Venegas E, Boeckx P (2015) Soil C storage controlled by interactions between geochemistry and climate. Nat Geosci 8:780–783
Dorrepaal E, Toet S, van Logtestijn RSP, Swart E, van der Weg MJ, Callaghan TV, Aerts R (2007) C respiration from subsurface peat accelerated by climate warming in the subartic. Nature 460:616–619
Eswaran H, Van Den Berg E, Reich P (1993) Organic Cin soils of the world. Soil Sci Soc Am J 57:192–194
FAI (Fertilizer Association of India) (2011) Fertilizer statistics. Fertilizer Association of India, New Delhi
Falloon P, Jones CD, Cerri CE, Al-Adamat R, Kamoni P, Bhattacharyya T, Easter M, Paustian K, Killian K, Coleman K, Milne E (2007) Climate change and its impact on soil and vegetation C storage in Kenya, Jordan, India and Brazil. Agric Ecosyst Environ 122:114–124
FAO (1994) Land degradation in South Asia: its severity, causes and effects upon the people. World soil resources reports 77. FAO, Rome, Italy
FAO (2015) Status of the world's soil resources. Main report. Food and Agriculture Organization of the United Nations (FAO), Rome, p 607
Ferrari A, Hagedorn F, Niklaus PA (2016) Experimental soil warming and cooling alters the partitioning of recent assimilates: evidence from a 14C-labelling study at the alpine treeline. Oecologia 181:25–37
Folger P (2009) The C cycle: implications for climate change and congress. Congressional Research Service, 13
Follain S, Minasny B, McBratney AB, Walter C (2006) Simulation of soil thickness evolution in a complex agricultural landscape at fine spatial and temporal scales. Geoderma 133:71–86
Fuchs M, Gál A, Michéli E (2010) Depth distribution of SOM stock in fine-textured soils of Hungary. Agrokém Talajt 59(1):93–98
Gebauer RLE, Ehleringer JR (2000) Water and nitrogen uptake patterns following moisture pulses in a cold desert community. Ecology 81:1415–1424
Gennadiyev AN, Zhidkin AP, Olson KR, Kachinskii VL (2010) Soil erosion under different land uses: assessment by the magnetic tracer method. Eurasian Soil Sci 43(9):1047–1054
Giardina CP, Litton CM, Crow SE, Asner GP (2014) Warming-related increases in soil CO_2 efflux are explained by increased below-ground C flux. Nat Clim Chang 4:822–827
Goddard MA, Mikhailova EA, Post CJ, Schlautman MA (2007) AtmosphericMg^{2+} wet deposition within the continental United States and implications for soil inorganic C sequestration. Tellus 59B:50–56
Goebel MO, Bachmann J, Woche SK, Fischer WR (2005) Soil wettability, aggregate stability, and the decomposition of soil organic matter. Geoderma 128:80–93
Graham MH, Haynes RF, Meyer JH (2002) Soil organic matter content quality: effects of fertilizer applications, burning and trash retention on a long-term sugarcane experiment in South Africa. Soil Biol Biochem 34:93–102
Guo YY, Amundson R, Gong P, Yu Q (2006) Quantity and spatial variability of soil C in the conterminous United States. Soil Sci Soc Am J 70:590–600
Heimsath AM, Chappell J, Spooner NA, Questiaux DG (2002) Creeping soil. Geology 30:111–114
Hernández J, del Pino A, Vance ED, Califra A, Del Giorgio F, Martínez L, GonzálezBarrios P (2016) Eucalyptus and Pinus stand density effects on soil C sequestration. For Ecol Manag 368:28–38
Hu YL, Zeng DH, Ma XQ, Chang SX (2016) Root rather than leaf litter input drives soil C sequestration after afforestation on a marginal cropland. For Ecol Manag 362:38–45

ICAR (2010) (Indian Council of Agricultural Research): state of Indian agriculture, 2012–2013, a report of Department of Agriculture and Cooperation, New Delhi, 9
ICRISAT (2010) ICRISAT report, International Crops Research Institute for the Semiarid Tropics. India: Patancheru; in the Red River Valley of North Dakota. J Environ Qual 22:305–310
Jandl R, Lindner M, Vesterdal L, Bauwens B, Baritz R, Hagedorn F, Johnson DW, Minkkinen K, Byrne KA (2007) How strongly can forest management influence soil C sequestration? Geoderma 137:253–268
Jenny H (1941) Factors of soil formation. McGraw-Hill, New York, p 281
Karale RL, Seshagiri Rao KV, Gaikwad ST, Venkatratnam L (1991) Soil degradation in India. In Biswas TD et al (eds) Soil-related constraints in crop production, Indian Society Soil Science Bulletin 15, New Delhi, India, pp 1–13
Kaupppi P, Sedjo R, Apps M, Cerri C, Fujimoro T, Janzen H (2001) Technological and economic potential of options to enhance, maintain and manage biological C reservoirs and geo-engineering. In: Metz B, Davidson O, Swart R, Pan J (eds) Climate change2001 – mitigation. Cambridge University Press, Cambridge, pp 301–343
Kerényi A (1991) Talajerózió, térképezés, laboratóriumi és szabadföldi kísérletek. Akadémiai Kiadó, Budapest, p. 219 (In Hungarian)
Kern JS, Johnson MG (1993) Conservation tillage impacts on national soil and atmospheric C levels. Soil Sci Soc Am J 57:200–210
Kinnell PIA (2001) Particle travel distances and bed and sediment compositions associated with rain-impacted flows. Earth Surf Process Landf 26:749–758
Kirkels FMSA, Cammeraat LH, Kuhn NJ (2014) The fate of soil organic C upon erosion, transport and deposition in agricultural landscapes – a review of different concepts. Geomorphology 226:94–105
Kleber M, Eusterhues K, Keiluweit M, Mikutta C, Mikutta R, Nico PS (2015) Mineral–organic associations: formation, properties, and relevance in soil environments. In: Advances in Agronomy. Elsevier, pp 1–140
Koven CD, Ringeval B, Friedlinstein P, Ciais P, Cadule P, Khvorostyanov D, Krinner G, Tarnocai C (2011) Permafrost C-climate feedbacts accelerate global warming. Proc Natl Acad Sci USA 108:14769–14774
Kramer MG, Chadwick OA (2016) Controls on C storage and weathering in volcanic soils across a high-elevation climate gradient on Mauna Kea, Hawaii. Ecology 97:2384–2395
Kumar S, Kadono A, Lal R, Dick W (2012) Long-term no-till impacts on organic C and properties of two contrasting soils and corn yields in Ohio. Soil Sci Soc Am J 76(5):1798–1809
Kumar S, Meena RS, Pandey A, Seema (2017a) Soil acidity management and an economics response of lime and sulfur on sesame in an alley cropping system. Int J Curr Microbiol App Sci 6(3):2566–2573
Kumar S, Meena RS, Yadav GS, Pandey A (2017b) Response of sesame (*Sesamum indicum* L.) to Sulphur and lime application under soil acidity. Int J Plant Soil Sci 14(4):1–9
Kumar S, Meena RS, Bohra JS (2018) Interactive effect of sowing dates and nutrient sources on dry matter accumulation of Indian mustard (*Brassica juncea* L.). J Oilseed Brassica 9(1):72–76
Laganière J, Angers DA, Paré D (2010) C accumulation in agricultural soils after afforestation: a meta-analysis. Glob Chang Biol 16:439–453
Lal R (1995) Global soil erosion by water and C dynamics. In: Lal R, Kimble J, Levine E, Stewart BA (eds) Soil management and greenhouse effect. Lewis Publishing, Boca Raton
Lal R (2004) Soil Carbon sequestration to mitigate "C" sequestration. Geoderma 123:1–22
Lal R (2006) Enhancing crop yields in the developing countries through restoration of the soil organic C pool in agricultural lands. Land Degrad Dev 17:197–209
Lal R, Kimble JM, Follett RF, Cole CV (1998) The potential of U. S. cropland to sequester C and mitigate the greenhouse effect. Ann Arbor Press, Chelsea
Lal R, Follett RF, Kimble J, Cole CV (1999) Managing US cropland to sequester C in soil. J Soil Water Conserv 54:374–381

Lance JC, McIntyre SC, Naney J, Rousseva W (1986) Measuring sediment movement at low erosion. Prog Environ Sci 1:307–326
Landi A, Mermut AR, Anderson DW (2003) Origin and rate of pedogenic Cate accumulation in Saskatchewan soils, Canada. Geoderma 117:143–156
Lavelle P (2000) Ecological challenges for soil science. Soil Sci 165:73–86
Lawrence CR, Harden JW, Xu X, Schulz MS, Trumbore SE (2015) Long-term controls on soil organic C with depth and time: a case study from the Cowlitz River Chrono sequence, WA USA. Geoderma 247:73–87
Liu W, Zhang Z, Wan S (2009) Predominant role of water in regulating soil and microbial respiration and their responses to climate change in a semiarid grassland. Glob Chang Biol 15:184–195
Long X, Ji J, Barrón V, Torrent J (2016) Climatic thresholds for pedogenic iron oxides under aerobic conditions: processes and their significance in paleoclimate reconstruction. Quat Sci Rev 150:264–277
Luizão RCC, Luizão FJ, Paiva RQ, Monteiro TF, Sousa LS, Kruijt B (2004) Variation of C and nitrogen cycling processes along a topographic gradient in a central Amazonian forest. Glob Chang Biol 10:592–600
Luo G, Ling N, Nannipieri P, Chen H, Raza W, Wang M, Guo S, Shen Q (2017) Long-term fertilization regimes affect the composition of the alkaline phosphor monoesterase encoding microbial community of a vertisol and its derivative soil fractions. Biol Fertil Soils 53:375–388
Maheswarappa HP, Nanjappa HV, Hegde MR (2011) Biddappa CC nutrient content and uptake by galangal (Kaempferia galanga L.) as influenced by agronomic practices as intercrop in coconut (Cocos nucifera L.) garden. J Spices Arom Crops 9:65–68
Maji AK, Reddy GPO, Sarkar D (2010) Degraded and wastelands of India status and spatial distribution (Eds. Virmani SM, Prasad R, Pathak PS). ICAR, Pusa, New Delhi, p 155
Mandal B (2011) The 29th professor JN Mukherjee-ISSS foundation lecture-soil organic C research in India – a way forward. J Indian Soc Soil Sci 59:S9
Mandal B, Majumder B, Bandyopadhyay PK, Hazra GC, Gangopadhyay A, Samantaray RN, Mishra AK, Chaudhury J, Saha MN, Kundu S (2007) The potential of cropping systems and soil amendments for C sequestration in soils under long-term experiments in subtropical India. Glob Chang Biol 13:357–369
Mandal B, Majumder B, Adhya TK, Bandyopadhyay PK, Gangopadhyay A, Sarkar D, Kundu MC, Choudhury SG, Hazra GC, Kundu S, Samantaray RN, Misra AK (2008) Potential of double-cropped rice ecology to conserve organic C under subtropical climate. Glob Chang Biol 14:2139–2151
Mann LK (1986) Changes in soil C storage after cultivation. Soil Sci 142:279–288
Martin MP, Wattenbach M, Smith P, Meersmans J, Jolivet C, Boulonne L, Arrouays D (2011) Spatial distribution of soil organic C stocks in France. Biogeo Sci 8:1053–1065
Masiello CA, Chadwick OA, Southon J, Torn MS, Harden JW (2004) Weathering controls on mechanisms of C storage in grassland soils. Glob Biogeochem Cycles 18:GB4023
McCorkle EP, Berhe AA, Hunsaker CT, Johnson DW, McFarlane KJ, Fogel ML, Hart SC (2016) Tracing the source of soil organic matter eroded from temperate forest catchments using C and nitrogen isotopes. Chem Geol 445:172–184
MEA (Millennium Ecosystem Assessment) (2005) Ecosystems and human well-being: biodiversity synthesis. In: Millennium ecosystem assessment. World Resources Institute, Washington, DC
Meena H, Meena RS (2017) Assessment of sowing environments and bio-regulators as adaptation choice for clusterbean productivity in response to current climatic scenario. Bangladesh J Bot 46(1):241–244
Meena RS, Yadav RS (2014) Phonological performance of groundnut varieties under sowing environments in hyper arid zone of Rajasthan, India. J Appl Nat Sci 6(2):344–348

Meena RS, Yadav RS (2015) Yield and profitability of groundnut *(Arachis hypogaea* L) as influenced by sowing dates and nutrient levels with different varieties. Legum Res 38(6):791–797
Meena RS, Yadav RS, Meena VS (2014) Response of groundnut (*Arachis hypogaea* L.) varieties to sowing dates and NP fertilizers under Western dry zone of India. Bangladesh J Bot 43(2):169–173
Meena RS, Dhakal Y, Bohra JS, Singh SP, Singh MK, Sanodiya P (2015a) Influence of bioinorganic combinations on yield, quality and economics of Mungbean. Am J Exp Agric 8(3):159–166
Meena RS, Meena VS, Meena SK, Verma JP (2015b) The needs of healthy soils for a healthy world. J Clean Prod 102:560–561
Meena RS, Meena VS, Meena SK, Verma JP (2015c) Towards the plant stress mitigate the agricultural productivity: a book review. J Clean Prod 102:552–553
Meena RS, Yadav RS, Meena H, Kumar S, Meena YK, Singh A (2015d) Towards the current need to enhance legume productivity and soil sustainability worldwide: a book review. J Clean Prod 104:513–515
Meena RS, Yadav RS, Reager ML, De N, Meena VS, Verma JP, Verma SK, Kansotia BC (2015e) Temperature use efficiency and yield of groundnut varieties in response to sowing dates and fertility levels in Western dry zone of India. Am J Exp Agric 7(3):170–177
Meena H, Meena RS, Singh B, Kumar S (2016a) Response of bio-regulators to morphology and yield of clusterbean [*Cyamopsis tetragonoloba* (L.) Taub.] under different sowing environments. J Appl Nat Sci 8(2):715–718
Meena RS, Bohra JS, Singh SP, Meena VS, Verma JP, Verma SK, Shiiag SK (2016b) Towards the prime response of manure to enhance nutrient use efficiency and soil sustainability a current need: a book review. J Clean Prod 112:1258–1260
Meena RS, Gogaoi N, Kumar S (2017a) Alarming issues on agricultural crop production and environmental stresses. J Clean Prod 142:3357–3359
Meena RS, Kumar S, Pandey A (2017b) Response of sulfur and lime levels on productivity, nutrient content and uptake of sesame under guava (*Psidium guajava* L.) based Agri-horti system in an acidic soil of eastern Uttar Pradesh, India. J Crop Weed 13(2):222–227
Meena RS, Meena PD, Yadav GS, Yadav SS (2017c) Phosphate solubilizing microorganisms, principles and application of microphos technology. J Clean Prod 145:157–158
Meena H, Meena RS, Lal R, Singh GS, Mitran T, Layek J, Patil SB, Kumar S, Verma T (2018a) Response of sowing dates and bio regulators on yield of clusterbean under current climate in alley cropping system in eastern U.P. Indian Legum Res 41(4):563–571
Meena RS, Kumar V, Yadav GS, Mitran T (2018b) Response and interaction of *Bradyrhizobium japonicum* and *Arbuscular mycorrhizal* fungi in the soybean rhizosphere: a review. Plant Growth Regul 84:207–223
Mikhailova EA, Post CJ (2006) Effect of land use on soil inorganic C stocks in the Russian Chernozem. J Environ Qual 35:1384–1388
Minasny B, McBratney AB, Malone BP, Wheeler I (2013) Digital mapping of soil C. Adv Agron 118:1–47
MNRE (2009) Annual Report of the Ministry of New and Renewable Energy, Government of India, New Delhi, India
MoA (1978) Indian agriculture in brief, 17th edn. Directorate of Economics and Statistics, Ministry of Agriculture, Department of Agriculture and Cooperation, Ministry of Agriculture and irrigation, Government of India, New Delhi, India
MoA (1985) Indian agriculture in brief, 20th edn. Directorate of Economics and Statistics, Ministry of Agriculture, Department of Agriculture and Cooperation, Ministry of Agriculture, Government of India, New Delhi, India
MoA (1994) Indian agriculture in brief 25th edn. Directorate of economics and statistics, Ministry of Agriculture, Department of Agriculture and Cooperation, Government of India, New Delhi, India
Monger HC, Matrinez-Rios JJ (2000) Inorganic C sequestration in grazing lands. In: Follett RF, Kimble J, Lal R (eds) The potential of U. S. grazing lands to sequester C and mitigate the greenhouse effect. CRC/Lewis Publisher, Boca Raton, pp 87–118

Montgomery DR, Dietrich WE (2002) Runoff generation in a steep, soil-mantled landscape. Water Resour Res 38:1168. https://doi.org/10.1029/2001WR000822
Müller-Nedebock D, Chaplot V (2015) Soil C losses by sheet erosion: a potentially critical contribution to the global C cycle. Earth Surf Process Landf 40:1803–1813
NAAS (2012) Management of crop residues in the context of conservation agriculture; policy paper no. 58. National Academy of Agricultural Sciences: New Delhi, India, p 12
National Bureau of Soil Survey and Land Use Planning (NBSS&LUP) (1994) Global assessment of soil degradation (GLASOD) guidelines; NBSS&LUP, Nagpur, India
National Bureau of Soil Survey and Land Use Planning (NBSS&LUP) (2005) Annual Report 2005, Nagpur
National Bureau of Soil Survey& Land Use Planning (NBSS&LUP) (2004) Soil Map (1:1 Million Scale); NBSS&LUP: Nagpur, India
NBSS&LUP: Nagpur, India, NCA. Report of the National Commission on Agriculture (1976) National Commission of Agriculture. Government of India, New Delhi, India, pp 427–472
NRSA (2000) Waste land atlas of India. Government of India, Balanagar, Hyderabad, India
NWDB (1985) Ministry of Environment and Forests, National Wasteland Development Board Guidelines for action. Government of India, New Delhi, India
Oakes EGM, Hughes JC, Jewitt GPW, Lorentz SA, Chaplot V (2012) Controls on a scale explicit analysis of sheet erosion. Earth Surf Process Landf 37:847–854
Oertel C, Matschullat J, Zurba K, Zimmermann F, Erasmi S (2016) Greenhouse gas emissions from soils – a review. Chemie der Erde-Geochem 76:327–352
Olson KR, Gennadiyev AN, Zhidkin AP, Markelov MV (2011) Impact of land use change and soil erosion in Upper Mississippi River Valley on soil organic C retention and greenhouse gas emissions. Soil Sci 176:449–458
Olson KR, Gennadiyev AN, Zhidkin AP, Markelov MV (2012) Impact of land use change, slope and erosion on soil organic C retention and storage USA. Soil Sci 177:269–278
Olson KR, Gennadiye AN, Zhidkin AP, Markelov MV, Golosov VN, Lang JM (2013a) Magnetic tracer methods to determine cropland erosion rates. Catena 104:103–110. https://doi.org/10.1016/j.catena.2012.10.015
Olson KR, Gennadiyev AN, Kovach RG, Lang JM (2013b) The use of fly ash to determine the extent of sediment transport on nearly level western Illinois landscapes. Soil Sci 178:24–28
Olson KR, Al-Kaisi MM, Lal R, Lowery B (2014a) Examining the paired comparison method approach for determining soil organic C sequestration rates. J Soil Water Conser 69:193A–197A. https://doi.org/10.2489/jswc.69.6.193A
Olson KR, Al-Kaisi MM, Lal R, Lowery B (2014b) Experimental consideration, treatments, and methods in determining soil organic C sequestration rates. Soil Sci Soc Am J 78:348–360
Olson KR, Gennadiyev AN, Kovach RG, Schumacher TE (2014c) Comparison of prairie and eroded agricultural lands on soil organic C retention (South Dakota). Open J Soil Sci 4:136–149. https://doi.org/10.4236/ojss.2014.4417
Papiernik SK, Lindstrom MJ, Schumacher TE, Schumacher JA, Malo DD, Lobb DA (2007) Characterization of soil profiles in a landscape affected by long-term tillage. Soil Tillage Res 93:335–345
Pathak H, Byjesh K, Chakrabarti B, Aggarwal PK (2011) Potential and cost of C sequestration in Indian agriculture: estimates from long-term field experiments. Field Crop Res 120:102–111
Prasad RN, Biswas PP, Soil Resources of India (2000) 50 Years of natural resource management (Eds. Singh GB, Sharma BR). Indian Council of Agricultural Research, New Delhi, India
Ram K, Meena RS (2014) Evaluation of pearl millet and mungbean intercropping systems in arid region of Rajasthan (India). Bangladesh J Bot 43(3):367–370
Rasmussen C (2006a) Distribution of soil organic and inorganic C pools by biome and soil taxa in Arizona. Soil Sci Soc Am J 70:256–265
Rasmussen C (2006b) Distribution of soil organic and inorganic C pools by biome and soil taxa in Arizona. Soil Sci Soc Am J 70:256–265

Riebe CS, Kirchner JW, Finkel RC (2004) Erosional and climatic effects on long-term chemical weathering rates in granitic landscapes spanning diverse climate regimes. Earth Planet Sci Lett 224:547–562

Riebe CS, Sklarm LS, Lukens CE, Shuster DL (2015) Climate and topography control the size and flux of sediment produced on steep mountain slopes. Proc Natl Acad Sci 112:15574–15579

Robert M (2002) La séquestration du Ce dans le sol pour une meilleure gestion des terres. Rapport sur les ressources en sols du monde.Organisation des Nations Unies pour l'Alimentation et l'Agriculture, Rome, 76 pp.

Rockstrom J, Karlberg L, Wani SP, Barron J, Hatibu N, Oweis T (2010) Managing water in rainfed agriculture – the need for a paradigm shift. Agric Water Manag 97:543–550

Royal Commission on Agriculture in India Report (1928) Agricole Publishing Academy. New Delhi, India, pp 75–76

Salemme R, Olson KR (2014) Land-use change effects on soil organic C and total soil nitrogen. Rates using Cesium-137. Soil Sci Soc Am J 50:1303–1309

Sartori F, Lal R, Ebinger MH, Eaton JA (2007) Changes in soil C and nutrient pools along a chronosequence of poplar plantations in the Columbia Plateau, Oregon. USA Agric Ecosyst Environ 118:6–2

Scarpone C, Schmidt MG, Bulmer CE, Knudby A (2016) Modelling soil thickness in the critical zone for southern British Columbia. Geoderma 282:59–69

Scherr SJ, Yadav S (1996) Land degradation in the developing world: implications for food, agriculture and the environment to 2020. IFPRI, Foof. Agril. and the environment discussion paper 14, Washington, DC, p 36

Schimel DS (1995) Terrestrial ecosystems and the C cycle. Glob Chang Biol 1:77–91

Sehgal J, Abrol IP (1994) Soil degradation in India. Status and the impact. Oxford IBH Publishing, New Delhi/Bombay/Calcutta, p 80

Sharma KL, Vittal KPR, Srinivas K, Venkateswarlu B, Neelaveni K (1994) Prospects of organic farming in dryland agriculture. In: Singh HP, Ramakrishna YS, Sharma KL, Venkateswarlu B (eds) Fifty years of dryland agricultural research in India. Central Research Institute for Dryland Agriculture, Hyderabad

Shen W, Reynolds JM, Hui D (2009) Responses of dryland soil respiration and soil C pool size to abrupt vs. gradual and individual vs. combined changes in soil temperature, precipitation, and atmospheric [CO_2]: a simulation analysis. Glob Chang Biol 15:2274–2294

Shi ZH, Fang NF, Wu FZ, Wang L, Yue BJ, Wu G (2012) Soil erosion processes and sediment sorting associated with transport mechanisms on steep slopes. J Hydrol 454:123–130

Sihag SK, Singh MK, Meena RS, Naga S, Bahadur SR, Gaurav YRS (2015) Influences of spacing on growth and yield potential of dry direct seeded rice (*Oryza sativa* L) cultivars. Ecoscan 9(1–2):517–519

Singh NT, Bandyopadhyay AK (1996) Chemical degradation leading to salt-affected soils and their management for agriculture and alternate uses. In Biswas TD, Narayanasamy G (eds) Soil management in relation to land degradation and environment. Indian Society Soil Science Bulletin 17, New Delhi, India, pp 89–101

Singh SK, Singh AK, Sharma BK, Tarafdar JC (2007) C stock and organic C dynamics in soils of Rajasthan. Indian J Arid Environ 68:408–421

Sombroe WG, Nachtergaeke FO, Hebel A (1993) Amounts, dynamics and sequestrations of C in tropical and subtropical soils. Ambio 22:417–426

Sreenivas K, Sujatha G, Sudhir K, Kiran DV, Fyzee MA, Ravisankar T, Dadhwal VK (2014) Spatial assessment of soil organic C density through random forests based imputation. J Indian Soc Remote Sens 42:577–587

Sreenivas K, Dadhwal VK, Kumar S, Harsha GS, Mitran T, Sujatha G, Ravisankar T (2016) Digital mapping of soil organic and inorganic C status in India. Geoderma 269:160–173

Srinivasarao CH, Venkateswarlu B, Dixit S, Kundu S, Gayatri Devi K (2011) Livelihood impacts of soil health improvement in backward and tribal districts of Andhra Pradesh. Central Research Institute for Dryland Agriculture, Hyderabad, p 119

Srinivasarao CH, Deshpande AN, Venkateswarlu B, Lal R, Singh AK, Kundu S (2012a) Grain yield and C sequestration potential of post monsoon sorghum cultivation in Vertisols in the semi-arid tropics of Central India. Geoderma 176:90–97

Srinivasarao CH, Venkateswarlu B, Lal R, Singh AK, Kundu S, Vittal KPR (2012b) Soil C sequestration and agronomic productivity of an Alfisol for a groundnut based system in a semiarid environment in South India. Eur J Agron 43:40–48. https://doi.org/10.1016/j.eja.05.001

Srinivasarao CH, Venkateswarlu B, Lal R, Singh AK, Kundu S, Vittal KPR (2012c) Sustaining agronomic productivity and quality of a Vertisolic soil (vertisol) under soybeans and flower cropping system in semi-arid Central India. Can J Soil Sci 92:771–785

Srinivasarao CH, Venkateswarlu B, Singh AK, Vittal KPR, Kundu S, Gajanan GN (2012d) Yield sustainability and C sequestration potential of groundnut–fingermillet rotation in Alfisols under semi-arid tropical India. Int J Agric Sustain 10:1–15

Srinivasarao CH, Venkateswarlu B, Lal R, Singh AK, Vittal KPR, Kundu S (2012e) Long-term effects of soil fertility management on C sequestration in a rice–lentil cropping system of the Indo-Gangetic plains. Soil Sci Soc Am J 76:168–178

Srinivasarao CH, Venkateswarlu B, Singh AK, Vittal KPR, Kundu S, Gajanan GN (2012f) Critical C inputs to maintain soil organic C stocks under long term finger millet (Eleusine coracana (L.) Gaertn) cropping on Alfisols in semiarid tropical India. J Plant Nutr Soil Sci 175(5):681–688. https://doi.org/10.1002/jpln.201000429

Srinivasarao CH, Venkateswarlu B, Lal R, Singh AK, Kundu S, Jakkula VS (2013a) C sink capacity and agronomic productivity of soils of semiarid regions of India. In: Lal R, Stewart BA (eds) Principles of sustainable soil management in agroecosystems. Advance soil science. CRC Press, Boca Raton, p 568

Srinivasarao CH, Venkateswarlu B, Lal R, Singh AK, Kundu S (2013b) Sustainable management of soils of dryland ecosystems of India for enhancing agronomic productivity and sequestering C. In: Sparks DL (ed) Advances in agronomy. Academic Press, Burlington, pp 253–329

Stavi I, Lal R (2011) Variability of soil physical quality in un eroded, eroded, and depositional cropland sites. Geomorphology 125:85–91

Sundermeier A, Reeder R, Lal R (2005) Soil C sequestration–fundamentals. Soil C fact sheet. The Ohio State University Extension, Columbus

Syers JK (1997) Managing soils for long-term productivity. Philos. Trans R Soc Lond Ser B Biol Sci 352:1011–1021

Tandon HLS (2004) Fertilizers in Indian agriculture—from 20th to 21st century. FDCO, New Delhi, p 240

Tendon HLS (1992) Assessment of soil nutrient depletion. In: Proceedings of the FADINAP regional seminar on fertilization and the environment, Chiangmai, Thailand, 7–11 September

Thompson JA, Pena-Yewtukhiw EM, Grove JH (2006) Soil landscape modelling across a physiographic region: topographic patterns and model transportability. Geoderma 133:57–70

Troll D (1965) Seasonal climates of the earth. In: Rodenwaldt E, Jusatz H (eds) Page 28 inworld maps of climatology. Springer, Berlin

Varma D, Meena RS, Kumar S (2017a) Response of mungbean to fertility and lime levels under soil acidity in an alley cropping system in Vindhyan Region, India. Int J Chem Stud 5(2):384–389

Varma D, Meena RS, Kumar S, Kumar E (2017b) Response of mungbean to NPK and lime under the conditions of Vindhyan Region of Uttar Pradesh. Legum Res 40(3):542–545

Velayutham M, Pal DK, Bhattacharyya T (2000) Organic C stock in soils of India. In: Lal R, Kimble JM, Stewarts BA (eds) Global climate change and tropical ecosystem. CRC Press, Boca Raton, pp 71–95

Velmurugan A, Kumar S, Dadhwal VK, Gupta MK (2014) Soil organic status of Indian forests. Indian Forester 140:468–477

Verma JP, Jaiswal DK, Meena VS, Meena RS (2015) Current need of organic farming for enhancing sustainable agriculture. J Clean Prod 102:545–547

Verma JP, Meena VS, Kumar A, Meena RS (2015a) Issues and challenges about sustainable agriculture production for management of natural resources to sustain soil fertility and health: a book review. J Clean Prod 107:793–794

Verma SK, Singh SB, Prasad SK, Meena RN, Meena RS (2015b) Influence of irrigation regimes and weed management practices on water use and nutrient uptake in wheat (*Triticum aestivum* L. Emend. Fiori and Paol.). Bangladesh J Bot 44(3):437–442

Vohra BB (1980) A policy for land and water, vol. 18. Department of Environment, Government of India: New Delhi, India, pp 64–70

Wang Y, Fu B, Lü Y, Song C, Luan Y (2010) Local-scale spatial variability of soil organic C and its stock in the hilly area of the Loess Plateau, China. Quat Res 73:70–76

Wang X, Cammeraat ELH, Cerli C, Kalbitz K (2014) Soil aggregation and the stabilization of organic C as affected by erosion and deposition. Soil Biol Biochem 72:55–65

West PC, Gibbs HK, Chad M, Wagner J, Barford CC, Carpenter SR, Foley JA (2010) Trading C for food: global comparison of C stocks vs. crop yields on agricultural land. Proc Natl Acad Sci USA 107:19645–19648

Yadav JSP (1996) Extent, nature, intensity and causes of land degradation in India. In Biswas TD, Narayanasamy G (eds) Soil management in relation to land degradation and environment, Indian Society soil Science Bulletin 17, New Delhi, India, pp 1–26

Yadav GS, Babu S, Meena RS, Debnath C, Saha P, Debbaram C, Datta M (2017a) Effects of godawariphosgold and single supper phosphate on groundnut (*Arachis hypogaea*) productivity, phosphorus uptake, phosphorus use efficiency and economics. Indian J Agri Sci 87(9):1165–1169

Yadav GS, Lal R, Meena RS, Babu S, Das A, Bhomik SN, Datta M, Layak J, Saha P (2017b) Conservation tillage and nutrient management effects on productivity and soil carbon sequestration under double cropping of rice in North Eastern Region of India. Ecol Indic. http://www.sciencedirect.com/science/article/pii/S1470160X17305617

Yadav GS, Lal R, Meena RS, Datta M, Babu S, Das, Layek J, Saha P (2017c) Energy budgeting for designing sustainable and environmentally clean/safer cropping systems for rainfed rice fallow lands in India. J Clean Prod 158:29–37

Yadav GS, Das A, Lal R, Babu S, Meena RS, Saha P, Singh R, Datta M (2018a) Energy budget and carbon footprint in a no-till and mulch based rice–mustard cropping system. J Clean Prod 191:144–157

Yadav GS, Das A, Lal R, Babu S, Meena RS, Patil SB, Saha P, Datta M (2018b) Conservation tillage and mulching effects on the adaptive capacity of direct-seeded upland rice (*Oryza sativa* L.) to alleviate weed and moisture stresses in the North Eastern Himalayan Region of India. Arch Agron Soil Sci. https://doi.org/10.1080/03650340.2018.1423555

Yoo K, Amundson R, Heimsath AM, Dietrich WE (2006) Spatial patterns of soil organic C on hillslopes: integrating geomorphic processes and the biological C cycle. Geoderma 130:47–65

Young CJS, Liu JA, Schumacher TE, Schumacher TC, Kaspar GW, McCarty D, Napton, Jaynes DB (2014) Evaluation of a model framework to estimate soil and soil organic C redistribution by water and tillage using 137Cs in two US Midwest soils. Geoderma 232(234):437–448. https://doi.org/10.1016/j.geoderma.2014.05.019

Zapata-Rios X, McIntosh J, Rademacher L, Troch PA, Brooks PD, Rasmussen C, Chorover J (2015) Climatic and landscape controls on water transit times and silicate mineral weathering in the critical zone. Water Resour Res 51:6036–6051

Restoration of Degraded Soil for Sustainable Agriculture

2

Muhammad Ashar Ayub, Muhammad Usman, Tehmina Faiz, Muhammad Umair, Muhammad Anwar ul Haq, Muhammad Rizwan, Shafaqat Ali, and Muhammad Zia ur Rehman

Abstract

Land degradation is a serious threat to agriculture which is adversely affecting the soil functions and productivity, while degraded soils stretch up to 6 billion ha worldwide. The population of the world is increasing day by day and agricultural land is declining due to degradation. It is estimated that 30% of forestry, 20% of agricultural land, and 10% of rangeland are severely affected by land degradation. Agriculture land is being degraded due to many reasons like deforestation, mining, misuse of fertilizers, and use of industrial water for irrigation purposes. This damage to ecosystem can be countered by adopting several soil restoration strategies. Major factors in land degradation which contribute to damaging the soil plant system are soil erosion, salt affectedness, decline in soil fertility and soil heavy metals contamination. Soil erosion can be minimized/controlled by afforestation, use of timber alternate, controlling the flow of water by growing cover crops, managing agricultural intensification and urban sprawl. Saline soils can be rehabilitated by growing salt-resistant crops, ploughing the salt-affected field deeply, and mixing of soil horizon. Chemically, reclamation of saline soils is also an option involving organic and inorganic amendments which can make salt-affected soils capable of giving a sustainable production. Different agronomic practices can also be followed to aid the soil rehabilitation and to increase crop productivity. Nutritional status of the soil can be improved by applying fertilizers, growing leguminous crops, green manuring, employing zero tillage practices, and crop rotation. Heavy metal contamination is one of the most severe

M. A. Ayub · M. Usman · T. Faiz · M. Umair · M. A. ul Haq · M. Zia ur Rehman (✉)
Institute of Soil and Environmental Sciences, University of Agriculture Faisalabad, Faisalabad, Pakistan
e-mail: ziasindhu1399@gmail.com; ziasindhu1399@uaf.edu.pk

M. Rizwan · S. Ali
Department of Environmental Sciences and Engineering, Government College University Faisalabad, Faisalabad, Pakistan

R. S. Meena (ed.), *Soil Health Restoration and Management*,
https://doi.org/10.1007/978-981-13-8570-4_2

degradation threats which can be minimized by using remediation techniques, organic and inorganic amendments, and phytoremediation. By adopting these strategies, degraded soils can be restored, and the world's agriculture economic losses due to land degradation can be minimized.

Keywords
Land degradation · Soil pollution · Soil erosion · Soil health · Restoration strategies

Abbreviations

4Rs	Right time, Right place, Right source, Right rate
Al	Aluminum
As	Arsenic
BET	Brunauer-Emmett-Teller
Ca	Calcium
Cd	Cadmium
CdS	Cadmium sulfide
CH_4	Methane
CNTs	Carbon nanotubes
Co	Cobalt
CO_2	Carbon dioxide
Cr	Chromium
Cu	Copper
DAP	Diammonium phosphate
DTPA	Diethylenetriaminepentaacetic acid
EDTA	Ethylenediaminetetraacetic acid
Fe	Iron
Fe-DTPA	Iron diethylenetriaminepentaacetic acid
Hg	Mercury
HMs	Heavy metals
IPNI	International Plant Nutrition Institute
Mg	Magnesium
MgO	Magnesium oxide
Mn	Manganese
MnO	Manganese oxide
Na-TNB	Sodium titanate nanobelt membrane
Ni	Nickel
Pb	Lead
Pd	Palladium
PGPB	Plant growth-promoting bacteria
pH_s	pH of saturated soil paste
Pt	Platinum

PVC	Polyvinyl chloride
PVDF	Polyvinylidene fluoride
Sb	Antimony
Th	Thorium
TiO	Titanium oxide
TiO_2	Titanium dioxide
U	Uranium
Zn	Zinc
ZnO	Zinc oxide

2.1 Introduction

Soil degradation occurs in many forms simply from erosion to soil contamination and brings the soil even toward complete unproductiveness under extreme conditions. The definition of soil degradation is very vast, and several associated causes are different natural and anthropogenic activities responsible for land degradation that results in decreased land productivity for agriculture and disturbing the natural balance of the whole ecosystem. Major factors deteriorating the soil quality are salinization, erosion, depletion of nutrients by exhaustive agricultural practices, and contamination with toxic metal ions (Gong et al. 2015; Oldeman et al. 1991; He et al. 2015; Žížala et al. 2017; Ashoka et al. 2017). Food security to meet the requirements of the ever-increasing population is a starving challenge for every country in the world, while soil degradation is negatively correlated with food security due to shrinking cultivated lands. Population census reflects that the world's population is going to increase from 7.6 billion in 2018 to 8.3 billion and 9.2 billion in 2030 and 2050, respectively (UN 2007; UNEP 2009; Cohen 2003). To feed this population, we need to rehabilitate degraded soil and stop further degradation of productive soils. Key points for sustainable production to feed the population are rehabilitation of existing degraded soils by using different amendments, good production from the degraded soils by growing resistant varieties, and controlling the factors governing degradation of agriculture soils with proper management. There are large variations among different assessment reports of the extent or intensity of degraded lands (Gibbs and Salmon 2015; Lal 2012; Bindraban et al. 2012; FAO and ITPS 2015; Meena and Meena 2017). According to different sources, figures about agricultural land degradation vary from 15% to 80% during the last two decades (FAO 2011; Bindraban et al. 2012; Oldeman et al. 1991; Gibbs and Salmon 2015; Bai et al. 2008; Pimentel and Burgess 2013; Hurni et al. 2015). Some recent estimates described the agricultural lands to be about 25% highly degraded, 44% slightly to moderately degraded, and only 8% being improved (Bindraban et al. 2012; FAO 2011). Reynolds et al. (2007) reported that dry lands are more prone to degradation which cover about 40% of the earth's land area and is related to the livelihood of 2.5 billion populations. Wood et al. (2000) reported that about 16% of the agricultural land reduces its yield due to soil erosion. Similarly, about 7% of world's arable land of arid or semiarid regions is salt affected and influences the socioeconomic balance of the world (Yensen 2008; Metternicht and Zinck 2003; Meena et al. 2015c). Nutrient depletion due to poor agricultural management and loss of soil fertility is another important issue in the soil degradation process. It is reported that about 135.3 million hectares of soil have been depleted from nutrients globally and are going to increase over time (FAO et al. 1994; Scherr and Yadav 1996; FAO, RAP 1999). Yield reduction and loss of biological diversity of soil due to the presence of certain pollutants are also sources of soil health decline under the definition of soil degradation. In this respect, soil quality can be assessed by a number of indicators like soil organic matter, soil structure, vegetation cover, soil strength, nutrients availability, bulk density, soil water infiltration, soil electrical conductivity, soil pH, microbial diversity, nematodes population, soil enzymatic activities, etc. (Schloter et al. 2003; Muñoz-rojas 2018; Yadav et al. 2018b). There are different measurements for the rehabilitation of affected soils according to the conditions prevailing there. It includes vegetation management and soil structure improvement for the control of erosion (Durán Zuazo and Rodriguez 2008). Salt-affected soils need some chemical amendments and good quality irrigation water with comprehensive drainage system (Murtaza et al. 2006). For maintenance of soil fertility, it requires proper fertilizer and crop management (Kraaijvanger and Veldkamp 2015), while effects of heavy metals can be alleviated by in situ immobilization using some immobilizing agents and phytoremediation (Rehman et al. 2016; Verma et al. 2015c). Introduction of transgenic and resistive plants for getting high biomass and good yield from such soils along with their restoration is also an appropriate way of managing this situation. In current era, it is necessary to consider the severe increase in soil degradation and to make comprehensive management practices to restore the soil productivity. In that regard, this chapter is an effort to comprehensively explain soil degradation scenario, factors responsible, and restoration practices to help concerned people worldwide to control soil degradation. This chapter is an effort to explain all leading causes of soil degradation and possible mechanisms to reclaim degraded soil and get sustainable agricultural production.

2.2 Modern Agriculture and Soil Degradation's Types, Causes, and Classification

Modern day agriculture is associated with land degradation in many ways (Fig. 2.1). The following describes in details the types, causes, and classification of degraded soils.

2.2.1 Types of Soil Degradation

Soil degradation is divided into four main types including physical, biological, chemical, and ecological degradation (Acosta et al. 2009). Physical degradation of soil generally destroys the structural features of the soil including pore continuity or geometry which makes the soil more vulnerable to compaction, crusting, surface

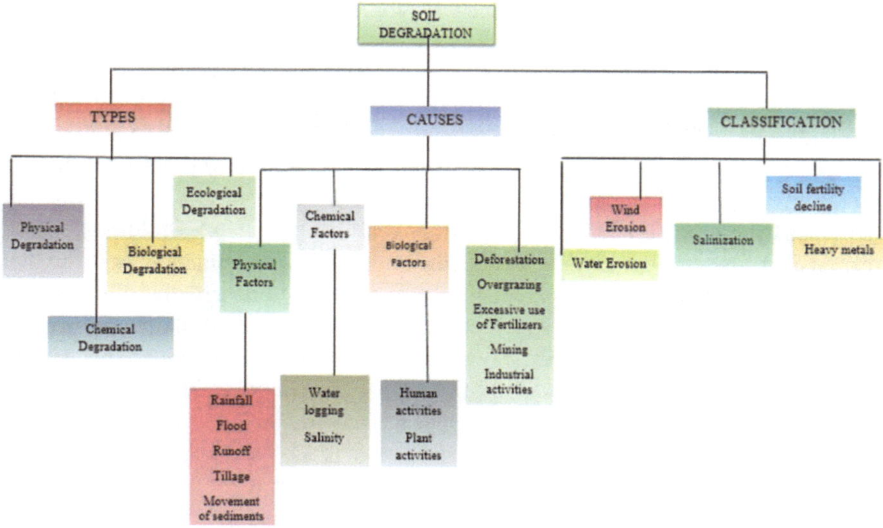

Fig. 2.1 Modern agriculture and soil degradation: types, causes, and classification

runoff, decreased water infiltration, water and wind erosion, more temperature changes, and increased tendency for desertification (Lal 2015). Chemical degradation of soil is characterized by salinization, acidification, nutrient decline, decreased cation exchange capacity, deficiency of Mg or Ca, toxicity of Mn or Al, or leaching of the essential plant nutrients (Paz-Ferreiro and Fu 2016). Contamination due to the industrial effluents or by-products also falls in the category of the chemical degradation of soil (Edelstein and Ben-Hur 2018). Biological degradation of soil includes the loss of soil biodiversity, increased emissions of greenhouse gases from soil to the atmosphere, and reduced carbon sink capacity of soil (Lehman and Taheri 2017). Due to the biological degradation, soil becomes the source of greenhouse gases (methane (CH_4) and carbon dioxide (CO_2)) emission instead of a carbon sink (Oertel et al. 2016; Meena and Yadav 2015). The combination of all the three categories including physical, chemical, and biological degradation refers to the ecological degradation. It disrupts the functions of the ecosystem including water infiltration, water purification, elemental cycling, and disturbance of the hydrological cycle and causes a decrease in the net productivity of the biome (Barrios 2007).

2.2.2 Causes of Soil Degradation

Many physical, chemical, and biological factors are involved in the degradation of the land (Khresat and Qudah 2006; Martínez et al. 2005; Kumar et al. 2017b). Physical factors responsible for land degradation include rainfall, floods, runoff, tillage, and movement of sediments which deteriorates the top fertile soil layer and reduces the soil quality. Such physical factors cause wind and water erosion, and their resultant forces change the structure and the composition of soil by moving the soil particles away from the soil surface (Lal 2015). These factors also reduce the level of organic matter and hence cause the decrease in soil fertility (Rajan et al. 2010). Decrease in the soil nutrients due to waterlogging or salinity is classified as the chemical factor involved in the soil degradation. Salts are accumulated in the soil and cause the nutrients to leach down which results in the decrease of soil quality. These chemical factors cause the irreversible decrease in the nutrients and the productive capacity like hardening of the aluminum and iron-rich clayish soils into hardpans (Stefanou and Papazafeiriou 2013; Yadav et al. 2018a). Biological factors responsible for soil degradation involve different plant and human activities which reduce the soil quality. The overgrowth of fungi and bacteria greatly influences the activity of the microbes in soil by biochemical reactions and ultimately reduces the crop yield. Human activities like poor management practices reduce the nutrient level of the soil and decrease the soil fertility (Montgomery 2007). Mainly, biological factors decrease the activity of microbes in the soil.

In addition to these factors, land degradation is also caused by deforestation, overgrazing, excessive fertilizer use, mining and industrial activities, etc. Increase in the population of livestock results in the overgrazing due to which soil quality is affected greatly. Overgrazing reduces the crop cover and breaks the particles of the soil which increases the erosion rates (Kairis et al. 2015).The misuse or excess use of fertilizers or pesticides also causes land degradation as they kill the organisms that secrete adhesive chemicals that bind the soil particles together. Excessive use of fertilizers also affects the biological activity and causes the toxicities of mineral nutrients and pollutants in the soil (Andreu and Pic 2004; Meena et al. 2016).

Deforestation causes land degradation by removal of the vegetative cover and trees (Varela et al. 2001). Vegetation binds the soil particles together and helps in soil formation. When vegetative cover is removed, it affects the soil properties like water holding capacity, soil aeration, and biological activity. Activities like slash-and-burn techniques and logging used by people who invade the forest areas for farming purposes make the soil less fertile and unproductive at the end. Mining and industrial activities affect the soil's physical, biological, and chemical properties (Swain et al. 2011; Datta et al. 2017a). Industries release the industrial waste and effluents into the environment—rivers, land, groundwater—which contaminates and reduces the quality of soil. Mining reduces the crop cover and releases hazardous chemicals (heavy metals) that cause the poisoning of the soil making it unproductive for other purposes.

2.2.3 Classification of Soil Degradation

Land degradation can be divided into five classes including:

(i) Water erosion
(ii) Wind erosion
(iii) Salinization

(iv) Soil fertility decline
(v) Heavy metal polluted soils

2.3 Soil Degradation Due to Erosion

Soil erosion is the loss of fertile upper layer of the soil. It is a worldwide problem that is an alarm for natural resources (Cerdà et al. 2009; Zhao et al. 2013; Dhakal et al. 2015). The changes occurring on the soils due to human-induced erosion are alarming. Soil erosion basically declines the soil quality which ultimately reduces the productivity of the natural ecosystems (Jiang and Zhang 2016; Kumar et al. 2017a). Due to the removal of the fertile layer of soil, the productive capacity of the soil is also reduced (Henry et al. 2014). The productivity of some soils has decreased up to 50% due to soil erosion. It also damages the diversity of animals, microbes, and plants present in the soil (Fitzpatrick et al. 2018). Detachment, transportation, and deposition are the three processes which occur during erosion (Kirkels et al. 2014; Kinnell 2005; Mahmoodabadi et al. 2014a, b). Detachment is the beginning process of the soil erosion (Kirkels et al. 2014) in which the aggregates are broken down into smaller particles (microaggregates). The flowing water and wind energy transport these particles and deposit them at the greater distances when the velocity of the wind and water decreases, and as a result the soil gets eroded.

2.3.1 Factors Affecting Soil Erosion

Soil erosion is affected by various environmental factors, among which the impacts of rainfall intensity, wind, topography, slope, and vegetation are prominent. Rainfall is a very important factor responsible for erosion. The higher the intensity of rainfall, the higher will be the rate of erosion because the rate of the soil particles detachment is increased due to the highly intensified rainfalls. The beating action of the raindrops causes the compaction and crusting of the soil which decreases the rate of infiltration and increases the surface runoff and ultimately increases erosion (Vanwalleghem et al. 2017). The length and the steepness of the slope also had a great influence on soil erosion. Erosion increases by increasing the steepness and the length of the slope. By consolidating the smaller fields to a larger one, the length of the slope increases; due to which, erosion potential increases. The impact of the steepness of the slope on soil erosion also depends on the land use and farming system (Balasubramanian 2017b; Meena et al. 2017b). Steeper and marginal lands which are converted from forests to agriculture to replace the already eroded land have high rate of erosion. In Jamaica and the Philippines, erosion rates are reported higher than 400 t/ha per year because in Jamaica about 52% of the land has slope larger than 20% and in Philippines almost 58% of the land has slope greater than 11% (Lal and Stewart 1990). Soil erosion due to the impact of wind energy is also very common. Due to wind energy, soil particles are displaced and settled down at the great distances. Wind energy is capable enough to drive the soil particles up to thousand miles away (Ravi et al. 2010).

Soil is more prone to erosion when it is exposed to wind or rainwater. Raindrops strike the exposed soil with great intensity, and the soil particles are easily displaced from the surface. As a result, a thin film of the soil is removed from the surface, and such type of erosion is termed as sheet erosion which is a very dominant form of land degradation (MacDonald and Larsen 2009). The influence of soil erosion due to rainfall is more on sloping land because on sloping land maximum soil moves downslope in the form of water splashes. Soil erosion and rainfall have a direct relationship. The greater the rainfall, the greater will be the rate of erosion (Van Dijk et al. 2002). Raindrops hit the bare soil and drive the soil particles away. Under the same vegetation and soil conditions, heavy rainfall for short period causes more erosion than low-intensity rainfall (Mohamadi and Kavian 2015; Yadav et al. 2017b). It is very difficult for humans to control the causes of precipitation, but harmful effects of rainfall can be reduced by the proper management of soil and crops (Tadesse et al. 2017). The loss of vegetation is another major cause of erosion. The rate of the erosion can be greatly reduced by the vegetation. When the vegetative cover of the soil increases up to 30%, the erosion starts decreasing. Soils having the vegetative cover (plant biomass) are less prone to erosion because the layer of the biomass present on the soil dissipates the raindrops and wind energy and the fertile topsoil remains safe (Swag 2002). The loss of vegetation has caused several problems. Soil is more prone to erosion if there is no vegetative cover present on the soil. Vegetation acts as a protective cover for the soil and protects the soil from being destroyed by the impact of raindrops or by the action of the wind. Stubble and plant residue present in the soil cut the speed of the wind at ground level. Different crop covers provide different kinds of protection, so suitable plant cover can be developed for the control of soil erosion. The efficiency of the vegetative cover depends on the extent of protection and the amount of rainfall during time (Hugo 2008). Loss of vegetation is widespread in the countries where there is large population and management practices are not proper to protect the soil. In some areas where there is scarcity of fuelwood, people use grasses and shrubs for burning purposes (Neelo et al. 2015). Such type of practices left the soil completely exposed to rain and wind energy, and erosion is caused more frequently in these soils.

Soil erosion due to natural adversities like landslides and earthquakes is beyond the human control. In landslides soil particles are displaced and move down the slope and are mainly accompanied by human activities like construction of buildings and roads and removal of trees (Amundson et al. 2015; Sihag et al. 2015). However, during earthquake events, a large amount of soil and crops in the hilly areas and the surrounding areas are affected. But the impact of the earthquakes is comparatively less because such type of event occurs relatively rare. The construction of infrastructures including roads, buildings, and parks increases the rate of erosion (Alexiew et al. 2015). The erosion rates from construction increases almost 20–500 t/ha per year. The erosion due to construction can be decreased if the natural vegetation regrows (IECA 1991). If the soil remains occupied by the construction sites, the area is completely lost for vegetation and remains eroded.

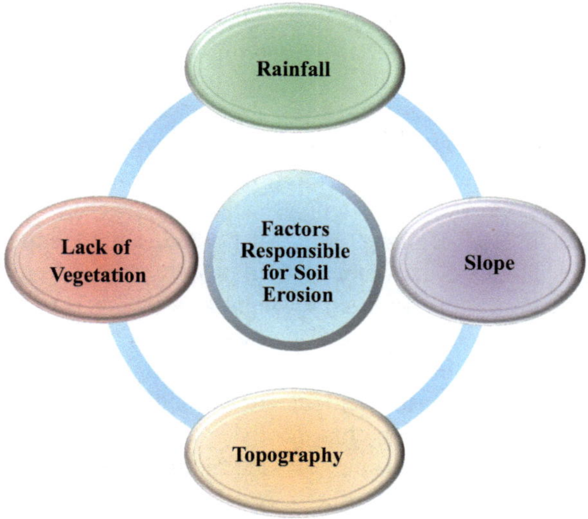

Fig. 2.2 Factors affecting soil erosion

Soil structure also plays a main role in soil erosion. Medium- to fine-textured soils and the soils with low organic matter and weak structure can easily be eroded. These soils are easily eroded because there is less infiltration rate of water and more runoff; therefore, the soil particles are displaced easily with water. Well-drained soils having deep and porous structure show more resistance to erosion. Large soil aggregates are difficult to erode both by wind and water (Jiao et al. 2009; Misra and Teixeira 2001; Meena et al. 2017a) (Fig. 2.2).

2.3.2 Effect of Soil Erosion on the Environment

2.3.2.1 Effects on Soil

The extent of soil erosion depends on the different characteristics of the soil profile (Pernar, et al. 2011). Soils are made up of different horizons each having variations in their properties. Upper layers of the soil contain more nutrients and also high water holding capacity (Vanwalleghem et al. 2017; Buragohain et al. 2017). Erosion deteriorates the quality of soil as it removes the nutrient-rich layer of the soil. It also affects the structure and even the texture of the soils by breaking down the aggregates, by moving the smaller size particles, and by the removal of organic matter. Continuous removal of the soil particles by the action of wind or water results in the textural change of the soil. Changes in the texture reduce the water holding capacity of soils, and ultimately the soil suffers from a severe drought condition. Erosion is also responsible for reducing the water quality as erosion breaks down the aggregates into smaller particles; these particles are driven away from the soil and enter different watercourses like rivers, dams, lakes, etc., and disrupt the ecosystem and also contaminate the drinking water. The entry of the sediments into watercourses may lead to the blockage of the drainage channels and streams and causes bank erosion, and ultimately water quality is reduced. It also threatens infrastructures like roads, buildings, etc. (Balasubramanian 2017).

2.3.2.2 Effects on Plants

Soil is the natural medium for plant growth and a natural resource for production of food. Plant growth is severely affected by soil erosion. Soil, when degraded by different factors, affects crop growth and yield (Brevik et al. 2015; Brevik et al. 2017; Blum 2013; Steffan et al. 2018; Dadhich and Meena 2014). Erosion causes nutrient loss, reduces the water holding capacity, and also reduces the soil layer thickness which is very important for plant growth (García-Díaz et al. 2017; Li et al. 2016). Poor tilling practices and excessive irrigation cause the loss of nutrients and make the soil less fertile for agricultural activities and natural vegetation. Hiel et al. (2018) stated that by using agricultural methods like leaving the organic matter in the soil and by ensuring the last year's crop residue remains in the soil can increase soil fertility. Erosion also affects the rooting depth, plant-available water reserves, and organic matter present in the soil. When the rate of erosion exceeds the formation of the new topsoil, it leads to the decrease in soil productivity and results in less agricultural yield and income (Hemant and Pani 2017; Yadav et al. 2017a). Drifting of the soil causes the loss of fertility and leads to the decline in growth and yield of the crops. Wind erosion causes the sandblasting of the seedlings, and as a result crops are completely damaged, and resowing of the seeds becomes necessary. Plants which are suffering from the sandblasting are more prone to disease which results in the decrease of the quality and yield of crops (Balasubramanian 2017).

2.3.3 Restoration of Eroded Soil with Different Management Practices

Restoration of eroded soils includes the many processes (Table 2.1) explained blow.

2.3.3.1 Soil Stabilization

Since the 1950s, different practices including terracing, natural vegetation, micro-catchments, check dams, etc., have been used for the conservation of soil and water (Tang 2004).

Terrace Farming

Terraces are the structures designed to control water erosion and are used for the stabilization of hillslopes (Wei et al. 2016; Meena et al. 2014). Field terraces are used to conserve moisture in the less rainfall areas and to control runoff in the areas having high rainfall. Terracing is also used to mitigate the off-site effects of erosion including the water quality decline, accumulation of soil particles in reservoirs, flooding, etc. (Vanmaercke et al. 2011). Soil and water conservations are more in terraced fields compared to the sloppy land. In China terracing reduces sediment

Table 2.1 Restoration of eroded soils with different management practices

Control practice	Characteristics	References
Terrace farming	Structures designed to control the water erosion and used for the stabilization of hilly slopes	Wei et al. (2016); Vanmaercke et al. (2011)
	Mitigate the off-site effects of erosion	
Vegetative cover	Physical barrier for the soil and changes the flow of the sediments on the soil surface	Van Dijk et al. (2002); Martínez et al. (2006); Rey (2003); Puigdefábregas (2005); Durán et al. (2006); Wainwright et al. (2002)
	Protects the soil from erosion by decreasing the runoff rate and increasing the rate of infiltration	
Deep ploughing	Used to break the plough or hard pan	Arachchi (2009); Alcántara et al. (2016))
	Increases the water retention of the soil, rate of infiltration, root penetration	
	Promotes the plant growth and yield	
Micro-catchment	Small structures like holes, basins, and bunds	Paricha et al. (2017)
	Designed to control the runoff and ultimately increase the infiltration rate	
Mulching	Agronomic practice	Jordán et al. (2011); Jiménez et al. (2016); Keizer et al. (2015); Mandal and Sharda (2013); Robichaud et al. (2013); Zhao et al. (2015); Cook et al. (2006); Mulumba and Lal (2008)
	Different materials like plant residues, straw, stones, gravel, etc., are spread on the soil	
	Improves the infiltration capacity and increases the intake of water and storage	
Temporary control measures	Topsoiling, hydroseeding, soil binders, silt fence, sediment pit, and riprap structure are used	Balasubramanian (2017)

formation and runoff by 53.0% and 48.9%, respectively (Chen et al. 2017). In Tanzania terracing decreased the soil loss by 96.3% (Wickama et al. 2014). In Palestine there was almost 20 times more erosion in non-terraced fields as compared to the terraced field (Hammad et al. 2006). The efficiency of terracing to conserve soil and water depends on a number of factors such as land use, soil texture, climate, land management, soil structure, etc. (Arnaez et al. 2007; Ehigiator and Anyata 2011; Liu et al. 2011; Park et al. 2014; Meena and Yadav 2014).

Vegetative Cover

Vegetation is a very important measure to protect the soil from being eroded. Vegetative cover protects the soil from erosion by decreasing the runoff rate (Rey 2003; Puigdefábregas 2005, Durán et al. 2006; Varma et al. 2017) and increasing the rate of infiltration (Wainwright et al.2002). Plants provide shelter to the soils and the soil is fixed with the roots of the plants (Gyssels et al. 2005; de Baets et al. 2007a, b). Vegetation behaves like a physical barrier for the soil and changes the flow of the sediments on the soil surface (Martínez et al. 2006). Vegetation plays a key role in controlling the erosion caused by to the impact of rainfall. Plant cover intercepts the rainfall and protects the soil from the impact of raindrops, but for long term it may increase the stability of soil aggregates and increase the infiltration rate (Hugo 2008). The efficiency of plant cover to reduce erosion depends on the quantity and type of the cover. Its effectiveness also depends on the extent of protection available for the different periods of time throughout the year. Winter crops or alfalfa provides the protection to the soil for the maximum time in a year (Balasubramanian 2017; Dhakal et al. 2016).

Deep Ploughing

Due to erosion, soils become compacted, and the soil compaction causes low water retention by the blockage of capillary edges and reduces the infiltration rates which ultimately affect the growth and yield of the crops (Singh et al. 2014; Meena et al. 2018). Due to the decrease in the water and nutrient supply, the fertility of soils may also decrease (Hamza and Anderson 2005). Deep ploughing is one of the best options to cope with the problem as it increases the water retention of the soil, rate of infiltration, and root penetration and thus promotes plant growth and yield (Arachchi 2009). In deep ploughing, moldboard plough is used to invert the soil, and the continuous cultivation results in the formation of a new surface horizon. It also improves the soil organic carbon as topsoil material is buried on the subsoil in deep ploughing. Since 1930 deep ploughing is done to the depth of about 60–120 cm and is currently used in different agricultural soils throughout the world. The practice of deep ploughing has become rare since 1970 but is still used to break the plough or hard pan and increases the rooting zone (Alcántara et al. 2016; Kumar et al. 2018).

Micro-catchment

Micro-catchments are the small structures including holes/basins designed to control the runoff and ultimately increase the infiltration rate and conserve water. These structures are made artificially by man. The micro-catchments are effectively introduced into the soils having good texture and can be exposed to runoff. The construction of the micro-catchments depends mainly on the crop requirements. Longitudinal micro-catchments are designed for field crops, while circular, square, or rectangular micro-catchments are mainly designed for trees or the pasture plants. The main benefit of the micro-catchment is having the high runoff yield as compared to the larger catchment (Paricha et al. 2017).

2.3.3.2 Temporary Control of the Sediments

Temporary control of the sediments involves the installation of temporary control measures which are completely removed when permanent control measures are introduced or vegetation is established (Measures 2011). Some common temporary measures include topsoiling, mulching, hydroseeding, soil binders, silt fence, sediment pit, and riprap structure. Sometimes, topsoil is placed over the subsoil, and it's the good medium for the growth of vegetation. This process is called topsoiling, and it provides a short-term protection from erosion (Balasubramanian 2017; Meena et al. 2018a).

Mulching is an agronomic practice in which a different material is spread on the soil for conservation purpose, and it also favors the plant growth (Jordán et al. 2011). Different materials used for mulching include plant residues, straw, stones, gravel, etc. (Jiménez et al. 2016; Keizer et al. 2015; Mandal and Sharda 2013; Robichaud et al. 2013; Zhao et al. 2015). Mulches protect the soil from the impact of raindrop and reduce the rate of soil and water loss in various environments (Sadeghi et al. 2015; Keesstra et al. 2016; Prosdocimi et al. 2016; Mwango et al. 2016; Dadhich et al. 2015). It breaks the concentration of nutrients and sediments present in the runoff water (Gholami et al. 2013). It also improves the infiltration capacity and increases the intake of water and storage (Cook et al. 2006; Mulumba and Lal 2008).

Hydroseeding is a process in which the suspension of seeds is used. Seeds are mixed with water and then sprayed on the area where revegetation must be done. This can control erosion temporarily and helps to establish vegetative cover quickly (Afzal et al. 2005). Soil binders are used to bind the soil particles and improve the resistance of the soil against pressure and water. It stabilizes the soil and control the erosion. Cement, lime (hydrated lime, quick lime, lime slurry), and fly ash are some common soil binders (Morgan 2014). Silt fence is also used to control erosion temporarily. It is a temporary barrier made up of very porous fabric and supported by a metal or wooden post in the soil, so it can easily be removed from the soil. Silt fence helps the soil to be retained on the disturbed surface. It controls the runoff and retains the sediments on the surface. A silt fence should be of appropriate material and placed in the soil properly (Mehenni et al. 2016; Service et al. 2002). A sediment pit is a temporary pit formed on the soil which is used to treat the sediment-loaded water and trap the water before it is discharged. A riprap is the structure in which the rocks are piled up along armor bridges, embankments, streambeds, shorelines, etc., to control runoff and ultimately prevent erosion (Balasubramanian 2017).

2.4 Soil Degradation Caused by Salinization

Salinization is developed due to the accumulation of soluble salts in soil. The salt accumulation in soil exceeds upto a range to which plant growth is reduced that leads to reduction in agricultural production, health of environment and also affects the economics (Metternicht 2017). At the initial stages, it affects the soil organisms that decrease the soil productivity, but at advance stages it damages all living organisms and plants. This condition ultimately transforms the productive and fertile soils into unproductive, decertified, and barren lands. Sodium salts cause dispersion in the clayey type of soil and convert it into degraded soil (Van Meter et al. 2011).

Soil health is directly related with the health of plants which are useful for the animals. Degraded soils also damage the flora and fauna of these areas. In the salt-affected soils, soil structure is better, but the plant growth is very poor because of the low uptake of water and nutrient in the plant body experiencing exosmosis (Li et al. 2009). Excess amount of salts can inhibit the growth of plants by reducing the uptake of water into the plants. In this condition, plants can also face the deficiency of nutrients (Pisinaras et al. 2010; Meena et al. 2015d).

2.4.1 Causes of Soil Salinity and Sources of Salts

Salinity is caused by natural processes as well as human activities. Natural salinity is caused by climate change, weathering of rocks, ion exchange, and equilibrium reactions of minerals that can control the chemistry of soil and water (Li et al. 2014). In those areas where precipitation is less than the evaporation, downward movement of water is low because these soluble salts cannot move out of the root zone and soil profile (Re and Sacchi 2017). So, the salts can precipitate in the soil and increase the salinity status of the soil. Irrigation water is the main source of salts containing soluble salts (Herbert et al. 2015).

Salinity that is caused by natural processes is also called primary salinity. In this process salts are accumulating in the soil from the weathering of rocks and from rainfall for longer period of time (Yan et al. 2015). During the rainy season, some rainwater evaporates from the soil, water bodies, and vegetation, and some rainwater infiltrates into the soil and underground water and some goes into rivers, streams, and oceans. But the rainwater contains small amount of salts and builds salinity by passage of time especially in the clayey soil and also in underground water (Kaushal et al. 2018; Pedrotti et al. 2015). In dry areas where rainfall is less, much amount of water evaporates from the soil and transpire from the plant, and large amount of salts accumulates in the land (Chongo and Kawanga 2015; Amiri et al. 2016; Yadav et al. 2017).

In the natural process chemical, physical, and biological weathering of parent rocks produces soluble salts and weathered material which is transported to groundwater and geologically deposits (Jagoutz 2006). Salinity is occurring due to the presence of carbonate minerals and feldspars in the parent material (Lottermoser 2010). Seawater contains greater amount of salts, and sea level rises day by day; seawater is also called saline water (Jagoutz 2006). Salinity is also caused by the evapotranspiration of water and less rainfall in the arid zone. Furthermore, winds in the coastal areas can move some amount of salt to adjacent normal areas (Kaushal et al. 2018; Verma et al. 2015b).

Anthropogenic activity can also cause salinization by applying the salt-rich water to the crop land. This type of salinity is called secondary salinity that can aggravate by the exploitation of coastal water and coastal aquifers causing intrusion of seawater, inappropriate agricultural and irrigation practices, and also low drainage rate into the soil (Cañedo-Argüelles et al. 2013; Hintz and Relyea 2017; Rina et al. 2013). In irrigation practices, excess amount of water used in arid climatic conditions in clayey soil types can also cause the salts to accumulate in the soil, and because of these, the land receives less amount of rain (Ayers et al. 2017). In some soils, having less drainage results in salt-affected soils under waterlogged conditions. Waterlogging prevents the leaching of salts which are imported by irrigated water.

2.4.2 Effects of Salts on Soil Fertility Status and Plants

Plant growth is affected by salts because they can increase the osmotic pressure which is interfering with the plant's nutrition. In high concentration, salts reduce the ability of plants to uptake water which can cause the deficiency of water (Ventura et al. 2014; Xiao et al. 2016). Reduction in water uptake can cause metabolic and some genotypic changes in plants, and this condition is called "wilting" (Grewal 2010). Moreover, plant growth is reduced by the nutrient imbalance and specific ion-toxicity under salt-affected conditions. Reduction in plant growth is time dependent, and this process takes place on two phases. In the first phase, the plant faces the deficiency of water, and in the second phase, the plant faces the toxicity of ions due to the high accumulation of salts in the plant roots and other parts (Lycoskoufis et al. 2005; Hapani and Marjadi 2015).

Salts also damage the physiological system of plant carbon metabolism and photosynthesis because of the less availability of CO_2 due to diffusion limitation and also reduction in the photosynthetic pigments (Bazihizina et al. 2012; Boursiac et al. 2005). Salts accumulated in the leaves can inhibit photosynthesis, and primarily salts reduce mesophyll and stomatal conductance to carbon dioxide and decrease the chlorophyll contents that can affect the light absorbance during the process of photosynthesis. Salt can also reduce the leave's surface area, and expansion due to this can reduce light interception (Hasaneen et al. 2009). Some ions that are more injurious can inhibit protein synthesis, photosynthesis, inactivation of enzymes and damage the other organelles like chloroplast. All these effects are generally important in the older leaves because they can transpire more and thus can accumulate more salt ions (Lima Neto et al. 2014).

Cell ion homeostasis and osmotic stress is caused by the salt accumulation in root zone. Cell ion homeostasis is induced by both the reduction in the uptake of essential nutrients like K^+, NO^{3-}, and Ca^{2+} and accumulation of Cl^- and Na^+. Plants which are grown under salt stress can also face specific ion toxicity by the accumulation of boron, sodium, and chloride in the transpiring leaves tissues and damage them. Those plants under salt stress have several nutrient deficiencies and imbalance caused by the higher concentration of sodium and chloride which are derived from ion competition, i.e., Na^+/K^+, Na^+/Ca^{2+}, Cl^-/NO^{3-}, and Ca^{2+}/Mg^{2+}, in the tissues of plants. Plants can also show the deficiency symptoms of nutrients under salt stress (Giuffrida et al. 2014; Wang et al. 2003; Newell 2013; Khan et al. 2000; Chen and Hoehenwarter 2015; Chartzoulakis et al. 2002).

Salt-affected soils decrease the marketable and increase the unmarketable yield of the crops by reducing the size and taste of fruits. In vegetable, it reduces the productivity of tubers and leaves leaving them of no commercial value (Hiwale 2015). Salts (mineral salts) can also cause some diseases in the vegetables such as blossom-end rot in pepper fruit, tomato, and eggplants due to calcium deficiency (Hu and Schmidhalter 2005; Koyro 2006; Meena et al. 2016a). But in some cases, salinity increases the dry matter content in fruits, total soluble solids, and contents of acid in tomato, melon cucumber, and sweet pepper. Also, it increases the antioxidant activity and carotenoids contents in tomato plants. Salt stress increases carotenoid and polyphenol contents and decreases concentration of nitrate ions and oxalic acid in spinach (Stavridou et al. 2017; Colla et al. 2012; Oyetunji and Imade 2015). Effect of salinity on crops is highly crop type, environment and soil type dependent.

2.4.3 Restoration Practices for Salt-Affected Soil

For the restoration of salt-affected soil, drainage conditions of soil must be considered because leaching of soluble salts is necessary during the restoration of salt affected soil (Amini et al. 2016; Thimmappa et al. 2017). For better drainage of water, the number of pores and size of pores are very important. Drainage can move the soluble salts out of the root zone, and it has the ability to move water throughout the plant's roots. Good quality irrigation water must be available in large quantity during the restoration of salt-affected soils (FAO 2005).

2.4.4 Remediation of Salt-Affected Soils

Salt-affected soil can be remediated by chemicals such as calcite, gypsum, calcium chloride, and other organic and inorganic amendments being used worldwide. These are of low cost, effective easy to use and have long lasting impacts. The physical, chemical, and biological properties of salt-affected soils are improved by the application of these amendments (Amini et al. 2016; Shaaban et al. 2013). All these amendments have their own characteristics, mechanisms and properties to improve the degraded soils. These are also improving the plant's growth and development. Gypsum is less costly and is useful in the remediation of salt-affected soil. It is mostly used in sodic and saline-sodic soils as an amendment but not for the saline soils. Leaves of plants grown in salt affected (Sodic or saline sodic) soils have high sodium to potassium ratio, but when gypsum is applied in the soil, this ratio decreases (Murtaza et al. 2011; Abou-Shady 2016; Öztürk et al. 2016; Ram and Meena 2014).

Organic matter is also used as an amendment; it does not only reduce the concentration of salts in the soil but also improve plant growth. Organic matter provides the exchange sites for the adsorption of ions from the soil solution. Organic matter undergoes biochemical oxidation and releases CO_2, plant food nutrients, and a variety of organic acids which can reduce salinization. Decomposition of organic matter leads to the formation of H_2CO_3 which reacts with $CaCO_3$ and gives $Ca(HCO_3)_2$ such as the following:

$$H_2CO_3 \quad + \quad CaCO_3 \rightarrow Ca(HCO_3)_2$$

Carbon dioxide and organic acids can dissolve the lime which is present in the soil and release soluble calcium ions to reduce sodium ion desorption ratio (Esteban et al. 2016; Rath and Rousk 2015).

2.4.5 Management of Salt-Affected Soils

Salt-affected soils can be managed by adopting different irrigation methods. The effect of soluble salt water and soil salinity can be mitigated and prevented by the artificial drainage. These are influenced by nutrient use efficiency and water use efficiency, salt distribution and accumulation, and salt leaching (Chesworth 2008). Water is applied through different irrigation methods such as surface drip irrigation, furrow irrigation, low-energy precision application, and subsurface drip irrigation (Jayasekera and Hall 2007). Surface irrigation and subsurface irrigation methods are better than other irrigation methods for salinity management because it is increasing nutrient use efficiency and water use efficiency (Qadir et al. 2008). An appropriate scheduling irrigation with drip irrigation and subsurface irrigation methods reduces the effect of salt via maintaining soil moisture level in the rhizosphere and leaching of salt steadily to the edge of wetting zone. Subsurface drip irrigation reduces the sodium and chloride accumulation and increases water use efficiency in some plants. In furrow irrigation method, wetting front can move the soluble salts in the soil. Salts accumulate between the furrows in the middle space by the moving water in adjacent furrow. Salts present in soil can also be reduced by adopting different agronomic practices like planting arrangement and seedbed preparation that can accumulate salts away from the zone of germinating seeds and roots of plants (Singh et al. 2016; Jamil et al. 2017).

Fertilizers are the major source of salts that can cause salinization in the soil which is affected by application of fertilizers, fertilizer characteristics, quality of irrigation water, and fertilizer application schedule. Excessive use of fertilizer must be avoided, and selection of the chloride-free, low-saline, and pure fertilizers should be prioritised. Nutritional requirements of plants must be taken into account while applying fertilizer, soil fertility status should be checked so we can get idea of present nutrient status of soil and remaining nutrients can be provided via exogenious fertilizer application (Mehdi et al. 2015; Ivits et al. 2013).

Fertilizers can be applied by mixing it with water (fertigation) and it can reduce the salinization stress on plants. This stress is reduced due to improvement in fertilizer use efficiency and increased site specific nutrient availability (Kaushal and Wani 2017). This process allows very low rate of fertilizer application and increases the frequency of fertilizer application that can supply nutrient to plant according to its requirement. During fertigation, the EC of water must be lower than the EC of the threshold level that can be tolerated by the plants. Fertigation with sulfuric acid and nitric acid is the rapid way to minimize the sodicity and salinity in arid region. Soil pH is decreased by nitric acid fertigation; this can increase the calcium dissolution in soil that can minimize the calcium-sodium ion competition. It can also decrease the salinity in root zone caused by chloride, because excess of chloride can be counterbalanced by nitrate. In some regions where rainfall is less (semiarid to arid regions) alkaline calcareous soils are found having high concentration of calcium carbonate in soil. For this, sulfuric acid must be applied in fertigation as this can solublise calcium carbonate producing calcium ions which replace adsorbed sodium making it leach down the soil profile and result in reclamation of saline sodic and sodic soils (Jamil et al. 2017; Singh et al. 2016).

Effect of salts can be mitigated by applying biofertilizers which can assist in production of growth hormones involved in growth improvement of plants under salinity stress. A formulated product that can contain one or more microbes involved in enhancement of plant growth by providing nutrient, making nutrients in available form (nutrient solublization), and/or production of growth hormones is known as biofertilizer. Endomycorrhizal fungi, ectomycorrhizal fungi, plant growth-promoting rhizobacteria, and many other microbes can improve plant growth, nutrient uptake, and plant tolerance to salt stress (Bech 2017; Singh et al. 2016; Meena et al. 2015). Chelating agent must be used to remediate the salts like Fe-DTPA (Andreottola et al. 2009).

2.5 Degraded Soils Bearing Low Fertility Status

Soil fertility relates with crop yield; fertile soil contains all major essential nutrients for the basic plant needs like nitrogen, phosphorus, and potassium; contains some other nutrients which are required to plants like Ca, Mg, S, Fe, Ni, Zn, Cu, Br, and Mo; and has some other beneficial nutrients for the plants. Fertile soils have some amount of organic matter which is responsible for improving soil structure, soil nutrient retention, and soil nutrient holding capacity and maintaining the pH of the soil. Unfortunately, all these nutrients are not present in the soil to fill up the crop requirements, and some soil has unfavorable conditions for the uptake of nutrient (Madejón et al. 2016).

2.5.1 Causes of Low Soil Fertility

Soil fertility is lost by the loss of the upper fertile layer of the soil that contains organic matter and nutrients. This process is referred to as soil erosion, where the soil is moved with water and air, resultantly decreasing soil fertility. This is related to human activities like deforestation, poor soil management, and overgrazing (Gascho 2005; Frossard et al. 2016). Complete process and mechanism of soil erosion are already explained in content 2.3 ("Soil Degradation Due to Erosion"). Extensive cropping can cause the removal of nutrients from the soil due to over usage and this process is called nutrient depletion or nutrient mining. Nutrient mining is severe threat to sustainable agriculture especially in small farmlands having extensive cropping and imbalanced use of fertilizer (Lehman and Taheri 2013; Wang et al. 2016).

Soil fertility is lost because of the physical degradation of the soil like the compaction of soil by heavy machinery, poor structure of soil, waterlogging, and crusting. Soil structure is a dynamic property of the soil that can be deteriorated or improved by agronomic practices. Poor cultural practices like tillage cause the deterioration of soil structure (Sainju et al. 2017; Dutta and Sen 2017). Compaction of soil reduces soil fertility by decreasing the large pore size that can reduce the root penetration in the soil and uptake of nutrients. Nitrogen is lost in the waterlogged soil in a form of nitrite, as nitrate leaching prevails in such condition. Organic matter is an important factor which can improve the physical, chemical, and biological properties of soil. Decreasing the level of organic matter in soil can reduce all these properties and reduce the nutrient holding capacity of soil by decreasing organic matter. Microbial activity is also reduced in soil because of the reduced organic matter causing hinderance in nutrient cycling as microbes are responsible for nutrient recycling and availability (Abuye and Achamo 2016).

Inefficient and poor managements of soil decrease the fertility status of the soil. Tremendously, soil fertility is decreased by improper crop rotation and excessive tillage practices. Soil fertility is reduced by the pollutants present in the soil in the form of heavy metals. Soil is being polluted by the indiscriminate use of chemicals and heavy metals containing pesticides and fertilizers. These pollutants reduce the activity of microbes in the soil which can ultimately damage the biological life of the soil. Salinization, alkalization, and acidification reduce soil fertility, which eventually lead to imbalance, toxicity, and deficiency of nutrients. All these factors are generally interrelated with soil degradation (Kong and Six 2010; Benaragama et al. 2016; Ghosh et al. 2016).

2.5.2 Consequences Associated with Less Soil Fertility

2.5.2.1 Loss of Agricultural Production

For agricultural production, land resource is a major factor because it supports both the food security and ecosystem (Jie et al. 2002; Zhao et al. 2015). In low fertile soil, degradation process can reduce the soil's ability to produce crops or biomass for livestock and human beings (El Baroudy 2011). Soil is non-renewable, it is a limited resource (Blum 2006), and it continuously undergoes degradation process (Bai et al. 2008) which reduces the fertility status of the soil. Due to this, the potential of soil is reduced for crop production, and it endangers the people of the country who completely depend on agricultural resources. Loss of fertility in degraded soil is the widely recognized form of degradation (Ravi et al. 2010; Varma et al. 2017a). Yield reduction is a serious problem in semiarid regions; it is true in these areas and is threatening the sustainability of agricultural production for many decades (Wang et al. 2009; Liu et al. 2010; Li et al. 2013; Vaezi and Bahrami 2014).

The on-farm declining of yield caused by the losses in land fertility is due to the degradation of land bearing low fertility (Eswaran et al. 2001; Mullan 2013). Low fertility on pot and field scale can reduce the production of agricultural crops between 30% and 90%, respectively (Mbagwu et al. 1984; Lal 1995). Degraded soil has low fertility and is a much more serious problem in the tropics than in temperate areas, and tropical soils are more degraded because of their climatic conditions and inherent properties (Asioet al. 2009). Globally, climate change enhances land degradation causing low soil fertility (Blum 2006; Seager et al. 2007). Approximately 6 million hectares of cultivated land worldwide annually undergo unproductive due to different processes of degradation (Asio et al. 2009).

2.5.2.2 Socioeconomic Issues Due to Less Productive Soil

Soil and land degradation is the loss of productive cover of soil due to soil erosion, salinization, compaction, and acidification. The extent of soil degradation and its remediation depend on which type of degradation process occurs such as salinization and soil erosion. Both are serious problems for the farmers who face the high management cost for cropping in the degraded land. Land degradation causes the loss of fertility of soil which can decline the productivity of crops, causing huge loss of economics and putting the farmer's livelihood and food security at risk (Bhattacharyya et al. 2015). Nutrient depletion is the initial or primary form of land degradation which causes the declining of crop productivity, and this is linked with poverty and hunger (Tully et al. 2015). In some countries, conserving the fertility of soil for longer period by adopting zero tillage practices and extensive monoculturing bring some economic benefits (Wingeyer et al. 2015).

Degraded soils force farmers to find new lands. Some new lands are not suitable for the crop production and require very high investment to improve the fertility status of the soil for it to become a productive land. Low fertile land due to degradation is a major threat to food security in poor countries. There is a strong relationship between poverty and land bearing low fertility (FAO and ITPS 2015). So, it is important to adopt different practices to improve soil fertility and health. More populated areas may face reduction in agricultural production and may also cause the loss of income of the people (FAO and ITPS 2015; Giampietro and Saltelli 2014; Meena et al. 2017c).

Soils, minerals, and forests are natural resources and have an economic value; it is known as natural capital. These natural resources are priced according to their cost of use, such as the logging cost of forests and the extracting cost of minerals. But in the case of soil, in classical economics, it is treated as land factor and priced at market value of farmland (Somda et al. 2002). Decrease in rural population due to migration might be a reason behind land resource conservation. For human survival, soil and water are the basic resources. Civilization rose due to the presence of abundant amount of good quality water and the high quality of land in the past nations. Civilization fell due to the mismanagement and exhausted use of land (Flower et al. 2012).

2.5.3 Practices to Increase Soil Fertility

2.5.3.1 Balance Use of Fertilizer

Balance use of fertilizer is the key for maintaining productivity of soil and nutrient use efficiency. It is not only meant for application of define amount of nutrient like nitrogen, phosphorus and potassium to soil in the form of fertilizer but also fertilizer application in balanced amount. It is a more than 150-year-old concept and it is very simple. According to this idea, crop requires all essential nutrients in adequate supply for optimum growth. The growth of crop is affected by the deficiency of any nutrient. In the recent concept, the rational use of organic manures and fertilizers supply nutrients for the production of agricultural crops in such a way that it should maintain the productivity of soil; minimally affect the environment by less leaching of nutrients; improve the quality of product; increase the crop yield; have the positive and synergetic interaction among the different factors of production such as water, seed, agrochemicals, etc.; and efficiently use fertilizers (Etissa et al. 2013; Kraaijvanger and Veldkamp 2015; Rego et al. 2003).

Balanced application of fertilizers also increases the fertilizer use efficiency that leads to enhanced yield of crop by improving the chemical, physical, and biological conditions of soil. It also encounter the nutrient removal; profitability and economic of fertilizers; investment ability of farmers; soil moisture and adverse conditions of soil like salinity, sodicity and acidity. It must also be ensured that the plants have tolerance to cold, pests, insects, diseases, and drought. Imbalanced fertilizer leads to mining of soil that causes sickness of soil and waste of resources uneconomically, but balanced fertilizers improve soil health (Poudel et al. 2001; Dejene 2003; Ferrández-Villena and Ruiz-Canales 2017).

2.5.3.2 Approaches to Manage Soil Fertility

In the management of soil fertility, 4Rs (right source, right amount, right application time, and right place) approach can be followed (Johnston and Bruulsema 2014). First, select the right source of fertilizers for the management of soil fertility status. Source of fertilizer is selected by the assessment of which essential nutrient is deficient in the soil. This type of information comes from diagnosing the site like soil and plant tissue testing and nutrient removal rate of harvested crops. International Plant Nutrition Institute (IPNI) has developed a system for the guidance of fertilizer source selection and has common use for small land holdings. The continuous application of urea or DAP for nitrogen and phosphorus nutrition can be lethal to crop health as it cause imbalance in N:P:K ratio and plant experience K deficiency. Nitrogenous fertilizer once applied in soil convert to nitrate (NO3-) or ammonium (NH4+) and application of nitrate and ammonium containing fertilizers is soil condition dependent. In flooded conditions nitrate leaching is an issue so ammonical form is recommended, while in alkaline and calcarious soils ammonium fixation is an issue so nitrate form of fertilizer is recommended (Lambers et al. 2011; Meena et al. 2015b).

Second, select the right rate of fertilizer; this information comes from the assessment of plant nutrition demand. Assess the proportion of nutrient uptake by the crop and the yield of that crop. The goals of yield are developed from the previous performance of crop; high hopes are not considered in this. Testing of soil, plant canopy sensors, and plant analysis help in the assessment of nutrient supply to the plants. There should be an assessment of overall nutrient status of soil either from organic source (compost, manure, biosolids, atmospheric deposition, and crop residues) or from chemical fertilizers. Also, consider the impact of soil resources; fertility of soil is declined if the outputs are greater than the inputs. This depends on the current status of soil (Shcherbak et al. 2014).

Third, select the right time of application; nutrients uptaken by crops is different in the growing period or season. Some nutrients are specific to crops at specific growth season of crops. Also assess the nutrient supply dynamic of soil, because the soil warms in the growing season and mineralization of organic matter is accelerated causing net accumulation of nutrients. Compaction of soil affects the application of fertilizer at different times, and it can also be considered that evaluating the field operation while applying the nutrients can delay the planting time of crops (Obour et al. 2015). Dynamics of nutrient loss are also recognized because the excessive rainfall exceeds the water storage capacity of soil. Mostly, in the early spring and late fall, the risk of runoff is higher.

Fourth, select the right place of fertilizer application; place the fertilizers where the plant can easily uptake through roots when needed. Chemistry of soil is also considered; some nutrients retained in the concentrating soil such as phosphorus or the small amount of soil also improves the availability of nutrients. Placement of fertilizer in subsurface technique covers the soil with crop residues that can conserve the water and nutrients of the soil (Zingore and Johnston 2013).

2.5.3.3 No Tillage or Zero Tillage Practices

Nutrient availability and presence in the soil depends on the physical and chemical properties of soil. Chemistry of soil is generally affected by tillage practices, such as pH, exchangeable cations, soil total nitrogen, and cation exchange capacity. Under no tillage practices, surface layer chemical properties are more favorable than in tilled soil. No tillage over a longer period of time is beneficial to structure and chemical properties enhancement and maintenance. In no tillage practices, plant residues left in the field or on the soil surface can increase the organic matter in the surface soil. In high tillage practices, the structure of soil is deteriorated, organic matter is reduced, there is higher mineralization, and there are more leaching losses of nutrients (Anaya and Huber-Sannwald 2015; Shahidi et al. 2014; Datta et al. 2017b).

In no tillage practices, accumulation of organic matter in few upper centimeters can change the pH of soil depending upon nature of organic matter. This can improve the chemical properties and also enhance the nutrient availability to plants. Tillage practices cannot directly affect the pH of soil, but it affects those factors that change the soil's pH. Exchangeable ions such as calcium, potassium, and magnesium are significantly higher in zero tillage soil than in ploughed soils (Shahidi et al. 2014).

2.6 Heavy Metals (HMs) in Soil Plant System

One of the key practices for restoration of degraded soils is to maintain soil health to accomplish maximal productivity. It is the basic factor in the restoration of degraded soils as it is the function of fertility status of soil which is determined by different indicators in which one is the remediation of suspected pollutants (NRCS 1996; USDA 2008). A lot of studies have been reported about the deleterious effects of heavy metals (HMs) on the soil system. High concentration of HMs in the soil has adverse effects on the soil quality and crop productivity besides contamination of food chain, which is a disturbing situation worldwide (Manta et al. 2002; Iqbal and Shah 2011; Zhou et al. 2014; Silva et al. 2016). Heavy metal ions compete for the uptake by plants with nutrient elements and cause nutrient disturbance. Their pollution shows toxicity for biotic life in soil, bioaccumulation in plants, and persistence over a longer period (Nriagu 1996). A number of studies have been conducted about their pervasiveness, and pollution in soil in relation to different sources and majority has been attributed to anthropogenic activities like unwise fertilizers usage, untreated industrial wastes, pesticides, transportation, and sewage irrigation (Sridhara Chary et al. 2008; Chabukdhara and Nema 2013; Ding et al. 2017; Tedoldi et al. 2017) besides pedogenic sources (Khan et al. 2010).

A performance-based soil quality index was proposed by Doran and Parkin (1994) to give an interpretation of soil function about environmental quality, sustainable production, and human health. They described soil quality index a function of these elements:

$$\text{Soil quality index} = f\left(\text{E1}, \text{E2}, \text{E3}, \text{E4}, \text{E5}, \text{E6}\right)$$

where

E1 = food quality, E2 = erosivity, E3 = ground water quality, E4 = surface water quality, E5 = air quality, and E6 = fiber and food production.

This approach is helpful to assess soil functions and propose the specific performance criteria for these elements in soil (Doran et al. 1997) like under heavy metal contamination. Indicators of the health of the soils which are being disturbed by heavy metals include soil enzyme activities which are directly related to soil microbes and plants (Badiane et al. 2001; Khan et al. 2007; Hinojosa et al. 2004; Filip 2002; Chaperon and Sauvé 2007; Meena et al. 2015a). Activities of different soil enzymes including b-glucosidase, protease, cellulose, invertase, arylsulfatase, acid and alkaline phosphatase, dehydrogenase, and urease are prone to heavy metal contamination (Kunito et al. 2001; Wang et al. 2008; Oliveira and Pampulha 2006; Effron et al. 2004). Similarly, detrimental effects of heavy metals on soil health can also be judged by the soil nematodes' diversity which reduced greatly under the presence of heavy metals (Gutiérrez et al. 2016). All these factors affect the productivity of soil, and their preponderance is the base for soil degradation.

2.6.1 Management Strategies for Soil Restoration

Strategies of restoring heavy metal-contaminated soil include checking HMs activity in soil and its penetration to plant tissues and food chain. This requires clean-up of irrigation water and agriculture soils through different chemical, physical, and biological methods (Edelstein and Ben-Hur 2018). It includes several techniques such as by adopting some farming practices to check heavy metal inclusion, by phytoremediation, by ex situ treatment, and by in situ stabilization. The in situ stabilization or immobilization of metals is a promising approach for remediation rather than ex situ as it is less laborious and cost-efficient. It can be enhanced by the addition of several amendments which have high surface charge, have the ability to form metal complexes, and have adsorption affinity as the associated parameters for toxicity or the bioavailability of heavy metals in contaminated soils are their exchangeable fraction and mobility in soil. Exogenous application of microbes and organic acids is also helpful in the mitigation of soil metal pollution as they support the plant growth and make the plant tolerant to stress besides metal detoxification (Etesami 2018; Kim et al. 2013). In the phytoremediation technique, hyper-accumulator plants are the efficient way of reclamation of HMs as they remove the heavy metals from the soil and enhance the soil properties, and it is easy to dispose of this biomass and cost-efficient (Rehman et al. 2017). A number of attempts have been made and still continued in this technique to make it more efficient and farmer friendly by applying genetic engineering. Details for these management practices are discussed below.

2.6.1.1 Conduct of Constructive Farming Practices

At farm level, contamination of heavy metals in soil and food can be managed by using certain strategies like irrigation water reclamation, wise use of fertilizers, check on pesticide's usage, and intercropping. Continuous application of untreated wastewaters for irrigation of soils puts a considerable amount of metal species to the soil. So, an ambient option for the reclamation of heavy metal-degraded soil is the treatment of irrigation water which is the prerequisite for management of these soils as it is the main source of contamination. Among the various techniques, a significant approach for the removal of HMs from water is the application of nano-adsorbents technology using different organic- and inorganic-based nano-materials. These nano-materials are highly efficient due to certain properties like high BET (Brunauer-Emmett-Teller) surface area, high dispersion ability, microporous structure, and being economic and environment friendly (Gupta et al. 2015; Li et al. 2003a, b). Each technique has its own specific efficiency and merits to remove pollutants; similarly, nano-adsorbents have great potential for the removal of metallic species from wastewater such as As, U, Cr, Cd, Cu, Hg, Pb, Ni, Pd, Sb, Pt, Th, etc. (Dubey et al. 2017). Different kinds of inorganic nano-materials (Anjum et al. 2016) for wastewater treatment have been investigated successfully including nano-adsorbents, i.e., TiO, MnO, MgO, Fe, ZnO, and carbon nanotubes (Gao et al. 2008; Tuzen and Soylak 2007; Gupta et al. 2011; Feng et al. 2012; Xu et al. 2008a, b); nano-membranes, i.e., electrospun PVDF, Na-TNB, multi-walled CNTs, and PVC

(Zhang et al. 2013a, b; Ho et al. 2012); and photocatalysts, i.e., TiO_2, ZnO, and CdS (Lin et al. 2014; Tristão et al. 2006; Zhu et al. 2009). Photocatalysts can remove heavy metals and other toxic pollutants from the environment. These can further be combined with biological process to enhance the HMs removal process (Luming and Xian 2008). On the other hand, there are some limitations to the use of nanomaterials due to their small size and the difficulty in separating them from aqueous solution which may lead to some secondary pollution (Ray et al. 2015). However, to overcome these kinds of challenges, efforts have been carried out to modify this technique such as the formation of multi-walled carbon nano-tubes (Tarigh and Shemirani 2013; Tang et al. 2012) and hydrous manganese oxide (Gupta et al. 2015). Some other nano-materials reported for metal removal from irrigation water contain chitosan (Rorrer et al. 1993), zeolite (Álvarez-Ayuso et al. 2003), sphagnum moss peat (Sharma and Forster 1993), bentonite (Viraraghavan and Kapoor 1994), seaweeds (Vijayaraghavan et al. 2005), red mud (Pradhan et al. 1999), fly ash (Rao et al. 2002), kaolinite (Yavuz et al. 2003), montmorillonite (Bhattacharyya and Gupta 2007), wollastonite (Sharma et al. 1990), sawdust (Memon et al. 2007), soya cake (Daneshvar et al. 2002), and alumina (Wu et al. 2000). All these materials can be used for the wastewater or irrigation water treatment or for soil reclamation.

To reduce soil contamination risk, excessive fertilization and pesticides application must be checked. It has been reported by Czarnecki and Düring (2015) that long-term application of increased mineral fertilizers raised the metal content level of soil; however, termination in fertilization for eight years reduced the soil concentration of different metals: Cd, Mn, Zn, Cu, and Pb by 82.6%, 48.5%, 56.9%, 54.2%, and 74.4%. Among them, phosphatic fertilizers are contributing to the metal contamination of soil like calcium phosphate and triple super phosphate which contains HMs in different concentrations depending on the source of rock phosphate. Some rock phosphates have been banned in some countries for agriculture purposes as they contain even more than 50 mg Kg^{-1} of Cd which is hazardous for soil health (Mortvedt and Beaton 1995). To control pests, biological and physical methods should be promoted rather than chemical control which is putting heavy metals to the soils. Recently, the double cropping pattern (growing of high metal-accumulating plants with crops of agricultural interest) was introduced to reduce the heavy metals in food chain and to enhance food quality and productivity besides soil reclamation (Wu et al. 2007; Whiting et al. 2001). Efficiency of this planting pattern depends on the metal species type and interactions between the coinciding plant species (Zou 2015). In this system, one plant improves the growth of the second plant by enhancing the nutrient availability to its companion species (Lu et al. 2012). Planting of hyper-accumulator *Solanum nigrum* with rye grass not only reduces the soil metal but also reduces the metal concentration in the rye grass to significant levels so that it may meet the standards of animal feed (Wei et al. 2006; Zou 2015). There are many other studies reported which are the same as the co-cropping of *Sedum alfredii* with maize that can reduce the Cd and Zn concentration in maize (Wu et al. 2007). Similarly, yield of maize is enhanced when co-cropped with *Kummerowia striata* rather than monoculture (Liu et al. 2012). So, an effective cropping system should be adopted to remediate the soil from heavy metals and to take care of food standards for human and animal consumption.

2.6.1.2 Application of Soil Conditioners

Several soil conditioners are present which influence the soil properties to remediate the metal-contaminated soils. In this respect certain microbes and organic acids are helpful besides ordinary amendments to boost the remediation process by making certain links with soil organic matter, exogenously applied materials, and soil properties (Vítková et al. 2015; Hou et al. 2017). Certain microbes directly affect the heavy metals in the soil and have direct influence on the HMs dynamics in soil. A number of soil conditioners are present in literature, but here the researchers focus on organic acids and microbes.

Exogenous application of organic acids is a good way to enhance the phytoremediation activity of plants in metal-contaminated soil. Organic acids are also released by the plant roots because of various environmental stresses (Jones 1998). Application of organic acids in soil results in modifications of physico-chemical properties, local rhizospheric conditions, and nature of soil organic matter (Bravin et al. 2012; Hinsinger et al. 2003). Several studies are present for organic acid-assisted remediation of heavy metals. Kim et al. (2013) reported that oxalic and succinic acid showed significant results for the removal of copper and zinc from soil, and respective metals were extracted in free ionic form and oxalate complexes. Similarly, some other organic acids were reported by Nascimento (2006) including DTPA, EDTA, oxalic acid, gallic acid, citric acid, and vanillic acid for desorption of cadmium, lead, copper, zinc, and nickel from soil to enhance phyto-extraction by plants; all of them showed significant results in metal removal from soil.

Microbes, especially plant growth-promoting bacteria (PGPB), play a vital role in reclaiming heavy metals in soils by many mechanisms (Rajkumar et al. 2012). There are a number of studies related to PGPB on phytoremediation and mobilization of heavy metals in soil (Ma et al. 2011; Sessitsch et al. 2013; Glick 2010) by various mechanisms like metal solubilization, decreasing the rhizospheric pH, increasing the root surface area of extracting plant, and enhancing the root exudates production by plants, thus enhancing the HMs removal process. However, some PGPB are known to have tolerance to heavy metals and promote plant growth by reducing the HMs uptake of plants which ensures safe food production. To mitigate HMs toxicity and damage to plants under heavy metal stress and to enhance growth, certain PGPB have been implicated (Rajkumar et al. 2010, 2012; Ma et al. 2016a, b). These microbes show a number of responses toward metal ions like metal biosorption, precipitation, bioaccumulation, complexation, enzymatic transformation, reduction, and oxidation, thereby reducing the HMs stress toward themselves and the plants (Rajkumar et al. 2012; Ma et al. 2016a; Sharma and Archana 2016). Usually, these kinds of microbes can cause both decrease and increase in the HMs bioavailability depending upon different plant interactions and soil conditions. There are comprehensive reports on the microbe-assisted phytoremediation of metal-contaminated soils (Ma et al. 2011; Glick 2010; Rajkumar et al. 2012; Sharma and Archana 2016; Sessitsch et al. 2013; Ullah et al. 2015).

2.6.1.3 Application of Soil Amendments

In situ immobilization of metals in soil by using different organic and inorganic substances is a nondisruptive, simple, and cost-efficient way for the remediation of degraded soils when it is difficult and costly to treat ex situ. It aims at boosting natural depletion processes like adsorption by charged surfaces, complexation, and precipitation which exist in soil to decrease the bioavailability and mobility of HMs in soil (Guo et al. 2006). These amendments do not aim for the decrease in the total content of heavy metals but assist in the reduction of mobile fraction of potentially toxic metals (Ench et al. 2000; Bolan and Duraisamy 2003; Oste et al. 2002; Adriano et al. 2004; Raicevic et al. 2005; Mora et al. 2005; Kumpiene et al. 2006, 2008). The demerit of this approach for reclamation is that the immobilized fraction or final product of this technique remains present in soil, although in inactive form (Raicevic et al. 2005). Various inexpensive treatment agents which are admixed to soil include alkaline materials for acidic soils (i.e., liming materials), phosphatic minerals, aluminosilicate mineral, hydroxides of iron and aluminum, alkaline biosolids, organic matter-based additives, and industrial by-products which have typical behavior for metal fixation (Negim 2009). Furthermore, the HMs fixation efficiency depends on the amount and time of amendment application. And the deciding factor for it is the physico-chemical nature of amendments and metals on site (Guo et al. 2006). These amendments reduce the bioavailable content of the metals in soil solution, thus decreasing the metal mobility, phytotoxicity, and uptake by plants. Based on nature, these are divided into organic and inorganic as discussed below.

Organic Amendments

Organic matter is the most important amendment for the cation exchange reactions and metal complexes formation to immobilize heavy metals. Metal-organic interaction can proceed both in soil solution and on the solid organic surfaces, which is present either naturally or added to soil (Silveira et al. 2003). Soluble organic molecules form stable complexes which are mobile in solution and bioavailable because of the protection of metal ions from adsorption on the soil's solid surfaces due to the organic ligands present (Alloway and Ayres 1997). With increase in pH, different functional groups (i.e., phenolic, carboxylic, alcoholic, carbonyl, etc.) of organic matter dissociate and increase the affinity for metal ions adsorption. The affinity of different metal ions for the complexation with organic matter is in the following order $Zn < Mn < Co < Ni < Pb < Fe < Cd < Cu$ (Adriano 2001). The extent of complex formation between organic ligands and metals ions is dependent on their interaction that exists between soluble organic ligands and metal-binding sites as it varies from metal to metal. A lot of biosolids have been applied to agricultural soils to treat metal contamination, and a number of studies are present in this regard. Fernández-Calviño et al. (2017) reported that pine bark application can be used to decrease the HMs release from mine soils and to enhance the metals' retention in acidic mine soils, thus minimizing the risk of transfer to healthy soils. Soil washing with some organic washing agents is a permanent technique for the removal of heavy metals which is influenced by the concentration of HMs, pH of the solution, and time of washing. Feng et al. (2018) have reported four washing agents including pineapple peel, broad beans straw, soybean straw, and tea residue of which pineapple peel showed the most significant results. Similarly, addition of the paper mills sludged to a contaminated soil, which is a source of carbonates, organic matter, and silicates, decreased the metal ions' availability (Calace et al. 2005). Biochar is also an emerging option for the remediation of heavy metal-contaminated soils to immobilize HMs like Cd, Pb, Ni, Mn, Sb, As, Zn, Cu, and Cr to reduce their deleterious effects on soil and availability for the plants. Efficiency of biochar for the remediation of HMs is related to its composition, porosity, pyrolysis temperature, residence time for pyrolysis, and feedstock (Lahori et al. 2017).

Inorganic Amendments

Application of inorganic amendments for this purpose is also a very effective approach to remediate the soil as most of the inorganic amendments are the industrial by-products and are available on large scale which make them cost-efficient; besides this, they also have environment-friendly properties. They provide more binding sites for the HMs immobilization which reduces their bioavailability. Different types of inorganic amendments that can be used for immobilization purpose include coal ash (Reviews et al. 2015), red mud (Panda et al. 2017), hydroxyapatite (He et al. 2013), liming material (Arunakumara et al. 2013), zeolites (Damian et al. 2013), fly ash (Dermatas and Meng 2003), slag, Ca-montmorillonite (Auboiroux et al. 1996), beringite (Vangronsveld et al. 1996), iron and aluminum oxides (See et al. 2014), bauxite residue (Lombi et al. 2002), portland cement (Li et al. 2001), phosphate salts (Seshadri et al. 2017), bentonite (Geebelen et al. 2002), silicon-containing materials (Anwaar et al. 2015), gravel sludge (Krebs et al. 1999), and nitrogenous amendments (Zhang et al. 2014). All these materials have specific affinity for different HMs and have different soil-working conditions. They reclaim HMs soil by making complexes, by changing pH of soil, by enhancing soil properties, by immobilizing, by supporting plant growth, etc. Different amendments have different ways of treatments; for example, fly ash has the ability to improve the properties of mine lands and degraded soils by decreasing the bulk density, reducing soil compaction, increasing water holding capacity, and precipitating the soluble metals; they have alkaline nature and can be used for acidic soils (Jala and Goyal 2006; Gorman et al. 2000). In addition to this, coal fly ash combined with peat increases the pH of soil, reduces the HMs mobility, allows the soil to revegetate, and decreases the metal toxicity to plants and microorganisms (Kumpiene et al. 2007). Phosphate and liming materials are the common amendments for the in situ stabilization to decrease the heavy metals in soil solution, to reduce their leachability and mobility, and to transform them from soluble to residual form, so that plant may not suffer (Bolan et al. 2003; Cao et al. 2003; Wang et al. 2001; Chen et al. 2003). Beringite is a modified aluminosilicate investigated for its ability of metal immobilization in soil based on ion exchange, chemical precipitation, and crystal growth. Combination of all these mechanisms explains well for its high sorption capacity for heavy metals. Zeolites can be used to enhance the soil quality as small amount of them can influence soils with high acid buffering capacity especially to decrease HMs uptake by plants (Oste et al. 2002). A number of studies are present for zeolites use as immobilization agents (Singh and Oste 2001; Mahabadi et al. 2007; Leggo and Ledésert 2001; Oste et al. 2002). Slags which include several oxides like calcium oxide, silicon oxide, phosphorus oxide, iron oxide, and other metal oxides are an industrial by-product and have been used for remediation of HMs-contaminated soils by adsorption on metal oxide surface and precipitation (Filho et al. 2004; Bes and Mench 2008; Pupatto et al. 2004; Kim et al. 2008).

2.6.1.4 Phytoremediation Approach for HMs Soils Reclamation

Phytoremediation technique is applicable to several reclamation treatments besides heavy metal-contaminated soils. It has high benefit to cost ratio as compared to all other treatment methods and is perceived as an efficient, eco-friendly, novel, and solar-driven technique with high public acceptance. Prasad (2003) have reported regarding its cost; it is as less as only 5% of other remediation techniques. Planting of vegetation on contaminated soils also prevents metal leaching and erosion (Chaudhry et al. 1998). Benefits of phytoremediation are threefolds including phyto-stabilization (risk containment), sustainable land management by phyto-extraction of heavy metals which improves soil quality, and phyto-extraction of valuable metals like Au, Tl, Ni, etc. (Vangronsveld et al. 2009; Verma et al. 2015a). It has no secondary pollution risk like other methods and is known as "green clean" technology (Pilon-Smits 2005), and the produced mass can be used for different commercial purposes to earn revenue. In this regard, high biomass-producing and fast-growing plants are opted. For this purpose, most efficient hyper-accumulators are being discovered to remediate polluted soils in a better way and short duration. Different molecular tools are being used for the better understanding of processes involved so that new avenues of biotechnology may open in this respect. Plants remediate land without affecting its topsoil, thus maintaining its fertility and utility (Mench et al. 2009).

Growing of Hyper-Accumulator Plants

Application of phytoremediation technology involves several processes like phyto-extraction, phyto-stabilization, phyto-volatilization, phyto-degradation, rhizodegradation, etc. All these are applicable by the introduction of efficient and native heavy metal hyper-accumulator plants (Cristaldi et al. 2017). So, the first step for this practice is the identification of certain hyper-accumulators that are fit for the specific polluted conditions as they are usually metal specific. There are some prerequisites for the selection of these; they should be high biomass-pertaining, fast-growing, and tolerant to toxic conditions and have good antioxidant system (Yao et al. 2012). The more biomass the plants have, the more are the extracting ratios of the HMs from the soil by the plants; similarly, it will be easy for it to stand the stressful environmental conditions due to HMs. Moreover, plants secrete root exudates which improve the soil quality and enhance the soil restoration process (Rehman et al. 2017). These plants also produce phyto-chelators to gain resistance against stress by HMs (Yadav 2010; Nagajyoti et al. 2010). A number of hyper-accumulator plants have been reported for different heavy metals like *Solanum nigrum* (Rehman et al. 2017), *Sedum alfredii* (Wu et al. 2007), *Sorghum bicolor* L. (Rizwan et al. 2017), *Crotalaria juncea* (Uraguchi et al. 2006), *Thlaspi caerulescens* (Milner and Kochian 2008), *Salix viminalis* L. (Mleczek et al. 2013), *Helianthus annuus* (Dhiman et al. 2017), Brassicaceae (Martos et al. 2016), *Brassica juncea* (Belimov et al. 2005), *Brassica chinensis* (Deng et al. 2011), *Salix* spp., *Salix babylonica* L., etc., and several others are present. All these plants are potential hyperaccumulators and can be used for the soil reclamation alone or combined with certain amendments which can enhance their ability to extract metals from soil.

Strive for Transgenic Plants

Phytoremediation is a sustainable and green solution for HMs-contaminated soils as it involves the plants for uptake, sequestration, volatilization, or detoxification of HMs from soil (Vegetal et al. 2005). This technique was born from the knowledge that plants have certain physiological properties against HMs stress. It was then followed by the selection of most efficient species and application of breeding methods to get most promising plants. Now transgenic plants have been produced for phytoremediation but still need to open avenues toward this way. No doubt hyperaccumulators have good ability to reclaim the soils, but many of them have certain limitations like low biomass production. This can be controlled by using genetic engineering in which transfer of hyperaccumulation genes to high biomass plants is done; the transfer of overexpression genes of other organisms like animals, yeast, and bacteria into plants to improve phytoremediation (Cherian and Oliveira 2005) is also a viable possibility. In this respect, several metal detoxification genes from bacteria and yeast which have been characterized functionally and genetically were transferred into plants to obtain genetically modified plants. The main approaches for this process involve transformation from other organisms (i.e., bacteria, etc.) and plants and overexpression genes of same plants (Maestri and Marmiroli 2011). Several genes from different organisms and targeted plants have been reported by many in the introduction of transgenic plants for phytoremediation purpose (Kawashima et al. 2004; Ellis et al. 2004; Wangeline et al. 2004; Song et al. 2003). It is a good way of obtaining efficient plants to further enhance the phytoremediation technology.

2.7 Summary and Future Perspectives

The soil is being disturbed by means of anthropogenic and natural degradation which results in the deterioration of soil properties and environmental sustainability. Erosion of soil is occurring due to poor soil structure, deforestation, loss of vegetation, and intense natural climatic factors. Prevalence of salts in soils is mainly caused by poor irrigation water, fallowing of land, and salty parent material, while loss of soil fertility is due to intensive cropping, unmanaged farming, and poor use of fertilizers. Untreated wastewaters and intense use of agricultural products like pesticides and fertilizers are adding up heavy metals to soil. All these processes are declining the soil properties either alone or simultaneously and affecting the agricultural yields and food quality. The challenge to feed the huge population of humans urges this situation to be resolved besides environment protection. There is a need to find out ways to increase and manage agricultural lands. To cope with this situation and to enhance soil rehabilitation, different management and reclamation procedures are required which can combat this need. So, in this respect, several practices have been considered which are efficient, environment supportive, and farmer oriented. These practices promote the control of erosion process, desalinization of lands, fertility enhancement, heavy metals remediation, etc. Different types of soil deterioration have respective set of approaches for reclamation which can be followed according to the prevailing conditions over the certain area. It includes soil stabilization through proper vegetation, farming practices, and different temporary structural managements for erosion control. Similarly, different amendments (i.e., gypsum, acids, organic matter, etc.) and provision of good quality water for salty soils are prerequisites to control soil salinization. Low fertility of soil can be managed by reducing the cultural practices and less growing of nutrient-exhaustive crops. Crop rotation is the most important agronomic practice in the management of fertility status of soil which is less fertile. Balance use of fertilizers also manages the fertility of the soil; it can also reduce the leaching of excess amount of nutrients to the underground water, and it is beneficial to the environment. Heavy metals reclamation can be done by in situ immobilization of metal ions using different immobilizing agents which can reduce the metal ions activity in soil to decrease the uptake and toxicity. They have different mechanisms to bind heavy metal ions in soil like complexes formation, adsorption, etc. Exogenous provision of organic acids and salutary microbes is also very good for soil structure improvement, plant growth, and metals removal. Phytoremediation in combination with metal mobilizers (i.e., sulfur, organic acids, chelators like EDTA, etc.) is a developing technique to remove metal ions permanently from soil. But there needs some care while using amendments as mobilizing agents in their rate of application because they can cause environmental pollution at higher rates. All these methodologies are involved in the restoration of soil properties in one or the other way and can be applied concurrently in a managed way in case of severe cases of soil degradation.

References

Abou-Shady A (2016) Reclaiming salt-affected soils using electro-remediation technology: PCPSS evaluation. Electrochim Acta 190:511–520

Abuye F, Achamo B (2016) Potential use of cyanobacterial bio-fertilizer on growth of tomato yield components and nutritional quality on grown soils contrasting pH. J Biol 6:54–62

Acosta JA, Faz Á, Martínez M (2009) Evaluation of land degradation in an area affected by different management systems, SE Spain. In: Land degradation and rehabilitation: dryland ecosystems. Papers presented at the Fourth International Conference on Land Degradation, Cartagena, Murcia, Spain, 12–17 September 2004. Catena Verlag, pp 81–96

Adriano DC (2001) Trace elements in terrestrial environments. Biogeochemistry, bioavailability, and risks of metals. Springer, New York

Adriano DC, Wenzel WW, Vangronsveld J, Bolan NS (2004) Role of assisted natural remediation in environmental cleanup. Geoderma 122:121–142

Afzal I, Basra SMA, Iqbal A (2005) The effects of seed soaking with plant growth regulators on seedling vigor of wheat under salinity. J Stress Physiol Biochem
Alcántara V, Don A, Well R, Nieder R (2016) Deep ploughing increases agricultural soil organic matter stocks. Glob Chang Biol 22:2939–2956
Alexiew D, Plankel A, Thomson G (2015) A geogrid-reinforced landslide stabilization: 20 years passed. In: 15th Asian Regional Conference on Soil Mechanics and Geotechnical Engineering, ARC 2015
Alloway BJ, Ayres DC (1997) Chemical principles of environmental pollution. Chapman & Hall London 395p. https://trove.nla.gov.au/work/7144673?q&versionId=45632709. Accessed 20 Aug 2018
Álvarez-Ayuso E, García-Sánchez A, Querol X (2003) Purification of metal electroplating waste waters using zeolites. Water Res 37:4855–4862. https://doi.org/10.1016/j.watres.2003.08.009
Amini S, Ghadiri H, Chen C, Marschner P (2016) Salt-affected soils, reclamation, carbon dynamics, and biochar: a review. J Soils Sediments 16:939–953
Amiri V, Nakhaei M, Lak R, Kholghi M (2016) Investigating the salinization and freshening processes of coastal groundwater resources in Urmia aquifer, NW Iran. Environ Monit Assess 188(188):233
Amundson R, Berhe AA, Hopmans JW, Olson C, Sztein AE, Sparks DL (2015) Soil and human security in the 21st century. Science:80
Anaya CA, Huber-Sannwald E (2015) Long-term soil organic carbon and nitrogen dynamics after conversion of tropical forest to traditional sugarcane agriculture in East Mexico. Soil Tillage Res 147:20–29
Andreottola G, Bonomo L, de Gioannis G, Ferrarese E, Muntoni A, Polettini A, Pomi R, Saponaro S (2009) Lab-scale feasibility tests for sediment treatment using different physico-chemical techniques. J Soils Sediments 10:142–150
Andreu V, Pic Y (2004) Determination of pesticides and their degradation products in soil: critical review and comparison of methods. Trends Analyt Chem 23:772–789
Anjum M, Miandad R, Waqas M, Gehany F, Barakat MA (2016) Remediation of wastewater using various nano-materials. Arab J Chem
Anwaar SA, Ali S, Ali S, Ishaque W, Farid M, Farooq MA, Najeeb U, Abbas F, Sharif M (2015) Silicon (Si) alleviates cotton (Gossypium hirsutum L.) from zinc (Zn) toxicity stress by limiting Zn uptake and oxidative damage. Environ Sci Pollut Res 22:3441–3450. https://doi.org/10.1007/s11356-014-3938-9
Arachchi LPV (2009) Soil & Tillage Research Effect of deep ploughing on the water status of highly and less compacted soils for coconut (Cocos nucifera L.) production in Sri Lanka. Soil Tillage Res 103:350–355
Arnaez J, Lasanta T, Ruiz-Flano P, Ortigosa L (2007) Factors affecting runoff and erosion under simulated rainfall in Mediterranean vineyards. Soil Tillage Res 93:324–334
Arunakumara KKIU, Walpola BC, Yoon MH (2013) Current status of heavy metal contamination in Asia's rice lands. Rev Environ Sci Biotechnol 12:355–377
Ashoka P, Meena RS, Kumar S, Yadav GS, Layek J (2017) Green nanotechnology is a key for eco-friendly agriculture. J Clean Prod 142:4440–4441
Asio VB, Jahn R, Perez FO, Navarrete IA, Abit SM Jr (2009) A review of soil degradation in the Philippines. Ann Trop Res 31:69–94
Auboiroux M, Baillif P, Touray JC, Bergaya F (1996) Fixation of Zn2+and Pb2+by a Ca-montmorillonite in brines and dilute solutions: preliminary results. Appl Clay Sci 11:117–126. https://doi.org/10.1016/S0169-1317(96)00014-2
Ayers JC, George G, Fry D, Benneyworth L, Wilson C, Auerbach L, Roy K, Karim MR, Akter F, Goodbred S (2017) Salinization and arsenic contamination of surface water in southwest Bangladesh. Geochem Trans 18:1–23
Badiane NNY, Chotte JL, Pate E, Masse D, Rouland C (2001) Use of soil enzyme activities to monitor soil quality in natural and improved fallows in semi-arid tropical regions. Appl Soil Ecol 18:229–238. https://doi.org/10.1016/S0929-1393(01)00159-7

Bai ZG, Dent DL, Olsson L, Schaepman ME (2008) Proxy global assessment of land degradation. Soil Use Manag 24:223–234

Balasubramanian A (2017) Soil Erosion- Causes and Effects. Cent Adv Stud Earth Sci 7. https://doi.org/10.13140/RG.2.2.26247.39841

Barrios E (2007) Soil biota, ecosystem services and land productivity. Ecol Econ 64(2):269–285

Bazihizina N, Barrett-Lennard EG, Colmer TD (2012) Plant growth and physiology under heterogeneous salinity. Plant Soil 354:1–9

Bech J (2017) Remediation of polluted soils. J Geochem Explor 174:1–3

Belimov AA, Hontzeas N, Safronova VI, Demchinskaya SV, Piluzza G, Bullitta S, Glick BR (2005) Cadmium-tolerant plant growth-promoting bacteria associated with the roots of Indian mustard (Brassica juncea L. Czern.). Soil Biol Biochem 37:241–250. https://doi.org/10.1016/j.soilbio.2004.07.033

Benaragama D, Shirtliffe SJ, Johnson EN, Duddu HSN, Syrovy LD (2016) Does yield loss due to weed competition differ between organic and conventional cropping systems. Weed Res. 56:274–283

Bes C, Mench M (2008) Remediation of copper-contaminated topsoils from a wood treatment facility using in situ stabilisation. Environ Pollut 156:1128–1138. https://doi.org/10.1016/j.envpol.2008.04.006

Bhattacharyya KG, Gupta SS (2007) Adsorptive accumulation of Cd(II), Co(II), Cu(II), Pb(II), and Ni(II) from water on montmorillonite: influence of acid activation. J Colloid Interface Sci 310(2):411–424

Bhattacharyya R, Ghosh B, Mishra P, Mandal B, Rao C, Sarkar D, Das K, Anil K, Lalitha M, Hati K, Franzluebbers A (2015) Soil degradation in India: challenges and potential solutions. Sustainability 7(4):3528–3570

Bindraban PS, van der Velde M, Ye L, van den Berg M, Materechera S, Kiba DI, Tamene L, Ragnarsdóttir KV, Jongschaap R, Hoogmoed M, Hoogmoed W, van Beek C, van Lynden G (2012) Assessing the impact of soil degradation on food production. Curr Opin Environ Sustain 4:478–488

Blum WEH (2006) Soil resources-The basis of human society and the environment. Die Bodenkultur 57:197–202

Blum WEH (2013) Soil and land resources for agricultural production: general trends and future scenarios-a worldwide perspective. Int. Soil Water Conser Res 1:1–1

Bolan NS, Duraisamy VP (2003) Role of inorganic and organic soil amendments on immobilisation and phytoavailability of heavy metals: a review involving specific case studies. Austr J Soil Res 41:533–555

Bolan NS, Adriano DC, Naidu R (2003) Role of phosphorus in (Im)mobilization and bioavailability of heavy metals in the soil-plant system. Rev Environ Contam Toxicol 177:1–44

Boursiac Y, Chen S, Luu D-T, Sorieul M, van den Dries N, Maurel C (2005) Early effects of salinity on water transport in Arabidopsis roots. Molecular and cellular features of aquaporin expression. Plant Physiol 139:790–805

Bravin MN, Garnier C, Lenoble V, Gérard F, Dudal Y, Hinsinger P (2012) Root-induced changes in pH and dissolved organic matter binding capacity affect copper dynamic speciation in the rhizosphere. Geochim Cosmochim Acta 84:256–268. https://doi.org/10.1016/j.gca.2012.01.031

Brevik EC, Cerdà A, Mataix-Solera J, Pereg L, Quinton JN, Six J, Van Oost K (2015) The interdisciplinary nature of soil. Soil 1:117–129

Brevik EC, Steffan JJ, Burgess LC, Cerdà A (2017) Links between soil security and the influence of soil on human health. In: Global Soil Security. Springer, Cham, pp 261–274

Buragohain S, Sharma B, Nath JD, Gogaoi N, Meena RS, Lal R (2017) Impact of ten years of biofertilizer use on soil quality and rice yield on an inceptisol in Assam, India. Soil Res. https://doi.org/10.1071/SR17001

Calace N, Campisi T, Iacondini A, Leoni M, Petronio BM, Pietroletti M (2005) Metal-contaminated soil remediation by means of paper mill sludges addition: chemical and ecotoxicological evaluation. Environ Pollut 136:485–492. https://doi.org/10.1016/j.envpol.2004.12.014

Cañedo-Argüelles M, Kefford BJ, Piscart C, Prat N, Schäfer RB, Schulz CJ (2013) Salinisation of rivers: an urgent ecological issue. Environ Pollut 173:157–167

Cao RX, Ma LQ, Chen M, Singh SP, Harris WG (2003) Phosphate-induced metal immobilization in a contaminated site. Environ Pollut 122:19–28. https://doi.org/10.1016/S0269-7491(02)00283-X

Cerdà A, Giménez-Morera AY, Bodí MB (2009) Soil and water losses from new citrus orchards growing on sloped soils in the western Mediterranean basin. Earth Surf Process Landf 34:1822–1830

Chabukdhara M, Nema AK (2013) Heavy metals assessment in urban soil around industrial clusters in Ghaziabad, India: probabilistic health risk approach. Ecotoxicol Environ Saf 87:57–64. https://doi.org/10.1016/j.ecoenv.2012.08.032

Chaperon S, Sauvé S (2007) Toxicity interaction of metals (Ag, Cu, Hg, Zn) to urease and dehydrogenase activities in soils. Soil Biol Biochem 39:2329–2338. https://doi.org/10.1016/j.soilbio.2007.04.004

Chartzoulakis K, Loupassaki M, Bertaki M, Androulakis I (2002) Effects of NaCl salinity on growth, ion content and CO2 assimilation rate of six olive cultivars. Sci Hortic (Amsterdam) 96:235–247

Chaudhry TM, Hayes WJ, Khan AG, Khoo CS (1998) Phytoremediation – focusing on accumulator plants that remediate metal-contaminated soils. Aust J Ecotoxicol 4:37–51

Chen Y, Hoehenwarter W (2015) Changes in the phosphoproteome and metabolome link early signaling events to rearrangement of photosynthesis and central metabolism in salinity and oxidative stress response in arabidopsis. Plant Physiol 169:3021–3033

Chen M, Ma LQ, Singh SP, Cao RX, Melamed R (2003) Field demonstration of in situ immobilization of soil Pb using P amendments. Adv Environ Res 8:93–102. https://doi.org/10.1016/S1093-0191(02)00145-4

Chen D, Wei W, Chen L (2017) Earth-science reviews effects of terracing practices on water erosion control in China: a meta- analysis. Earth Sci Rev 173:109–121

Cherian S, Oliveira MM (2005) Transgenic plants in phytoremediation: recent advances and new possibilities. Environ Sci Technol 39:9377–9390. https://doi.org/10.1021/es051134l

Chesworth W (2008) Management of irrigation-induded Salt-affected soils. Fao., p 611

Chongo N, Kawanga C (2015) Inclusive stakeholder participation for sustaining dry sanitation solution in Peri-Urban areas: a Madimba Community Experience, Zambia. In: 5th International Dry Toilet Conference

Cohen JE (2003) Human population: the next half century. Science 302(80):1172–1175

Colla G, Rouphael Y, Rea E, Cardarelli M (2012) Grafting cucumber plants enhance tolerance to sodium chloride and sulfate salinization. Sci Hortic (Amsterdam). 135:177–185

Cook HF, Valdes GSB, Lee HC (2006) Mulch effects on rainfall interception, soil physical characteristics and temperature under Zea mays L. Soil Till Res 91:227–235

Cristaldi A, Conti GO, Jho EH, Zuccarello P, Grasso A, Copat C, Ferrante M (2017) Phytoremediation of contaminated soils by heavy metals and PAHs. A brief review. Environ Technol Innov 8:309–326

Czarnecki S, Düring RA (2015) Influence of long-term mineral fertilization on metal contents and properties of soil samples taken from different locations in hesse, Germany. Soil 1:23–33. https://doi.org/10.5194/soil-1-23-2015

Dadhich RK, Meena RS (2014) Performance of Indian mustard (*Brassica juncea* L.) in response to foliar spray of thiourea and thioglycollic acid under different irrigation levels. Indian J Ecol 41(2):376–378

Dadhich RK, Meena RS, Reager ML, Kansotia BC (2015) Response of bio-regulators to yield and quality of Indian mustard (*Brassica juncea* L. Czernj. and Cosson) under different irrigation environments. J App Nat Sci 7(1):52–57

Damian F, Damian G, Lácátuşu R, Postolache C, Iepure G, Jelea M, Násui D (2013) The heavy metals immobilization in polluted soils from Romania by the natural zeolites use. Carpathian J Earth Environ Sci 8:231–250

Daneshvar N, Salari D, Aber S (2002) Chromium adsorption and Cr(VI) reduction to trivalent chromium in aqueous solutions by soya cake. J Hazard Mater 94:49–61. https://doi.org/10.1016/S0304-3894(02)00054-7

Datta R, Baraniya D, Wang YF, Kelkar A, Moulick A, Meena RS, Yadav GS, Ceccherini MT, Formanek P (2017a) Multi-function role as nutrient and scavenger off reeradical in soil. Sustain MDPI 9:402. https://doi.org/10.3390/su9081402

Datta R, Kelkar A, Baraniya D, Molaei A, Moulick A, Meena RS, Formanek P (2017b) Enzymatic degradation of lignin in soil: a review. Sustain MDPI (9):1–18. https://doi.org/10.3390/su9071163

de Baets S, Poesen J, Knapen A, Barbera GG, Navarro JA (2007a) Root characteristics of representative Mediterranean plant species and their erosion-reducing potential during concentrated runoff. Plant Soil. 294:169–183

de Baets S, Poesen J, Knapen A, Galindo P (2007b) Impact of root architecture on the erosion-reducing potential of roots during concentrated flow. Earth Surf Proc Land 32:1323–1345

De Mora AP, Ortega-Calvo JJ, Cabrera F, Madejón E (2005) Changes in enzyme activities and microbial biomass after "in situ" remediation of a heavy metal-contaminated soil. Appl Soil Ecol 28:125–137. https://doi.org/10.1016/j.apsoil.2004.07.006

Dejene A (2003) Integrated natural resources management to enhance food security – the case for community-based approaches in Ethiopia. Environ Nat Resour Work Pap

Deng Z, Cao L, Huang H, Jiang X, Wang W, Shi Y, Zhang R (2011) Characterization of Cd- and Pb-resistant fungal endophyte Mucor sp. CBRF59 isolated from rapes (Brassica chinensis) in a metal-contaminated soil. J Hazard Mater 185:717–724. https://doi.org/10.1016/j.jhazmat.2010.09.078

Dermatas D, Meng X (2003) Utilization of fly ash for stabilization/solidification of heavy metal contaminated soils. Eng Geol 70:377–394. https://doi.org/10.1016/S0013-7952(03)00105-4

Dhakal Y, Meena RS, De N, Verma SK, Singh A (2015) Growth, yield and nutrient content of mungbean (*Vigna radiata* L.) in response to INM in eastern Uttar Pradesh, India. Bangladesh J Bot 44(3):479–482

Dhakal Y, Meena RS, Kumar S (2016) Effect of INM on nodulation, yield, quality and available nutrient status in soil after harvest of green gram. Leg res 39(4):590–594

Dhiman SS, Zhao X, Li J, Kim D, Kalia VC, Kim IW, Kim JY, Lee JK (2017) Metal accumulation by sunflower (Helianthus annuus L.) and the efficacy of its biomass in enzymatic saccharification. PLoS One 12. https://doi.org/10.1371/journal.pone.0175845

Ding F, He Z, Liu S, Zhang S, Zhao F, Li Q, Stoffella PJ (2017) Heavy metals in composts of China: historical changes, regional variation, and potential impact on soil quality. Environ Sci Pollut Res 24:3194–3209. https://doi.org/10.1007/s11356-016-8057-3

Doran JW, Parkin TB (1994) Defining and assessing soil quality. Defining soil quality for a sustainable environment Madison, WI. Soil Sci Soc Am 35:3–21

Doran JW, Sarrantonio M, Liebig MA (1997) Soil health and sustainability. Adv Agron 56:1–54

Dubey S, Banerjee S, Upadhyay SN, Sharma YC (2017) Application of common nano-materials for removal of selected metallic species from water and wastewaters: a critical review. J Mol Liq 240:656–677

Durán Zuazo V, Rodriguez C (2008) Soil-erosion and runoff prevention by plant covers: a review. In: Sustainable agriculture. Springer, Dordrecht, pp 785–811. https://doi.org/10.1007/978-90-481-2666-8_48

Durán ZVH, Francia MJR, Rodríguez PCR, Martínez RA, Cárceles RB (2006) Soil erosion and runoff prevention by plant covers in a mountainous area (SE Spain): implications for sustainable agriculture. Environmentalist 26:309–319

Dutta S, Sen D (2017) Application of SWAT model for predicting soil erosion and sediment yield. Sustain Water Resour Manag:1–22

Edelstein M, Ben-Hur M (2018) Heavy metals and metalloids: Sources, risks and strategies to reduce their accumulation in horticultural crops. Sci Hortic (Amsterdam) 234:431–444

Effron D, De La Horra AM, Defrieri RL, Fontanive V, Palma RM (2004) Effect of cadmium, copper, and lead on different enzyme activities in a native forest soil. Commun Soil Sci 35:1309–1321. Plant Anal. https://doi.org/10.1081/CSS-120037548

Ehigiator OA, Anyata BU (2011) Effects of land clearing techniques and tillage systems on runoff and soil erosion in a tropical rain forest in Nigeria. J Environ Manag. 92:2875–2880

El Baroudy AA (2011) Monitoring land degradation using remote sensing and GIS techniques in an area of the middle Nile Delta Egypt. Catena 87:201–208

Ellis DR, Sors TG, Brunk DG, Albrecht C, Orser C, Lahner B, Wood KV, Harris HH, Pickering IJ, Salt DE (2004) Production of Se-methylselenocysteine in transgenic plants expressing selenocysteine methyltransferase. BMC Plant Biol 4. https://doi.org/10.1186/1471-2229-4-1

Ench MJM, Anceau AM, Angronsveld JV, Mench MJ, Manceau a VJ, Clijsters H, Mocquot B (2000) Capacity of soil amendments in lowering the phytoavailability of sludge-borne zinc. Agronomie 20:383–397. https://doi.org/10.1051/agro:2000135

Esteban W, Pacheco P, Tapia L, Bastías E (2016) Remediation of salt and boron-affected soil by addition of organic matter: an investigation into improving tomato plant productivity. Idesia (Arica) 34:25–32

Eswaran H, Lal R, Reich PF (2001) Land degradation: an overview. Responses to Land degradation, pp 20–35

Etesami H (2018) Bacterial mediated alleviation of heavy metal stress and decreased accumulation of metals in plant tissues: mechanisms and future prospects. Ecotoxicol Environ Saf 147:175–191

Etissa E, Dechassa N, Alamirew T, Alemayehu Y, Drechsel P, Heffer P, Magen H, Mikkelsen R, Wichelns D, Tekeste AN, Mohamed A, Shahabz M, Haile W, Lule M, Dubois T, Coyne D, Kisitu D, Kamusiime H, Bbemba J, Etissa E, Dechassa N, Alamirew T, Alemayehu Y, Crop P, Series S, Gübbük H, Pekmezci M, Drechsel P, Heffer P, Magen H, Mikkelsen R, Wichelns D, Mohamed A, Njukwe E, Tenkouano A, Amah D, Sadik K, Nyine M, Dubois T, Tanji K, Grattan S, Grieve C, Harivandi A, Shaw D, Sheikh B, Wu L, Shahabz M, Tekeste AN, Rahman S, Biswas N, Hassan M, Ahmed G, Tkassahun H, Engda TA, Collick A, Oumer HA, Bayabil HK, Zewdie T, Solomon D, Nicholson CF, Steenhuis TS, Banata AT, MLT C, Giesen LF, Araya G, Pérez-Cotapos MLS, VERGARA RL, Manca M, Tohme RA, Holmberg SD, Bressmann T, Lirio DR, Román JS, Solís RG, Thakur S, Rao SN, Modelado EL, La ADE, Durante C, Tradición UNA, En M, Espejo EL, Fuentes DELAS, De Yucatán UA, Lenin CM, Cian LF, Douglas MJ, Plata L, Héritier F, In SE, Ababa A, Abebe TF, Alamirew T, Abegaz F, Youngquist J, English B, Scharmer R, Chow P, Shook R, Haileslassie A, Priess JA, Veldkamp E, Lesschen JP, Alemu Lema* and Abera Degebassa Zeway Fisheries Resources Research Center, Zeway, Ethiopia. Accepted 3 September 2012, Haile W, NDP D, Etissa E, Dechassa N, Alamirew T, Alemayehu Y, Yoseph T, Shiferaw W, Sorsa Z, Simon T, Shumbullo A, Solomon W, Blomme G, Price N, Coyne D, Lepoint P, Nicolas N, Ndayihazamaso P, Production B, Equatoria H (2013) Household fertilizers use and soil fertility management practices in vegetable crops production: the case of central rift valley of Ethiopia. Int J Agric Biol 2:47–55

Fahnestock P, Lal R, Hall GF (1996) Land use and erosional effects on two Ohio alfisols: II. Crop yields. J Sustain Agric 7:85–100

FAO (The Food and Agriculture Organization of the United Nations) (2011) The state of the World's land and water resources for Food and Agriculture (SOLAW)—managing systems at risk; Food and Agriculture Organization of the United Nations: Rome, Italy; Earthscan: London, UK, Available online: http://www.fao.org/docrep/017/i1688e/i1688e.pdf

FAO and ITPS (2015) Status of the World's Soil Resources (SWSR); Main Report; Food and Agriculture Organization of the United Nations and Intergovernmental Technical Panel on Soils: Rome, Italy, Available online: ftp://extftp.fao.org/nr/Data/Upload/SWSR_MATTEO/Main_report/Pdf/web_Soil_Report_Main_001.pdf

FAO Food and Agriculture Organisation of the United Nations (2005) Management of irrigation-induced salt-affected soils. Cisau, Iptrid, Agll, Fao 97:61–85

FAO, RAP (1999) Poverty alleviation and food security in Asia: land resources. RAP Publication 1999/2. Bangkok. http://www.fao.org/DOCREP/
FAO, UNDP, UNEP (1994) Land degradation in south Asia: its severity, causes and effects upon the people
Feng L, Cao M, Ma X, Zhu Y, Hu C (2012) Superparamagnetic high-surface-area Fe 3O 4 nanoparticles as adsorbents for arsenic removal. J Hazard Mater 217:439–446. https://doi.org/10.1016/j.jhazmat.2012.03.073
Feng C, Zhang S, Li L, Wang G, Xu X, Li T, Zhong Q (2018) Feasibility of four wastes to remove heavy metals from contaminated soils. J Environ Manag 212:258–265. https://doi.org/10.1016/j.jenvman.2018.01.030
Fernández-Calviño D, Cutillas-Barreiro L, Paradelo-Núñez R, Nóvoa-Muñoz JC, Fernández-Sanjurjo MJ, Álvarez-Rodríguez E, Núñez-Delgado A, Arias-Estévez M (2017) Heavy metals fractionation and desorption in pine bark amended mine soils. J Environ Manag 192:79–88. https://doi.org/10.1016/j.jenvman.2017.01.042
Ferrández-Villena M, Ruiz-Canales A (2017) Advances on ICTs for water management in agriculture. Agric Water Manag 183:1–3
Filho MPB, Zimmermann FJP, Silva OFD (2004) Influence of calcium silicate slag on soil acidity and upland rice grain yield. Cienc Agrotec 28:323–331
Filip Z (2002) International approach to assessing soil quality by ecologically-related biological parameters. Agric Ecosyst Environ 88:169–174. https://doi.org/10.1016/S0167-8809(01)00254-7
Fitzpatrick CR, Copeland J, Wang PW, Guttman DS, Kotanen PM, Johnson MT (2018) Assembly and ecological function of the root microbiome across angiosperm plant species. Proc Natl Acad Sci 201717617
Flower KC, Cordingley N, Ward PR, Weeks C (2012) Nitrogen, weed management and economics with cover crops in conservation agriculture in a Mediterranean climate. F Crop Res. 132:63–75
Frossard E, Buchmann N, Bünemann EK, Kiba DI, Lompo F, Oberson A, Tamburini F, Traoré OYA (2016) Soil properties and not inputs control carbon: nitrogen: phosphorus ratios in cropped soils in the long term. Soil 2:83–99
Gao C, Zhang W, Li H, Lang L, Xu Z (2008) Controllable fabrication of mesoporous MgO with various morphologies and their absorption performance for toxic pollutants in water. Cryst Growth Des 8:3785–3790
García-Díaz A, Bienes R, Sastre B, Novara A, Gristina L, Cerdà A (2017) Nitrogen losses in vineyards under different types of soil groundcover. A field runoff simulator approach in central Spain. Agric Ecosyst Environ 236:256–267
Gascho GJ (2005) Soil fertility decline in the tropics: with case studies on plantations. 149-151.
Geebelen W, Vangronsveld J, Adriano DC, Carleer R, Clijsters H (2002) Amendment induced immobilization of lead in a lead-spiked soil: Evidence from phytotoxicity studies. Water Air Soil Pollut 140:261–277
Gholami L, Sadeghi SHR, Homaee M (2013) Straw mulching effect on splash erosion, runoff and sediment yield from eroded plots. Soil Sci Soc Am J 77:268–278
Ghosh BN, Meena VS, Alam NM, Dogra P, Bhattacharyya R, Sharma NK, Mishra PK (2016) Impact of conservation practices on soil aggregation and the carbon management index after seven years of maize-wheat cropping system in the Indian Himalayas. Agric Ecosyst Environ 216:247–257
Giampietro M, Saltelli A (2014) Reply to Goldfinger et al. (2014) Footprint facts and fallacies: a response to Giampietro and Saltelli (2014) Footprints to nowhere. Ecol Indic 46:260–263
Gibbs HK, Salmon JM (2015) Mapping the world's degraded lands. Appl Geogr 57:12–21
Giuffrida F, Graziani G, Fogliano V, Scuderi D, Romano D, Leonardi C (2014) Effects of Nutrient and NaCl salinity on growth, yield, quality and composition of pepper grown in soilless closed system. J Plant Nutr 37:1455–1474
Glick BR (2010) Using soil bacteria to facilitate phytoremediation. Biotechnol Adv 28:367–374

Gong L, Ran Q, He G, Tiyip T (2015) A soil quality assessment under different land use types in Keriya river basin, Southern Xinjiang, China. Soil Tillage Res 146:223–229. https://doi.org/10.1016/j.still.2014.11.001

Gorman JM, Sencindiver JC, Horvath DJ, Singh RN, Keefer RF (2000) Erodibility of fly ash used as a topsoil substitute in mineland reclamation. J Environ Qual 29:805–811

Grewal HS (2010) Water uptake, water use efficiency, plant growth and ionic balance of wheat, barley, canola and chickpea plants on a sodic vertosol with variable subsoil NaCl salinity. Agric Water Manag 97:148–156

Guo G, Zhou Q, Ma LQ (2006) Availability and assessment of fixing additives for the in situ remediation of heavy metal contaminated soils: a review. Environ Monit Assess 116:513–528. https://doi.org/10.1007/s10661-006-7668-4

Gupta VK, Agarwal S, Saleh TA (2011) Synthesis and characterization of alumina-coated carbon nanotubes and their application for lead removal. J Hazard Mater 185:17–23

Gupta VK, Tyagi I, Sadegh H, Ghoshekand RS, Makhlouf ASH, Maazinejad B (2015) Nanoparticles as adsorbent; a positive approach for removal of noxious metal ions: a review. Sci Technol Dev 34:195

Gutiérrez C, Fernández C, Escuer M, Herrera RC, Rodríguez MEB, Carbonell G, Martín JAR (2016) Effect of soil properties, heavy metals and emerging contaminants in the soil nematodes diversity. Environ Pollut 213:184–194. https://doi.org/10.1016/j.envpol.2016.02.012

Gyssels G, Poesen J, Bochet E, Li Y (2005) Impact of plant roots on the resistance of soils to erosion by water: a review. Prog Phys Geog 29:189–217

Hammad AA, Børresen T, Haugen E (2006) Effects of rain characteristics and terracing on runoff and erosion under the Mediterranean. Soil Till Res 87:39–47

Hamza MA, Anderson WK (2005) Soil compaction in cropping systems: A review of the nature, causes and possible solutions. Soil Tillage Res 82:121–145

Hapani P, Marjadi D (2015) Salt tolerance and biochemical resposes as a stress indicator in plants to salinity: a review. CIBTech J Biotechnol ISSN

Hasaneen MNA, Younis ME, Tourky SM (2009) Plant growth, metabolism and adaptation in relation to stress conditions XXIII. Salinity-biofertility interactive effects on growth, carbohydrates and photosynthetic efficiency of lactuca sativa. Plant Omics 2:197–205

He M, Shi H, Zhao X, Yu Y, Qu B (2013) Immobilization of Pb and Cd in contaminated soil using nano-crystallite hydroxyapatite. Procedia Environ Sci 18:657–665

He Z, Shentu, Yang X, Baligar, Zhang T, Stoffella (2015) Heavy metal contamination of soils: sources, indicators, and assessment. J Environ Indic 9:17–18

Hemant K, Pani P (2017) Effects of soil erosion on agricultural productivity in semi-arid regions. J Agron 32:165–184

Henry J, Aherne J, Starkloff T, Stolte J, Bejan A, Wikipedia F, Asia E (2014) Testing your soil why and how to take a soil-test sample. Agron Soil 4:1–12

Herbert ER, Boon P, Burgin AJ, Neubauer SC, Franklin RB, Ardon M, Hopfensperger KN, Lamers LPM, Gell P, Langley JA (2015) A global perspective on wetland salinization: ecological consequences of a growing threat to freshwater wetlands. Ecosphere 6:1–43

Hiel M-P, Barbieux S, Pierreux J, Olivier C, Lobet G, Roisin C, Garré S, Colinet G, Bodson B, Dumont B (2018) Impact of crop residue management on crop production and soil chemistry after seven years of crop rotation in temperate climate, loamy soils. PeerJ 6:e4836. https://doi.org/10.7717/peerj.4836

Hinojosa MB, Carreira JA, Ruíz RG, Dick RP (2004) Soil moisture pre-treatment effects on enzyme activities as indicators of heavy metal-contaminated and reclaimed soils. Soil Biol Biochem 36:1559–1568. https://doi.org/10.1016/j.soilbio.2004.07.003

Hinsinger P, Plassard C, Tang C, Jaillard B (2003) Origins of root-mediated pH changes in the rhizosphere and their responses to environmental constraints: A review. Plant Soil 248:43–59

Hintz WD, Relyea RA (2017) A salty landscape of fear: responses of fish and zooplankton to freshwater salinization and predatory stress. Oecologia 185:147–156

Hiwale S (2015) Sustainable horticulture in semiarid dry lands, pp 1–393

Ho HL, Chan WK, Blondy A, Yeung KL, Schrotter JC (2012) Experiment and modeling of advanced ozone membrane reactor for treatment of organic endocrine disrupting pollutants in water. Catal Today 193:120–127

Hou J, Li M, Mao X, Hao Y, Ding J, Liu D, Xi B, Liu H (2017) Response of microbial community of organic matter impoverished arable soil to long-term application of soil conditioner derived from dynamic Rapid fermentation of food waste. PLoS One 12. https://doi.org/10.1371/journal.pone.0175715

Hu Y, Schmidhalter U (2005) Drought and salinity: a comparison of their effects on mineral nutrition of plants. J Plant Nutr Soil Sci 168:541–549

Hugo V (2008) Review article soil-erosion and runoff prevention by plant covers. A review. 28:65–86

Hurni H, Giger M, Liniger H, Mekdaschi Studer R, Messerli P, Portner B, Schwilch G, Wolfgramm B, Breu T (2015) Soils, agriculture and food security: the interplay between ecosystem functioning and human well-being. Curr Opin Environ Sustain 15:25–34

IECA (1991) Erosion control – a global perspective. Proceedings of Conference XXII, International Erosion Control Association, Steamboat Springs, CO.

Iqbal J, Shah MH (2011) Distribution, correlation and risk assessment of selected metals in urban soils from Islamabad, Pakistan. J Hazard Mater 192:887–898. https://doi.org/10.1016/j.jhazmat.2011.05.105

Ivits E, Cherlet M, Tóth T, Lewińska KE, Tóth G (2013) Characterisation of productivity limitation of salt-affected lands in different climatic regions of Europe using remote sensing derived productivity indicators. L Degrad Dev 24:438–452

Jagoutz E (2006) Salt-induced rock fragmentation on Mars: the role of salt in the weathering of Martian rocks. Adv Sp Res 38:696–700

Jala S, Goyal D (2006) Fly ash as a soil ameliorant for improving crop production – a review. Bioresour Technol 97:1136–1146

Jamil M, Hussain SS, Qureshi MA, Mehdi SM, Nawaz MQ (2017) Impact of sowing techniques and nitrogen fertilization on castor bean yield in salt affected soils. J Anim Plant Sci 27:451–456

Jayasekera S, Hall S (2007) Modification of the properties of salt affected soils using electrochemical treatments. Geotech Geol Eng 25:1–10

Jiang C, Zhang L (2016) Ecosystem change assessment in the Three-river Headwater Region, China: patterns, causes, and implications. Ecol Eng. 93:24–36

Jiao J, Zou H, Jia Y, Wang N (2009) Research progress on the effects of soil erosion on vegetation. Acta Ecol Sin. https://doi.org/10.1016/j.chnaes.2009.05.001

Jie C, Jing-Zhang C, Man-Zhi, Zi-tong G (2002) Soil degradation: a global problem endangering sustainable development. J Geogr Sci 12:243–252

Jiménez MN, Fernández-Ondoño E, Ripoll MA, Castro-Rodríguez J, Huntsinger L, Navarro FB (2016) Stones and organic mulches improve the Quercus Ilex L. afforestation success under Mediterranean climatic conditions. Land Degrad Dev 27:357–365

Johnston AM, Bruulsema TW (2014) 4R nutrient stewardship for improved nutrient use efficiency. Procedia Eng 83:365–370

Jones DL (1998) Organic acids in the rhizosphere - A critical review. Plant Soil 205:25–44

Jordán A, Zavala LM, Muñoz-Rojas M (2011) Mulching, effects on soil physical properties. In: Gliński J, Horabik J, Lipiec J (eds) Encyclopedia of agrophysics. Springer, Dordrecht, pp 492–496

Kairis O, Karavitis C, Salvati L, Kounalaki A (2015) Exploring the impact of overgrazing on soil erosion and land degradation in a dry Mediterranean agro-forest landscape (Crete, exploring the impact of overgrazing on soil erosion and land degradation in a dry Mediterranean agro-forest landscape). Arid Land Res Manag 29:360–374

Kaushal M, Wani SP (2017) Nanosensors: frontiers in precision agriculture. In: Nanotechnology: an agricultural paradigm, pp 279–291

Kaushal SS, Likens GE, Pace ML, Utz RM, Haq S, Gorman J, Grese M (2018) Freshwater salinization syndrome on a continental scale. Proc Natl Acad Sci 115:574–583

Kawashima CG, Noji M, Nakamura M, Ogra Y, Suzuki KT, Saito K (2004) Heavy metal tolerance of transgenic tobacco plants over-expressing cysteine synthase. Biotechnol Lett 26:153–157. https://doi.org/10.1023/B:BILE.0000012895.60773.ff

Keesstra S, Pereira P, Novara A, Brevik EC, Azorin-Molina C, Parras-Alcántara L, Jordán A, Cerdà A (2016) Effects of soil management techniques on soil water erosion in apricot orchards. Sci Total Environ 551:357–366

Keizer JJ, Martins MAS, Prats SA, Santos LF, Vieira DCS, Nogueira R, Bilro L (2015) Assessing the performance of a plastic optical fibre turbidity sensor for measuring post-fire erosion from plot to catchment scale. Soil 1:641–650

Khan MA, Ungar IA, Showalter AM (2000) Effects of salinity on growth, water relations and ion accumulation of the subtropical perennial halophyte, Atriplex griffithii var. stocksii. Ann Bot 85:225–232

Khan S, Cao Q, Hesham AE-L, Xia Y, He J-Z (2007) Soil enzymatic activities and microbial community structure with different application rates of Cd and Pb. J Environ Sci (China) 19:834–840. https://doi.org/10.1016/S1001-0742(07)60139-9

Khan S, Rehman S, Zeb Khan A, Amjad Khan M, Tahir Shah M (2010) Soil and vegetables enrichment with heavy metals from geological sources in Gilgit, northern Pakistan. Ecotoxicol Environ Saf 73:1820–1827. https://doi.org/10.1016/j.ecoenv.2010.08.016

Khresat SA, Qudah EA (2006) Formation and properties on aridic soils of Azraq Basin in northeastern Jordan. J Arid Environ 64:116–136

Kim DH, Shin MC, Choi HD, Il SC, Baek K (2008) Removal mechanisms of copper using steelmaking slag: adsorption and precipitation. Desalination 223:283–289. https://doi.org/10.1016/j.desal.2007.01.226

Kim JO, Lee YW, Chung J (2013) The role of organic acids in the mobilization of heavy metals from soil. Ksce J Civ Eng 17:1596–1602. https://doi.org/10.1007/s12205-013-0323-z

Kinnell PIA (2005) Raindrop impact induced erosion processes and prediction: a review. Hydrol Process 19:2815–2844

Kirkels FMSA, Cammeraat LH, Kuhn NJ (2014) The fate of soil organic carbon upon erosion, transport and deposition in agricultural landscapes—a review of different concepts. Geomorphology 226:94–105

Kong AYY, Six J (2010) Tracing root vs. residue carbon into soils from conventional and alternative cropping systems. Soil Sci Soc Am J 74:1201–1210

Koyro HW (2006) Effect of salinity on growth, photosynthesis, water relations and solute composition of the potential cash crop halophyte Plantago coronopus (L.). Environ Exp Bot 56:136–146

Kraaijvanger R, Veldkamp T (2015) Grain productivity, fertilizer response and nutrient balance of farming systems in Tigray, Ethiopia: a multi-perspective view in relation to soil fertility degradation. L Degrad Dev 26:701–710

Krebs R, Gupta SK, Furrer G, Schulin R (1999) Gravel sludge as an immobilizing agent in soils contaminated by heavy metals: A field study. Water Air Soil Pollut 115:465–479. https://doi.org/10.1023/A:1005167004828

Kumar S, Meena RS, Pandey A, Seema (2017a) Soil acidity management and an economics response of lime and sulfur on sesame in an alley cropping system. Int J Current Microb App Sci 6(3):2566–2573

Kumar S, Meena RS, Yadav GS, Pandey A (2017b) Response of sesame (*Sesamum indicum* L.) to sulphur and lime application under soil acidity. Int J Plant Soil Sci 14(4):1–9

Kumar S, Meena RS, Bohra JS (2018) Interactive effect of sowing dates and nutrient sources on dry matter accumulation of Indian mustard (*Brassica juncea* L.). Journal of Oilseed Brassica 9(1):72–76

Kumpiene J, Ore S, Renella G, Mench M, Lagerkvist A, Maurice C (2006) Assessment of zerovalent iron for stabilization of chromium, copper, and arsenic in soil. Environ Pollut 144:62–69. https://doi.org/10.1016/j.envpol.2006.01.010

Kumpiene J, Lagerkvist A, Maurice C (2007) Stabilization of Pb- and Cu-contaminated soil using coal fly ash and peat. Environ Pollut 145:365–373. https://doi.org/10.1016/j.envpol.2006.01.037

Kumpiene J, Lagerkvist A, Maurice C (2008) Stabilization of As, Cr, Cu, Pb and Zn in soil using amendments – a review. Waste Manag 28:215–225. https://doi.org/10.1016/j.wasman.2006.12.012
Kunito T, Saeki K, Goto S, Hayashi H, Oyaizu H, Matsumoto S (2001) Copper and zinc fractions affecting microorganisms in long-term sludge-amended soils. Bioresour Technol 79:135–146. https://doi.org/10.1016/S0960-8524(01)00047-5
Lahori A, Guo Z, Zhang Z, Li R, Mahar A, Mk A, Shen F, Ta S, Kumbhar F, Wang P, Jiang S (2017) Use of biochar as an amendment for remediation of heavy metal-contaminated soils: prospects and challenges. Pedosphere 27:991–1014. https://doi.org/10.1016/S1002-0160(17)60490-9
Lal R (1995) Erosion-crop productivity relationships for soils of Africa. Soil Sci Soc Am J 59:661–667
Lal R (2012) Research needs for credible data on soil resources and degradation. In: Lal R, Stewart BA (eds) World soil resources and food security. CRC Press, Boca Raton, pp 539–546
Lal R (2015) Restoring soil quality to mitigate soil degradation. Sustainability 7:5875–5895
Lal R, Stewart BA (1990) Soil degradation. Springer, New York
Lambers H, Finnegan PM, Laliberte E, Pearse SJ, Ryan MH, Shane MW, Veneklaas EJ (2011) Phosphorus nutrition of proteaceae in severely phosphorus-impoverished soils: are there lessons to be learned for future crops? Plant Physiol 156:1058–1066
Leggo PJ, Ledésert B (2001) Use of organo-zeolitic fertilizer to sustain plant growth and stabilize metallurgical and mine-waste sites. Mineral Mag 65:563–570. https://doi.org/10.1180/002646101317018398
Lehman RM, Taheri WI (2013) Soil microorganisms can reduce P loss from cropping systems. In: Sustainable agriculture reviews. Springer, Cham, pp 15–36
Lehman RM, Taheri WI (2017) Soil microorganisms can reduce P loss from cropping systems. In: Sustainable agriculture reviews. Springer, Cham, pp 15–36
Li XD, Poon CS, Sun H, Lo IMC, Kirk DW (2001) Heavy metal speciation and leaching behaviors in cement based solidified/stabilized waste materials. J Hazard Mat A82:215–230
Li YH, Ding J, Luan Z, Di Z, Zhu Y, Xu C, Wu D, Wei B (2003a) Competitive adsorption of Pb2+, Cu2+ and Cd2+ ions from aqueous solutions by multiwalled carbon nanotubes. Carbon NY 41:2787–2792. https://doi.org/10.1016/S0008-6223(03)00392-0
Li YH, Wang S, Luan Z, Ding J, Xu C, Wu D (2003b) Adsorption of cadmium(II) from aqueous solution by surface oxidized carbon nanotubes. Carbon NY 41:1057–1062. https://doi.org/10.1016/S0008-6223(02)00440-2
Li FR, Zhao WZ, Liu JL, Huang ZG (2009) Degraded vegetation and wind erosion influence soil carbon, nitrogen and phosphorus accumulation in sandy grasslands. Plant Soil 317:79–92
Li Z, Huang J, Zeng G, Nie X, Ma W, Yu W, Guo W, Zhang J (2013) Effect of erosion on productivity in subtropical red soil hilly region: a multi-scale spatio-temporal study by simulated rainfall. PloS One 8:77838. https://doi.org/10.1371/journal.pone.0077838
Li Q, Zhou J, Zhou Y, Jia R (2014) Salinization of deep groundwater in plain areas of Xinjiang: causes and countermeasures. Desalin Water Treat 52:2724–2733
Li Z, Nie X, Chang X, Liu L, Sun L (2016) Characteristics of soil and organic carbon loss induced by water erosion on the loess plateau in China. PLoS One 11:e0154591
Lima Neto MC, Lobo AKM, Martins MO, Fontenele AV, Silveira JAG (2014) Dissipation of excess photosynthetic energy contributes to salinity tolerance: a comparative study of salt-tolerant Ricinus communis and salt-sensitive Jatropha curcas. J Plant Physiol 171:23–30
Lin S-T, Thirumavalavan M, Jiang T-Y, Lee J-F (2014) Synthesis of ZnO/Zn nano photocatalyst using modified polysaccharides for photodegradation of dyes. Carbohydr Polym 105:1–9. https://doi.org/10.1016/j.carbpol.2014.01.017
Liu XB, Zhang XY, Wang YX, Sui YY, Zhang SL, Herbert SJ, Ding G (2010) Soil degradation: a problem threatening the sustainable development of agriculture in Northeast China. Plant Soil Environ 56:87–97
Liu SL, Wang C, Zhang XL, Yang JJ, Qiu Y, Wang J (2011) Soil and water conservation effect of different terrace configurations in land consolidation project. J Soil Water Conserv 25:59–62

Liu Z, Gu C, Chen F, Yang D, Wu K, Chen S, Jiang J, Zhang Z (2012) Heterologous expression of a Nelumbo nucifera phytochelatin synthase gene enhances cadmium tolerance in Arabidopsis thaliana. Appl Biochem Biotechnol 166:722–734. https://doi.org/10.1007/s12010-011-9461-2

Lombi E, Zhao FJ, Zhang G, Sun B, Fitz W, Zhang H, McGrath SP (2002) In situ fixation of metals in soils using bauxite residue: chemical assessment. Environ Pollut 118:435–443. https://doi.org/10.1016/S0269-7491(01)00294-9

Lottermoser BG (2010) Mine Wastes (third edition): Characterization, treatment and environmental impacts. DOI. https://doi.org/10.1007/978-3-642-12419-81-400

Lu N, Maruo T, Johkan M, Hohjo M, Tsukagoshi S, Ito Y, Ichimura T, Shinohara Y (2012) Effects of supplemental lighting within the canopy at different developing stages on tomato yield and quality of single-truss tomato plants grown at high density. Environ Control Biol 50:1–11

Luming MA, Xian ZW (2008) Enhanced biological treatment of industrial wastewater with bimetallic zero-valent iron. Environ Sci Technol 42:5384–5389. %U http://search.ebscohost.com.proxy.gre

Lycoskoufis IH, Savvas D, Mavrogianopoulos G (2005) Growth, gas exchange, and nutrient status in pepper (Capsicum annuum L.) grown in recirculating nutrient solution as affected by salinity imposed to half of the root system. Sci Hortic (Amsterdam) 106:147–161

Ma Y, Prasad MNV, Rajkumar M, Freitas H (2011) Plant growth promoting rhizobacteria and endophytes accelerate phytoremediation of metalliferous soils. Biotechnol Adv 29:248–258

Ma Y, Oliveira RS, Freitas H, Zhang C (2016a) Biochemical and molecular mechanisms of plant-microbe-metal interactions: relevance for phytoremediation. Front Plant Sci 7. https://doi.org/10.3389/fpls.2016.00918

Ma Y, Rajkumar M, Zhang C, Freitas H (2016b) Beneficial role of bacterial endophytes in heavy metal phytoremediation. J Environ Manage 174:14–25

MacDonald LH, Larsen IJ (2009) Runoff and erosion from wildfires and roads: effects and mitigation, ch 9. L Restor to Combat Desertif Innov Approaches, Qual Control Proj Eval

Madejón P, Alaejos J, García-Álbala J, Fernández M, Madejón E (2016) Three-year study of fast-growing trees in degraded soils amended with composts: effects on soil fertility and productivity. J Environ Manag 169:18–26

Maestri E, Marmiroli N (2011) Transgenic plants for phytoremediation. Int J Phytoremediation 13:264–279. https://doi.org/10.1080/15226514.2011.568549

Mahabadi AA, Hajabbasi MA, Khademi H, Kazemian H (2007) Soil cadmium stabilization using an Iranian natural zeolite. Geoderma 137:388–393. https://doi.org/10.1016/j.geoderma.2006.08.032

Mahmoodabadi M, Ghadiri H, Yu B, Rose C (2014a) Morpho-dynamic quantification of flow-driven rill erosion parameters based on physical principles. J Hydrol 514:328–336

Mahmoodabadi M, Ghadiri H, Rose C, Yu B, Rafahi H, Rouhipour H (2014b) Evaluation of GUEST and WEPP with a new approach for the determination of sediment transport capacity. J Hydrol 513:413–421

Mandal D, Sharda VN (2013) Appraisal of soil erosion risk in the Eastern Himalayan region of India for soil conservation planning. Land Degrad Dev 24:430–437

Manta DS, Angelone M, Bellanca A, Neri R, Sprovieri M (2002) Heavy metals in urban soils: a case study from the city of Palermo (Sicily), Italy. Sci Total Environ 300:229–243. https://doi.org/10.1016/S0048-9697(02)00273-5

Martínez FJ, Sierra M, Sierra C, Roca A (2005) Assessing sustainable use of land under olive cultivation in Alcala la Real (Jaen, Spain) using GIS. Adv Geo Ecol 36:75–84

Martínez RA, Durán ZVH, Francia FR (2006) Soil erosion and runoff response to plant cover strips on semiarid slopes (SE Spain). Land Degrad Dev 17:1–11

Martos S, Gallego B, Sáez L, López-Alvarado J, Cabot C, Poschenrieder C (2016) Characterization of zinc and cadmium hyperaccumulation in three noccaea (Brassicaceae) populations from non-metalliferous sites in the eastern pyrenees. Front Plant Sci 7. https://doi.org/10.3389/fpls.2016.00128

Mbagwu JSC, Lal R, Scott TW (1984) Effects of artificial de-surfacing on alfisols and ultisols in Southern Nigeria: II. Changes in soil physical properties 1. Soil Sci Soc MJ 48:834–838

Measures PC (2011) Temporary and permanent control measures, (June), pp 1–22
Meena H, Meena RS (2017) Assessment of sowing environments and bio-regulators as adaptation choice for clusterbean productivity in response to current climatic scenario. Bangladesh J Bot 46(1):241–244
Meena RS, Yadav RS (2014) Phonological performance of groundnut varieties under sowing environments in hyper arid zone of Rajasthan, India. J App Nat Sci 6(2):344–348
Meena RS, Yadav RS (2015) Yield and profitability of groundnut *(Arachis hypogaea* L) as influenced by sowing dates and nutrient levels with different varieties. Legum Res 38(6):791–797
Meena RS, Yadav RS, Meena VS (2014) Response of groundnut (*Arachis hypogaea* L.) varieties to sowing dates and NP fertilizers under Western Dry Zone of India. Bangladesh J Bot 43(2):169–173
Meena RS, Dhakal Y, Bohra JS, Singh SP, Singh MK, Sanodiya P (2015) Influence of bioinorganic combinations on yield, quality and economics of mungbean. Am J Exp Agri 8(3):159–166
Meena RS, Meena VS, Meena SK, Verma JP (2015a) The needs of healthy soils for a healthy world. J Clean Prod 102:560–561
Meena RS, Meena VS, Meena SK, Verma JP (2015b) Towards the plant stress mitigate the agricultural productivity: a book review. J Clean Prod 102:552–553
Meena RS, Yadav RS, Meena H, Kumar S, Meena YK, Singh A (2015c) Towards the current need to enhance legume productivity and soil sustainability worldwide: a book review. J Clean Prod 104:513–515
Meena RS, Yadav RS, Reager ML, De N, Meena VS, Verma JP, Verma SK, Kansotia BC (2015d) Temperature use efficiency and yield of groundnut varieties in response to sowing dates and fertility levels in Western Dry Zone of India. Am J Exp Agric 7(3):170–177
Meena H, Meena RS, Singh B, Kumar S (2016) Response of bio-regulators to morphology and yield of clusterbean [*Cyamopsis tetragonoloba* (L.) Taub.] under different sowing environments. J App Nat Sci 8(2):715–718
Meena RS, Bohra JS, Singh SP, Meena VS, Verma JP, Verma SK, Shiiag SK (2016a) Towards the prime response of manure to enhance nutrient use efficiency and soil sustainability a current need: a book review. J Clean Prod 112:1258–1260
Meena RS, Gogaoi N, Kumar S (2017a) Alarming issues on agricultural crop production and environmental stresses. J Clean Prod 142:3357–3359
Meena RS, Kumar S, Pandey A (2017b) Response of sulfur and lime levels on productivity, nutrient content and uptake of sesame under guava (*Psidium guajava* L.) based agri-horti system in an acidic soil of eastern Uttar Pradesh, India. J Crop Weed 13(2):222–227
Meena RS, Meena PD, Yadav GS, Yadav SS (2017c) Phosphate solubilizing microorganisms, principles and application of microphos technology. J Clean Prod 145:157–158
Meena H, Meena RS, Lal R, Singh GS, Mitran T, Layek J, Patil SB, Kumar S, Verma T (2018) Response of sowing dates and bio regulators on yield of clusterbean under current climate in alley cropping system in eastern U.P, Indian. Legum Res 41(4):563–571
Meena RS, Kumar V, Yadav GS, Mitran T (2018a) Response and interaction of *Bradyrhizobium japonicum* and *Arbuscular mycorrhizal* fungi in the soybean rhizosphere: A review. Plant Growth Regul 84:207–223
Mehdi SM, Sarfraz M, Ilyas M, Amjad Qureshi M, Zaka MA (2015) Integrated nutrient management using P-fixation factor in rice-wheat cropping system under salt affected conditions. Int J Agric Biol 17:643–647
Mehenni A, Cuisinier O, Masrouri F (2016) Impact of lime, cement, and clay treatments on the internal erosion of compacted soils. J Mater Civ Eng. https://doi.org/10.1061/(ASCE)MT.1943-5533.0001573
Memon SQ, Memon N, Shah SW, Khuhawar MY, Bhanger MI (2007) Sawdust-a green and economical sorbent for the removal of cadmium (II) ions. J Hazard Mater 139:116–121. https://doi.org/10.1016/j.jhazmat.2006.06.013
Mench M, Schwitzguébel JP, Schroeder P, Bert V, Gawronski S, Gupta S (2009) Assessment of successful experiments and limitations of phytotechnologies: contaminant uptake, detoxifica-

tion and sequestration, and consequences for food safety. Environ Sci Pollut Res 16:876–900. https://doi.org/10.1007/s11356-009-0252-z
Metternicht G (2017) Soils: salinization. In: International encyclopedia of geography: people, the earth, environment and technology. https://doi.org/10.1002/9781118786352.wbieg1044.1-10
Metternicht GI, Zinck JA (2003) Remote sensing of soil salinity: potentials and constraints. Remote Sens Environ 85:1–20
Milner MJ, Kochian LV (2008) Investigating heavy-metal hyperaccumulation using Thlaspi caerulescens as a model system. Ann Bot 102:3–13
Misra RK, Teixeira PC (2001) The sensitivity of erosion and erodibility of forest soils to structure and strength. Soil Tillage Res. https://doi.org/10.1016/S0167-1987(01)00155-6
Mleczek M, Gąsecka M, Drzewiecka K, Goliński P, Magdziak Z, Chadzinikolau T (2013) Copper phytoextraction with willow (Salix viminalis L.) under various Ca/Mg ratios. Part 1. Copper accumulation and plant morphology changes. Acta Physiol Plant 35:3251–3259. https://doi.org/10.1007/s11738-013-1360-4
Mohamadi MA, Kavian A (2015) Effects of rainfall patterns on runoff and soil erosion in field plots. Int Soil Water Conser Res 3:273–281
Montgomery DR (2007) Soil erosion and agricultural sustainability. Proc Natl Acad Sci 104:13268–13272
Morgan RPC (2014) Soil erosion and conservation
Mortvedt J, Beaton J (1995) Heavy metal and radionuclide contaminants in phosphate fertilizers. In: Phosphorous in the global environment. Wiley, Chichester, pp 93–106
Mullan D (2013) Soil erosion under the impacts of future climate change: assessing the statistical significance of future changes and the potential on-site and off-site problems. Catena 109:234–246
Mulumba LN, Lal R (2008) Mulching effects on selected soil physical properties. Soil Tillage Res 98:106–111
Muñoz-rojas M (2018) Soil quality indicators: critical tools in ecosystem restoration. Curr Opin Environ Sci Heal 5:47–52. https://doi.org/10.1016/j.coesh.2018.04.007
Murtaza G, Ghafoor A, Qadir M (2006) Irrigation and soil management strategies for using saline-sodic water in a cotton–wheat rotation. Agric Water Manag 81:98–114. https://doi.org/10.1016/j.agwat.2005.03.003
Murtaza B, Murtaza G, Zia-ur-Rehman M, Ghafoor A, Abubakar S, Sabir M (2011) Reclamation of salt-affected soils using amendments and growing wheat crop. Soil Environ 30:130–136
Mwango SB, Msanya BM, Mtakwa PW, Kimaro DN, Deckers J, Poesen J (2016) Effectiveness of mulching under miraba in controllingsoil erosion, fertility restoration and crop yield in the Usambara mountains, Tanzania. Land Degrad Dev 27:1266–1275
Nagajyoti PC, Lee KD, Sreekanth TVM (2010) Heavy metals, occurrence and toxicity for plants: a review. Environ Chem Lett 8:199–216. https://doi.org/10.1007/s10311-010-0297-8
Nascimento CWA d (2006) Organic acids effects on desorption of heavy metals from a contaminated soil. Sci Agric 63:276–280. https://doi.org/10.1590/S0103-90162006000300010
Neelo J, Kashe K, Teketay D, Masamba W (2015) Ethnobotanical survey of woody plants in Shorobe and Xobe villages, northwest region of Botswana. Ethnobot Res App 14:367–379
Negim O (2009) New technique for soil reclamation and conservation. In: Situ stabilization of trace elements in contaminated soils. Earth Sciences Université Sciences et Technologies - Bordeaux I, English <tel-00408020>
Newell N (2013) Review: effects of soil salinity on plant growth. Nicole Newell Plant Physiol
Nrcs (1996) Indicators for soil quality evaluation. Soil Qual Inf Sheet:1–2
Nriagu JO (1996) A history of global metal pollution. Science. https://doi.org/10.1126/science.272.5259.223
Obour PB, Dadzie FA, Kristensen HL, Rubæk GH, Kjeldsen C, Saba CKS (2015) Assessment of farmers' knowledge on fertilizer usage for peri-urban vegetable production in the Sunyani Municipality, Ghana. Resour Conserv Recycl 103:77–84
Oertel C, Matschullat J, Zurba K, Zimmermann F, Erasmi S (2016) Greenhouse gas emissions from soils—a review. Chem Erde Geochem 76:327–352

Oldeman L, Hakkeling R, Sombroek W (1991) World map of the status of human-induced soil degradation: an explanatory note. Global Assessment of Soil Degradation (GLASOD), Agris
Oliveira A, Pampulha ME (2006) Effects of long-term heavy metal contamination on soil microbial characteristics. J Biosci Bioeng 102:157–161. https://doi.org/10.1263/jbb.102.157
Oste LA, Lexmond TM, Van Riemsdijk WH (2002) Metal immobilization in soils using synthetic zeolites. J Environ Qual 31:813–821. https://doi.org/10.2134/jeq2002.8130
Oyetunji OJ, Imade FN (2015) Effect of different levels of NaCl and Na2SO4 salinity on dry matter and ionic contents of cowpea (Vigna unguiculata L. Walp.). Afr J Agric Res. https://doi.org/10.5897/AJAR2014.9463
Öztürk M, Ashraf M, Aksoy A, Ahmad MSA, Hakeem KR (2016) Plants, pollutants and remediation, pp 1–404
Panda I, Jain S, Das SK, Jayabalan R (2017) Characterization of red mud as a structural fill and embankment material using bioremediation. Int Biodeterior Biodegrad 119:368–376. https://doi.org/10.1016/j.ibiod.2016.11.026
Paricha AMP, Sethi KC, Gupta V, Pathak A, Chhotray SK (2017). Soil water conservation for microcatchment water harvesting systems. ISBN: 9780998900001
Park JY, Yu YS, Hwang SJ, Kim C, Kim SJ (2014) SWAT modeling of best management for Chungju dam watershed in South Korea under future climate change scenarios. Paddy Water Environ 12:65–75
Paz-Ferreiro J, Fu S (2016) Biological indices for soil quality evaluation: perspectives and limitations. Land Degrad Dev 27:14–25
Pedrotti A, Chagas RM, Ramos VC, do Nascimento Prata AP, Tadeu Lucas AA, dos Santos PB (2015) Causes and consequences of the process of soil salinization. Rev Eletronica EM Gest Educ E Tecnol Ambient 19:1308–1324
Pilon-Smits E (2005) Phytoremediation. Annu Rev Plant Biol 56:15–39. https://doi.org/10.1146/annurev.arplant.56.032604.144214
Pimentel D, Burgess M (2013) Soil erosion threatens food production. Agriculture 3:443–463. https://doi.org/10.3390/agriculture3030443
Pisinaras V, Tsihrintzis VA, Petalas C, Ouzounis K (2010) Soil salinization in the agricultural lands of Rhodope District, northeastern Greece. Environ Monit Assess 166:79–94
Poudel D, Horwath W, Mitchell J, Temple S (2001) Impacts of cropping systems on soil nitrogen storage and loss. Agric Syst. 68:253–268
Pradhan J, Das SNS, Thakur RSR (1999) Adsorption of hexavalent chromium from aqueous solution by using activated red mud. J Colloid Interface Sci 217:137–141. https://doi.org/10.1006/jcis.1999.6288
Prasad MNV (2003) Phytoremediation of metal-polluted ecosystems: hype for commercialization. Russ J Plant Physiol 50:686–700
Prosdocimi M, Jordán A, Tarolli P, Keesstra S, Novara A, Cerdà A (2016) The immediate effectiveness of barley straw mulch in reducing soil erodibility and surface runoff generation in Mediterranean vineyards. Sci Total Environ 547:323–330
Puigdefábregas J (2005) The role of vegetation patterns in structuring runoff and sediment fluxes in drylands. Earth Surf Process Landf 30:133–147
Pupatto JGC, Bull LT, Crusciol CAC (2004) Soil chemical attributes, root growth and rice yield according to slag application. Pesquisa Agropecuaria Brasileria 39:1213–1218
Qadir M, Qureshi AS, Cheraghi SAM (2008) Extent and characterisation of salt-affected soils in Iran and strategies for their amelioration and management. L Degrad Dev 19:214–227
Raicevic S, Kaludjerovic-Radoicic T, Zouboulis AI (2005) In situ stabilization of toxic metals in polluted soils using phosphates: Theoretical prediction and experimental verification. J Hazard Mater 117:41–53. https://doi.org/10.1016/j.jhazmat.2004.07.024
Rajan K, Natarajan A, Kumar K, Badrinath M, Gowda R (2010) Soil organic carbon—the most reliable indicator for monitoring land degradation by soil erosion. Curr Sci 99:823–827
Rajkumar M, Ae N, Prasad MNV, Freitas H (2010) Potential of siderophore-producing bacteria for improving heavy metal phytoextraction. Trends Biotechnol 28:142–149

Rajkumar M, Sandhya S, Prasad MNV, Freitas H (2012) Perspectives of plant-associated microbes in heavy metal phytoremediation. Biotechnol Adv 30:1562–1574

Ram K, Meena RS (2014) Evaluation of pearl millet and mungbean intercropping systems in Arid region of Rajasthan (India). Bangladesh J Bot 43(3):367–370

Rao M, Parwate AV, Bhole AG (2002) Removal of Cr6+ and Ni2+ from aqueous solution using bagasse and fly ash. Waste Manag 22:821–830. https://doi.org/10.1016/S0956-053X(02)00011-9

Rath KM, Rousk J (2015) Salt effects on the soil microbial decomposer community and their role in organic carbon cycling: a review. Soil Biol Biochem 81:108–123

Ravi S, Breshears DD, Huxman TE, D'Odorico P (2010) Land degradation in drylands: Interactions among hydrologic–aeolian erosion and vegetation dynamics. Geomorphology 116:236–245

Ray PZ, Shipley HJ, Fu F et al (2015) Inorganic nano-adsorbents for the removal of heavy metals and arsenic: a review. RSC Adv 5:29885–29907. https://doi.org/10.1039/C5RA02714D

Re V, Sacchi E (2017) Tackling the salinity-pollution nexus in coastal aquifers from arid regions using nitrate and boron isotopes. Environ Sci Pollut Res 24:13247–13261

Rego TJ, Nageswara Rao V, Seeling B, Pardhasaradhi G, Kumar Rao JVDK (2003) Nutrient balances – A guide to improving sorghum- and groundnut-based dryland cropping systems in semi-arid tropical India. F Crop Res 81:53–68

Rehman MZ, Rizwan M, Ali S, Fatima N, Yousaf B, Naeem A, Sabir M, Raza H, Sik Y (2016) Ecotoxicology and Environmental Safety Contrasting effects of biochar, compost and farm manure on alleviation of nickel toxicity in maize (Zea mays L .) in relation to plant growth , photosynthesis and metal uptake. Ecotoxicol Environ Saf 133:218–225. https://doi.org/10.1016/j.ecoenv.2016.07.023

Rehman MZ, Rizwan M, Ali S, Ok YS, Ishaque W, Saifullah NMF, Akmal F, Waqar M (2017) Remediation of heavy metal contaminated soils by using Solanum nigrum: a review. Ecotoxicol Environ Saf 143:236–248

Reviews E, Yao ZT, Ji XS, Sarker PK, Tang JH, Ge LQ, Xia MS, Xi YQ (2015) A comprehensive review on the applications of coal fly ash. Earth-Science Rev 141:105–121. https://doi.org/10.1016/j.earscirev.2014.11.016

Rey F (2003) Influence of vegetation distribution on sediment yield in forested marly gullies. Catena 50:549–562

Reynolds JF, Stafford Smith DM, Lambin EF, Turner BL, Mortimore M, Batterbury SPJ, Downing TE, Dowlatabadi H, Fernández RJ, Herrick JE, Huber-Sannwald E, Jiang H, Leemans R, Lynam T, Maestre FT, Ayarza M, Walker B (2007) Ecology: global desertification: building a science for dryland development. Science (80-.) 316:847–851

Rina K, Datta PS, Singh CK, Mukherjee S (2013) Isotopes and ion chemistry to identify salinization of coastal aquifers of sabarmati river basin. Curr Sci 104:335–344

Rizwan M, Ali S, Qayyum MF, Ok YS, Zia-ur-Rehman M, Abbas Z, Hannan F (2017) Use of Maize (Zea mays L.) for phytomanagement of Cd-contaminated soils: a critical review. Environ Geochem Health 39:259–277. https://doi.org/10.1007/s10653-016-9826-0

Robichaud PR, Lewis SA, Wagenbrenner JW, Ashmun LE, Brown RE (2013) Post-fire mulching for runoff and erosion mitigation. Part I: Effectiveness at reducing hillslope erosion rates. Catena 105:75–92

Rorrer GL, Hsien TY, Way JD (1993) Synthesis of porous-magnetic chitosan beads for removal of cadmium ions from wastewater. Ind Eng Chem Res 32:2170–2178. https://doi.org/10.1021/ie00021a042

Sadeghi SHR, Gholami L, Homaee M, KhalediDarvishan A (2015) Reducing sediment concentration and soil loss using organic and inorganic amendments at plot scale. Soild Earth 6:445–455

Sainju UM, Lenssen AW, Allen BL, Stevens WB, Jabro JD (2017) Soil residual nitrogen under various crop rotations and cultural practices. Zeitschrift fur Pflanzenernahrung und Bodenkd 180:187–198

Scherr S, N Yadav (1996) Land degradation in the developing world: implications for food, agriculture, and the environment to 2020

Schloter M, Dilly O, Munch JC (2003) Indicators for evaluating soil quality. Agric Ecosyst Environ 98(03):255–262. https://doi.org/10.1016/S0167-8809, 00085-9

Seager R, Ting M, Held I, Kushnir Y, Lu J, Vecchi G, Huang HP, Harnik N, Leetmaa A, Lau NC, Li C (2007) Model projections of an imminent transition to a more arid climate in southwestern North America. Science 316:1181–1184

See TP, Pandikumar A, Ngee LH, Ming HN, Hua CC (2014) Magnetically separable reduced graphene oxide/iron oxide nanocomposite materials for environmental remediation. Catal Sci Technol 4:4396–4405. https://doi.org/10.1039/C4CY00806E

Service F, Mountain R, Robichaud PR, Brown RE (2002) Silt Fences: an economical technique for measuring hillslope soil erosion. Analysis

Seshadri B, Bolan NS, Choppala G, Kunhikrishnan A, Sanderson P, Wang H, Currie LD, Tsang DCW, Ok YS, Kim K (2017) Potential value of phosphate compounds in enhancing immobilization and reducing bioavailability of mixed heavy metal contaminants in shooting range soil. Chemosphere 184:197–206. https://doi.org/10.1016/j.chemosphere.2017.05.172

Sessitsch A, Kuffner M, Kidd P, Vangronsveld J, Wenzel WW, Fallmann K, Puschenreiter M (2013) The role of plant-associated bacteria in the mobilization and phytoextraction of trace elements in contaminated soils. Soil Biol Biochem 60:182–194

Shaaban M, Abid M, RAI A-S (2013) Amelioration of salt affected soils in rice paddy system by application of organic and inorganic amendments. Plant, Soil Environ 59:227–233

Shahidi BMR, Dyck M, Malhi SS (2014) Carbon dioxide emissions from tillage of two long-term no-till Canadian prairie soils. Soil Tillage Res 144:72–82

Sharma RK, Archana G (2016) Cadmium minimization in food crops by cadmium resistant plant growth promoting rhizobacteria. Appl Soil Ecol 107:66–78

Sharma DC, Forster CF (1993) Removal of hexavalent chromium using sphagnum moss peat. Water Res 27:1201–1208. https://doi.org/10.1016/0043-1354(93)90012-7

Sharma YC, Gupta GS, Prasad G, Rupainwar DC (1990) Use of wollastonite in the removal of Ni(II) from aqueous solutions. Water Air Soil Pollut 49:69–79. https://doi.org/10.1007/BF00279511

Shcherbak I, Millar N, Robertson GP (2014) Global metaanalysis of the nonlinear response of soil nitrous oxide (N2O) emissions to fertilizer nitrogen. Proc Natl Acad Sci 111:9199–9204

Sihag SK, Singh MK, Meena RS, Naga S, Bahadur SR, Gaurav YRS (2015) Influences of spacing on growth and yield potential of dry direct seeded rice (*Oryza sativa* L.) cultivars. Ecoscan 9(1-2):517–519

Silva FBVD, do Nascimento CWA, Araújo PRM, da Silva LHV, da Silva RF (2016) Assessing heavy metal sources in sugarcane Brazilian soils: an approach using multivariate analysis. Environ Monit Assess 188. https://doi.org/10.1007/s10661-016-5409-x

Silveira MLA, Alleoni LRF, Guilherme LRG (2003) Biosolids and heavy metals in soils. Sci Agric 60:793–806. https://doi.org/10.1590/S0103-90162003000400029

Singh BR, Oste L (2001) In situ immobilization of metals in contaminated or naturally metal-rich soils. Environ Rev 9:81–97. https://doi.org/10.1139/er-9-2-81

Singh N, Dhillon BS, Sandhu PS (2014) Alleviation of soil physical constraints for improved crop production: a review. Ecol Environ Conserv 20:29–36

Singh YP, Mishra VK, Singh S, Sharma DK, Singh D, Singh US, Singh RK, Haefele SM, Ismail AM (2016) Productivity of sodic soils can be enhanced through the use of salt tolerant rice varieties and proper agronomic practices. F Crop Res. 190:82–90

Somda J, Nianogo AJ, Nassa S, Sanou S (2002) Soil fertility management and socio-economic factors in crop-livestock systems in Burkina Faso: a case study of composting technology. Ecol Econ. 43:175–183

Song WY, Sohn EJ, Martinoia E, Lee YJ, Yang YY, Jasinski M, Forestier C, Hwang I, Lee Y (2003) Engineering tolerance and accumulation of lead and cadmium in transgenic plants. Nat Biotechnol 21:914–919. https://doi.org/10.1038/nbt850

Sridhara Chary N, Kamala CT, Samuel Suman Raj D (2008) Assessing risk of heavy metals from consuming food grown on sewage irrigated soils and food chain transfer. Ecotoxicol Environ Saf. https://doi.org/10.1016/j.ecoenv.2007.04.013

Stavridou E, Hastings A, Webster RJ, Robson PRH (2017) The impact of soil salinity on the yield, composition and physiology of the bioenergy grass Miscanthus × giganteus. GCB Bioenergy 9:92–104
Stefanou S, Papazafeiriou A (2013) Effects of iron and aluminum oxides and clay content on penetration resistance of five Greek soils. Eurasian J Soil Sci 2:122–130
Steffan JJ, Brevik EC, Burgess LC, Cerdà A (2018) The effect of soil on human health: an overview. Eur J Soil Sci 69:159–171
SWAG (2002) Principles of erosion and sediment control. http://www.watershedrestoration.org/erosion.htm. (8/10/2002)
Swain KB, Goswami S, Das M (2011) Impact of mining on soil quality: a case study from hingula opencast coal mine, Angul District, Orissa. Vistas Geol Res 10:77–81
Tadesse L, Suryabhagavan KV, Sridhar G, Legesse G (2017) Land use and land cover changes and Soil erosion in Yezat Watershed, North Western Ethiopia. Int Soil Water Conser Res 5:85–94
Tang KL (2004) Soil and water conservation in China. Science Press, Beijing (in Chinese)
Tang WW, Zeng GM, Gong JL, Liu Y, Wang XY, Liu YY, Liu ZF, Chen L, Zhang XR, Tu DZ (2012) Simultaneous adsorption of atrazine and Cu (II) from wastewater by magnetic multi-walled carbon nanotube. Chem Eng J 211–212:470–478. https://doi.org/10.1016/j.cej.2012.09.102
Tarigh GD, Shemirani F (2013) Magnetic multi-wall carbon nanotube nanocomposite as an adsorbent for preconcentration and determination of lead (II) and manganese (II) in various matrices. Talanta 115:744–750. https://doi.org/10.1016/j.talanta.2013.06.018
Tedoldi D, Chebbo G, Pierlot D, Branchu P, Kovacs Y, Gromaire MC (2017) Spatial distribution of heavy metals in the surface soil of source-control stormwater infiltration devices – inter-site comparison. Sci Total Environ 579:881–892. https://doi.org/10.1016/j.scitotenv.2016.10.226
Thimmappa K, Singh YP, Raju R (2017) Reclamation of sodic soils in India: an economic impact assessment. In: Bioremediation of salt affected soils: an Indian perspective. Springer, Cham, pp 257–274
Tristão JC, Magalhães F, Corio P, Sansiviero MTC (2006) Electronic characterization and photocatalytic properties of CdS/TiO2 semiconductor composite. J Photochem Photobiol A Chem 181:152–157. https://doi.org/10.1016/j.jphotochem.2005.11.018
Tully K, Sullivan C, Weil R, Sanchez P (2015) The state of soil degradation in Sub-Saharan Africa: baselines, trajectories, and solutions. Sustainability 7:6523–6552
Tuzen M, Soylak M (2007) Multiwalled carbon nanotubes for speciation of chromium in environmental samples. J Hazard Mater 147:219–225. https://doi.org/10.1016/j.jhazmat.2006.12.069
Ullah A, Heng S, Munis MFH, Fahad S, Yang X (2015) Phytoremediation of heavy metals assisted by plant growth promoting (PGP) bacteria: a review. Environ Exp Bot 117:28–40
UN (United Nations) (2007) Population Newsletter Number 83. http://www.un.org/esa/population/publications/popnews/Newsltr83.pdf. Accessed on 2 July 2009
UNEP (United Nations Environmental Programme) (2009) The environmental food crisis the environment's role in averting future food crises a UNEP rapid response assessment. UNEP, GRID-Arendal. http://www.grida.no/publications/rr/food-crisis/. Accessed on 12 May 2010
Uraguchi S, Watanabe I, Yoshitomi A, Kiyono M, Kuno K (2006) Characteristics of cadmium accumulation and tolerance in novel Cd-accumulating crops, Avena strigosa and Crotalaria juncea. J Exp Bot 57:2955–2965. https://doi.org/10.1093/jxb/erl056
USDA (2008) Soil quality indicators. Usda 0–1. https://doi.org/10.1016/j.jhazmat.2011.07.020
Vaezi AR, Bahrami HA (2014) Relationship between soil productivity and erodibility in rainfed wheat lands in Northwestern Iran. J Agric Sci Technol 16:1455–1466
Van Dijk AIJM, Bruijnzeel LA, Rosewell CJ (2002) Rainfall intensity–kinetic energy relationships: a critical literature appraisal. J Hydrol 261:1–23
Van Meter RJ, Swan CM, Snodgrass JW (2011) Salinization alters ecosystem structure in urban stormwater detention ponds. Urban Ecosyst 14:723–736
Vangronsveld J, Colpaert JV, Van Tichelen KK (1996) Reclamation of a bare industrial area contaminated by non-ferrous metals: physico-chemical and biological evaluation of the durability of soil treatment and revegetation. Environ Pollut 94:131–140. https://doi.org/10.1016/S0269-7491(96)00082-6

Vangronsveld J, Herzig R, Weyens N, Boulet J, Adriaensen K, Ruttens A, Thewys T, Vassilev A, Meers E, Nehnevajova E, van der Lelie D, Mench M (2009) Phytoremediation of contaminated soils and groundwater: Lessons from the field. Environ Sci Pollut Res 16:765–794. https://doi.org/10.1007/s11356-009-0213-6

Vanmaercke M, Poesen J, Maetens W, de Vente J, Verstraeten G (2011) Sediment yield as a desertification risk indicator. Sci Total Environ 409:1715–1725

Vanwalleghem T, Gómez JA, Amate JI, de Molina MG, Vanderlinden K, Guzmán G, Laguna A, Giráldez JV (2017) Impact of historical land use and soil management change on soil erosion and agricultural sustainability during the Anthropocene. Anthropocene 17:13–29

Varela ME, De Blas E, Benito E (2001) Physical soil degradation induced by deforestation and slope modification in a temperate-humid environment. Land Degrad Dev 12:477–484

Varma D, Meena RS, Kumar S (2017) Response of mungbean to fertility and lime levels under soil acidity in an alley cropping system in Vindhyan Region, India. Int J Chem Stu 5(2):384–389

Varma D, Meena RS, Kumar S, Kumar E (2017a) Response of mungbean to NPK and lime under the conditions of Vindhyan Region of Uttar Pradesh. Legum Res 40(3):542–545

Vegetal B, Cie F, Lisboa U, Grande C, Cherian S, Oliveira MM (2005) Critical review transgenic plants in phytoremediation: recent advances and new possibilities. Environ Sci Technol 39:9377–9390

Ventura Y, Myrzabayeva M, Alikulov Z, Omarov R, Khozin-Goldberg I, Sagi M (2014) Effects of salinity on flowering, morphology, biomass accumulation and leaf metabolites in an edible halophyte. AoB Plants 6:1–11

Verma JP, Jaiswal DK, Meena VS, Meena RS (2015a) Current need of organic farming for enhancing sustainable agriculture. J Clean Prod 102:545–547

Verma JP, Meena VS, Kumar A, Meena RS (2015b) Issues and challenges about sustainable agriculture production for management of natural resources to sustain soil fertility and health: a book review. J Clean Prod 107:793–794

Verma SK, Singh SB, Prasad SK, Meena RN, Meena RS (2015c) Influence of irrigation regimes and weed management practices on water use and nutrient uptake in wheat (*Triticum aestivum* L. Emend. Fiori and Paol.). Bangladesh J Bot 44(3):437–442

Vijayaraghavan K, Jegan J, Palanivelu K, Velan M (2005) Biosorption of cobalt(II) and nickel(II) by seaweeds: batch and column studies. Sep Purif Technol 44:53–59. https://doi.org/10.1016/j.seppur.2004.12.003

Viraraghavan T, Kapoor A (1994) Adsorption of mercury from wastewater by bentonite. Appl Clay Sci 9:31–49. https://doi.org/10.1016/0169-1317(94)90013-2

Vítková M, Komárek M, Tejnecký V, Šillerová H (2015) Interactions of nano-oxides with low-molecular-weight organic acids in a contaminated soil. J Hazard Mater 293:7–14. https://doi.org/10.1016/j.jhazmat.2015.03.033

Wainwright J, Parsons AJ, Schlesinger WH (2002) Hydrology–vegetation interactions in areas of discontinuous flow on a semi-arid bajada, Southern New Mexico. J Arid Environ 51:319–338

Wang YM, Chen TC, Yeh KJ, Shue MF (2001) Stabilization of an elevated heavy metal contaminated site. J Hazard Mater 88:63–74. https://doi.org/10.1016/S0304-3894(01)00289-8

Wang W, Vinocur B, Altman A (2003) Plant responses to drought, salinity and extreme temperatures: towards genetic engineering for stress tolerance. Planta 218:1–14

Wang Y, Li Q, Shi J, Lin Q, Chen X, Wu W, Chen Y (2008) Assessment of microbial activity and bacterial community composition in the rhizosphere of a copper accumulator and a non-accumulator. Soil Biol Biochem 40:1167–1177. https://doi.org/10.1016/j.soilbio.2007.12.010

Wang Z, Liu B, Wang X, Gao X, Liu G (2009) Erosion effect on the productivity of black soil in Northeast China. Sci China Ser D Earth Sci 52:1005–1021

Wang W, Dong W, Chen S, Li J, Chen T, Hu C (2016) Effect of continuously applying controlled-release fertilizers on nitrogen balance and utilization in winter wheat-summer maize cropping system. Nongye Gongcheng Xuebao/Transactions Chinese Soc Agric Eng 32:135–141

Wangeline AL, Burkhead JL, Hale KL, Lindblom SD, Terry N, Pilon M, Pilon-Smits EAH (2004) Overexpression of ATP Sulfurylase in Indian Mustard. J Environ Qual 33:54. https://doi.org/10.2134/jeq2004.5400

Wei S, Zhou Q, Koval PV (2006) Flowering stage characteristics of cadmium hyperaccumulator Solanum nigrum L. and their significance to phytoremediation. Sci Total Environ 369:441–446. https://doi.org/10.1016/j.scitotenv.2006.06.014

Wei W, Chen D, Wang LX, Daryanto S, Chen LD, Yu Y, Lu YL, Sun G, Feng TJ (2016) Global synthesis of the classifications, distributions, benefits and issues of terracing. Earth Sci Rev 159:388–403

Whiting SN, Leake JR, McGrath SP, Baker AJM (2001) Hyperaccumulation of Zn by Thlaspi caerulescens can ameliorate Zn toxicity in the rhizosphere of cocropped Thlaspi arvense. Environ Sci Technol 35:3237–3241. https://doi.org/10.1021/es010644m

Wickama J, Okoba B, Sterk G (2014) Effectiveness of sustainable land management measures in West Usambara highlands, Tanzania. Catena 118:91–102

Wingeyer AB, Amado TJ, Pérez-Bidegain M, Studdert GA, Varela CHP, Garcia FO, Karlen DL (2015) Soil quality impacts of current South American agricultural practices. Sustainability 7:2213–2242

Wood S, Sebastian K, Scherr S (2000) Pilot analysis of global ecosystems: agroecosystems, a joint study by International Food Policy Research Institute and World Resources Institute: Washington, DC, USA, Available online: http://www.wri.org/sites/default/files/pdf/page_agroecosystems.pdf. Accessed on 15 Nov 2015

Wu CH, Lo SL, Lin CF (2000) Competitive adsorption of molybdate, chromate, sulfate, selenate, and selenite on γ-Al2O3. Coll Surf A Physic Chem Eng Asp 166:251–259. https://doi.org/10.1016/S0927-7757(99)00404-5

Wu QT, Wei ZB, Ouyang Y (2007) Phytoextraction of metal-contaminated soil by Sedum alfredii H: effects of chelator and co-planting. Water Air Soil Pollut 180:131–139. https://doi.org/10.1007/s11270-006-9256-1

Xiao Y, Sun J, Liu F, Xu T (2016) Effects of salinity and sulphide on seed germination of three coastal plants. Flora Morphol Distrib Funct Ecol Plants 218:86–91

Xu D, Tan X, Chen C, Wang X (2008a) Removal of Pb(II) from aqueous solution by oxidized multiwalled carbon nanotubes. J Hazard Mater 154:407–416. https://doi.org/10.1016/j.jhazmat.2007.10.059

Xu Z, Gu Q, Hu H, Li F (2008b) A novel electrospun polysulfone fiber membrane: Application to advanced treatment of secondary bio-treatment sewage. Environ Technol 29:13–21. https://doi.org/10.1080/09593330802008412

Yadav SK (2010) Heavy metals toxicity in plants: An overview on the role of glutathione and phytochelatins in heavy metal stress tolerance of plants. South African J Bot 76:167–179

Yadav GS, Babu S, Meena RS, Debnath C, Saha P, Debbaram C, Datta M (2017) Effects of godawariphosgold and single supper phosphate on groundnut (*Arachis hypogaea*) productivity, phosphorus uptake, phosphorus use efficiency and economics. Indian J Agric Sci 87(9):1165–1169

Yadav GS, Lal R, Meena RS, Babu S, Das A, Bhomik SN, Datta M, Layak J, Saha P (2017a) Conservation tillage and nutrient management effects on productivity and soil carbon sequestration under double cropping of rice in North Eastern Region of India. Ecol Indi. http://www.sciencedirect.com/science/article/pii/S1470160X17305617

Yadav GS, Lal R, Meena RS, Datta M, Babu S, Das LJ, Saha P (2017b) Energy budgeting for designing sustainable and environmentally clean/safer cropping systems for rainfed rice fallow lands in India. J Clean Prod 158:29–37

Yadav GS, Das A, Lal R, Babu S, Meena RS, Saha P, Singh R, Datta M (2018a) Energy budget and carbon footprint in a no-till and mulch based rice–mustard cropping system. J Clean Prod 191:144–157

Yadav GS, Das A, Lal R, Babu S, Meena RS, Patil SB, Saha P, Datta M (2018b) Conservation tillage and mulching effects on the adaptive capacity of direct-seeded upland rice (*Oryza sativa* L.) to alleviate weed and moisture stresses in the North Eastern Himalayan Region of India. Arch Agron and Soil Sc. https://doi.org/10.1080/03650340.2018.1423555

Yan X, Ding J, Li X, Zhang Z, Ma E, Zeng X, Wang G (2015) Effect of salt dust storm migration pathways on degradation of Ebinur lake wetland. Shengtai Xuebao/Acta Ecol Sin 35:5856–5865

Yao Z, Li J, Xie H, Yu C (2012) Review on remediation technologies of soil contaminated by heavy metals. Procedia Environ Sci 16:722–729. https://doi.org/10.1016/j.proenv.2012.10.099

Yavuz O, Altunkaynak Y, Güzel F (2003) Removal of copper, nickel, cobalt and manganese from aqueous solution by kaolinite. Water Res 37:948–952. https://doi.org/10.1016/S0043-1354(02)00409-8

Yensen NP (2008) In: Khan MA, Weber DJ (eds) Halophyte uses for the twenty-first century BT-ecophysiology of high salinity tolerant plants. Springer, Dordrecht, pp 367–396

Zhang F, Ge Z, Grimaud J, Hurst J, He Z (2013a) Long-term performance of liter-scale microbial fuel cells treating primary effluent installed in a municipal wastewater treatment facility. Environ Sci Technol 47:4941–4948. https://doi.org/10.1021/es400631r

Zhang S, Wang R, Zhang S, Li G, Zhang Y (2013b) Development of phosphorylated silica nanotubes (PSNTs)/polyvinylidene fluoride (PVDF) composite membranes for wastewater treatment. Chem Eng J 230:260–271. https://doi.org/10.1016/j.cej.2013.06.098

Zhang F, Wan X, Zhong Y (2014) Nitrogen as an important detoxification factor to cadmium stress in poplar plants. J Plant Interact 9:249–258. https://doi.org/10.1080/17429145.2013.819944

Zhao G, Mu X, Wen Z, Wang F, Gao P (2013) Soil erosion, conservation, and eco-environment changes in the Loess Plateau of China. Land Degrad Dev 24:499–510

Zhao X, Wu P, Gao X, Persaud N (2015) Soil quality indicators in relation to land use and topography in a small catchment on the Loess Plateau of China. Land Degrad Dev 26:54–61

Zhou L, Yang B, Xue N, Li F, Seip HM, Cong X, Yan Y, Liu B, Han B, Li H (2014) Ecological risks and potential sources of heavy metals in agricultural soils from Huanghuai Plain, China. Environ Sci Pollut Res 21:1360–1369. https://doi.org/10.1007/s11356-013-2023-0

Zhu H, Jiang R, Xiao L, Chang Y, Guan Y, Li X, Zeng G (2009) Photocatalytic decolorization and degradation of Congo Red on innovative crosslinked chitosan/nano-CdS composite catalyst under visible light irradiation. J Hazard Mater 169:933–940. https://doi.org/10.1016/j.jhazmat.2009.04.037

Zingore S, Johnston A (2013) The 4R Nutrient Stewardship in the context of smallholder Agriculture in Africa. Agro-ecological Intensification of Farming Systems in the East and Central African Highlands, edited by: Vanlauwe B, Blomme G, Van Asten P, Earthscan, UK, pp 77–84.

Žížala D, Zádorová T, Kapička J, Žížala D, Zádorová T, Kapička J (2017) Assessment of soil degradation by erosion based on analysis of soil properties using aerial hyperspectral images and ancillary data, Czech Republic. Remote Sens 9:28. https://doi.org/10.3390/rs9010028

Zou XL (2015) Phytoextraction of heavy metals from contaminated soil by co-cropping solanum nigrum L. with ryegrass associated with endophytic bacterium. Sci Technol 50:1806–1813. https://doi.org/10.1080/01496395.2015.1014058

An Effective Organic Waste Recycling Through Vermicompost Technology for Soil Health Restoration

3

Abdullah Adil Ansari, Lydia Ori,
and Yvonne Indrani Ramnarain

Abstract

Globally the increase in food productivity was the major concern during the last century that started with green revolution with the use of chemical fertilizers and pesticides to grow the different crops for feeding millions of growing population. Seventy-five percent of agriculture sector is dependent on chemical fertilizers during the last 20 years for the production of agricultural produce. This had positive effects initially for many years, but during the last decade, the negative impacts of excessive usage have resulted in low crop productivity, increased infestation of pests and diseases, soil degradation, and consequently the adverse effect on the environmental parameters. This has led the advent of use of organic agriculture by using different organic amendments, biopesticides, and biocontrol measures by researchers and farmers in many different countries. This has created selective markets for organic produce as well. This is also substantiated by increased solid waste produced at various levels that mainly includes organic waste to the tune of 46% globally which is incorporated in the soil and water causing pollution. These organic wastes can be recycled by processes like vermicomposting which can produce nutrient-rich organic fertilizers which are enriched source of beneficial microbes. The large-scale vermicomposting is being practiced in many countries like India, Canada, Italy, Japan, Malaysia, the Philippines, the USA, and Nepal. Vermicomposting is an efficient way of using organic waste materials like various plant litter matter, manure, and other solid

A. A. Ansari (✉)
University of Guyana, Georgetown, Guyana
e-mail: abdullah.ansari@uog.edu.gy

L. Ori
Anton de Kom University of Suriname, Paramaribo, Suriname

Y. I. Ramnarain
Department of Agricultural Research, Marketing and Processing, Ministry of Agriculture, Animal Husbandry and Fisheries, Paramaribo, Suriname

R. S. Meena (ed.), *Soil Health Restoration and Management*,
https://doi.org/10.1007/978-981-13-8570-4_3

wastes to convert through epigeic and anecic species of earthworms into useful organic fertilizer vermicompost which when applied to soil ecosystem can enhance plant growth productivity and ameliorate soils. Such technologies can be economically sustainable and viable. Vermicompost contains micronutrients; microbes from diverse groups like bacteria, fungi, and actinomycetes; phytohormones; soil enzymes; and humic acids and is free of pest/pesticides and diseases. The nutrient quality of vermicompost varies and depends on the substrates used (moisture %, 60–70; aeration, 50%; temperature, 18–35° C; pH, 6.5–7.5; nitrogen %, 0.8–3.0; phosphate %, 0.5–1.7; and potassium %, 0.5–1.6). The application of vermicompost in cultivation of crops is beneficial not only in terms of yield but plays an important role in soil amelioration in terms of structure and nutrient management. The researchers, farmers, industries, and practitioners across many countries have reported success and recorded numerous benefits with the use organic inputs like vermicompost to cultivate various crops like paddy, wheat, eggplant, okra, and tomatoes.

Keywords
Earthworms · Organic agriculture · Organic input · Plant productivity · Vermicomposting

Abbreviations

%	Percentage
C: N	Carbon: nitrogen
Ca	Calcium
Cu	Copper
EC	Electrical conductivity
ESP	Exchangeable sodium percentage
Fe	Ferrous
Kg/m2	Kilogram/meter2
Mg	Magnesium
Mg	Milligram
Mn	Manganese
NPK	Nitrogen, phosphorus, potassium
Zn	Zinc

3.1 Introduction

Plants on earth have been providing food, promoting health, and supporting the ecosystem. The soil is critical for plant growth that harbors important organisms (Ismail 2005). Soil microorganisms influence ecosystem by contributing to plant nutrition, plant health, and soil structure and soil fertility. Soil microorganisms are integral part of biogeochemical cycles (Kirk et al. 2004; Meena and Meena 2017). Microbes stimulate plant growth by increasing nutrient availability (Ismail 2005).

Modern agriculture is based on chemical fertilizers and pesticides which is not sustainable in long terms and causes irreversible damages to the soil ecological balances across the globe. The excessive and indiscriminate usage has resulted in contamination of groundwater and atmospheric pollution. High levels of nitrate residues leach into the soil and find its way into groundwater. It has also adversely affected the crop productivity in terms of quantity and quality (persistence of pesticide residues), thereby causing reduction by 15–18% (Swietlik 1992; IFDC 2013; Asma et al. 2016). There is also increase in cost of production by 33% and increase in greenhouse gases by 60% (Swietlik 1992; IFDC 2013; Asma et al. 2016; Rahman and Zang 2018; Ashoka et al. 2017). Overuse of chemical fertilizers and pesticides in agricultural lands over long period of time has resulted in poor soil health with combined effect on crop production and increase incidences of pests and diseases. The global usage of nitrogen and phosphorus fertilizers as recorded by Lu and Tian (2017) in 2013 is as follows:

Nitrogen fertilizers: China (31%), India (15%), the USA (11%), Brazil (3%), Pakistan (2%), and others (37%)
Phosphorus fertilizers: China (27%), India (13%), Brazil (11%), the USA (10%), Canada (2%), and others (37%)

These concerns have led to greater economic impact on farmers (Swietlik 1992; IFDC 2013; Asma et al. 2016; Rahman and Zang 2018). Over the last few years, the problems associated with food security have led to thinking and application of technologies for organic agriculture by soil management techniques and microbial innovations (use of biofertilizers, biopesticides, and organic fertilizers). The soil stability is maintained by formation of aggregate particles with humification and aeration process through integrated soil management. The practice of adding organic input into soil is essential to achieve stable soil structures (Asma et al. 2016; Rahman and Zang 2018; Meena et al. 2017b).

Biofertilizers are essential component of organic farming that facilitate soil microorganisms in increasing the availability and uptake of mineral nutrients, thereby increasing the nutrient status (Ismail 2005; Kumar et al. 2017b). Application of vermicompost assists in increasing microbial activity, enzymes, and humic acids which helps to increase the aggregation and stability of soils causing improvement in aeration. It is prepared through the inoculation of earthworms to organic waste materials. The vermicomposted material has excellent biochemical properties that improve the soil qualities in terms of porosity, aeration, and nutrient availability (Ansari 2008).

Biofertilizers like vermicompost increase useful microorganisms that recycle the organic substances in the soil and release the available nutrients slowly to the plants (Kirk et al. 2004; Ismail 2005; Ansari 2008; Meena et al. 2018b). The process of composting consists of two phases – breakdown and building up phase. Breakdown involves decomposition of biodegradable substances (organic waste) into degraded substances into simpler form like amino acids and subsequently into nitrates and nitrites. Building up phase consists of formation of complex humic acids and finally transforms to fine humus substance in composting material at the final stages. The processes are carried out by aerobic microflora like *actinomycetes*, *Azotobacter*, and *Nitrosomonas*.

Vermicomposting is an efficient process as it is faster than the conventional composting methods that take longer time (about 3 months). Conversion of organic waste results in superior quality of end material (vermicompost) which contains earthworm cast and available nutrients encapsulated in mucus formed through the processing of composting material mixed with fine soil particles when processed through the earthworm gut systems. The plants are able to absorb the macro- and micronutrients easily slowly from soil results in optimum quantity (Kale 1998; Ansari and Ismail 2001; Meena et al. 2015d). These nutrients from organic source result in effective growth of plant productivity.

Use of organic inputs has become an essential component of organic agriculture. Vermicomposting plays important role in realizing the agriculture without the use of chemical fertilizers. Vermicomposting is a simple process and has excellent properties without causing any damage to the plants. The process comprises the use of earthworms for the degradation of organic waste through vermin, stabilizing it in association with microorganisms. Organic waste is processed by aerobic microorganisms and is consumed by earthworms which pass through gut where it is mixed with coelomic fluid and mucus. It is also processed through microflora in the gut that also enhances the antibacterial properties. The vermicast is mixed with the composted material that is free of any pathogens. There is also a process of thermophilic digestion in the initial phase of composting that utilizes the thermophilic microorganisms. The diversity of useful microorganisms increases in the vermicompost (Ansari 2008, 2012; Meena et al. 2017c; Buragohain et al. 2017). Vermicomposting is an effective method of reducing organic waste in order to produce vermicompost which is the transformation of garbage to valuable product with high market value that can substitute the chemical fertilizers in agriculture in an effective way by improving the soil health in terms of enhancement of soil microbial community with effective recycling of minerals ensuring safe environment and human health (Asma et al. 2016). Vermicomposting is the process of producing compost by utilizing earthworms to turn organic waste into high-quality compost that consists mainly of worm cast in addition to decayed organic matter (Ismail 2005; Devi and Prakash 2015; Verma et al. 2015). Vermicomposting helps to convert the organic wastes (agro-wastes, animal manure, and domestic refuse) into highly nutrient fertilizers for plant and soil (Gajalakshmi and Abassi 2003; Yadav et al. 2018b). Vermicompost is soft light and dark black material like peat with all the properties of standard compost like structure, aeration, and moisture enriched with useful microbes (Ismail 2005; Edwards et al. 2011; Meena et al. 2015b).

Organic farming through the use of vermicompost which is enriched with NPK and micronutrients and essential microbes like nitrogen-fixing and phosphate-solubilizing bacteria and actinomycetes is an alternative to chemical fertilizer-based agriculture (Sinha et al. 2011). Vermicompost not only enhances plant growth but also plays a role in plant protection without causing any harm to the plants (Ansari and Jaikishun 2011; Meena and Yadav 2015).

The organic waste generated has been a serious problem affecting the environment and has caused a global concern. These concerns can be addressed by solving the problems with sustainable solutions using organic waste recycled and composted materials like vermicompost. Researchers across many countries have carried out researches on organic waste management through various methods and technologies to produce composted materials. The organic solid waste is a major problem in fast-growing urbanized areas in many developing countries. The recycling of these organic wastes through vermicomposting to produce organic fertilizers like vermicompost is highly cost effective and critical for farming practices (Ansari 2008; Meena et al. 2015c). Reduction in the use of chemical fertilizers/pesticides and reduced environmental hazards are the critical benefits of recycling organic waste (Nath et al. 2009; Varma et al. 2017b).

Soil is defined as the unconsolidated mineral material on the immediate surface of the earth that serves as a natural medium for the growth of land plants (Donahue et al. 1990). Soils are formed from hard rocks and other organic residues. Soil formation is determined by the climate, living organisms, nature of parent material, topography, and time period and is brought about by the following processes (Brady 1999):

1. Weathering and organic matter breakdown
2. Translocation of inorganic and organic materials up and down the soil profile
3. Accumulation of soil material in horizons in the soil profile

The physical properties of soil include texture, structure, density, porosity, consistency, temperature, color, and water content. According to the US Department of Agriculture, soils are classified into 12 textural classes based on the percentage of soil separates – sand, silt, and clay: sands, loamy sands, sandy loam, loam, silt loam, silt, sandy clay loam, silty clay loam, clay loam, sandy clay, silty clay, and clay. The chemical properties of soil include mineral solubility, nutrient availability, pH, cation exchange, and buffering action. The biological properties of soil are due to the activity of soil dwellers like rodents, worms, insects, bacteria, fungi, actinomycetes, algae, and protozoa, which, by churning the soil and mixing organic residues, alter the physical conditions of the soil (Donahue et al. 1990; Sihag et al. 2015).

Soil is a living system that contains essential nutrients like NPK and macro- and micronutrients. It also contains microbial community supported by macroinvertebrates. Soil undergoes physical and biological processes that maintain the structure and stability with different physiological processes like respiration, nutrient recycling, and decomposition. The addition of chemical fertilizers and pesticides used in crop productivity provides available nutrients and promotes plant growth. The chemical fertilizers and pesticides have been an important component of modern agriculture for the past decade. But over the years, due to persistent and excessive use, it has resulted in maneuvering the soil physical, chemical, and biological properties. It has caused the hardening/compaction of soil, reduction in soil fertility, and pesticide residue in soil, thus changing the nature of soil in terms of its properties (soil degradation). Compaction is an important indication of soil degradation which is affecting the modern agriculture. Soil porosity is reduced leading to increase in bulk density. One of the major reasons is frequent and excessive use of chemical fertilizers over longer period of time (Mari et al. 2008; Varma et al. 2017a). Soil compaction results in poor soil aeration and percolation of water, soil erosion, restricting the root growth, and excessive runoff. The groundwater recharge and mineral movement are also affected (Blanco et al. 2002; Batey 2009). Across the globe, the fertile soils have been transformed into desertified lands, problematic soils like alkaline/saline/sodic soils, and acidic soils due to indiscriminate modern agricultural practices.

Large extents of land in India and other parts of the world are affected by salinity and alkalinity due to major degradation processes like salinization, water logging, chemical impairment, and desertification (Dagar and Singh 1994; Dagar et al. 1994; Datta et al. 2017b). Twenty-seven subgroups of salt-affected soils have been identified based on clay mineralogy, presence or absence of calcic horizon, textural variations, and groundwater conditions (Murthy et al. 1980). In India these are classified as saline soils, alkali (sodic) soils, and saline-alkali soils. Saline soils contain a concentration of neutral soluble salts with the electrical conductivity (EC) of a saturated extract of the soil being more than 4 dSm^{-1}. The exchangeable sodium percentage (ESP) is less than 15, and the pH is less than 8.5 because the salts are neutral; the chlorides and sulfates of the base-forming cations dominate. Saline-alkali soils contain appreciable quantities of neutral soluble salts and enough sodium. ESP is greater than 15 and EC of the saturated soil extract is more than 4 dSm^{-1}. The pH is 8.5 or less.

Sodic soil (Fig. 3.1) does not contain any significant amount of soluble salts. The detrimental effects these soils have on plants are not only due to the toxicity of Na^+, HCO_3^-, and OH^- ions but also due to reduced water infiltration and aeration. The pH is always above 8.5, often rising to 10.0 and above due to hydrolysis of Na_2CO_3.

These problematic soils have also been reclaimed using organic amendment methods. The use of organic amendments, such as composts, biosolids, straw, sawdust, manures, and all crop residues (Wallace and Terry 1998), enhances bioremediation by improving physical properties, including infiltration, aeration, and water holding capacity as well as conditions for microbial growth, such as aeration, nutrients, and available organic carbon (Bollag and Bollag 1995). The effectiveness of an organic material in reclaiming alkali soils depends on the amount of CO_2 produced and conditions resulting through a drop in the redox potential (Swarup 1994). Mulches and composts are important soil amendments (Figs. 3.2 and 3.3). Organic mulches are carbon-containing material derived from sources such as sawdust, wood chips, hay, straw, and leaves (Wallace and Terry 1998) and are applied to the

Fig. 3.1 Alkaline/sodic soils

Fig. 3.2 Soil profile of sodic soil

soil surface, whereas composts are humus-like products of an engineered process (Stratton et al. 1995). Many organic materials like rice straw (Midmore 1983), grass clippings (Castle et al. 1995), hay (Borland and Weinstein 1989), uncomposted leaves (Borland and Weinstein 1989), and manure (Castle et al. 1995) have been evaluated as effective mulches.

Fig. 3.3 Soil profile of bio-remediated sodic soil using organic amendments

3.2 Earthworms in Soil Fertility and Nutrient Management

Soil fertility and nutrient recycling are the essential roles of earthworms in soil. Organic nutrients are processed by earthworms that are up taken by plants efficiently for plant growth. Soil fertility is affected in terms of transformation in terms of physical, chemical, and biological properties. The earthworm cast released in the earthworm burrows in the soil enriches it with diverse microflora. Earthworms are categorized into three ecological groups:

1. Epigeics (*Eisenia foetida*, *Eudrilus eugeniae*) – These are earthworms dwelling on the surface of the soil and are efficient in feeding on detritus material like leaf litter. They are phytophagous and do not affect the soil structure directly as they do not move into the soil.
2. Anecics (*Lampito mauritii*) – These earthworms have the ability to recycle leaf litter (detritus) along with the soil and are found in the upper layers of the soil. They are geophytophagous in nature. They also produce cast on surface of the soil and bring about changes in the soil structure in terms of nutrient movement, moisture retention, and soil porosity.
3. Endogeic earthworms (*Octochaetona thurstoni*) – These earthworms are found dwelling in the deeper layers of the soil and feed only on the soil and derive nutrients from the organically rich soil. They do not play a role in composting (Ismail 2005).

3.3 Vermicomposting and Organic Agriculture

Organic agriculture involves the use of organic inputs (organic manures, biofertilizers, and biopesticides/biocontrol agents) for the cultivation of crops (Thampan 1995; Meena et al. 2016a). Use of vermicompost for plant growth is one of the important organic fertilizers that is being used in many countries involved in organic agriculture. Vermicomposting is a controlled and sustainable method of recycling biodegradable organic waste to produce useful end product vermicompost through the use of epigeic and anecic earthworms. Vermicompost is an organic fertilizer that is enriched with nitrogen (1.2–6.1% more than farmyard manure), phosphates (1.8–2.0% more than farmyard manure), and potassium (0.5–0.75% more than farmyard manure). There is presence of phytohormones like auxins and cytokinins, enzymes, vitamins, and essential microbes like *Actinomycetes*, *Azotobacter*, *Nitrosomonas*, protozoans, and fungi (Ismail 1997). It is considered to be good soil conditioner that is being used by practitioners of organic agriculture in many developed and developing countries. The quality of vermicompost depends on the starting organic material (enriched with nutrients) fed to the earthworms that is transformed into cast which is present in vermicompost and can be readily available to plants when added (Ismail 1997).

The beneficial effects of vermicompost are as follows (Hussain and Abbasi 2018):

1. Vermicompost contains higher nutrient content than the other composts due to increased mineralization and humification process through earthworm activity.
2. Vermicompost contains higher available nutrients like NPK, soluble calcium, and trace elements necessary for plant growth.
3. There is presence of humic acid in vermicompost that is responsible for slow release of nutrients.
4. Vermicompost is rich in organic matter that increases the soil porosity, aeration, and reduce bulk density.
5. Contain higher concentration and diversity of microbiota that play a role in production of plant growth regulators, enzymes, and phytohormones (auxins and cytokinins).

3.4 Types of Earthworms for Vermicomposting

Commonly used earthworms in vermicomposting and vermiculture are *Bimastos parvus*, *Dendrobaena rubida*, *D. veneta*, *Eisenia foetida* (Fig. 3.4), *E. hortensis* and *Eudrilus andrei*, *E. eugeniae* (Fig. 3.6), *Amynthas diffringens*, *A. morrisi*, *Lampito*

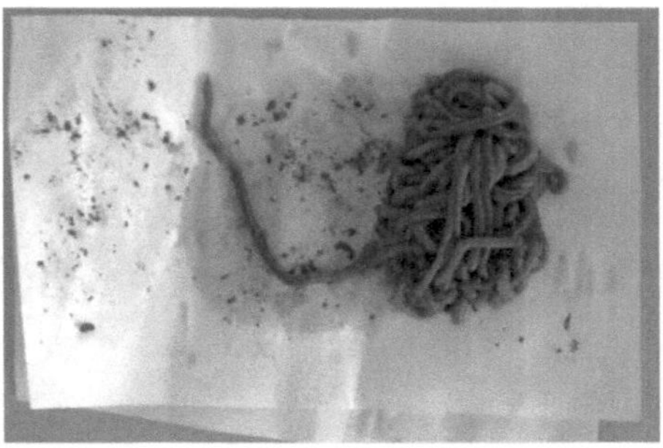

Fig. 3.4 *Eisenia foetida*

mauritii, *Metaphire anomala*, *M. birmanica*, *Perionyx excavatus* (Fig. 3.5), *P. sansibaricus*, *Megascolex megascolex*, *Pontoscolex corethrurus*, *Octochaetona serrata*, *O. surensis*, *Pheretima elongata*, and *P. posthuma* (Munnoli et al. 2010). On a large scale, very few of the abovementioned earthworms are used in vermicomposting (Domínguez and Edwards 2011; Datta et al. 2017a), because they differ in their ability to degrade the litter matter. Epigeics like *Eisenia foetida* and *Eudrilus eugeniae* are often used in the process of vermicomposting to turn food waste, agriculture waste, and manures into a valuable soil amendment called vermicompost (Fig. 3.6).

Fig. 3.5 *Perionyx excavatus*

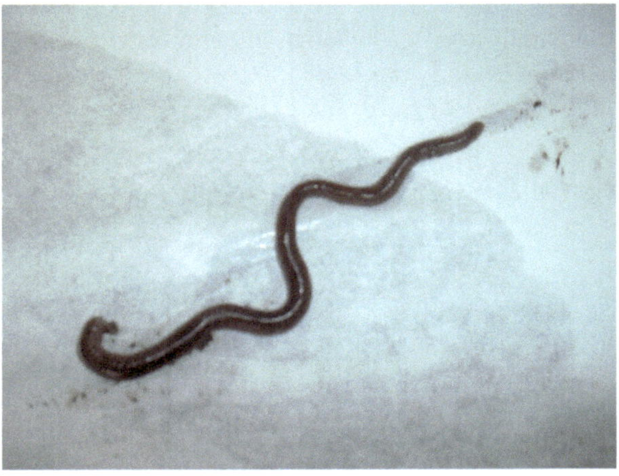

Fig. 3.6 *Eudrilus eugeniae*

3.5 Types of Waste Material Used for Vermicomposting

There are different types of organic waste that can be recycled through vermicomposting. These can be grouped as follows (Hussain and Abbasi 2018):

- *Agricultural waste*: Agricultural fields – stubble waste, husk, straw, and farmyard manure.
- Stems, leaf matter, fruit rind, pulp, and stubble. But be careful while handling an all-citric waste.
- *Animal waste*: Dung, urine, and biogas slurry.
- *Urban solid waste*: Kitchen waste from household and restaurants, waste from market yards and places of worship, and sludge from sewage treatment plants).
- *Agro industries*: Food processing units – peel, rind, and unused pulp of fruits and vegetables, fine bagasse, press mud and seed husk, stems, leaves, and flowers after extraction of oil.

Culture, breeding, and multiplication of epigeic earthworms and their use in organic waste recycling have become an important practical and scientific tool (Sinha et al. 2010). The technology of vermicomposting gives a great opportunity to effectively manage household, agricultural (rice straw, grass clippings, coffee husk), and agro-industrial waste (vegetable market solid waste) (Mane and Raskar 2012; Okwor et al. 2012; Yadav et al. 2018a). According to Suthar (2009), the degradable materials are used to obtain vermicompost in this process.

Large-scale organic waste (cattle dung, grass clippings, and water hyacinth) recycled through partial digestion (anaerobic and aerobic processes) through bio-dung composting for 30 days followed by vermicomposting (*Eisenia foetida*) for next 25 days can be a successful process to produce nutrient-rich vermicompost (Ansari 2008; Ansari and Rajpersaud 2012; Meena et al. 2018a).

3.6 Effect of Environmental Factors on Earthworms and Vermicomposting

Temperature, moisture, and pH are important factors, as earthworms have known to have tolerance level to these environmental parameters. If these limits are exceeded, the earthworms move to safe areas in the wastes, leave it, or die, and as a result the processing of organic waste slows down (Ansari 2012).

3.6.1 Temperature

The activity, metabolism, growth, respiration and reproduction, fecundity, and growth period from hatching to sexual maturity of earthworms are greatly influenced by temperature. According to Dominguez and Edwards (2011), during the process of vermicomposting, the temperature range should be between 0 °C and

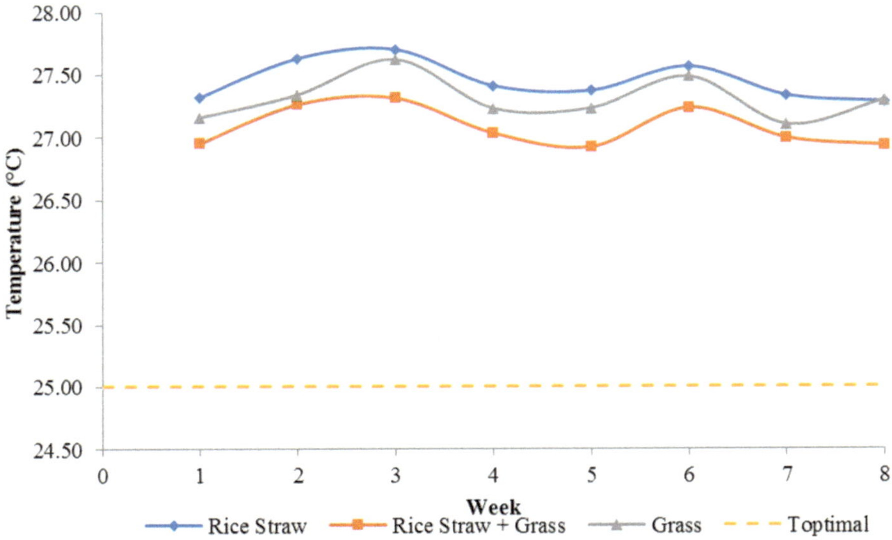

Fig. 3.7 Temperature changes during vermicomposting

35 °C, but the optimal temperature is 25 °C. Munnoli (2007) suggested a temperature range of 25–35 °C for *Eisenia foetida* in vermibeds (Fig. 3.7).

3.6.2 Moisture

Dominguez and Edwards (2011) reported that during the process of vermicomposting, for a rapid growth of *Eisenia foetida*, the moisture should be between 80% and 90%, and the optimal moisture is 85%. Water forms 75–90% of the body weight of the earthworms, and therefore it is important to keep the optimum level of moisture in in vermicomposting units (Fig. 3.8). This facilitates the progression of composting without any hindrance to earthworm activity and microbial degradation in vermi-reactors.

In natural soil-based ecosystem, the earthworms move to safer areas with more moisture in case of loss of soil moisture in the dwelling areas. When there is drastic loss of moisture in the soil, the earthworm has reasonable ability to adjust and survive through water loss from the body (Munnoli et al. 2010; Kumar et al. 2018).

3.6.3 pH

The pH fluctuates in the range of 5–9 during the progress of vermicomposting (Fig. 3.9) and reaches to neutral at the end of the process when the vermicompost is ready for harvest (Dominguez and Edwards 2011). This may occur due to the production of CO_2 and the organic acids produced during microbial metabolism.

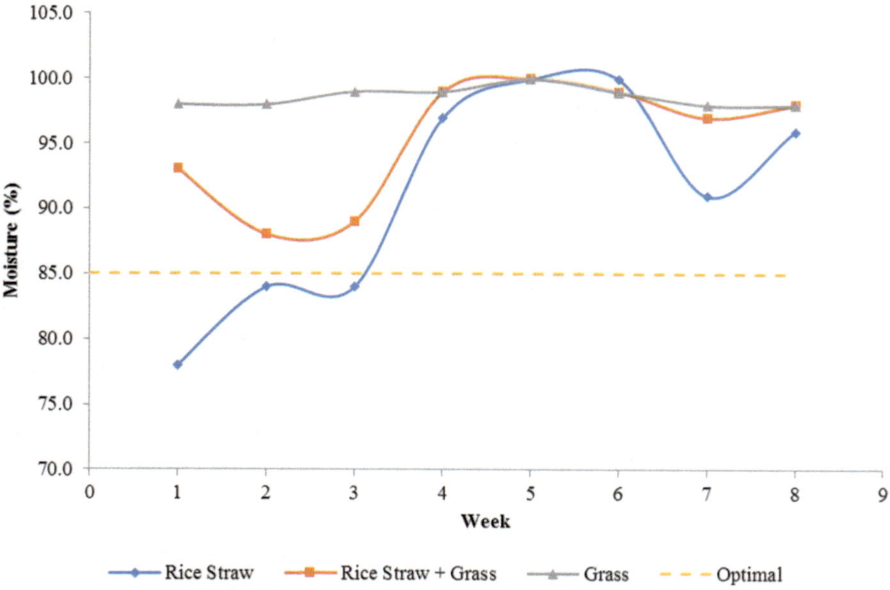

Fig. 3.8 Moisture changes during vermicomposting

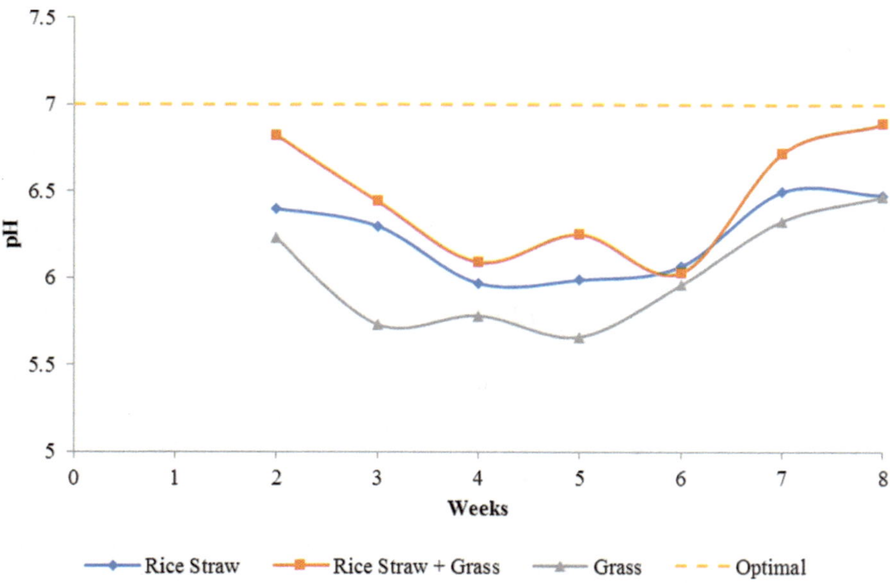

Fig. 3.9 Changes in pH during vermicomposting

Several researchers reported that most species of earthworms have an optimum pH of about 7.0 (Narayan 2000; Pagaria and Totwat 2007; Suthar 2008; Panday and Yadav 2009; Yadav et al. 2017c). Edwards (1988) noted that *Eisenia foetida* have a tolerance pH range from 4.0 to 7.0.

3.7 Criteria of Vermicompost Maturity

A mature vermicompost is dark black and is composed of fine granular material like peat with excellent physical, chemical, and biological properties that could impact the soil in terms of structure, porosity, and aeration (Dominguez and Edwards 2011; Meena et al. 2014). During vermicomposting, the earthworms feed on the detritus material that passes through earthworm gut where it mixes with gut microflora and is churn, degraded along with the activity of microbes present in the composting material. The material that is excreted is cast which mixes with material undergoing composting which is further enhanced by continuous activity of microbes and finally converted to a fine dark sweet-smelling vermicompost which is stabilized and becomes homogeneous. This is a very good source of desired level of plant growth nutrients with negligible levels of contaminants. The micronutrients are present in available form at higher levels when compared with the starting material (Edwards and Bohlen 1996).

3.8 Evaluation of Compost Stability

According to Dominguez and Edwards (2011), the most common method to evaluate the compost stability are C:N ratio, humic substances, absence of plant inhibitors, and human pathogens. Below a specific C:N ratio and absence of human pathogens are important aspects in determining the stability of vermicompost.

3.8.1 Carbon/Nitrogen Ratio

One of the important parameters commonly used to see the progression of organic material undergoing the process of vermicomposting process is C:N ratio, and there is distinct variation in C:N ratio depending on the type of starting organic material in the process. The ideal C:N ratio at the time of harvesting mature vermicompost is 10, but it is hardly achieved due to presence of various recalcitrant organic compounds and some of the materials undergoing poor decomposition. Therefore the acceptable C:N ratio ranges between 14 and 20 in vermicomposts at maturity (Dominguez and Edwards 2011; Dadhich and Meena 2014).

3.8.2 Absence of Human Pathogens

Absence of adequate information on the presence of potential human pathogens in vermicomposts produced from different organic waste materials like sewage-based biosolids and different animal manures is major obstacle in acceptance of vermicomposting as an alternative means of organic solid waste management. Based on researches, vermicompost is considered hygienic if does not contain *Salmonella*, human viruses, infective parasitic helminthic eggs, and no more than 5×10^4 fecal coliforms and 5×10^5 fecal streptococci per 100 grams of sample (Dominguez and Edwards 2011).

3.9 Chemical Properties of Vermicompost

The process of vermicomposting is an aerobic method of decomposition, and there is complete involvement of mesophilic microbial activity. There are complex food webs associated with vermicomposting systems, and there are several nutrient elements which undergo changes to different chemical forms finally modified to stable organic compounds important in nutrient dynamics and plant growth regulators (Dominguez and Edwards 2011; Dhakal et al. 2015). Below the pH and soluble salt concentrations are discussed.

3.9.1 pH

The pH of soil or potting mixture to grow the plants is critical for different aspects of soil fertility and plant growth parameters. The pH of vermicompost varies based on the type of raw materials used in the process of vermicomposting. Vermicompost produced from cattle dung has pH range of 6.0–6.7. The pH should be regulated between 6 and 7 during vermicomposting to get the final vermicompost which is suitable for the optimum release of plant nutrients (NPK and other micronutrients) and effective absorption by plants (Edwards and Bohlen 1996).

3.9.2 Soluble Salt Concentrations

The growth media or soil conditioners should have the soluble salt concentration (electrical conductivity) should be less than 100–200 mS/m for seedlings and sensitive plants. For established plants, it should be less than 200–300 mS/m. The effective vermicomposting process normally ensures that the salt contents are low which is as a result of activity due to earthworms. If the salt content is greater than 0.5%, then the earthworms are affected in their activity (Edwards and Arancon 2004).

3.10 Nutrient Contents in Vermicompost

3.10.1 Total Organic Carbon

Organic matter in vermicompost determines the total organic carbon. From the initial process of vermicomposting to the final stages, the organic carbon content changes which is an important indicator of how the process is proceeds and stabilization occurs (Edwards et al. 2011; Meena et al. 2016b).

3.10.2 Total Nitrogen

The range of total nitrogen content in vermicomposts varies and depends on the type of organic input processed through vermicomposting. It varies from 0.1% to 4% and can be increased further by manipulation of nitrogen-rich starting material and is an important parameter in determining the quality of vermicompost in terms of usage for various crop productions (Edwards and Bohlen 1996).

3.10.3 Carbon/Nitrogen Ratio

Carbon/nitrogen ratio is critical in vermicomposting process, and it normally decreases from the initial phase to final stages where it tends to reduce, indicting the stabilization of vermicompost. At this stage the C:N ratio should be below 20–22 (Edwards and Bohlen 1996).

3.10.4 Total Phosphorus and Potassium

The mature vermicompost when stabilized should have the optimum macronutrients – phosphates and potassium – which are an indication of quality. Phosphates in general should be more than 0.5%. It is necessary to specify the total content of P and K in finished vermicompost as it is an indication of overall macronutrient value. Generally, P contents of more than 0.5% is desirable. Seedlings and some plants that are sensitive to phosphates may require less than 0.1% (Edwards and Bohlen 1996).

3.10.5 Total Calcium and Magnesium

Calcium and magnesium are critical for plant growth, and it is important to check these in mature vermicompost which should contribute to overall nutrient status (Edwards and Bohlen 1996).

3.10.6 Micronutrients (Manganese, Copper, Zinc, and Iron)

The micronutrients are also essential for plant growth and are present in optimal quantity in vermicompost. It is possible in some situations that there may be excess of micronutrient element in vermicompost that may be toxic to specific plants (Edwards and Bohlen 1996).

3.11 Vermicomposting Process

Vermicomposting is a method of composting by utilization of epigeic and anecic earthworms at an optimum population in recycling of organic waste in order to produce an effective organic fertilizer vermicompost. Vermicompost is enriched with macro- and micronutrients in optimum levels of organic matter and organic acids like humic acid along with inoculum of useful microbiota. It can be applied in wide-ranging soil conditions as soil conditioner and for cultivation of variety of crops at different rates based on the requirements of each crop. The use of earthworms is essential in vermicomposting as they tend to convert the organic material with the aid of gut micoflora into fine earth like material. The process is beneficial and cost-effective and utilizes the organic waste material. It is in turn also helping in ecosystem services like reducing the organic waste to useful organic fertilizer that can substitute the chemical fertilizers. The technology is environmentally friendly and is the best of reducing the organic garbage to zero waste.

Vermicomposting can be done in many ways. The vermicomposting units can be set up in tanks, boxes, containers, basket, drums, and pits in the soil. The depth of the units should be less than 1 meter or 1 foot. These should be provided with appropriate drainage at the bottom (with appropriate perforations). The other dimensions can vary based on suitability and quantity of vermicompost production required. The ideal size of the tank or pit could be 2 m × 1 m × 0.75 m which can produce approximately 100 kg of vermicompost per harvest. The units should be set up in shade (outside), or shade house can be constructed.

The temperature and moisture are key factors in regulating the units and should always be maintained at optimum levels. The organic materials should be added gradually and in small quantity say, for example, three additions per week or every day if the quantity is very small. The material should be turn at least once in a week. The units should be sprinkled with water every day to maintain the optimum moisture levels of 80%. The units should be left to dry unless it is time for harvest. At the time of harvest, the units should not be sprinkled with water for 2–3 days, and then the vermicompost can be harvested. Initial period to set up the unit takes 50–60 days. This is for stabilization of the units and to achieve the appropriate density of earthworms in units. The units can be inoculated with the number of epigeic earthworms (approximately 100–500). The number depends on the size of the units. When the optimum population is reached (50–60 days), then start loading thee units with organic waste at regular intervals. The vermicompost can be harvested every 40–45 days.

The basic layer at the bottom of the unit is filled with broken bricks/pebbles/blue metal (5–6 cm thick). This is followed by addition of layer of coarse sand (10 cm thick). Then layer is created with good loamy soil (10 cm thick). The units should be sprinkled with water so that all the layers are moistened with the drainage system at the bottom the unit working in perfect condition. This is followed by organic materials like dried grass/leaves/plant litter and cattle dung to make a layer of 5–6 cm. The epigeic earthworms (100–200) can then be inoculated in the units. After the initial phase which is first 50–60 days, the materials should be added once per week. The vermicomposting units (Fig. 3.10) are stabilized after the initial phase following which the loading of organic waste can be optimized to three times per week. Then the vermicompost can be harvested every 40–45 days (Fig. 3.11). At the time of harvest the addition of water to units can be stopped for 2–3 days so that the earthworms can move to the bottom layers. A heap is created within the unit and removed so that earthworms are least disturbed. The vermicompost is packed in

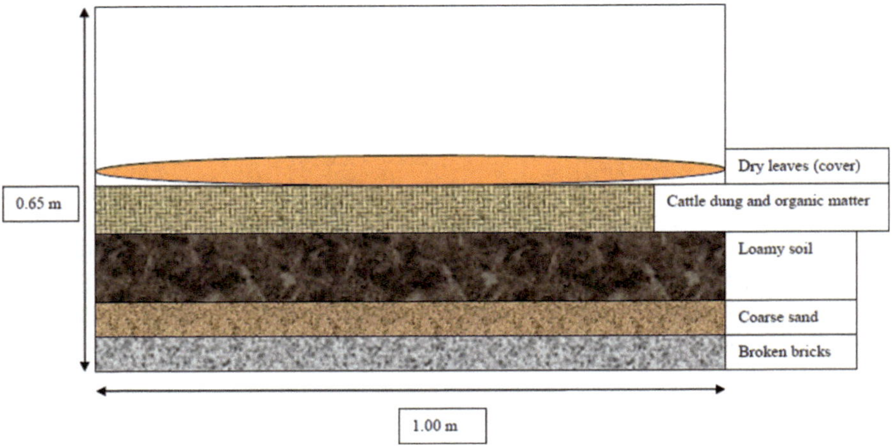

Fig. 3.10 Setup of vermicomposting unit

Fig. 3.11 Vermicompost at harvest

bags with proper aeration and optimum moisture level to sustain the microbial population. It can be directly applied to the soil at the required rate of the specific crop. It can be stored up to a period to 3 months in proper bags without any nutrient loss.

3.12 Use of Vermicompost as Soil Amendment and in Crop Productivity

Vermicompost serves as a nutrient-rich natural fertilizer; improves the physical, chemical, and biological properties of soil; and reduces the use of chemical fertilizers (Kale 1998; Nath et al. 2009; Ansari and Jaikishun 2011; Meena et al. 2015e). It also increases the amount of readily available water and induction of N, P, and K exchange, which results in better growth of the plants (Ismail 2005). Application of vermicompost to soil reduces the bulk density, thereby increasing the soil porosity and aeration which facilitates the movement of nutrients. It also helps in soil aggregate formation and better root development during plant growth. Soil pH is maintained by humic acid component of vermicompost that acts as buffer and facilitates the slow release of nutrients. Thus use of vermicompost as organic amendment improves and restores the physical, chemical, and biological properties of soil (Hussain and Abbasi 2018; Yadav et al. 2017b).

A samba rice cultivation study reveals that the addition of vermicompost has significant positive effects on the soil physical, chemical properties and plant growth parameters (Tharmaraj et al. 2011). The application of vermicompost increases the soil properties such as organic matter, total nitrogen, phosphorus, potassium, sulfur, zinc, and boron contents; grain and straw yields of rapeseed also increased significantly, when increasing the dose of vermicompost.

Vermicompost has shown to improve plant growth significantly (Lalitha et al. 2000), and it has positive influence on yield parameters of potato (*Solanum tuberosum*), spinach (*Spinacia oleracea*), turnip (*Brassica campestris*), okra (*Abelmoschus esculentus*) (Ansari, 2008; Ansari and Sukhraj 2010; Varma et al. 2017b), and black gram (*Vigna mungo*) (Tharmaraj et al. 2010). According to Sinha et al. (2011) in experiments with corn and wheat crops, tomato and egg-plants vermicompost has displayed excellent growth performances in terms of height of plants, color and texture of leaves, appearance of flowers and fruits, seed ears as compared to chemical fertilizers, and the conventional compost. The application of vermicompost also increased the shelf life and total soluble solids in tomato fruits (Ansari 2012).

Vermicompost can be applied to different crops, and some of the recommended dosages are as follows:

- Rice/wheat/sugarcane: 4–5 tonnes per hectares
- Vegetables (okra, brinjal, tomatoes): 500 g per plant or 2–3 tonnes per hectares
- Potted plants: 500 grams to 1 kg depending on stage

3.13 Recent Experiments on Vermicomposting in Guyana and Suriname

3.13.1 Experiment 1

Experiment 1 was conducted on large-scale vermicomposting of cattle dung, rice straw, and grass clippings and was carried out during the year 2014 at the National Agricultural Research and Extension Institute (NARIE). Four tanks with dimension of 2.1 × 2.1 × 1.0 m^2 were set up (Fig. 3.12) based on structural guidelines of Vermitech and inoculated with 625 earthworms (*Eisenia foetida*) each. The vermicompost was harvested after 60 days and was subjected to physicochemical analysis. Promix which is widely used in agricultural system in Guyana was also analyzed for chemical parameters. Earthworm population was also recorded. The results indicated that the vermicompost productivity (Figs. 3.13 and 3.14) was to the tune of 1376.42 kg (Table 3.1) and numbers of earthworms were 29,110 (9.6/kg/m^2). Comparative analysis showed that vermicompost is better than promix in terms of total nitrogen, calcium potassium, and phosphates (Table 3.2) whereas equally potent in other nutrients indicating that vermicompost can be better substitute in nursery management in terms of quality as well as cost implications. Based on the experiment, vermicompost was found to be enriched with nutrients and has potential to ameliorate the different soil conditions. The potential effect and benefits of vermicomposting on environment are considerable. Thus, this technique may prove to be beneficial to the soil enrichment with reduction in the use of synthetic fertilizers.

3.13.2 Experiment 2

Experiment was carried out during 2014–2015 at Anton de Kom University of Suriname, Paramaribo, with the objective of exploring the vermicomposting process. The research was done in different stages: building of a vermicompost station at the

Fig. 3.12 Vermicomposting unit at the NARIE

Fig. 3.13 Raw material in vermicomposting units

Fig. 3.14 Mature vermicompost

Table 3.1 Vermicompost production matrix

Unit (grass clippings + cattle dung)	No. of earthworms introduced (*Eisenia foetida*)	Earthworm population/2.2 m^2	Vermicompost harvest kg/2.2 m^2
T1	500	22701.0	771.6
T2	1000	52289.8	1631.3
T3	500	26942.0	1416.1
T4	500	14507.0	1686.7
Average	625	29110.0	1376.4

Table 3.2 Nutrient analysis

Parameters	Vermicompost	Promix
pH	6.36	6.2
Organic carbon (%)	5.18	6.35
Total nitrogen (%)	0.91	0.65
C/N ratio	5.69	9.77
Available phosphate (mg/kg)	102	80
Calcium (mg/kg)	1.66	0.48
Magnesium (mg/kg)	10.55	2.78
Potassium (mg/kg)	6.71	5.41
Manganese (ppm)	40.60	54.20
Iron (ppm)	5.11	3.68
Copper (ppm)	0.22	0.57
Zinc (ppm)	28.80	29.70

University compound; import of a compost earthworm, *Eisenia foetida*, from Guyana; and production of vermicompost using dry grass clippings, rice straw, and cow manure. Vermicomposting (Vermitech pattern) was done using *Eisenia foetida* with three treatments [T1 (rice straw), T2 (rice straw + grass), and T3 (grass)] in the units. During the process, the temperature, humidity, and pH were measured for all the three treatments. The population of earthworms, the production of vermicompost, and chemical and microbial analyses of the vermicompost were recorded after a hundred and forty (140) days. The results were collected and analyzed using Sigma Plot 12.0 tools. Results indicated that for all the three treatments, the temperature was in range of 0–35 °C, the humidity was between 80 and 100%, and the pH fluctuated in the range of 5.5–7.0 and stabilized to near neutral on the 60th day. The number of earthworms was counted using the hand count method. The vermicomposting results showed that the dry grass clippings and rice straw along with cow manure were successfully processed to vermicompost during the period of 60 days and had a dark color, mull-like soil odor and were homogeneous. The combination of rice straw and grass had the highest production of 105 kg, followed by grass and rice straw with 102.5 kg and 87 kg, respectively (Table 3.3). The harvested vermicompost had an excellent nutrient status (Table 3.4), which was confirmed by the chemical analyses and had all the essential macro and micro plant nutrients like N, P, K, Ca, Mg, Mn, Cu, Zn, and Fe, indicating the achievement of getting an environmentally friendly enriched nutrient fertilizer for the agriculture sector.

3.14 Impact of Organic Inputs in the Soil

3.14.1 Significance of Vermicompost as Organic Input

Organic inputs like vermicompost facilitate the process of humification in the soil with enhanced microbial activity and enzyme productivity that improves the soil stability with aggregate formation, porosity, aeration, and nutrient cycling (Perucci

Table 3.3 Harvest data of vermicompost

Units (composition)	Rice straw (T1)	Rice straw + grass (T2)	Grass (T3)
Total mass of feed (kg) initially	168	168	168
Average harvest per unit (kg)	10.90	13.80	9.80
1st harvested vermicompost (kg)	43.50	55	39
Production of vermicompost (%)	***25.90***	***32.70***	***23.20***
Total mass of feed (kg) secondly	96	96	96
Average harvest per unit (kg)	10.90	12.50	15.90
2nd harvested vermicompost (kg)	43.50	50	63.50
Production of vermicompost (%)	***45.30***	***52.10***	***66.20***
Total harvested vermicompost (kg)	**87**	**105**	**102.50**

Table 3.4 Physicochemical properties of raw material and vermicompost

Parameter	Dry grass clippings	Cow manure	Vermicompost
pH-H_2O	6.50	6.20	6.50
EC (mS/cm)	3.00	5.72	3.71
Total organic carbon (%)	42.96	21.02	18.53
Total-N (%)	1.88	1.57	1.36
C/N ratio	23:1	13:1	13:1
Total-P (%)	0.26	0.78	0.58
Total-K (%)	1.23	0.86	0.56
Total-Zn (ppm)	118	921	611
Total-Mn (ppm)	235	633	544
Total-Cu (ppm)	6.80	34.8	26.90
Total-Fe (%)	0.18	1.62	1.56

1990). Organic matter added in the soil binds the nutrients like calcium, magnesium, and potassium in the form of soil colloids with humic acids that are critical for plant growth (Haynes 1986). Addition of organic matter in soil causes improvement in soil which is indicated by microbial biomass and activity of enzymes (Perucci 1990). It is also reported that casting present in vermicompost contains plant growth promoters like auxins and cytokinins (Krishnamoorthy and Vajranabhaiah 1986; Yadav et al. 2017a).

Application of vermicompost as organic amendment is one of the better composts that contains earthworm cast and contains high available nitrogen and microbial inoculum (Satchell and Martin 1984; Satchell et al. 1992; Verma et al. 2015b). This enhances microbial activity like stimulating nitrogen-fixing bacteria and actinomycetes (Kale 1998; Borken et al. 2002; Meena et al. 2015a). Several researchers have reported that application of organic inputs like vermicompost increases crop yield and productivity which is as a result of available nutrients like phosphorus and potassium in optimum levels. These also facilitate microorganisms phosphatases in soil (Ozores-Hampton et al. 1994; Kumar et al. 2017a). The synergistic role of the microorganisms in vermicompost-amended soil has positive effects in nutrient management (Buchanan and Gliessman 1990; Dhakal et al. 2016).

3.14.2 Effect of Vermicompost on Plant Growth Productivity

Several researchers (Sharma and Mittra 1991; Ismail 1997; Meena et al. 2017a) have reported increase in crop yield with reference to wheat-paddy cropping system. The readily available nutrients in vermicompost cause the positive effect on the plant growth parameters (Ozores-Hampton et al. 1994; Rajkhowa et al. 2000; Dadhich et al. 2015). Plants absorption is very effective due to microbial activity in soil amendment like vermicompost. This solubilizes the phosphorus (Mishra and Banger, 1986; Singh et al. 1987). The cost of cultivation is reduced, and higher income has been recorded in various organic farming trails like wheat, paddy peanut (*Arachis hypogaea*), and eggplant (*Solanum melongena*) cultivation through Vermitech in comparison with chemical fertilizers (Ismail 1997). With the experiences across the globe, countries like India and New Zealand have proved that organic farming is efficient in terms of cost, sustainability, and environmental impact (Reganold et al. 2001; Ram and Meena 2014).

Significant results have also been reported on the researches on the effect of earthworms and vermicompost on cultivation of vegetable crops like tomato (*Lycopersicum esculentum*), brinjal (*Solanum melongena*), and okra (*Abelmoschus esculentus*) (Ismail 1997). Application of organic input like vermicompost on vegetables and other crops has been successful in several extension trials by farmers (Ismail 1997). Such practices can contribute to higher quality food production (Ouédraogo et al. 2001; Verma et al. 2015a). Application of organic inputs like vermicompost contributes to slow release on nutrients and also contains plant growth promoters like gibberellin, cytokinin, and auxins (Lalitha et al. 2000; Meena and Yadav 2014).). Vermicompost-treated plots were reported to show higher productivity of potato tubers attributed to increase in availability of phosphorus (Erich et al. 2002). Other researchers have reported significant yield of spinach and onions by application of vermicompost and vermiwash in soil which facilitated the release of exchangeable nutrients in soil (Cook et al. 1980; Tiwari et al. 1989). During the recent years, concern has been raised about excessive use of chemical fertilizers and pesticides causing damages to soil and environment, thereby affecting the crop productivity and human health. So alternative agricultural practices with the use of organic inputs like vermicompost can be very promising and suitable for sustainable organic agriculture for better future (Reganold et al. 2001).

3.15 Conclusion

Soil is an integral system for natural ecosystem like agricultural fields that require nutrient resources. Intensive use of agricultural chemicals over the years worldwide has caused negative effects on soil like soil degradation and acidification. The organic agriculture can be substantiated with the use of organic fertilizers like vermicompost toward sustainable crop productivity and stable soil bio-systems. Organic agriculture can be integrated with not only vermicompost production and application on various crop system but can be integrated with other organic composting technologies based on different microbial inoculum, biopesticide production and use, and biocontrol mechanisms. These would be in the near future can help us to achieve sustainable goal toward food security with least impact on the environment.

References

Ansari AA (2008) Effect of vermicompost on the productivity of potato (*Solanum tuberosum*), spinach (*Spinach oleracea*) and turnip (*Brassica campestris*). World J Agric Sci 4(3):333–336

Ansari AA (2012) Permutations and combinations of organic waste-vermitechnology. Lambert Academic Publishing, p 76

Ansari AA, Ismail SA (2001) A case study on organic farming in Uttar Pradesh. J Soil Biol 27:25–27

Ansari A, Jaikishun S (2011) Vermicomposting of sugarcane bagasse and rice straw and its impact on the cultivation of *Phaseolus vulgaris* L. in Guyana. S Am J Agric Technol 7(2):225–234

Ansari A, Rajpersaud J, (2012) Physicochemical changes during vermicomposting of water hyacinth (*Eichhornia crassipes*) and grass clippings. ISRN Soil Sci

Ansari AA, Sukhraj K (2010) Effect of Vermiwash and Vermicompost on soil parameters and productivity of okra (*Abelmoschus esculentus*) in Guyana. Pak J Agric Res 23(3–4):137–142

Ashoka P, Meena RS, Kumar S, Yadav GS, Layek J (2017) Green nanotechnology is a key for eco-friendly agriculture. J Clean Prod 142:4440–4441

Asma F, Ahmad L, Singh P (2016) Vermicomposting: an effective technique to recycle waste into valuable organic fertilizer: a review. J Pure Appl Microbiol 10(2):1–7

Batey T (2009) Soil compaction and soil management – a review. Soil Use Manag 25(4):335–345

Blanco CH, Granter CH, Anderson SH, Alberts EE, Ghidey F (2002) Saturated hydraulic conductivity and its impact on simulated runoff for claypan soils. Soil Sci Soc Am J 66:1596–1602

Bollag JM, Bollag WB (1995) Soil contamination and the feasibility of biological remediation. In: Skipper HD, Turco RF (eds) Bioremediation science and application, American Society of Agronomy special pub no. 35. Madison, WI

Borken W, Muhs A, Beese F (2002) Changes in microbial and soil properties following compost treatment of degraded temperate forest soils. Soil Biol Biochem 34:403–412

Borland J, Weinstein G (1989) Mulch: is it always beneficial? Grounds Maint 24:10–120

Brady NC (1999) The nature and properties of soils, 10th edn. Macmillan, New York, 621 pp

Buchanan RA, Gliessman SR (1990) The influence of conventional and compost fertilization on phosphorus use efficiency by broccoli in a phosphorus deficient soil. Am J Altern Agric 5:38

Buragohain S, Sharma B, Nath JD, Gogaoi N, Meena RS, Lal R (2017) Impact of ten years of bio-fertilizer use on soil quality and rice yield on an inceptisol in Assam, India. Soil Res. https://doi.org/10.1071/SR17001

Castle WL, Minassian V, Menge JA, Lovatt CJ, Pond E, Johnson E, Guillement F (1995) Urban and agricultural wastes for use as mulches on avocado and citrus and for delivery of microbial biocontrol agents. J Hortic Sci 70:315

Cook AG, Critchley BR, Critchley U (1980) Effects of cultivation and DDT on earthworm activity in a forest soil in the sub-humid tropics. J Appl Ecol 17:21–29

Dadhich RK, Meena RS (2014) Performance of Indian mustard (*Brassica juncea* L.) in response to foliar spray of thiourea and thioglycollic acid under different irrigation levels. Indian J Ecol 41(2):376–378

Dadhich RK, Meena RS, Reager ML, Kansotia BC (2015) Response of bio-regulators to yield and quality of Indian mustard (*Brassica juncea* L. Czernj. and Cosson) under different irrigation environments. J Appl Nat Sci 7(1):52–57

Dagar JC, Singh NT (1994) Agroforestry options in the reclamation of problem soils. In: Thampan PK (ed) Tree and tree farming. Peekay Tree Crops Development Foundation, Cochin, pp 63–103

Dagar JC, Singh NT, Singh G (1994) Agroforestry options for degraded and problematic soils in India. In: Singh P, Pathak PS, Roy MM (eds) Agroforestry systems for sustainable land use. Oxford & IBH Publishing, New Delhi, pp 96–120
Datta R, Baraniya D, Wang YF, Kelkar A, Moulick A, Meena RS, Yadav GS, Ceccherini MT, Formanek P (2017a) Multi-function role as nutrient and scavenger off reeradical in soil. Sustain MDPI 9:402. https://doi.org/10.3390/su9081402
Datta R, Kelkar A, Baraniya D, Molaei A, Moulick A, Meena RS, Formanek P (2017b) Enzymatic degradation of lignin in soil: a review. Sustain MDPI 9:1163. https://doi.org/10.3390/su9071163, 1–18
Devi J, Prakash M (2015) Microbial population dynamics during vermicomposting of three different substrates amended with cowdung. Int J Curr Microbiol Appl Sci 4(2):1086–1092
Dhakal Y, Meena RS, De N, Verma SK, Singh A (2015) Growth, yield and nutrient content of mungbean (*Vigna radiata* L.) in response to INM in eastern Uttar Pradesh, India. Bangladesh J Bot 44(3):479–482
Dhakal Y, Meena RS, Kumar S (2016) Effect of INM on nodulation, yield, quality and available nutrient status in soil after harvest of green gram. Legum Res 39(4):590–594
Domínguez JJ, Edwards CA (2011) Biology and ecology of earthworms species used for vermicomposting. In: Edwards CA, Arancon NQ, Sherman RL (eds) Vermiculture technology: earthworms, organic waste and environmental management. CRC Press, Boca Raton, pp 27–40
Donahue RL, Miller RW, Shickluna JC (1990) Soils. an introduction to soils and plant growth. Prentice-Hall of India, New Delhi, 667 pp
Edwards CA (1988) Breakdown of animal, vegetable and industrial organic waste by earthworms. Agric Ecosyst Environ 24:21–31
Edwards CA, Arancon N (2004) Interactions among organic matter, earthworms and microorganisms in promoting plant growth. In: Magdoff F, Weil R (eds) Functions and management of organic matter in agro-ecosystems. CRC Press, Boca Raton, pp 327–376
Edwards CA, Bohlen PJ (1996) Biology and ecology of earthworms, 3rd edn. Chapman and Hall, London, p 426
Edwards CA, Subler S, Arancon N (2011) Quality criteria for vermicomposts. In: Edwards CA, Arancon NQ, Sherman RL (eds) Vermiculture technology: earthworms, organic waste and environmental management. CRC Press, Boca Raton, pp 287–301
Erich MS, Fitzgerald CB, Porter GA (2002) The effect of organic amendments on phosphorus chemistry in a potato cropping system. Agric Ecosys Environ 88(1):79–88
Gajalakshmi S, Abassi S (2003) Earthworms and vermicomposting. Indian J Biotechnol 3:486–494
Haynes RJ (1986) The decomposition process mineralization, immobilisation, humus formation and degradation. In: Haynes RJ (ed) Mineral nitrogen in the plant-soil system. Academic Press, NewYork
Hussain N, Abbasi SA (2018) Efficacy of the vermicomposts of different organic wastes as "clean" fertilizers: state of the art. Sustainability 10:1205): 1–1205):63
International Fertilizer Development Centre (IFDC) (2013) Fertilizer Deep placement (FDP). Muscle Shoals AL USA
Ismail SA (1997) Vermicology: the biology of earthworms. Orient Longman Press, Hyderabad, p 92
John P. Reganold, Jerry D. Glover, Preston K. Andrews, Herbert R. Hinman, (2001) Sustainability of three apple production systems. Nature 410 (6831):926–930
Ismail SA (2005) The earthworm book. Other India Press Mapusa, Goa, p 101
Kale RD (1998) Earthworm Cinderella of organic farming. Prism Book, Bangalore, p 88
Kirk JL, Beandette LA, Hart M, Moutoglis P, Klironomos JN, Lee H, Trevors JT (2004) Methods of studying soil microbial diversity. J Microbiol Methods 58:169–188
Krishnamoorthy RV, Vajranabhaiah SN (1986) Biological activity of earthworm casts: an assessment of plant growth promoter levels in the casts. Proc Indian Acad Sci (Anim Sci) 95:341–351

Kumar S, Meena RS, Pandey A, Seema (2017a) Soil acidity management and an economics response of lime and sulfur on sesame in an alley cropping system. Int J Curr Microb Appl Sci 6(3): 2566–2573

Kumar S, Meena RS, Yadav GS, Pandey A (2017b) Response of sesame (*Sesamum indicum* L.) to sulphur and lime application under soil acidity. Int J Plant Soil Sci 14(4):1–9

Kumar S, Meena RS, Bohra JS (2018) Interactive effect of sowing dates and nutrient sources on dry matter accumulation of Indian mustard (*Brassica juncea* L.). J Oilseed Brassica 9(1):72–76

Lalitha R, Fathima K, Ismail SA (2000) Impact of biopesticides and microbial fertilizers on productivity and growth of *Abelmoschus esculentus*. Vasundhara Earth 1 & 2:4–9

Lu C, Tian H (2017) Global nitrogen and phosphorous fertilizer use for agriculture production in the past half century: shifted hot spots and nutrient imbalance. Earth Syst Sci Data 9:181–192

Mane TT, Raskar S (2012) Management of Agriculture Waste from market yard through vermicomposting. Res J Recent Sci 1:289–296

Mari GR, Changying J, Zhou J (2008) Effect of soil compaction on soil physical and nitrogen, phosphorous, potassium uptake in wheat plants. J Trans CSAE 24(1):74–79

Meena H, Meena RS (2017) Assessment of sowing environments and bio-regulators as adaptation choice for clusterbean productivity in response to current climatic scenario. Bangladesh J Bot 46(1):241–244

Meena RS, Yadav RS (2014) Phonological performance of groundnut varieties under sowing environments in hyper arid zone of Rajasthan, India. J Appl Nat Sci 6(2):344–348

Meena RS, Yadav RS (2015) Yield and profitability of groundnut (*Arachis hypogaea* L) as influenced by sowing dates and nutrient levels with different varieties. Legum Res 38(6):791–797

Meena RS, Yadav RS, Meena VS (2014) Response of groundnut (*Arachis hypogaea* L.) varieties to sowing dates and NP fertilizers under Western dry zone of India. Bangladesh J Bot 43(2):169–173

Meena RS, Dhakal Y, Bohra JS, Singh SP, Singh MK, Sanodiya P (2015a) Influence of bioinorganic combinations on yield, quality and economics of Mungbean. Am J Exp Agric 8(3):159–166

Meena RS, Meena VS, Meena SK, Verma JP (2015b) The needs of healthy soils for a healthy world. J Clean Prod 102:560–561

Meena RS, Meena VS, Meena SK, Verma JP (2015c) Towards the plant stress mitigate the agricultural productivity: a book review. J Clean Prod 102:552–553

Meena RS, Yadav RS, Meena H, Kumar S, Meena YK, Singh A (2015d) Towards the current need to enhance legume productivity and soil sustainability worldwide: a book review. J Clean Prod 104:513–515

Meena RS, Yadav RS, Reager ML, De N, Meena VS, Verma JP, Verma SK, Kansotia BC (2015e) Temperature use efficiency and yield of groundnut varieties in response to sowing dates and fertility levels in Western dry zone of India. Am J Exp Agric 7(3):170–177

Meena H, Meena RS, Singh B, Kumar S (2016a) Response of bio-regulators to morphology and yield of clusterbean [*Cyamopsis tetragonoloba* (L.) Taub.] under different sowing environments. J Appl Nat Sci 8(2):715–718

Meena RS, Bohra JS, Singh SP, Meena VS, Verma JP, Verma SK, Shiiag SK (2016b) Towards the prime response of manure to enhance nutrient use efficiency and soil sustainability a current need: a book review. J Clean Prod 112:1258–1260

Meena RS, Gogaoi N, Kumar S (2017a) Alarming issues on agricultural crop production and environmental stresses. J Clean Prod 142:3357–3359

Meena RS, Kumar S, Pandey A (2017b) Response of sulfur and lime levels on productivity, nutrient content and uptake of sesame under guava (*Psidium guajava* L.) based Agri-horti system in an acidic soil of eastern Uttar Pradesh, India. J Crop Weed 13(2):222–227

Meena RS, Meena PD, Yadav GS, Yadav SS (2017c) Phosphate solubilizing microorganisms, principles and application of microphos technology. J Clean Prod 145:157–158

Meena H, Meena RS, Lal R, Singh GS, Mitran T, Layek J, Patil SB, Kumar S, Verma T (2018a) Response of sowing dates and bio regulators on yield of clusterbean under current climate in alley cropping system in eastern U.P. Indian Legum Res 41(4):563–571

Meena RS, Kumar V, Yadav GS, Mitran T (2018b) Response and interaction of *Bradyrhizobium japonicum* and *Arbuscular mycorrhizal* fungi in the soybean rhizosphere: a review. Plant Growth Regul 84:207–223

Midmore DJ (1983) The use of mulch for potatoes in the hot tropics. *CIP Circulare* (March), International Potato Center, Lima, Peru

Mishra MM, Banger KC (1986) Rock phosphate comprising: transformation of phosphorus forms and mechanisms of solubilization. Biol Agric Hortic 3:331

Munnoli PM (2007) Management of industrial organic solid wastes through vermiculture biotechnology with special reference to microorganisms. PhD thesis, Goa University, India, pp 1–334

Munnoli PM, daSilva T, Bhosle S (2010) Dynamics of the soil-earthworm-plant relationship: a review. Dynamics Soil, Dynamics Plant, pp 1–21

Murthy RS, Herkerpur LR, Bhattacharjee JC (1980) The taxanomy of salt affected soils of the Indian sub-continent. In: International symposium in salt affected soils. CSSRI, Karnal India, pp 67–76

Narayan J (2000) Vermicomposting of biodegradable wastes collected from Kuvempu University campus using local and exotic species of earthworm. In: Proceedings of a National conference on industry and environment, 28th to 30th December 1999, Karad India, pp 417–419

Nath G, Sing K, Singh D (2009) Chemical analysis of vermicomposts/vermiwash of different combinations of animal, agro and kitchen wastes. Aust J Basic Appl Sci 3(4):3671–3676

Okwor AC, Ebenebe CI, Anizoba MA (2012) Biodegradation of domestic organic waste using earthworm (*Eudrilus eugenia):* a veritable tool for agricultural and environmental sustainability. Int J Agric Biosci 1(1):39–41

Ouédraogo E, Mando A, Zombré NP (2001) Use of compost to improve soil properties and crop productivity under low input agricultural system in West Africa. Agric Ecosys Environ 84:259–266

Ozores-Hampton M, Schaffer B, Bryan HH, Hanlon EA (1994) Nutrient concentrations, growth and yield of tomato and squash in municipal solid-waste-amended soil. Hortic Sci 29:785

Pagaria P, Totwat KL (2007) Effects of press mud and spent wash in integration with s with phosphogypsum on metallic catión build up in the calcareous sodic soils. J Indian Soc Soil Sci 55(1):52–57

Panday SN, Yadav A (2009) Effect of vermicompost amended alluvial soil on growth and metabolic responses of rice (*Oryza sativa* L) plants. J Eco-friendly Agric 4(1):35–37

Perucci P (1990) Effect of the addition of municipal solid-waste compost on microbial biomass and 541 enzyme activities in soil. Biol Fertil Soils 10:221

Rahman KM, Zang D (2018) Effects of fertilizer broadcasting on the excessive use of inorganic fertilizers and environmental sustainability. Sustainability 10(759):1–15

Rajkhowa DJ, Gogoi AK, Kandal R, Rajkhowa KM (2000) Effect of vermicompost on Greengram nutrition. J Indian Soc Soil Sci 48:207–208

Ram K, Meena RS (2014) Evaluation of pearl millet and mungbean intercropping systems in arid region of Rajasthan (India). Bangladesh J Bot 43(3):367–370

Reganold JP, Glover JD, Andrews PK, Hinman HR (2001) Sustainability of three apple production systems. Nature 410(6831):926–930

Satchell JE, Martin K (1984) Phosphatase activity in earthworm species. Soil Biol Biochem 560(16):191

Satchell JE, Marti K, Krishnamoorthy RV (1992) Stimulation of microbial phosphatase 562 production by earthworm activity. Soil Biol Biochem 16:195

Sharma AR, Mittra BN (1991) Effect of different rates of application of organic and nitrogen fertilisers in a rice- based cropping system. J Agric Sci 117:313

Sihag SK, Singh MK, Meena RS, Naga S, Bahadur SR, Gaurav YRS (2015) Influences of spacing on growth and yield potential of dry direct seeded rice (*Oryza sativa* L) cultivars. Ecoscan 9(1–2):517–519

Singh CP, Singh YP, Singh M (1987) Effect of different carbonaceous compounds on the transformation of soil nutrients. II. Immobilization and mineralization of phosphorus. Biol Agric Hortic 4:301

Sinha R, Agarwal S, Chauhan K, Chandran V, Soni B (2010) Vermiculture technology: reviving the dreams of sir Charles Darwin for scientific use of earthworms in sustainable development programs. Technol Invest 1:155–172

Sinha K, Valani D, Soni B, Chandran V (2011) Earthworm vermicompost: a sustainable alternative to chemical fertilizers for organic farming. Agriculture issues and policies. Nova Science Publishers, New York, p 71

Stratton ML, Barker AV, Rechcigl JE (1995) Compost. In: Rechcigl JE (ed) *Soil amendments and environment quality*. Lewis Publishers, Boca Raton

Suthar S (2008) Development of a novel epigeic-anecic-based polyculture vermireactor for efficient treatment of municipal sewage water sludge. Int J Environ Waste Manag 2(1/2):84–101

Suthar S (2009) Bioremediation of agricultural wastes through vermicomposting. Biorem J 13(1):21–28

Swarup A (1994) Chemistry of salt affected soils and fertility management. In: DLN R, Singh NT, Gupta RK, Tyagi NK (eds) Salinity management for sustainable agriculture. CSSRI, Karnal, pp 18–40

Swietlik D (1992) Causes and consequences of over fertilization in orchards. HortTechnology 2:112–132

Thampan PK (1995) Perspectives on organic agriculture. In: Thampan PK (ed) Organic agriculture. Peekay Tree Crops Development Foundation, Cochin, pp 1–38

Tharmaraj K, Ganesh P, Kolanjinathan K, Suresh Kumar R, Anandan A (2010) Influence of vermicompost and vermiwash on physicochemical properties of black gram cultivated soil. Int J Recent Sci Res 3:077–083

Tharmaraj K, Ganesh P, Kolanjinathan K, Suresh Kumar R, Anandan A (2011) Influence of vermicompost and vermiwash on physic chemical properties of rice cultivated soil. Curr Bot 2(3):18–21

Tiwari SC, Tiwari BK, Mishra RR (1989) Microbial populations, enzyme activities and nitrogen-phosphorus-potassium enrichments in earthworm casts and in the surrounding soil of a pineapple plantation. Biol Fertil Soils 8:178–182

Varma D, Meena RS, Kumar S (2017a) Response of mungbean to fertility and lime levels under soil acidity in an alley cropping system in Vindhyan region, India. Int J Chem Stud 5(2):384–389

Varma D, Meena RS, Kumar S, Kumar E (2017b) Response of mungbean to NPK and lime under the conditions of Vindhyan region of Uttar Pradesh. Legum Res 40(3):542–545

Verma JP, Jaiswal DK, Meena VS, Meena RS (2015) Current need of organic farming for enhancing sustainable agriculture. J Clean Prod 102:545–547

Verma JP, Meena VS, Kumar A, Meena RS (2015a) Issues and challenges about sustainable agriculture production for management of natural resources to sustain soil fertility and health: a book review. J Clean Prod 107:793–794

Verma SK, Singh SB, Prasad SK, Meena RN, Meena RS (2015b) Influence of irrigation regimes and weed management practices on water use and nutrient uptake in wheat (*Triticum aestivum* L. emend. Fiori and Paol.). Bangladesh J Bot 44(3):437–442

Wallace A, Terry RE (1998) Handbook of soil conditioners. Marcel Dekkar, New York, p 594

Yadav GS, Babu S, Meena RS, Debnath C, Saha P, Debbaram C, Datta M (2017a) Effects of godawariphosgold and single supper phosphate on groundnut (*Arachis hypogaea*) productivity, phosphorus uptake, phosphorus use efficiency and economics. Indian J Agric Sci 87(9):1165–1169

Yadav GS, Lal R, Meena RS, Babu S, Das A, Bhomik SN, Datta M, Layak J, Saha P (2017b) Conservation tillage and nutrient management effects on productivity and soil carbon sequestration under double cropping of rice in north eastern region of India. Ecol Indic. http://www.sciencedirect.com/science/article/pii/S1470160X17305617

Yadav GS, Lal R, Meena RS, Datta M, Babu S, Das LJ, Saha P (2017c) Energy budgeting for designing sustainable and environmentally clean/safer cropping systems for rainfed rice fallow lands in India. J Clean Prod 158:29–37

Yadav GS, Das A, Lal R, Babu S, Meena RS, Saha P, Singh R, Datta M (2018a) Energy budget and carbon footprint in a no-till and mulch based rice–mustard cropping system. J Clean Prod 191:144–157

Yadav GS, Das A, Lal R, Babu S, Meena RS, Patil SB, Saha P, Datta M (2018b) Conservation tillage and mulching effects on the adaptive capacity of direct-seeded upland rice (*Oryza sativa* L.) to alleviate weed and moisture stresses in The North Eastern Himalayan Region of India. Arch Agron Soil Sci. https://doi.org/10.1080/03650340.2018.1423555

Sustainable Management and Restoration of the Fertility of Damaged and Contaminated Lands and Soils

4

Martin Banov, Svetla Rousseva, and Pavel Pavlov

Abstract

The material is related to the description of the existing methods of technical and biological recultivation in order to restore the fertility of the damaged lands and soils by the traditional method – using humus soil material. New, innovative technologies are recommended to smoothly restore soil fertility by using waste products. The two stages of reclamation – technical and biological – are presented, and the advantages and disadvantages of the various methods of their implementation are examined. Various technological solutions have been justified to restore the fertility of damaged terrain by using reclamation substrates, which are mixtures of geological and waste materials in different proportions depending on the physicochemical characteristics of the individual components. Also used is a biological substrate with different components, mixed in volume ratios – geological materials with mild to heavy sandy-clayey mechanical composition, organic matter (humus), low nutrients for plants and lack of toxic components and compost materials. Technological developments for the recultivation of landfills by using sludge from wastewater treatment plants (WWTPs), which compensate for the shortage of humus materials, have been considered. The characteristics of the sediments reveal that they are an organic mass rich in macro- and trace elements and can be used as a fertilizer and source for enrichment of soils with organic matter and nutrients in reclamation activities.

M. Banov (✉)
Department of Genesis, Geography and Soil Classification, Institute of Soil Science, Agrotechnologies and Plant Protection, Sofia, Bulgaria
e-mail: banovm@abv.bg

S. Rousseva
Department of Soil Erosion, Institute of Soil Science, Agrotechnologies and Plant Protection, Sofia, Bulgaria

P. Pavlov
University of Mining and Geology "St. Ivan Rilski", Sofia, Bulgaria

R. S. Meena (ed.), *Soil Health Restoration and Management*,
https://doi.org/10.1007/978-981-13-8570-4_4

Keywords
Methods of technical and biological recultivation · Damaged lands · Soils · Organic matter · Sludge from wastewater treatment plants (WWTPs)

Abbreviation

Ag	Argentum
Al	Aluminium
Al_2O_3	Aluminium oxide
As	Arsenicum
Au	Aurum
BDS	Bulgarian State Standard
CaO	Calcium oxide
Cd	Cadmium
Co	Cobalt
Cr	Chromium
Cu	Copper
Fe met.	Iron
Fe	Iron
$FeO+Fe_2O_3$	Iron(II) oxide + iron(III) oxide
Ga	Gallium
Ge	Germanium
K_2O	Potassium oxide
MgO	Magnesium oxide
Mn	Manganum
MnO	Manganum oxide
MPC	Maximum permissible concentration
Na_2O	Sodium oxide
Ni	Nickel
$N\text{-}NH_4$	Ammonium
$N\text{-}NO_3$	Nitrate
P_2O_5	Diphosphorus pentoxide
P_2O_5	Phosphorus pentoxide
Pb	Plumbum
pH (H_2O)	Acid water
pH (KCL)	Acid potassium chloride
pH	Acid
S	Sulphur
Sb	Stibium
Se	Selenium
Se	Selenium
SiO_2	Silicon dioxide
S_2O_4	Sulfite

Te	Tellurium
TiO_2	Titanium dioxide
WWTP	Wastewater treatment plant
WWTPs	Wastewater treatment plants
Zn	Zinc

4.1 Introduction

The intense processing of mineral resources has negative impact on the change of environmental components. During mining activities soil cover is destroyed, hydrological conditions are changed, and new relief forms are created, which disturbs the ecological balance and changes the natural landscape. Massive environmental disturbances are also caused by the processing of black, coloured and inert materials, the construction of linear and engineering facilities, urban agglomerations, etc. (Banov et al. 2010; Buragohain et al. 2017;Meena et al. 2018).

Progress of industry, construction, development of coalfields, etc. lead to a continuous reduction of the effective land fund, which is the main means of production in agriculture and forestry.

Disturbed terrains, such as mounds of mines, sand and gravel quarries, gullies, uncorrected river beds, landfills, etc., are such that they are not suitable for agricultural use. They have lost partly or thoroughly their economic purpose and are in practice a source of negative impact on the natural environment in result of the destruction and/or contamination of the soil cover through the production activity of man. The morphological structure of the natural soils' profile, as well as their composition and properties, is disturbed.

The analysis of the global experience shows that the reclamation of disturbed terrains is a complex problem determined by a number of factors: geo-climatic, mining and technical conditions, mechanical and agrochemical properties of soil and geological materials that are extracted, the system of development of the field, etc. (Zheleva and Bozhinova 2009; Zheleva-Bogdanova 2010; Ivanov and Banov 2008; Misheva et al. 2007; Tsolova et al. 2010; Kumar et al. 2017a). This requires that the technological solutions for restoring the fertility of the disturbed terrains should be tailored to each country's specificities and coordinated efforts of specialists from different scientific and development organizations and institutes should be applied.

The main task in the reclamation of lands disturbed by mining or other activities involves restoration of soil fertility and recovery of the disturbed land for use by creating an ecologically balanced system. Recultivation is one of the most radical methods of restoration and improvement of the disturbed terrains and their recovery into the full (cultivated) land fund.

In general, the recultivation of disturbed and destroyed land is a system of engineering, ecological, meliorative, agro-technical, forestry and other activities, aimed at restoration of technogenically disturbed terrains, of soil cover and of fertility, in order to create objects with different purposes – agricultural, forestry, sanitary, recreational, etc.

Presently, the restoration of the fertility of the disturbed lands is linked with use of a humus soil layer, which is applied on a technically reclaimed terrain of 40 cm. thickness.

At the same time the analysis and the estimates of the humus amounts available for humus recultivation in a number of disturbed and contaminated territories are insufficient. For this reason, it is necessary to look for additional raw material sources, which could replace the missing quantities of humic materials.

A possible solution to the problem is the usage of waste products from certain industries which, instead of being used for appropriate purposes, are dispersed around the factories, hampering the latter's functioning and polluting the environment. Examples are the sludge from biological treatment of wastewater, the solid and liquid manure from livestock farms, the various kinds of sawdust, plant leftovers, solid household waste, etc.

The lack of sufficient humus and the presence of waste products suitable for secondary processing and use require search for new variants and development of new technologies for recultivation through the use of an innovative approach, which will be presented in this paper.

The main objective of this development is to present the existing methods of technical and biological recultivation for restoring fertility of disturbed lands and soils by the traditional method – through the use of humic soil material – and to recommend innovative technologies for humusless restoration of soil fertility by using waste products.

4.2 State of the Question and Principal Schemes for Recultivation

The principal scheme, by which the recultivation of the disturbed lands in the Republic of Bulgaria is performed, is exemplified by two methods:

- Spreading of humic material, 40 cm thick, over technically constructed lands with geological or waste materials with appropriate physicochemical characteristics
- Direct utilization of land recultivated with geological or waste materials

The first method, by use of humic material, is more prospective, as recovery of soil fertility is implemented more quickly, to a much greater extent, and results in good yield stability. This method significantly improves the physical, chemical and biological properties of the recultivated lands. The humus' content regulates the aeration of the substrates in a positive direction, giving them looseness and porosity. This greatly facilitates their processing and enables normal nutrition and growth of plants.

In the absence of sufficient amounts of humus, the second method of recultivation is recommended.

In some cases, it is possible to replace the humus layer with a substrate of organic or inorganic components that has precisely defined analytical characteristics and bioproductive capabilities. The method is mainly applied in the cases of agricultural recultivation by direct utilization of the terrains recultivated with geological and waste materials, which is characteristic for the larger percentage of the sites. In these cases, before the implementation of the unearthing and the extraction activities, there should be made an assessment for the suitability for recultivation of the varying compositions and properties of geological materials, which cover the mineral resource. The assessment is made in regard of the following indicators: morphological features; mechanical and microaggregate composition; organic matter content (humus); general and absorbable forms of nitrogen, phosphorus and potassium; environmental reaction (acid (pH) in water); carbonates; sorption capacity and exchange cations; general chemical composition; content of microelements; relative and bulk density; etc.

In a most general aspect, the assessment shows the prospects for selective extraction and arrangement of geological materials with respect to the accelerated reclamation of the environment. The selective approach to the construction of recultivated lands is an opportunity to improve their characteristics and properties and to reduce the costs of recultivation. Meanwhile, it avoids the danger of deposition of materials with toxic or adverse properties within the superficial part of the dumps. This approach has not found large application in the practice yet since the recultivated lands are characterized by highly heterogeneous composition, which hinders the development and cost-effective implementation of activities for recovery and increase of their productivity.

In order to estimate the improvement of the properties and the fertility of the recultivated lands within the application of this approach, numerous studies of recultivated and envisaged to recultivation areas have been carried out on the territory of the whole country (Banov and Hristov 1996a, b; Banov and Marinkina 1997; Banov and Markov 1999; Gencheva 1995; Garbuchev et al. 1975; Hristov et al. 1996; Banov 1999; Dimitrova et al. 1998; Marinkina and Banov 2000; Meena and Meena 2017). A large number of geological, waste and soil materials have been studied in relation to their suitability for technical and biological recultivation. The results obtained have been tested in vegetative and field settings. A number of methodological instructions and technological developments have been prepared both in terms of specific sites and of national importance (Banov and Pavlov 2014, 2015; Pavlov and Banov 2014; Pavlov et al. 2015a, b, c; Banov et al. 2015; Ashoka et al. 2017).

Within the practice two successive stages of recultivation – technical and biological – can be distinguished. At the technical stage, principally, there is implemented the transportation and arrangement of geological materials, as well as the construction of the terrain, according to the requirements of biological recultivation.

The biological recultivation of the constructed recultivated terrains is performed by cultivation of certain crops, application of strictly dosed fertilizer norms, use of specific agro-technical equipment, etc., according to the purpose of the terrain's use.

The conducted research and practical recultivation activities in a number of countries as well as in Bulgaria show that the technogenically recultivated lands are characterized by low natural fertility. An additional disadvantage is the presence of an increased content of heavy metals and other synthetic inorganic and organic substances, sometimes in excess of the maximum permissible concentration (MPC).

The great variety found in morphological, physical and physicochemical characteristics of the materials that make up the profile of the recultivated terrain requires the development of particular (according to the specific conditions) technological solutions for the restoration of the fertility of these sites. In this regard, the present guidance is intended to describe and analyse some modern/cutting-edge methods of recultivation of damaged lands and soils.

In principle, there are three basic methods for technical recultivation of disturbed terrains:

1. Spreading of humus material, 40 cm thick, over technically constructed lands with geological or waste materials with appropriate physicochemical characteristics. It should be noted that the humus layer should be spread no earlier than on the second year of construction of the dumps. This is due to the settling of the loose materials, resulting in the formation of dents on the surface of the embankments. The latter are aligned, by which humic material is saved.
2. Direct utilization of terrains recultivated with geological or waste materials suitable for physicochemical purposes, without use of a humus layer.
3. In the surface layer of the technically constructed recultivated areas, meliorants are inserted in order to improve the quality of the geological and waste materials used.

The conducted long-term laboratory, vegetative and field studies have shown that recultivation activities are accompanied by various technical, practical and financial difficulties and problems, namely:

1. The use of humus material in the technical construction of recultivated terrains leads to a relatively fast recovery of soil fertility but is connected with spending of significant financial resources.
2. The biological stage of recultivation is very long, accompanied by planting of a large number of cultures, which require the implementation of strictly defined systems of fertilization, plant protection and agro-technology.
3. The two stages of recultivation (the technical and the biological) require spending of large capital investments.
4. The technically constructed recultivated areas are characterized by different unfavourable features – heavy mechanical composition, low natural fertility, heavy metal content above MPC, toxic levels of the reaction (pH), etc.
5. There is a significant discrepancy between the time limits for the implementation of the biological recultivation regulated by Ordinance No. 26 on Recultivation of Disturbed Terrains, Improvement of Low Productive Lands and Removal and Utilisation of the Humus Layer and the real terms for which the recovery of soil fertility of the damaged terrain is completed. According to the requirements of the normative documents, the duration of the biological recultivation is 5 years,

while the actual time reaches in some cases (humuslessly recultivated lands) up to 12–16 years.
6. There is no methodology for assessing the fertility of recultivated soils, which hinders the process of returning the lands to their real owners.
7. Technically built tailings dams during the first years of the mining operations require additional technical and biological activities to restore their agro-soil potential.

The solution to these problems is possible through the development and implementation of new, efficient and rational technological solutions to restore the fertility of anthropogenic soils. This is due to the fact that the available quantities of humic materials are in exhaustion, while at the same time large amounts of waste materials of different origins accumulate. For these and other reasons, recently various organic residues have found extremely wide use in the recultivation of lands disturbed by mining of mineral resources. The high content of essential nutrient macro- and microelements in the residue, and especially of carbon accessible for plants, provides a high meliorative effect, especially valuable in the case of impaired balance of the organic matter.

4.3 Innovative Methodological Approaches to Development of Technological Variants for Recovery of Terrain Tertility

In recent years, increasing attention has been paid to the search for new, efficient and rational technological solutions to restore the fertility of anthropogenic lands and soils. This is caused by the lack of sufficient quantities of humus materials and by the tendency for subsequent use of the materials disposed at landfills, which at a certain stage in the technological development become a raw material for the production of precious metals and power.

On this occasion, there has been developed a technology for the recultivation of terrains used as landfills for disposal of waste products from thermal power plants and in particular cinder dumps for storage of ashes. The location, climatic characteristics of the area, humidity, type and concentration of pollutants, acidity of the environment as well as the oxygen access are the basic indicators that have been considered in the development of the technology. The technology offers different approaches to recultivation depending on the suitability of geological and soil materials for formation of surface layers and the need for introducing enhancers to maintain and improve the soil fertility. The technology suggests technical and biological recultivation of the terrain by preparing a recultivation substrate and spreading it on the terrain used as a landfill, insertion of mineral fertilizers, terrain treatment and grassing with appropriate plant species. As a recultivation substrate, mixtures of geological and waste materials are used in different proportions depending on the physicochemical characteristics of the individual components. The characteristics of the product of concern enable ashes to be attributed to the limited number of materials suitable for use in agriculture. On the other hand, this is an opportunity for its future use and utilization. The advantages of the technology include the fact that the complex of events takes place directly on the terrain. The technology is protected by Patent Invention No. 66155/25.11.2011 entitled "Method for Restoration of Terrains Used as Landfills for Disposal of Waste Products".

Another technology for recultivation of damaged and contaminated lands and soils is directed to the use of a biological substrate with the following components mixed in different volume ratios: geological materials, characterized by light to heavy sandy-clayey mechanical composition, without reserves of organic matter (humus) and with low nutrient content for plants and lack of toxic components, and compost under the trade name Compovet B-4. To obtain the biological substrate, the geological materials are mixed with Compovet B-4 in ratios 2:1, 3:1, 4:1 or 5:1. The resulting biological substrate can be used as a substitute for humus materials for the technical and biological recultivation of terrains that have been utilized as mine dumps, mining sites, tailings ponds and others. The advantages of the biological substrate are connected with the fact that it can be used irrespective of the physicochemical characteristics of the materials constituting the terrains, its quality characteristics are constant and the frequent lack of humus materials and organic fertilizers for technical and biological recultivation in many cases makes it without alternative. The technology is protected by a certificate for Utility Model Registration No. 947/30.11.2007 – "Biological Substrate".

The efficiency of recultivation measures considerably increases after a preselection and evaluation of the plants suitable for phytorecultivation. A particularly suitable method for biological recultivation and creation of meadow pasture lands is the method of introduction of grasses that in natural conditions grow in similar areas and meet the following general requirements: to be drought-resistant, to have a feed value, to form a deep root system, to remain in good condition at low temperatures, to grow quickly without creating a powerful turf and to secure protection of the slowly growing species (Dinev et al. 2012a, b).

The negative impact of anthropogenic activity on the environment determines the need to continue the research to find new, rational and, most of all, cost-effective methods for restoration of damaged lands, which can be easily put into practice.

4.3.1 Use of Sludge of Wastewater Treatment Plants (WWTPs)

The shortage of humus materials and the presence of a large amount of waste of different origins require the restoration of fertility of the damaged terrain without the use of a humus soil layer, i.e. new technological options are proposed. Due to these and other reasons, recently the sludge from wastewater treatment plants has found extremely wide use in the recultivation of so-called damaged lands – disturbed, polluted, low-productive, etc.

The use of sludge makes it possible to overcome a number of disadvantages related to the recultivation of disturbed soils, namely:

(a) Balanced nutrition of the plants grown is ensured and the organic reserve of soils increases.
(b) The content of organic matter greatly facilitates the treatment of recultivated terrains and allows normal development of the plants' root system.
(c) The physical and the physical-mechanical characteristics of the recultivated soils are improved.
(d) The term for implementation of the biological stage of recultivation is reduced.
(e) The necessary investments for the implementation of the recultivation activities are substantially reduced.
(f) The process of returning the land to their real owners is facilitated.

Sludges are organic mass rich in macro- and microelements. The dry matter content in them (45–75%) classifies them as biomass, suitable for transporting and spreading by mobile transport. The values of the plant nutrient macroelements (nitrogen, phosphorus and potassium) show that sludges have the appropriate agrochemical properties. They can be used as fertilizers and sources for enrichment of soils with organic matter in recultivation activities.

Sludges are rich in soil-friendly microorganisms. Inserted to the soil, they become a source for formation of humus and absorbable nutrients for plants.

Given the existing shortage of organic sources in the country and the relatively low content (1–3%) of humus in our soils, the use of sludge as a fertilizer and a means of recultivation of disturbed terrains is highly recommended.

The studies conducted on their composition and the results of their practical testing prove their suitability for use in the recultivation of disturbed terrains.

Disturbed terrains can be found in the vicinity of almost every settlement. Within the Sofia metropolitan area only, thousands of decares of agricultural land have been destroyed and turned into swamps and lakes. Its amount increases annually as the construction needs of sand and gravel grow. The restoration of these lands' fertility, as well as their return into the arable land fund, is a problem that is gaining increasing importance every year. In these cases, sludge from wastewater treatment plants (WWTPs) can be used by:

- Mixing of the superficial non-fertile layer of the recultivated terrains with the sludge in an acceptable ratio, in order to improve the structure and to increase sustainably the soil fertility
- Creating a self-contained surface layer of sludge

As noted above, there are two successive stages of recultivation – technical and biological.

In the specific case, sludge could be used for technical recultivation of disturbed terrains, which are located near WWTP. They can be composted with other waste materials and used in the recultivation of disturbed and contaminated terrains. During the biological stage of recultivation, it is recommended to cultivate agricultural crops in compliance with the requirements of the legislation or with the grass and forest vegetation. Depending on the guidelines for the future use of recultivated terrains, a range of 60–80–100 t/dka of sludge is recommended. The rate may vary and is determined on a case-by-case basis, by designing a project and assessing in detail the state of the disturbed terrain.

The requirements related to the use of sludge in the reclamation of disturbed, contaminated or low-productive terrains are related to:

(a) Identification of the reasons for the terrain's disturbance
(b) Lack of enough humic soil layer
(c) Physicochemical characterization of the soils in the area, including the waste materials
(d) Future purpose of the recultivated terrain
(e) Characteristics of the sludges
(f) Presence of earth masses in the area and assessment of their properties and characteristics
(g) Distance to the treatment plant, etc.

According to the specific conditions of the terrain, technological variants, including the use of sludge, are recommended.

4.3.2 Reclamation of Landfill Sites: Technology Solutions

One of the practical examples for application of WWTP sludge is the recultivation of solid waste dumps, such as the one near the village of Yana, Sofia District (Scheme 4.1).

As indicated in the identification of the recultivation measures, the soil and climatic conditions in the area as well as the technology of disturbance and contamination of land/soils are essential. The study of these factors shows the following features:

Climatic characteristics – The industrial site of the dump falls into the temperate continental climate subarea of the European continental climate area or, more particularly, in the climatic region of the high fields in Western Central Bulgaria.

Table 4.1 data shows that the average annual temperature is relatively low and that the coldest months of the year are January and February; in some cases, the temperature drops to −20 °C. Sharp drops in temperatures in the spring and the autumn are also not excluded. The months of July and August are the warmest.

There is a danger of late spring and early autumn frosts in the area during the second half of May and the first ten days of September.

The distribution of rainfall over the year by months and seasons is uneven – sharply expressed continental climate. The rainfall is highest in June and lowest in February and September. The difference between the summer maximum and the winter minimum is more than twice. The average annual rainfall is below the country average (Table 4.2).

Winds predominate in the west, east and northwest.

Geological characteristics – The territory of the dump falls in the Eastern Sofia Field and belongs to the Sofia Valley. The average altitude of the Sofia Field is

Scheme 4.1 Yana, tailings pond

500–560 m, and the surrounding mountains rise above 800–1000 m. The relief is flat without hill formations.

Soil characteristics – The soil cover is formed exclusively over Pliocene and Quaternary materials (sands, gravel, sandy clays, clays, etc.).

The soil-forming factors (climate, relief, soil-forming rocks, vegetation, etc.) in the site's area determine the formation of the following soil differences: leached

Table 4.1 Average monthly and annual air temperature

Months												Average
January	February	March	April	May	June	July	August	September	October	November	December	Annually
−1,8	−0,3	4,6	10,3	14,4	18,0	20,9	19,7	16,1	10,5	5,1	0,0	9,8

Table 4.2 Average monthly and annual rainfall in mm

Months												Average
January	February	March	April	May	June	July	August	September	October	November	December	Annual amount
33	30	39	56	83	85	67	54	48	57	49	39	640

vertisols, gleyic chernozems, eutric fluvisols, dystric fluvisols, eutric gleysols and histosols. The so-established soil differences are characterized by the following physicochemical properties and indicators:

- Leached vertisols are characterized by thickness of the humus horizon of 66–86 cm. The total thickness of the soil profile is about 101–117 cm. The mechanical composition of these soils is medium clayey. They are medium humic in terms of humus content. The soil response ranges from slightly acidic to slightly alkaline. Carbonates are observed below 101–117 cm.

Gleyic chernozems (typic and calcic) – They are formed on the alluvial material, fluvial and torrential, under the influence of meadow vegetation. They are medium sandy-clayey to heavy sandy-clayey in mechanical composition in the surface horizon. The physical clay content varies from 41.3 to 47.2%. The content of organic matter (humus) defines them as low humic to medium humic – from 1.73 to 3.61% humus in the arable layer. The carbonate content is high and ranges from 0.27 to 18.77%. There is no presence of carbonates in the surface part (0–26 cm) of the profile of the typical meadow chernozems. The reaction of the medium changes from neutral to slightly alkaline.

Eutric fluvisols – The area of the site is characterized by eutric fluvisols, temporarily over-moisturizing at the surface with high groundwater level and thin humus layer (up to 0.30–0.40 m). They are slightly to medium sandy-clayey in mechanical composition but contain a high percentage of large particles. The soils are medium humic with an organic content of 1.10–1.70%. The carbonate content is quite diverse, which affects the reaction of the medium (pH in water). The latter changes from slightly alkaline to neutral and acidic in the arable layer. The underlying horizons are characterized by a slightly alkaline reaction of the environment.

Dystric fluvisols – They are characterized by a layered structure. They are formed on a variety of alluvial and partly deluvial material. The profile of these soils is represented by humus horizon of varying thickness (26–150 cm). After the arable layer or the humus horizon follow layers of sands and gravel. The mechanical composition of the soils is very diverse, ranging from light to medium sandy-clayey to medium clayey, depending on the type of soil-forming sediments. According to the humus content, the soils are medium humic (from 1.40 to 4.23%). The carbonate content is high, but there are areas where carbonates are missing. The soil reaction is neutral and alkaline, and for non-carbonaceous areas, it is acidic.

Eutric gleysols – Their development is associated with constant over-moisturizing and accumulation of organic matter from the meadow marsh vegetation. Their profile is composed of two horizons – humic and gleyic. By mechanical composition, the soils are sandy-clayey in the surface horizon. The depth of the profile shows a change in the mechanical composition, and in the gley horizon, it is already medium sandy-clayey.

The humus horizon of these soils is rich in organic matter – from 3.32 to 10.20%. Carbonates are present right from the surface, the amount being high – over 50%.

The soil reaction is alkaline. Eutric gleysols are slightly salinated as a result of the capillary rise and surface evaporation of the mineralized groundwater.

For the proper and accurate planning of recultivation activities and the creation of a recultivation substrate with the use of sludge from WWTP, a study has been implemented of the materials that make up the dump. It has been found that there are steel slags in the body of the dump, which are characterized by a volatile chemical composition.

Average samples show that the major oxides change to the following ranges:

Calcium oxide	**(CaO)**	30÷44%
Magnesium oxide	**(MgO)**	1.0÷4.0%
Silicon dioxide	**(SiO_2)**	15÷33%
Aluminium oxide	**(Al_2O_3)**	1.0÷3.0%
Manganum oxide	**(MnO)**	8÷25%
Iron(II) oxide + iron(III) oxide	**(FeO + Fe_2O_3)**	10÷15%
Phosphorus pentoxide	**(P_2O_5)**	1÷1.5%
Iron	**(Fe met.)**	15÷21%

In conclusion it can be said that the materials from the solid waste dump are not suitable for carrying out biological recultivation, which requires their covering with appropriate physicochemical or geological or soil materials. In order to find such materials, samples have been taken from the following sites:

1. Depot for storage of humic and soil materials on the territory of the Sofia Airport
2. Geological and soil materials collected during the construction of building sites on the city of Sofia's territory
3. Geological and soil materials deposited on the solid waste dump in the village of Yana

The materials studied cannot be accurately assigned to certain soil taxonomic units because of the anthropogenic impact on them and the substantial disturbance of their natural mechanical, physical and chemical properties. The area of distribution is in an intensely developed industrial area, which requires a complex assessment of their characteristics and properties. This requires the determination of the basic biogenic properties of the soil and geological materials as well as the permissible contaminants with heavy metals and radionuclides.

In connection with the abovementioned, the samples collected have been analysed in relation to the following indicators:

(a) Distribution of mechanical fractions (mechanical composition)
(b) Content of plant nutrients – nitrogen, phosphorus and potassium
(c) Content of organic matter (humus)
(d) Reaction of the medium (pH in water and potassium chloride)
(e) Total quantity of carbonates
(f) Hygroscopic moisture, heavy metal content – copper (Cu), zinc (Zn), plumbum (Pb), cadmium (Cd), cobalt (Co), nickel (Ni) and chromium (Cr)

The data presented in Table 4.3 show that by mechanical composition the individual materials studied are characterized by varying results, which confirms the fact that they have a strong anthropogenic impact on them. Four groups of materials are determined with respect to the distribution of mechanical fractions. The first group is represented by samples from the depot at the Sofia Airport, which have a physical clay content ranging from 24.9% to 32.7%, which defines them as light to medium sandy-clayey.

Geological and soil materials collected during the construction of building sites on the territory of Sofia are grouped in two categories – loose sand, where the fractions of the large and small sand prevail and the content of physical clay is about 3–4% (samples 5 and 6), and materials distinguished by relatively evenly distributed sand and dust fractions where the physical clay content is about 21% (samples 7 and 8).

Samples of the geological and soil materials disposed on the solid waste dump in the village of Yana have widely variable content of physical clay – from 25.2% to 55.2% – which is the result of different origins and times of disposal.

The results of the agrochemical and chemical analyses of the samples (Table 4.4) show that they can be divided into three main groups with the following features:

(a) Soil reaction of samples from the Sofia Airport depot is slightly alkaline. The content of organic matter (humus) determines the studied materials as poorly stocked. The amount of the major biogenic elements is unbalanced; as for phosphorus and nitrogen, they are in the poor grade of stock, while for potassium it is in the very good grade.
(b) The geological and soil materials collected during the construction of building sites on the city of Sofia's territory are characterized by a clearly expressed alkaline reaction and poor organic content. Phosphorus and nitrogen are almost absent, and the potassium content is at the lower limit of the average stock.
(c) Alkaline reaction is characteristic also for the samples of the materials deposited on the solid waste dump in the village of Yana. The organic matter content classifies them in the category of the low-humic. Nitrogen is low in stock; phosphorus and potassium are good and very good in stock.

The analytical results for the content of heavy metals (Table 4.5) in the studied soils and geological materials have been assessed according to the requirements of Ordinance No. 3 on the Norms on the Permissible Content of Harmful Substances in Soil and the Appendix thereto. Considering that the soil reaction in aqueous suspension of samples is > 7.00, the amount of heavy metals is below the acceptable limits of maximum permissible concentration (MPC).

The conclusions made show that soil and geological materials from all the three sites can be used for technical and biological recultivation purposes.

Table 4.3 Distribution of mechanical fractions in % to dry air condition in geological and soil materials

No.	Sample	Hyg. moisture %	Size of particles in mm							
			Total > 1	1–0.25	0.25–0.05	0.05–0.01	0.01–0.005	0.005–0.001	<0.001	Total < 0.01
1.	Sofia Airport depot	4.02	12.6	20.0	21.4	13.6	12.2	7.5	13.0	32.7
2.	Sofia Airport depot	3.08	10.4	17.1	19.6	27.9	5.1	4.7	15.2	25.0
3.	Sofia Airport depot	3.53	9.3	18.4	19.2	27.6	5.6	4.9	14.9	25.4
4.	Sofia Airport depot	3.98	9.5	18.9	18.9	26.8	6.3	5.1	13.5	24.9
5.	Geol. material – city of Sofia	1.41	0.5	57.4	34.8	3.9	1.0	1.5	0.9	3.4
6.	Geol. material – city of Sofia	1.39	0.7	55.3	36.7	3.8	1.2	1.4	1.4	4.0
7.	Geol. material – city of Sofia	4.43	2.4	20.4	39.2	16.6	12.4	5.0	4.0	21.4
8.	Geol. material – city of Sofia	3.23	2.0	19.9	40.1	16.1	12.0	5.4	4.3	21.7
9.	Geol. materials disposed on the dump	3.12	9.4	18.1	19.6	27.9	5.2	4.9	15.1	25.2
10.	Geol. materials disposed on the dump	4.10	22.6	31.0	0.4	13.6	12.5	7.2	12.7	32.4
11.	Geol. materials disposed on the dump	3.43	12.5	24.2	0.6	9.5	13.3	9.4	30.5	53.2
12.	Geol. materials disposed on the dump	2.51	18.1	39.7	3.3	7.3	11.1	4.3	16.2	31.6
13.	Geol. materials disposed on the dump	3.03	8.1	26.5	7.9	12.4	12.9	8.4	23.8	45.1
14.	Geol. materials disposed on the dump	3.84	5.8	20.7	5.7	12.6	16.3	7.0	31.9	55.2

Table 4.4 Agrochemical and chemical indicators of the studied geological and soil materials

No.	Sample	pH		$N\text{-}NH_4$	$N\text{-}NO_3$	P_2O_5	K_2O	Carbonates %	Humus %
		H_2O	KCL	mg/kg soil		mg/100 g soil			
1.	Sofia Airport depot	7.5	6.9	6.99	4.33	4.8	20.6	0.0	1.32
2.	Sofia Airport depot	7.4	6.7	5.72	4.21	5.1	21.6	0.0	1.40
3.	Sofia Airport depot	7.5	6.8	6.54	5.30	5.5	19.6	0.0	1.27
4.	Sofia Airport depot	7.5	6.8	6.20	4.65	5.1	20.6	0.0	1.30
5.	Geol. material – city of Sofia	8.5	7.5	3.34	1.99	2.4	16.7	0.0	1.17
6.	Geol. material – city of Sofia	8.4	5	3.21	2.25	2.1	15.6	0.0	1.20
7.	Geol. material – city of Sofia	8.4	7.0	4.33	2.33	1.6	13.3	0.0	0.21
8.	Geol. material – city of Sofia	8.5	7.3	3.53	2.36	1.1	12.6	0.0	0.19
9.	Geol. materials disposed on the dump	7.8	5.9	4.23	3.56	8.5	20.6	0.0	1.58
10.	Geol. materials disposed on the dump	8.7	7.3	5.99	4.99	25.3	23.1	0.0	2.16
11.	Geol. materials disposed on the dump	8.4	7.0	5.87	2.39	20.3	27.0	0.0	1.72
12.	Geol. materials disposed on the dump	7.9	6.1	4.33	3.56	8.5	20.6	0.0	1.54
13.	Geol. materials disposed on the dump	8.6	7.2	5.90	4.89	9.2	21.1	0.0	2.16
14.	Geol. materials disposed on the dump	7.6	7.0	4.33	1.66	18.9	23.8	0.0	1.32

Table 4.5 Contents of heavy metals in the studied geological and soil materials

No.	Sample	Cu	Zn	Pb	Cd	Co	Ni	Cr
		mg/kg	mg/kg	mg/kg	mg/kg	mg/kg	mg/kg	mg/kg
1.	Sofia Airport depot	19,8	38,0	13,8	<0,50	9,00	14,3	17,8
2.	Sofia Airport depot	17,6	36,0	11,3	<0,50	9,20	15,1	16,3
3.	Sofia Airport depot	18,9	37,5	12,6	<0,50	8,90	15,3	18,2
4.	Sofia Airport depot	19,2	38,1	14,2	<0,50	9,20	16,5	17,6
5.	Geol. material – city of Sofia	3,50	14,3	<2,00	<0,50	<2,50	4,50	5,50
6.	Geol. material – city of Sofia	2,60	13,3	<2,00	<0,50	<2,50	3,60	6,10
7.	Geol. material – city of Sofia	4,30	17,2	<2,00	<0,50	<2,50	4,30	6,20
8.	Geol. material – city of Sofia	3,40	15,1	<2,00	<0,50	<2,50	4,20	5,60
9.	Geol. materials disposed on the dump	43,3	123	61,5	<0,50	19,8	42,0	35,0
10.	Geol. materials disposed on the dump	25,5	46,0	7,75	<0,50	8,50	10,0	15,8
11.	Geol. materials disposed on the dump	28,9	98,0	8,95	<0,50	9,50	7,50	16,5
12.	Geol. materials disposed on the dump	33,1	83,5	7,65	<0,50	8,50	7,50	14,7
13.	Geol. materials disposed on the dump	38,7	111	9,15	<0,50	8,00	10,0	15,5
14.	Geol. materials disposed on the dump	41,5	105	8,25	<0,50	8,50	9,50	14,5

The scheme of recultivation activities is as follows:

1. Loading soil and geological materials with layer thickness of 30 cm.
2. A layer of soil and geological materials (recultivating substrate) with a thickness of 0.20 m is spread on the surface of the so-prepared terrain. As organic fertilizer, it is suggested to use stabilized sludge from a wastewater treatment plant (WWTP) in the city of Sofia, in the amount of 10 m^3/dka.

4.3.3 Reclamation of Land Disturbed During Construction of Tailings Ponds

Tailings ponds are specific sites built upon the processing and enrichment of ore minerals. Principally they are located in natural decreases of the terrain and occupy large areas – 1000 decares or more.

The destruction of soil differences is related to the filling of tailings ponds with waste products. In practice, these areas are not recoverable, with the exception of the surface area (the shoreline and the lake) and the embankments of the tailings pond walls, which fertility, however, can be restored only to some extent.

An additional adverse effect on soils is caused by their dust pollution due to wind erosion. Areas on which tailings ponds are built are unusable for agricultural purposes and are not subject to recultivation in the sense of restoring soil fertility. They are technogenic objects which, because of the law restrictions, cannot be used for agricultural production of food and feed. In this case, it is only possible to integrate the territory of the tailings pond into the surrounding landscape and to include it in the general natural landscape by carrying out specific recultivation measures. The character of the relief in the tailings ponds area favours the development of erosion processes, which however are further provoked by the technogenic interference of man. The executed forest fellings contribute to the deepening of the destructive processes. Deforested areas left without the natural protection of the vegetation are subject to the active development of erosion processes. A significant part of the humus-accumulative horizon is taken away, as well as the whole soil profile at some places.

4.3.4 Reclamation of Elshitsa Tailings Pond

The site and its associated territories are situated in the vicinity of the Elshitsa Village, Panagyurishte Municipality (Scheme 4.2). The tailings pond is of a sputtered type and is constructed as a result of copper ore processing on an area of 108.5 ha.

Climatically the area where the tailings pond is built falls into the southern part of the moderate zone, adjacent to the subtropical Mediterranean climate. This determines the climate as moderate continental to transition to the Mediterranean. The average temperature in January is around 0–1.5 °C below zero. Continuous retention of temperatures above 10 °C comes around the middle of April. The hottest month is July with an average temperature of about 23.5–25 °C, as during extremely

Scheme 4.2 "Elshitsa" Tailings Pond

strong warming the maximum temperature can reach 40–42 °C. The average temperature for October is about 11–13 °C, with the first autumn frosts occurring at the end of this month.

Wind has substantial impact on the dispersion of harmful substances and pollution of the environment near the source of pollution. The study shows that the area is characterized by a slight movement of air masses with an average monthly speed of 1.2–1.8 m/sec. The transfer of air masses occurs mainly from the northeast and northwest to the southeast (79.9% of the wind time).

In geological and structural terms, the area in which the tailings pond is located is made up of Precambrian and Upper Chalk materials (intrusive and sub-volcanic rocks such as granites, granodiorites and their porphyrite varieties, dacites and andesites and their agglomerate tufts). The area and its adjacent plots are characterized by copper-porphyry ore outcrops and pyrite and copper-pyrite mineralizations of volcanogenic-hydrothermal origin. The geological setting predetermines the formation of fissure groundwater in the area. Major aquifers are the cracked rock formations of metamorphic (gneiss), volcanogenic-sedimentary (andesites, dacites, rhyodacites) and intrusive (granite, granodiorite) rocks. Groundwater is distributed at a depth of about 100–150 m and the underground drainage module is in the most cases below 0.1 l/s/km^2. By chemical composition, the predominant types of water are sodium-calcium-sulphate and calcium-magnesium-sulphate.

The soil cover in the area is represented by several soil types (Koinov et al. 1968; Yadav et al. 2018), classified according to the Bulgarian Soil Classification (Penkov et al. 1992). A correlation with the World Reference Base for Soil Resources or WRBSR (IUSS Working Group WRB 2006) has also been made. Soil types are represented by:

Leached chromic forest soils – leptic luvisols. They have a thin illuvial horizon, passing into a solid soil-forming rock. The humus horizon's thickness is 25–30 cm, and the total soil profile ranges from 55 to 65 cm. By mechanical composition, the soils are slightly sandy-clayey. The soil reaction in the surface horizon is strongly acidic.

Chromic forest soils – leptic cambisols (chromic). They are formed on weathered products of acidic rocks. They are characterized by a nearby positioned bedrock. As to mechanical composition, they are sandy-clayey with a high content of skeletal fragments. The total depth of the soil profile is 50–60 cm. The soil reaction ranges from very strong to slightly acidic.

Chromic-podzolic (pseudopodzolic) forest soils – planosols (chromic). They are formed on Quaternary deposits under the influence of forest vegetation. They are characterized by a clearly differentiated profile – a well-defined podzolic and illuvial horizon and a clear differentiation in the mechanical composition. The illuvial horizon is reddish-coloured due to the hydroxides of iron and aluminium. The mechanical composition is slightly sandy-clayey. This mechanical composition causes poor water permeability.

The vegetation is represented by forest communities of oaks: *Quercus cerris* L., *Quercus frainetto* Ten. and *Quercus dalechampii* Ten. In some places there are small remains of field ash (*Fraxinus oxycarpa* Bieb. ex Willd.), field elm (*Ulmus minor* Mill.) and greyish oak (*Quercus pedunculiflora* C. Koch.). There are also artificially created populations of black and white pine trees (*Pinus nigra*, *Pinus sylvatica*). Of the shrub formations, the most common are dog rose (*Rosa canina* L.), blackberry (*Rubus caesius* L.) and hawthorn (*Crataegus monogyna* Jacq.). Grass formations represented mainly by cereal grasses (Poaceae) are also spreading in this area.

The anthropogenic impacts on the extraction and processing of copper ore on the landscape are expressed mainly in relief change resulting from the constructed roads, facilities, embankments and tailings ponds.

The construction of the technogenic structures serving mining activities has had an impact on the forest vegetation in the area, which has been greatly reduced. According to the factors that determine the biocenotic value of the terrestrial communities (degree of anthropogenic interference in the formation of biocenosis, tolerance to anthropogenic impacts, species diversity and uniqueness of the communities), the studied formations can be described as biocenoses under strong anthropogenic influence, partly created by man (artificial plantations, agricultural vegetation) with a high degree of tolerance and lack of rare floristic elements.

The negative impact of the tailings pond on the environmental components is a result of the toxic characteristics of the industrial tailings products disposed. The results of the implemented laboratory analyses of the tailings are presented in Table 4.6. The toxicity of the residue is determined by the extremely high content of sulphur, copper and lead, which exceed the concentrations characteristic for the soils (published in Ordinance 3/2008). The copper (Cu) content exceeds 27 times that in natural soils, plumbum (Pb) content 16 times and sulphur (S) content about 10 times (Treikyashki and Hristov 1982). Apart from their direct toxicity to aquatic and terrestrial environments, heavy metals degrade slowly and have a long-lasting lethal effect on vegetation.

The high sulphide content (denoted by S_S in Table 4.6) gives rise to sulphuric acid formation in amounts sufficient to cause the well-known problem of the acid mine drainage. The strong acidification of the medium (pH <4.0) increases the mobility and toxicity of hydrogenium (H^+), aluminium (Al) and manganum (Mn); decreases the digestibility of N, P, K and Mo; and strongly inhibits the plants' growth (Berg and Vogel 1973; Bengtson et al. 1973; Foy 1974; Safaya 1978; and others). It is known that the kinetics of the oxidation process are substantially influenced by the degree of dispersion of sulphide minerals, their simultaneous presence, the biochemical activity of thionic bacteria, the physical properties of geological materials containing them, the degree of aeration, the mechanical and mineral composition and the hydrothermal mode of the locality (Marinkina 1999; Kumar et al. 2017).

The relative stability of sulphides in tailings ponds, where the oxidation conditions are different than those in the dumps (Jambor 1994), is as follows: pyrrhotite <sphalerite = galenite <pyrite = arsenopyrite < chalcopyrite < magnetite (the most resistant). Each of these minerals is a source of a number of elements – impurities which, during the oxidation processes, are displaced from the sulphide minerals grid (e.g. pyrite and chalcopyrite contain arsenicum (As), nickel (Ni), cobalt (Co), selenium (Se), tellurium (Te), etc., while sphalerite contains Iron (Fe), cadmium (Cd), gallium (Ga), germanium (Ge), etc.). Some of these elements go directly into the rivers (with the drainage and the slope outflows), while others can take part in the secondary sedimentation phases such as sulphates, Fe oxides and hydroxides

Table 4.6 Chemical composition of the tailings from the tailings pond "Elshitsa"

Elements														
SiO_2	Al_2O_3	CaO	MgO	Na_2O	K_2O	TiO_2	S_T	S_{SO4}	S_S	Cu	Fe	Pb	Ag	Au
%													g/t	
69,32	12,41	1,10	2,06	1,80	2,75	0,34	2,90	0,10	2,80	0,16	4,50	0,08	1,36	0,20

which have a nickel (Ni) content of 0.7%, copper (Cu) content of 1.2%, arsenicum (As) content of 4.7% and plumbum (Pb) content of 6.7% (Benvenuti et al. 1997; Yadav et al. 2017b). When the physicochemical conditions change, these elements can be mobilized into the environment again.

Several geochemical trends have been identified in the behaviour of the various elements that can outline the risk to the environment. Typically, the content of cadmium (Cd), copper (Cu), zinc (Zn) and Iron (Fe) increases in depth due to the solubility of their oxidation products. Therefore, they are the first to migrate into river waters and are most often the determining criterion for the extent of pollution in a given region. The content of other elements such as argentum (Ag), aurum (Au), stibium (Sb), plumbum (Pb) and arsenicum (As) diminishes in depth due to the insolubility of their altered products (Boulet and Larocque 1998; Meena et al. 2016). This means that these elements will be present in the mineral phase and will represent a long-term potential source of pollution.

When the tailings enter the soils under the influence of wind, the underground or the slope drainage leads to a change in their characteristics, resulting in a strong decrease in pH values and overdose of precautionary concentrations for copper (Cu) and cadmium (Cd) (Ordinance 3/2008).

The studies carried out show that during the long-term mining and enrichment activities and under the influence of the dust pollution with tailings, the adjacent soils located on an area of 19.1 ha are characterized by significant deviations from the characteristics of the zonal soils. The data in Table 4.7 show that the rootstock layer (0–40 cm) of the chromic cambisols is anthropogenically contaminated with cadmium and copper. Contamination exceeds the maximum permissible concentration (MPC) regulated in Ordinance 3 by 54–79% for copper and by 67% for cadmium. There is also a reported decrease in the pH of the medium (Table 4.8) near the tailings pond due to the high content of sulphide minerals in the tailings and the acidity generated during the oxidation process.

The mobility of zinc also increases in the acidic conditions but does not exceed the precautionary levels in the studied soils. The soil's pH is a controlling factor for the digestibility of the soil manganum, whose solubility decreases with a decrease in acidity. It can be expected that a pH response below 5.0 will create conditions for increasing the content of easily reducible manganum, which causes the occurrence of toxic symptoms in plants. The established accumulation of manganum in subhorizons of the studied soils is most likely related to the processing and mechanical redistribution of manganum compounds.

In conclusion, it can be stated that the soils in the area of Elshitsa tailings pond are heavily polluted with copper and cadmium (exceeding by over 50% of the MPC) and pose a risk to the environment and human health. They also form an area with increased manganum content, which can affect the quality and safety of agricultural production likewise.

Table 4.7 Content of heavy metals (mg/kg) in soils in the area of Elshitsa tailings pond

Profile cut	Sample depth (cm)	Cu	Pb	Zn	Mn	Cd
Field near the tailings pond	A + B	14,5	13,0	23,5	97,5	1,0
	0–22					
	BC	143,5	24,5	61,0	993,5	1,5
	22–40					
	CD	154,5	23,5	69,5	–	2,0
	42–65					
Field far from the tailings pond	A arable	22,0	19,5	15,5	181,0	1,0
	0–21					
	BC	14,5	20,5	25,5	423,5	1,0
	21–47					
	C	16,5	17,0	25,5	145,5	1,5
	47–80					
Natural chromic forest soil near the tailings pond	A	123,5	29,0	107,0	1000,0	2,5
	0–25					
	BC	121,0	32,5	95,0	1050,0	2,5
	25–44					
	C	98,0	21,0	77,5	750,0	2,0
	44 – ↓					
Precautionary concentration		60	45	160	–	0,6
Maximum permissible concentration		80	60	200	–	1,5
Intervention concentration		500	500	900	–	12

[a]Precautionary concentration is the harmful substance content of the soil in mg/kg, the excess of which does not lead to disturbance of the soil functions or to danger to the environment and human health

[b]Intervention concentration is the harmful substance content in the soil in m/kg, the excess of which leads to disturbance of the soil functions or to danger to the environment and human health

Table 4.8 Chemical characteristics of the soils in the area of Elshitsa tailings pond

Profile cut	Sample depth (cm)	pH in H_2O	Total N (%)	P_2O_5 (mg/100 g)	Digestible K_2O (mg/100 g)
A field near the tailings pond	A + B	4,80	0,055	4,0	20,00
	0–22				
	BC	4,85	–	–	–
	22–40				
	CD	4,80	–	–	–
	42–65				
A field far from the tailings pond	A arable	5,60	0,028	2,0	24,50
	0–21				
	BC	5,70	–	–	–
	21–47				
	C	5,65	–	–	–
	47–80				
Natural chromic forest soil near the tailings pond	A	5,35	0,116	1,5	30,00
	0–25				
	BC	5,80	0,080	1,0	26,00
	25–44				
	C	6,30	0,026	1,0	27,50
	44 – ↓				

Technical and biological recultivation of the Elshitsa tailings pond and the adjacent contaminated areas

The tailings pond recultivation plan has been developed in accordance to the requirements of Ordinance No. 26 on Recultivation of Disturbed Terrains, Improvement of Low Productive Lands and Removal and Utilisation of the Humus Layer and Ordinance No. 8 of 24.08.2004 on Terms and Requirements for Construction and Exploitation of Landfills and Other Facilities and Installations for Utilisation and Disposal of Waste and provides for the implementation of the following stages:

Stage 1. Creation of conditions for mechanized implementation of the recultivation activities

As the tailings pond's surface is not sufficiently thickened, it was necessary to construct temporary roads reaching the interior of the site. This was done by laying a geogrid and then filling soil materials to protect the heavy machinery from sinking.

Construction of temporary roads

Stage 2. Sealing of the tailings pond's surface

In the next step, encapsulation of the tailings pond's surface was performed with soil cement polymer in order to form a superficial shielding layer on the POP and to protect the surface water from penetration into the tailings.

Sealing of the surface of Elshitsa tailings pond with soil cement polymer

Stage 3. Melioration

After the encapsulation, stabilized sludge from the Plovdiv wastewater treatment plant was spread out.

Spreading of stabilized sludge from the Plovdiv wastewater treatment plant

Stage 4. Grassing

During the last stage of recultivation, the surface is grassed with grass mixtures.

The measures taken in respect to the soils contaminated by tailings are directed to the following:

Cleaning of the soil surface from the blown tailings. The dredging was carried out at a depth of about 15 cm.

Melioration of the contaminated soils with lime materials at a rate of 200 kg/dka to reduce the harmful effects of the acidic reaction of the environment.

Organic matter (manure) input at a rate of 10 t/dka in order to provide a nutrient medium for plants. After the spreading of the manure, deep ploughing was made at a depth of no less than 20–25 cm.

Sowing of grass mixtures

In the selection of the grass species composition, the specific soil-climatic and temperature conditions in the area of the site, the altitude, the type of terrain, etc. were considered.

Status of the Site Following the Implementation of the Recultivation Activities

The combined use of sludge from WWTP in Plovdiv, after encapsulation of the surfaces against dust pollution with soil cement polymer and properly selected plant propagating material, has formed a tight sealing coat (upper shielding layer), which reduces the amount of rainwater leachate to zero, diminishes the piezometric surface and sharply reduces the amount of drainage water and the quality of surface drainage.

The encapsulation and recultivation activities have led to a cessation of dusting, stabilization of slopes and creation of sustainable perennial herbaceous vegetation consisting of *Onobrychis adbus*, *Dactylis glomerata* and *Bromus inermis* Leyss.

Elshitsa tailings pond before the biological recultivation

Elshitsa tailings pond after the biological recultivation

Regardless of the successfully conducted recultivation events, in accordance to the requirements of Ordinance No. 8, the reclaimed terrains (surface and slopes of the Elshitsa tailings pond) are not envisaged for use with agricultural purpose. This is due to the fact that by the moment of recultivation there are amounts of heavy

Table 4.9 Chemical and agrochemical characteristics of sludge from wastewater treatment plant in Plovdiv and maximum permissible concentration (MPC) of heavy metals

Indicators	Belt filter presses	Desiccant fields	Depot	MPC
Absolute dry matter %	35,89	31,34	84,19	–
Ash substances %	40,51	48,95	68,06	–
Ammoniacal N %	0,05	0,01	0,02	–
General N %	1,50	1,80	1,07	–
General P_2O_5%	0,64	0,24	0,57	–
General K_2O %	0,12	0,14	0,11	–
Na %	0,08	0,06	0,05	–
Ca %	2,05	2,35	2,90	–
Zn mg/kg	3122	3162	1673	3000
Cu mg/kg	523	310	214	1600
Mn mg/kg	317	319	249	–
Co mg/kg	6	5	1	–
Cd mg/kg	63	29	4	30
Ni mg/kg	449	125	252	350
Pb mg/kg	21	105	72	1000
Cr mg/kg	1468	1291	561	500
pH (H_2O)	6,8	6,9	6,7	–

metals exceeding the MPC for soils in the sludge from the WWTP in Plovdiv (Table 4.9).

The analysis of the plant production from the site indicates that the values of heavy metals are in safe concentrations (Table 4.10).

Elshitsa tailings pond after biological recultivation. External dump built up after extraction of copper ore (at the bottom)

Table 4.10 Content of heavy metals in the plant production from the Elshitsa tailings pond

Element	Value mg/kg	MPC mg/kg
Cu	20,7	280
Pb	1,68	80
Zn	76,8	370
Cd	1,31	3,0
Ni	3,0	70

4.3.5 Recultivation of Sedmochislenitsi Tailings Pond

Similar technological solutions are suitable for application in the recultivation of tailings ponds built up within the ore mining process. An example of this is the project for rehabilitation of the Sedmochislenitsi tailings pond, located 7 km westwards from Vratsa (Scheme 4.3). It is designed for disposal of industrial waste from the enrichment of the lead-zinc ore from the mine of the same name, located 2 km away. It is built in the bed of the small mountain river of Medna with a designed length of about 450 m and covers an area of about 35,000 m^2. The collection volume of the tailings pond is about 500,000 m^3.

Contemporary state of "Sedmochislenitsi" tailings pond

During the period 2003–2004, a team of the Institute of Soil Science "Nikola Pushkarov", together with colleagues from the University of Sassari, Sardinia, carried out a survey of the territory disturbed and contaminated by the waste materials deposited in the "Sedmochislenitsi" tailings pond.

Scheme 4.3 Sedmochislenitsi tailings pond

The main objectives of the survey were related to:

(a) Assessment of the waste (tailings) and the soil contamination caused by the accumulation of heavy metals as a result of the accelerated erosion of the available tailings
(b) Assessment of the spatial distribution (three-dimensional) of the heavy metals in the area of the site and along the Leva River flow
(c) Development of technological solutions for the restoration of the environmental and landscape components

During the working process, samples were collected from the deposited tailings and from the terrains located along the river of Leva downriver from the tailings pond site.

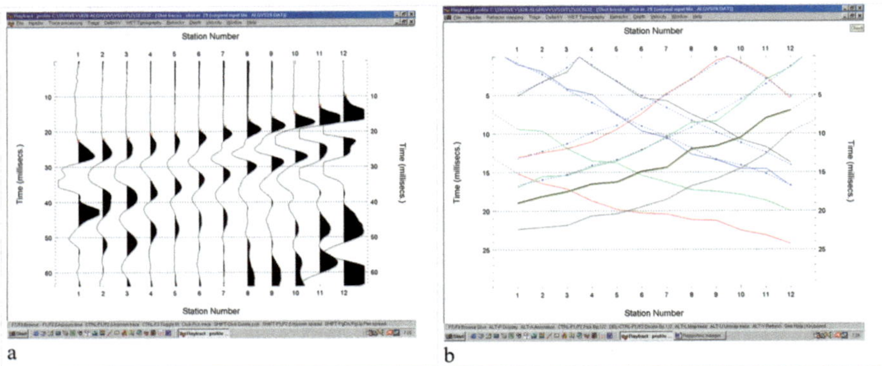

Fig. 4.1 Seismic determination of the tailings quantity

By special seismic testing methods, it was found that around 300,000 m^3 of waste materials (tailings) are still available in the tailings pond body (Fig. 4.1).

The analytical results obtained (Table 4.11) show that tailings deposited in the tailings pond are characterized by a neutral to slightly alkaline reaction of the medium and a low organic carbon content. The analytical determination of 43 pcs. of surface samples (0–10 cm) from the tailings (Table 4.12) shows that the heavy metal concentration of zinc (Zn), plumbum (Pb), copper (Cu), arsenicum (As), cadmium (Cd) and stibium (Sb) is extremely high. The deep sampling and the analytical measurements for the vertical distribution of the studied elements indicate a significant accumulation of Pb, Zn, Sb and Cu in the lower layers of the deposited materials (Table 4.13). The content of cadmium (Cd), chromium (Cr) and nickel (Ni) does not show significant differences by depth of the profile studies.

The laboratory analyses of samples collected along the Leva River flow show anthropogenic impact (heavy metal content exceeding the background values), resulting from the washing of the deposited tailings during the annual spills of the water flow.

As a result of the conducted field survey, analytical laboratory measurements and comparative characterizations, there were designed activities for inclusion of the territory of Sedmochislenitsi tailings pond to the surrounding landscape, which can be outlined in the following:

1. Construction of a correction of the water flow passing through the tailings pond body

   The activity aims to derive the slope waters out of the site territory. This way the erosion processes, which would lead to the transportation of the waste and contamination of the adjacent terrains, will be stopped.

   Additionally, outflow-drainage ditches will be built along the tailings pond body.
2. Surface levelling of the tailings pond

   In the course of this activity, an inclination of the surface part of the site will be formed towards the slope, where the rainwater will be collected in the already constructed outflow-drainage ditches. This way the waste material will gather at the top of the tailings pond.
3. Construction of an earth embankment barrier wall

Table 4.11 Physicochemical characteristics of collected samples

Profile	pH in H_2O	Organic carbon (g/kg)
Pr.-1-A	7,7	12
Pr.-1-B	7,4	11
Pr.-1-C	7,5	11
Pr.-2-A	7,8	12
Pr.-2-B	7,6	14
Pr.-2-C	7,3	13
Pr.-2-D	7,0	9
Pr.-3-A	7,3	11
Pr.-3-B	7,3	7
Pr.-3-C	7,7	6
Pr.-3-D	7,5	8
Pr.-3-E	7,2	9
Pr.-4-A	7,5	18
Pr.-4-B	7,5	23
Pr.-4-C	7,9	12
Pr.-4-D	8,0	13
Pr.-4-E	7,6	13
Pr.-5-A	7,3	12
Pr.-5-B	7,5	13
Pr.-5-C	7,2	13
Pr.-5-D	8,0	11
Pr.-5-E	7,3	12
Pr.-5-F	7,2	11
Pr.-6-A	7,6	10
Pr.-6-B	7,7	13
Pr.-6-C	7,7	11
Pr.-6-D	7,6	11
Pr.-6-E	7,7	13
Pr.-6-F	7,7	15
Pr.-7-A	7,6	11
Pr.-7-B	7,7	11
Pr.-7-C	7,8	11
Pr.-8-A	7,8	8
Pr.-8-B	**7,7**	**9**
Pr.-8-C	7,8	7
Pr.-9-A	7,9	9
Pr.-9-B	7,9	37
Pr.-9-C	7,9	34
Pr.-10–1-0-10	7,9	12
Pr.-10–3–0-10	7,7	6
Pr.-10–4–0-10	7,7	15
Pr.-10–5–0-10	7,6	10
Pr.-11–1–0-10	7,8	39
Pr.-11–3–0-10	7,1	17
Pr.-11–4–0-10	7,8	9
Pr.-11–5–0-10	7,8	8

(continued)

Table 4.11 (continued)

Profile	pH in H_2O	Organic carbon (g/kg)
Pr.-12–1–0-10	6,1	104
Pr.-12–3–0-10	7,1	124
Pr.-12–4–0-10	7,8	21
Pr.-12–5–0-10	7,8	30
Pr.-12–6–0-10	7,6	43
Pr.-13–1–0-10	7,9	13
Pr.-13–3–0-10	7,6	26
Pr.-13–4–0-10	7,5	12
Pr.-13–5–0-10	7,9	26

Table 4.12 Content of heavy metals at depth of 0–10 cm

	As	Cd	Cr	Cu	Ni	Pb	Zn	Sb
	mg/Kg	mg/Kg	mg/Kg	mg/Kg	mg/Kg	mg/Kg	mg/Kg	mg/Kg
Min	274,36	4,06	5,06	148,85	50,52	304,84	406,46	4,89
Max	1028,33	23,16	18,29	2692,79	132,10	2887,04	6121,18	28,19
Average	648,98	13,33	11,26	750,83	93,48	1556,94	3173,00	12,19

4. Construction of a clay screen

   The activity aims to cease the drainage water's contact with the deposited waste materials.
5. Construction of a drainage system.
6. Loading a layer of geological materials

   The thickness of the layer should be no less than 1.5 m, thereby providing the opportunity for growth in depth of the roots of the vegetation planted as well as protecting the drainage layer and the clay screen from damage.

   The geological materials must meet the following requirements:

   – Should not contain toxic amounts of heavy metals, salts, petroleum products, etc.
   – Should be suitable as physicochemical characteristics for arboreal and herbaceous vegetation
   – If possible, should be located close to the site in order to reduce transport costs
7. Spreading of a superficial recultivation layer

   Spreading of a mixture of geological materials and sludge from the WWTP in Vratsa in a ratio of 1:1.

   By implementing this activity, favourable conditions will be provided for the cultivation of appropriate arboreal and herbaceous vegetation.
8. Growing of arboreal and herbaceous vegetation

   In order to fully integrate the disturbed and polluted area into the surrounding landscape, it is necessary for the technically constructed terrain to be grassed and forested.

   In the selection of plant species, account should be taken of the prevalent vegetation in the site's area, which should be used for afforestation and grassing. In this way full integration and full balance between the technogenic terrain and the surrounding area will be achieved.

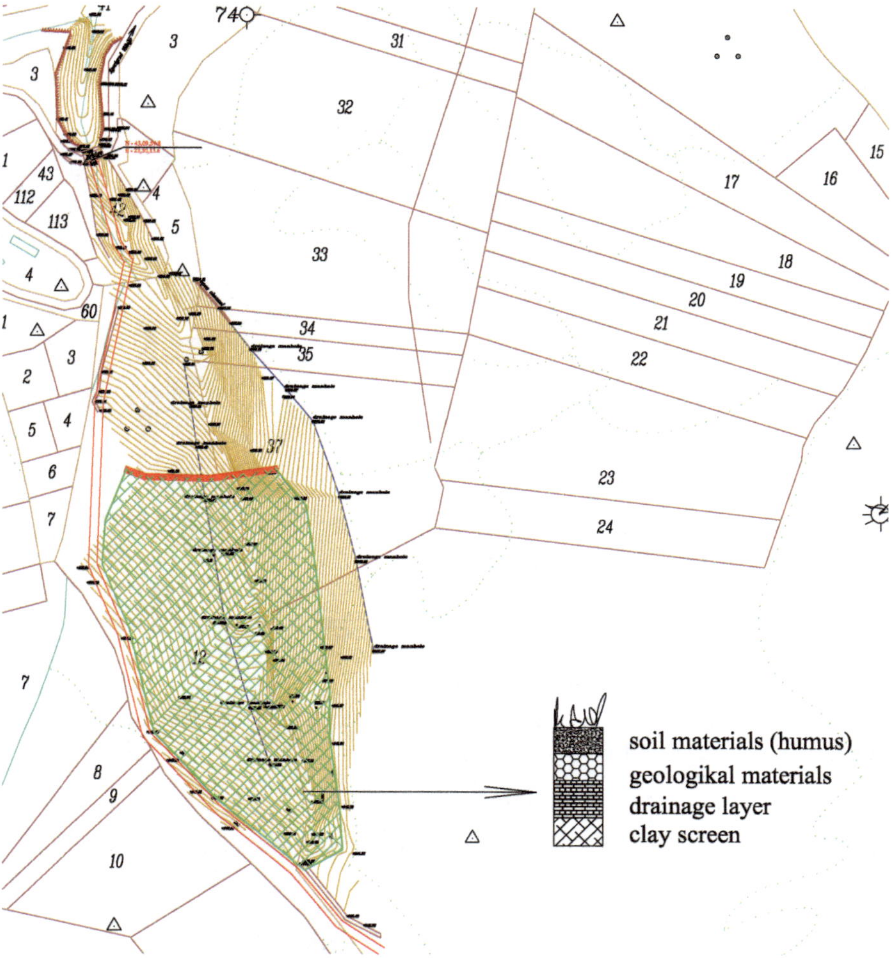

Schematic presentation of the planned activities for recultivation and reclamation of Sedmochislenitsi tailings pond

4.4 Conditions for the Environmentally Sound Use of Sludge on Potentially Suitable Terrains

4.4.1 Specification of the Current Land Use

This can be done at the respective municipalities, where the materials from the land commissions are kept. There is description of each land property as well as its location, and there is a sketch with the category of land and its neighbours. This is an important condition in regard to establishing a contact with landowners and concluding agreements on the use of sludge.

Table 4.13 Content of heavy metals by depth of profile

	As	Cd	Cr	Cu	Ni	Pb	Zn	Sb
Sample depth	mg/Kg	mg/Kg	mg/Kg	mg/Kg	mg/Kg	mg/Kg	mg/Kg	mg/Kg
0–10	606	8	<5	216	91	949	1313	11,5
10–20	508	7	<5	196	81	934	1279	10,4
20–30	510	5	<5	191	82	979	1224	10,2
30–40	611	6	<5	215	92	1151	1425	15,4
40–50	510	6	<5	225	92	1142	1448	15,6
50–60	510	7	<5	248	82	1102	1664	14,1
60–70	614	9	<5	269	82	1238	1657	14,5
70–80	509	11	<5	318	71	1212	1813	14,6
80–90	509	16	<5	516	71	1425	1832	10,9
90–100	507	15	<5	504	71	1430	1846	9,9
100–110	508	15	<5	494	81	1219	1808	9,9
110–120	508	11	<5	427	81	975	1838	10,3

4.4.2 Specification of the Capacities of the Areas and the Sites as Well as the Term of Their Use

A preliminary conversation is held with the chairman of the cooperative where the selected area or site is located. It provides information on the history of the field, including the type and amount of mineral or other fertilizers used, the previously grown culture, the treatments, the crop rotation, etc. The terms of lease, the use of sludge, etc. are determined.

4.4.3 Designation of Access, Transport Links and Distances

During the selection of a cultivated area, the area's distance to the treatment plant must be taken into account, on one hand, as well as the existing transport links, on the other. A more favourable condition is when the roads do not pass through central districts or major motorways. The greater distance is associated with additional fuel and lubricant costs, etc.

4.4.4 Determination of the Topographical Status of the Area, the Site Relief and the Soil Conditions

This is an important issue, including the flatness of the terrain, the presence of hilly, eroded territories, the proximity to communication links, etc. Soil conditions determine the content of organic matter, the presence of background heavy metal content, the stock of basic nutrients, etc.

In order to fully address the issue of the environmentally sound use of Sofia City's wastewater treatment plant sludge in agriculture as a fertilizer or for recultivation, a project should be developed, based on the legal norms including sludge assessment, soil type, area dimensions, transport communications, technological variants of sludge utilization, technological regulations on crop cultivation, economic and ecological effect, etc.

4.5 Contemporary Guidelines for the Use of Recultivated Areas

In the presence of sufficiently large arrays with a certain relief configuration, geological substrates with appropriate physicochemical characteristics and humic materials with high natural fertility, it is possible to construct recultivated areas that satisfy the initial requirements for conducting purposeful agriculture.

In this respect, an alternative for cultivation is provided by oil plants (Banov and Tsolova 2007; Yadav et al. 2017a). The latter, apart from being unpretentious in terms of soil and climatic conditions, find a good market among the European Union countries. The reasons for this are numerous, namely:

- Favourable soil and climatic conditions in the country for the production of oil-bearing and essential oil crops.
- During the plant processing, the heavy metals accumulated in the root and foliage mass do not separate into the obtained oils, which allows for the inclusion of the recultivated lands into the agricultural land of the country without additional costs for their melioration and improvement.
- Protection of the recultivated terrains from the development of erosive processes.
- Profitability – good yields are obtained at a relatively low cost per area unit.
- Existence of long-standing traditions in the cultivation and processing of essential oil plants.
- Significant reduction of the wild-growing species in their natural habitats.
- Demands of international client companies to cultivated species and varieties

In connection with the accession of Bulgaria to the EU, the expansion of oil crops can be assessed as a new strategic direction in the national agrarian program. It is about a promising agricultural production and a relatively empty market niche in the European and world markets. Undoubtedly, the realization of such projects is a direct response to a priority objective for the country's development in a near future. In this respect, the technogenically recultivated farmland built up within mining and ore processing can be seen as a unique opportunity to produce suitable bio-oil crops with economically significant and safe energy reproducibility.

The abovementioned allows to consider as an example the opportunities for rapeseed and lavender cultivation on recultivated lands on the territory of the Republic of Bulgaria.

Lavender is an essential oil plant that occupies an important place in the production of a range of products, primarily in perfumery, painting and ceramics, mainly as a component of soaps, cosmetics, varnishes and paints.

Meanwhile it is also an excellent honey-bearing plant; the honey obtained from it is characterized by excellent taste and healing qualities.

The specific geographic, climatic and soil conditions in the Republic of Bulgaria contribute to the formation of several specialized belts for growing the plant, such as those of Karlovo, Kalofer, Strelcha, Zelendol, Rozino and Razlog. In smaller areas and volumes, lavender is grown in other parts of the country as well. Continuous selection and improvement of the quality of production lead to gradual expansion of the lavender range here and today it is cultivated in numerous locations in Bulgaria. The latest studies of leading experts prove the good adaptability of the plant in the region of Northeastern Bulgaria, mainly around Alfatar and Dulovo.

The most suitable soils for its cultivation are those with a neutral and alkaline reaction (pH in water of 6–8). Plants do not tolerate shallow underground water, as on unlevelled plots where water is retained they quickly dry out and perish.

The world production of lavender oil is about 1400 t per year, with a relative share of 75% for lavandin, 20% for lavender and 5% for d'aspic sort.

The major application of lavender oil is in perfumery as a component of lotions, toilet waters and perfumes, most often in combination with other components.

The stabilization of the Bulgarian cosmetics industry and the opening of new foreign market niches imply the increase of the necessary quantities of it.

Another prospective culture is **rapeseed**. Its seeds have a high oil content. Rapeseed oil belongs to the group of technical semi-drying oils. In industry, it has a wide range of applications: in metallurgy and paint, varnish, leather, textile and soap production as well as production of nitroglycerine. In the refinement process the oil is also used for industrial purposes. An important by-product of oil production is the oilcake. It contains on average 33% protein and 10% fat and is used to feed livestock.

The major spread area of rapeseed in Bulgaria is North Bulgaria, mainly Vratsa, Pleven and Tyrnovo districts.

According to its climate requirements, the winter rapeseed, which is prevalent in Bulgaria, is close to the cereal crops. It survives winter frosts provided it is well rooted in the autumn. At a young age and early in the spring, it may freeze.

Winter rapeseed can also be grown on weak and shallow soils and even on marl soils. It finds the most favourable development on chernozems. For the poorer soils it is necessary to apply mineral fertilization or such with manure. Rapeseed is not suitable for skeletal, gravel or sandy soils as well as for salinized ones.

This gives us a reason to recommend, in order to restore the fertility of technogenically reclaimed lands and their return to the country's arable land fund, to grow essential oil crops such as rapeseed and lavender. This is an appropriate solution for the following reasons:

1. The soil-climatic conditions in the Republic of Bulgaria are favourable for the development of this type of plants. General and preliminary assumptions are made that the large technogenically disturbed areas around the open mines in Bulgaria can also be used for their cultivation.

2. Rapeseed and lavender are characterized by well-developed root systems and therefore are suitable for planting as anti-erosion crops, facilitating also the accelerated development of soil-forming processes in technogenically disturbed areas.
3. In the processing of plants, the heavy metals accumulated in the roots, stems and leaves do not concentrate in the extracted oils. This way an ecologically clean bioproduct is obtained from contaminated sites. This makes it possible to include the technogenically recultivated lands in the country's arable land without extra costs for their melioration and improvement.
4. Essential oil crops present a serious alternative for the diversification of crops in the conditions of implementation of the sustainable agriculture principles in the Republic of Bulgaria.

4.6 Monitoring of Reclaimed Land and Soils

The recultivation substrate and the crop production need to be periodically monitored, and it is only when the relevant regulatory requirements are met that the latter should be used for animal feed and for food.

4.6.1 Essence of Monitoring

Monitoring is a systematic observation of parameters, which allows forecasting and regulation of the ongoing processes in ecosystems. It mainly includes the following monitoring parameters: air and air emissions, aquatic environment and related processes, soil, physical impacts, biological and chemical impacts and human impact (standard of living, state of the local economy, working conditions including exposure to dangerous substances, etc.).

4.6.2 Required Monitoring Techniques and Procedures

1. Tool equipment – to be adequate to the monitored processes
2. Applicable standards – according to the Bulgarian State Standard (BDS) and ISO
3. Data quality control – selection and enforcement of standardized samples
4. Requirements for personnel and personnel training

4.6.3 Accountability

1. To the local and national structures
2. To the international structures

4.6.4 Reason for Conducting Soil Monitoring

Soil monitoring is a systematic observation, assessment and forecast of the quality changes of soil from a given territory over time as a result of anthropogenic activity and/or natural processes. It has a strong geographic component and can therefore be considered as a specialized geographic information system (GIS) for early detection of adverse or undesirable changes in soil quality in order to regulate the processes that lead to pollution, degradation or demolition of the soil cover.

The major issues that determine the need for comprehensive soil information to create a certain policy in the area of the recultivation site are:

- The impact of climate change
- The impact of pollution (diffusely and in specific points)
- The impact of agriculture
- The soil degradation – desertification, erosion, salinization, pollution and acidification
- The knowledge about the background soil conditions

In order to provide adequate information, it is important to:

- Identify the priorities of the soil problems.
- Identify the available data and the missing ones.
- Provide appropriate information from other aspects of the environment and the economic impacts.
- Provide advantages for more comparable measurements of changes as well as their assessment and calculation over a large geographic area.

Compared to the systems for monitoring and control of the state of atmospheric air, of surface waters and of groundwater, soil monitoring is relatively limited and insufficiently developed. Soils present a less dynamic and better-defined buffer system than air and water. Tracking changes in soil quality does not require frequent sampling and analysis. Soils however vary spatially and in depth, requiring application of a specific sampling system. Moreover, they accumulate for continuous periods the adverse effects of the anthropogenic activity (e.g. accumulation of pollutants, compounds that cause acidification and/or salinization, which directly or indirectly affects the quality of the other environmental components that are in contact with them). For this reason, it is assumed that systematic monitoring of soil quality is now as necessary as for the other environmental components.

Where information about local or diffuse pollution caused by heavy metals or organic pollutants is available, it must be ensured that the data are representative of the site, the soil type and the land use mode. The assessment should be based on conditional soil care, based on modern technical knowledge. As a consequence, a concept for the assessment procedure should also be developed. The following four points outline the four major steps in the soil assessment process:

(a) Identification of relevant indicators and priorities
(b) Evaluation of the indicators, using data collected through regular monitoring or specific surveys
(c) Ratification of the results achieved
(d) Comparison with planned values, threshold values and previous values (if any)

4.7 Conclusion

The reclamation of disturbed terrains is a complex problem determined by a number of factors: geo-climatic, mining and technical conditions, mechanical and agro-chemical properties of the soil and geological material extracted, the system of development of the field, etc. This requires that the technological solutions for restoring the fertility of the disturbed terrains should be tailored to each country's specificities and coordinated efforts of specialists from different scientific and development organizations and institutes should be applied.

The main task in the reclamation of lands disturbed by mining or other activities involves restoration of soil fertility and reclamation of the disturbed land for use by creating an ecologically balanced system. Recultivation is one of the most radical methods of restoration and improvement of the disturbed terrains and their recovery into the full (cultivated) land fund.

The conducted long-term laboratory, vegetative and field studies have shown that recultivation activities are accompanied by various technical, practical and financial difficulties and problems, namely:

1. The use of humus material in the technical construction of recultivated terrains leads to a relatively fast recovery of soil fertility but is connected with spending of significant financial resources.
2. The biological stage of recultivation is very long, accompanied by planting of a large number of cultures, which require the implementation of strictly defined systems of fertilization, plant protection and agro-technology.
3. The two stages of recultivation (the technical and the biological) require spending of large capital investments.
4. The technically constructed recultivated areas are characterized by different unfavourable features – heavy mechanical composition, low natural fertility, heavy metal content above MPC, toxic levels of the reaction (pH), etc.
5. There is a significant discrepancy between the time limits for the implementation of the biological recultivation regulated by Ordinance No. 26 on Recultivation of Disturbed Terrains, Improvement of Low Productive Lands and Removal and Utilisation of the Humus Layer and the real terms for which the recovery of soil fertility of the damaged terrain is completed. According to the requirements of the normative documents, the duration of the biological recultivation is 5 years, while the actual time reaches in some cases (humuslessly recultivated lands) up to 12–16 years.

6. There is no methodology for assessing the fertility of recultivated soils, which hinders the process of returning the lands to their real owners.
7. Technically built tailings dams during the first years of the mining operations require additional technical and biological activities to restore their agro-soil potential.

The solution to these problems is possible through the development and implementation of new, efficient and rational technological solutions to restore the fertility of anthropogenic soils. This is due to the fact that the available quantities of humic materials are in exhaustion, while at the same time large amounts of waste materials of different origins accumulate. For these and other reasons, various organic residues recently have found extremely wide use in the recultivation of lands disturbed by mining of mineral resources. The high content of essential nutrient macro- and microelements in the residue, and especially of carbon accessible for plants, provides a high meliorative effect, especially valuable in the case of impaired balance of the organic matter.

The use of sludge makes it possible to overcome a number of disadvantages related to the recultivation of disturbed soils, namely:

(a) Balanced nutrition of the plants grown is ensured and the organic reserve of soils increases.
(b) The content of organic matter greatly facilitates the treatment of recultivated terrains and allows normal development of the plants' root system.
(c) The physical and the physical-mechanical characteristics of the recultivated soils are improved.
(d) The term for implementation of the biological stage of recultivation is reduced.
(e) The necessary investments for the implementation of the recultivation activities are substantially reduced.
(f) The process of returning the land to their real owners is facilitated.

In connection with the abovementioned, some particular technological solutions were developed and implemented (the "Elshitsa" tailings pond) in order to restore the fertility of lands and soils disturbed by the mining industry, by construction of tailings ponds, landfills for solid and household waste, etc.

Individual Variants of Reclamation Include

Corrections of Water Currents Passing Through the Recultivated Terrain
The activity aims to derive the slope waters out of the site territory. This way the erosion processes will be stopped.

Surface Levelling of the Tailings Pond
In the course of this activity, an inclination of the surface part of the site will be formed towards the slope, where the rainwater will be collected in the already constructed outflow-drainage ditches.

Construction of a Clay Screen

The event aims to cease the drainage contact with the deposited waste materials.

Construction of a Drainage System

Loading a Layer of Geological Materials

The thickness of the layer may vary from 0.50 m to 1.5 m, depending on the type of vegetation grown. Geological materials must meet the following requirements:

(a) Do not contain toxic amounts of heavy metals, salts, petroleum products, etc.
(b) Should be suitable as physicochemical characteristics for arboreal and herbaceous vegetation
(c) If possible, should be located near the site in order to reduce transport costs

Spreading a Surface Recultivation Layer

The surface reclamation layer can be constructed by:

(a) A mixture of soil and geological materials (humus)
(b) A mixture of geological materials and sludge from WWTPs in a ratio of 1:1
(c) Creation of a self-contained surface layer of sludge

Growing of Herbaceous and Arboreal Vegetation

Within the selection of plant species, account should be taken of the prevalent vegetation in the site's area, which should be used for afforestation and grassing. In this way full integration and full balance between the technogenic terrain and the surrounding area will be achieved.

References

Ashoka P, Meena RS, Kumar S, Yadav GS, Layek J (2017) Green nanotechnology is a key for eco-friendly agriculture. J Clean Prod 142:4440–4441

Banov M (1999) Requirements concerning reclamation of lands disturbed by surface coal mining. J Balkan Ecol 2(1):81–84

Banov M, Hristov Bl (1996a) Nyakoi printsipi za klasificatsiya, nomenklatura i diagnostika na rekultivirani pochvi bez humusno sadarzhanie ot raiona na "Maritsa-iztok". Pochvoznanie, agrohimiya i ekologiya (3):26–30

Banov M, Hristov B (1996b) Teoretichni i prakticheski problemi pri rekultivatsiya na narushenite zemi v Balgariya. Konferentsiya "Reformi i ekologichna politika v selskoto stopanstvo" 2:7–50

Banov M, Marinkina V (1997) Soil-forming changes in reclaimed lands. I. Microbiological changes. Problems concerning genesis, evolution, use and protection of soil from the Eastern Region of Romania, Bucharest

Banov M, Markov E (1999) Savremenni problemi pri rekultivatsiya na zemite v raiona na Iztocnomarishkiya kamenovaglen baseyn. Pochvoznanie, agrohimiya i ekologiya (4–5):132–135

Banov M, Pavlov P (2014) Izsledvane i vazstanovyavane na tereni, narusheni pri promishlena deynost. Annual of the University of Mining and Geology "St. Ivan Rilski", Part II, Mining and Mineral Processing. Izdatelska kashta "Sv. Ivan Rilski", ISSN 1312-1820, pp 195–199

Banov M, Pavlov P (2015) Fiziko-himichna harakteristika na nasipishta s geologichni materiali, izgradeni pri dobiv na medna ruda. Annual of the University of Mining and Geology "St. Ivan Rilski", Part II, Mining and Mineral Processing. Izdatelska kashta "Sv. Ivan Rilski", ISSN 1312-1820, 58:38–43

Banov M, Tsolova V (2007) Perspektivi za otglezhdane na tehnicheski kulturi na tehnogenno rekultivirani pochvi v Balgariya. Pochvoznanie, agrohimiya i ekologiya, ISSN 0861-9425, XLI(2):10–16
Banov M, Tsolova V, Ivanov P, Hristova M (2010) Anthropogenically disturbed soils and methods for their reclamation. Agric Sci Technol 2(1):33–39
Banov M, Pavlov P, Blagiev M (2015) Restoration of land damaged in mining activities. In: Proceedings of the 24th international mining congress of Turkey, pp 1425–1431
Bengtson GW, Allen SE, Mays DA, Zarger TG (1973) Use of fertilizers to speed pine establishment on reclaimed coal-mine spoil in Northeastern Alabama: I. Green-house experiments. In: Ecology and reclamation of devastated land, vol 2. Gordon and Breach, New York, pp 199–225
Benvenuti M, Mascaro I, Corsini F, Lattanzi P, Parrini P, Tanelli G (1997) Mine waste dumps and heavy metal pollution in abandoned mining district of Boccheggiano (Southern Tuscany, Italy). Environ Geol 30(3/4):238–243
Berg WA, Vogel WG (1973) Toxicity of acid coal-mine spoils to plants. Ecology and reclamation of devastated land, vol 1. Gordon and Breach, New York
Boulet MP, Larocque ACL (1998) A comparative mineralogical and geochemistrical study of sulfide mine tailings at two sites in New Mexico. USA Environ Geol 33(2/3):130–142
Buragohain S, Sharma B, Nath JD, Gogaoi N, Meena RS, Lal R (2017) Impact of ten years of bio-fertilizer use on soil quality and rice yield on an inceptisol in Assam, India. Soil Res 56:49. https://doi.org/10.1071/SR17001
Dimitrova A, Banov M, Hristov Bl (1998) Topsoil storage effects on microbial activity. Ninth Congress of the Bulgarian Microbiologists 2:216–219
Dinev N, Banov M, Kutev V, Filcheva-Konisheva E, Georgiev B, Mitova I (2012a) Monitoring na pochvite i ustoychivo upravlenie na zemite v zemlishtata na s. Yana, s. Gorni Bogrov, raion "Kremikovtsi". ISBN 978-954-92592-4-7. Ambrosia NT OOD, 80
Dinev N, Banov M, Mitova T, Kutev V, Mitova I, Georgiev B, Filcheva-Konisheva E (2012b) Ustoychivo izpolzvane i vazstanovyavane na pochvite v raiona na Sofia. ISBN 978-954-92592-3-0. Ambrosia NT OOD, 104
Foy CD (1974) Differential responses of plant genotypes to micronutrients. In: Micronutrients in agriculture. Soil Sci Soc Am Inc, Medison, WI, pp 384–418
Garbuchev Iv, Lichev Sp, Treikyashki P, Kamenov P (1975) Prigodnost na substratite za rekultivatsiya na zemite v raiona na Maritsa-iztok. Sofia
Gencheva S (1995) Klasifikatsiya i nyakoi osobenosti na antropogennite pochvi. Dissertation
Hristov Bl, Banov M, Dimitrova A (1996) Mikrobiologichni harakteristiki na rekultivirani pochvi ot rayona na SO "Maritsa-iztok". Pochvoznanie, agrohimiya i ekologiya (1):37–39
IUSS Working Group WRB (2006) World reference base for soil resources (2006), World Soil Resources Reports No. 103, 2nd edn. FAO, Rome, p 128
Ivanov P, Banov M (2008) Klasifikatsiya na rekultivirani pochvi v zavisimost ot tipa zemepolzvane. Pochvoznanie, agrohimiya i ekologiya XLII(3):11–17
Jambor JL (1994) Mineralogy of sulfide-rich tailings and their oxidation products. In: Environmental geochemistry of sulfide mine-wates, MAC short cours, 59
Koinov V, Trashliev Hr, Yolevski M, Andonov T, Ninov N, Hadjijanakiev A, Angelov E, Boyagiev T, Fotakiewa E, Krastanov Sl, Staikov I (1968) Soil map of Bulgaria, Scale 1: 400000. (Bg)
Kumar S, Meena RS, Pandey A, Seema (2017) Soil acidity management and an economics response of lime and sulfur on sesame in an alley cropping system. Int J Current Microb and App Sci 6(3):2566–2573
Kumar S, Meena RS, Yadav GS, Pandey A (2017a) Response of sesame (*Sesamum indicum* L.) to sulphur and lime application under soil acidity. Int J Plant and Soil Sci 14(4):1–9
Marinkina V (1999) Prouchvane I vazmozhnosti za rekultivatsiya na sulfidosadarzhashti materiali, polucheni pri otkrit vagledobiv. Dissertation for graduation with the educational and scholarly grade "Doctor". Agricultural Academy, Institute of Soil Science, Agrotechnologies and Plant Protection "N. Pushkarov", Sofia, 231

Marinkina V, Banov M (2000) Comparative estimation of suitability for reclamation of geological materials at "Maritza-Iztok" and "Chukurovo" coal mine districts. First national conference on humus substances and soil tillage, 93–95

Meena H, Meena RS (2017) Assessment of sowing environments and bio-regulators as adaptation choice for clusterbean productivity in response to current climatic scenario. Bangladesh J Botany 46(1):241–244

Meena H, Meena RS, Singh B, Kumar S (2016) Response of bio-regulators to morphology and yield of clusterbean [*Cyamopsis tetragonoloba*(L.) Taub.] under different sowing environments. J App and Nat Sci 8(2):715–718

Meena H, Meena RS, Lal R, Singh GS, Mitran T, Layek J, Patil SB, Kumar S, Verma T (2018) Response of sowing dates and bio regulators on yield of clusterbean under current climate in alley cropping system in eastern U.P. Indian Leg Res 41(4):563–571

Misheva L, Tsolova V, Banov M, Poynarova M, Hadzhiyanakiev Y, Zlatev A, Kozhuharova S, Zhelyazkova S, Vasileva Z (2007) Vazmozhnosti za izpolzvane na zemi i tereni, zamarseni s radionuklidi i tezhki metali. Monography. Publ. National Centre for Agricultural Sciences & Institute of Soil Science "N. Pushkarov", Sofia, 39

Pavlov P, Banov M (2014) Narusheniya na okolnata sreda, predizvikani ot podzemen dobiv na polezni izkopaemi. Proceedings of the VI International Geomechanics Conference, ISSN 1314–6467, 482–486

Pavlov P, Banov M, Ivanov Pl (2015a) Possibilities for reclamation by using suitable geological and waste materials. Proceedings of the 24th international mining congress of Turkey, 1420–1424

Pavlov P, Banov M, Tondera D (2015b) Izsledvane na geolozhki materiali za tselite na rekultivatsiyata. Annual of the University of Mining and Geology "St. Ivan Rilski", Part II, Mining and Mineral Processing. Izdatelska kashta "Sv. Ivan Rilski", ISSN 1312-1820, 58:44–48

Pavlov P, Banov M, Totev L (2015c) Environmental consequences of copper mining. Technische Universitat Bergakademie Freiberg "Freiberger Geotechnik-Kolloquium 2015", ISBN 978-3-86012-517-5, 177–182

Penkov M, Donov V, Boyadzhiev T, Andonov T, Ninov N, Yolevski M, Antonov G, Gencheva S (1992) Klasifikatsiya i diagnostika na pochvite v Balgariya vav vrazka sas zemerazdelyaneto. Zemizdat, Sofia, 151

Safaya NM (1978) Delimation of minerals stresses in mine spoils and sceeming plants for adaptability. In: Ecology and coal resource development: based on the international congress for energy and the ecosystem, vol 2. Pergamon Press, Grand Yorks, ND

Treikyashki P, Hristov Bl (1982) Grupirovka na geologichnite materiali po prigodnost za biologichna rekultivatsiya. Sbornik s materiali ot III Natsionalna konferentsiya po pochvoznanie, Sofia

Tsolova V, Banov M, Hristova M, Kolchakov V (2010) Klasifikatsiya na rekultivirani pochvi ot raiona na mini "Maritsa-iztok". Pochvoznanie, agrohimiya i ekologiya, Nauchno spisanie na Selskostopanska academia XLIV(4):5–11

Yadav GS, Lal R, Meena RS, Babu S, Das A, Bhomik SN, Datta M, Layak J, Saha P (2017a) Conservation tillage and nutrient management effects on productivity and soil carbon sequestration under double cropping of rice in north eastern region of India. Ecol Indic. http://www.sciencedirect.com/science/article/pii/S1470160X17305617

Yadav GS, Lal R, Meena RS, Datta M, Babu S, Das, Layek J, Saha P (2017b) Energy budgeting for designing sustainable and environmentally clean/safer cropping systems for rainfed rice fallow lands in India. J Clean Prod 158:29–37

Yadav GS, Das A, Lal R, Babu S, Meena RS, Patil SB, Saha P, Datta M (2018) Conservation tillage and mulching effects on the adaptive capacity of direct-seeded upland rice (*Oryza sativa* L.) to alleviate weed and moisture stresses in the North Eastern Himalayan Region of India. Arch Agron Soil Sci 64:1254. https://doi.org/10.1080/03650340.2018.1423555

Zheleva El, Bozhinova P (2009) Metodi za remediatsiya na zamarseni pochvi/mesta. Ekologiya i badeshte (1):34–56

Zheleva-Bogdanova El (2010) Rekultivatsiya na narusheni tereni. PSSE. Sofia, 411

Relevance of Microbial Diversity in Implicating Soil Restoration and Health Management

5

Sunita Devi and Ruchi Soni

Abstract

Soil comprises three interconnected factors responsible for its fertility including physical, chemical, and biological. The soil fertility depends upon the diversity of living microorganisms in the soil and their interaction with other physico-chemical components, which accounts for their higher complexity and dynamic behavior. It has been documented as the well-understood component for soil fertility. Along with maintaining the soil fertility, soil microorganisms also impart essential roles in the nutrient biogeochemical cycles that are the fundamentals of life on the earth. A small amount of soil exhibits a great deal of microbial diversity, which includes bacteria, actinomycetes, fungi, algae, and protozoa. Bacteria comprise dominating population in the soil followed by actinomycetes, fungi, algae, and protozoa. It has been reported that one gram of soil may contain 10^9–10^{10} prokaryotes including bacteria-archaea and actinomycetes, 10^4–10^7 protists, ~100 m of fungal hyphae, and 10^8–10^9 viruses. The rhizosphere, a narrow zone influenced by plant roots, provides an active habitat for abundant microbes and is considered as one of the most complex ecosystems on the earth. To improve soil health and plant growth performance, it is important to know about the occurrence of diverse microbes and their behavior and role in the rhizosphere microbiome. Moreover, the ability of root exudates for mediating plant–microbe and plant–microbiome interactions could maintain agricultural practices sustainable. This chapter explores the utility and functioning of soil microbial diversity in terms of its agricultural relevance and subsequent increased crop production so that the growing world population scenario could conquer.

S. Devi
Department of Basic Science, College of Forestry, Dr. YSP University of Horticulture & Forestry Nauni, Solan, Himachal Pradesh, India

R. Soni (✉)
Regional Centre of Organic Farming, Nagpur, Maharashtra, India
e-mail: ruchihpau1987@gmail.com

R. S. Meena (ed.), *Soil Health Restoration and Management*,
https://doi.org/10.1007/978-981-13-8570-4_5

The microbial population has not been promoted effectively in agricultural practices till date because several beneficial soil microbes are still not explored. So, the chapter insights the various modern molecular tools that will provide an opportunity to discover new species currently unknown to science.

Keywords
Microbial diversity · Metagenome · Plant growth · Plant nutrition · Soil health management

Abbreviations

SOM	Soil organic matter
DOM	Dissolved organic matter
C	Carbon
CLPP	Community level physiological profile
SCSU	Sole-carbon-source utilization
CSUP	Carbon substrate utilization profiles
TTC	Triphenyl tetrazolium chloride
DGGE	Denaturing gradient gel electrophoresis
MPN	Most probable number
FAME	Fatty acid methyl ester
PLFA	Phospholipid fatty acid
DNA	Deoxyribonucleic acid
ARDRA	Amplified ribosomal DNA restriction analysis
Co	DNA concentration
t	Incubation time
MATE	Multidrug and toxic compound extrusion
PGPR	Plant-growth-promoting rhizobacteria
QS	Quorum sensing
IVET	In vivo expression technology
AMF	Arbuscular mycorrhizal fungi

5.1 Introduction

Soil is an active/incarnate and natural entity; it is essential for the proper and balanced functioning of agroecosystems. It comprises complex and natural ecosystem for numerous and diverse kinds of microbes including bacteria, actinomycetes, fungi, protists, and algae (Muller et al. 2016). Their comparative population count studies in some soils interpret that prokaryotic biomass exceeds greater than 5 tons per hectare, whereas fungal biomass varies from 1 to 15 tons (Brady and Weil 2014;

Yadav et al. 2018b). The soil with adequate precise surface area of 10^1 m^2 g^{-1} possessing bacterial density of 10^{10} cells g^{-1} and 1000 cells per colony estimated to be consistently disseminated over the surface would produce average spacing of 500 μm radius between adjacent colonies (Raynaud and Nunan 2014). The significance of soil microbial diversity in maintaining fertility and the correlation of biological accomplishments with soil physicochemical properties along with their role in sustainability of the farming systems is well studied nowadays (Madsen 2005; Manlay et al. 2007; Ashoka et al. 2017). Microbial diversity is considered as a major parameter for measuring the soil health (Nielsen and Winding 2002). These biological indicators including microbial biomass, microbial diversity, soil respiration, and different enzyme activities in the soil are the basics of soil health (Brussard et al. 2007). Biotic activities mediated by microbes are responsible for degradation of rocks and minerals and subsequent formation of the soil. It provides the medium for the growth and development of these microbes besides accomplishing all their nutritional requirements. Some physical characteristics and amount of soil organic matter (SOM) regulate the dynamics of microbial diversity, which further indicates the soil health and plant growth (Cynthia et al. 2016). SOM while contributing to soil fertility by maintaining microbial populations and increasing water retention also imparts a huge impact on the growth and development of the soil vegetation. The dynamics of microbial communities in the soil and rhizosphere may vary, and consequently, plant–microbe interactions are highly specific and defined depending on co-evolutionary pressures. The plant exudes root exudates, which behave as signalling molecules and are utilized by soil microbes as substrates (Huang et al. 2014; Meena and Meena 2017). Plant–microbe interactions in the rhizosphere facilitate a wide range of dynamic ecosystem processes coupled with carbon repossession and nutrient cycling. The microbial diversity plays a significant role in plant growth and development besides maintaining soil health. In the past decades, genomics studies revealed that the genomes of soil bacteria are comparatively larger and contain more number of genes than those bacteria genomes of aquatic or clinical environments (Land et al. 2015). The largest whole bacterial genome of 14.8 Mb and 11599 genes belongs to the *Myxobacterium sorangium* cellulosum (Han et al. 2013). Distinctive soil genera including *Pseudomonas* and *Streptomyces* generally possess genomes >6 Mb. Plants exhibit a large number of interactions with soil microbes, which comprises diverse ecological opportunities including competitive, commensal, neutral, and mutualistic. Based on molecular characterization and sequence analysis studies, it has been reported that the soil and rhizospheric niches act as hotspots of microbial diversity and ecological richness, with plant roots accomplishing diversified microbial taxa (Bulgarelli et al. 2013; Kumar et al. 2017a). The progress in new culture-dependent and molecular culture-independent approaches has presented a new era for soil and rhizosphere microbiome. Specially, preparation and selection of multifarious libraries generated from the soil metagenome through gene-mining provides opportunities to completely understand the vast genetic and metabolic variation of soil microorganisms. This chapter provides an overview of the role of microbial diversity in maintaining soil health, plant growth, and crop productivity along with implementation of culture-dependent and culture-independent molecular approaches to explore and exploit the diversity of microbial communities to improve soil health.

5.2 Soil as Natural Medium for Microbes and Plant

Soil acts as a storehouse for diverse soil-dweller microbes in addition to the nutrients and energy required by plants. These microbes transform organic matter present in the soil and release nutrients from it. Soils sustain a wide range of microbial species and a one gram may contain around 10^3–10^5 taxa with approx. 1.5×10^{10} bacterial density (Torsvik and Ovreas 2002). Schmidt and Eickhorst (2014) reported that the soil medium comprised of 10^{-1} and 10^2 m^2 g^{-1}sandy to clay soils with high specific surface area that endure high bacterial abundance in soil and moderately large sections of soil surfaces devoid of bacterial population.

Soil comprises three-phase absorbent medium components composed of solid, liquid, and gaseous phases (Sharma 2005) as depicted in Fig. 5.1. The organic constituent of soil consists of solid SOM and dissolved organic matter (DOM) (Liebeke et al. 2009). In general, SOM derived mainly from plant residues comprises stable organic fraction humus, and active organic matter contains nutrients in available form. Active organic matter acts as natural medium for microbes; for instance, bacterial community uses simpler organic compounds, while fungi absorb complex compounds like wood, fibrous plant residues, and soil humus as their nutritional sources (Rashida et al. 2016; Datta et al. 2017a). Decomposition of SOM is carried out by soil microbes by secreting extracellular enzymes. Microbes release enzymes to obtain carbon (C) or other diminishing nutrients and to transform the soil's most copious substrates into simpler forms (DeForest et al. 2012). The second useful component of DOM is a combination of organic aromatic compound derivatives of amino acids, fulvic acids, and lignin, and some are derived from both plant material

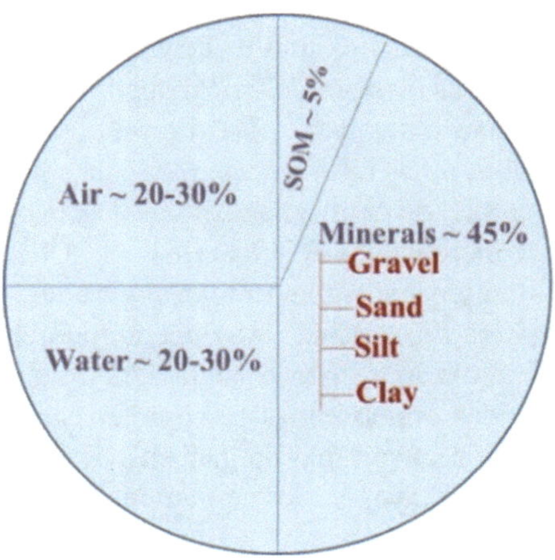

Fig. 5.1 Assessment of soil components

and dead bacteria, including oligomeric and monomeric sugar derivatives along withC14 and C54 fatty acids (Kalbitz et al. 2000).

Many researchers have revealed soil as a valuable medium for the microbial growth. Taylor (1951) reported that a medium containing only aqueous soil extract and agar provided much higher viable counts of microbes from soil than any other common medium. Soil extract, a commonly known culture medium has been acknowledged to support varied culturable as well as unculturable microbes since many years (Lima-Perim et al. 2016; Meena and Yadav 2015). Lochhead and Chase (1943) tentatively concluded that soil extract medium containing a wide range of amino acids, vitamins, and various beneficial growth-promoting substances facilitated the growth of bacterial culture. A study done by Pham and Kim (2016) revealed that application of transwell membranes along with soil extract led to isolation and growth of various soil bacteria, resistant to traditional cultivation, and this novel method was efficient for cultivation of uncultivable new bacterial species through adaptation to laboratory culturing conditions. Furthermore, microbial diversity composition is also based on various ecological factors such as moisture, O_2 availability, pH, and temperature (Eilers et al. 2012; Dadhich et al. 2015). Growth of microbes may also depend on variation in these factors which leads to formation of new microbial groups adapted to that particular modified environment possessing altered functional properties. Detailed knowledge of the essential components of the soil would be beneficial to analyze the adaptation of the microbial population to ecological niches.

5.3 Significance of Microflora in Soil Health Management

Agriculturist spontaneously acquainted the importance of healthy soils in terms of color, smell, softness, and taste to describe the physical conditions and ultimately performance of plant growth and crop productivity. Soil health disturbances can be due to physical, chemical, or biological activities comprising in the soil. All types of soil disturbances lessen the habitats for soil microorganisms and lead to diminished soil food networks. Lack of sufficient microflora in the soil reduces the probability of any cropping system along with increasing disease and pest risks and further lowering the crop productivity. Managing soil health is essential to maximizing profitability and microbial diversity contributing to improving soil health and function, reducing agriculture input costs, and increasing profitability of the crop.

Chemical transformation undergoing in the soil habitat involves active participation of diverse microbes. They are actively involved in maintaining soil fertility by facilitating the nutrient biogeochemical cycles like carbon and nitrogen that improve plant growth. Decomposition of the organic matter arriving in the soil and recycling of nutrients is one of the exclusive examples of it (Baldock 2007; Khatoon et al. 2017; Meena et al. 2016a). Certain groups of soil microbes increases the amount of nutrients in the soil such as nitrogen-fixing bacteria that convert nitrogen gas present in the atmosphere into soluble nitrogenous compounds and help plant its growth and development and mycorrhizal fungi increase the availability of mineral nutrients like phosphorus to plants (Nouri et al. 2014; Soni et al. 2017; Varma et al. 2017). Such type of soil beneficial microbes colonize the plant roots, improve the fertility eminence of the soil, and ultimately help in plant growth and development and are termed as biofertilizers which are continuously gaining high attention to be used as efficient microbial inoculants in the field of agriculture (Bhardwaj et al. 2014). Another group of soil microbes that has been known to produce compounds including vitamins and plant growth hormones help in maintaining plant health and promoting crop productivity are called as phytostimulators. Both plants and microorganisms depend on soil for their nutrition leading to change in soil characteristics through organic litter deposition and various metabolic activities, respectively. Microorganisms are involved in plant growth and development through direct mechanisms including manipulation of hormone signaling and protection against phytopathogens. Root exudate metabolites are responsible for plant–microbe communication (Fig. 5.2).

Normally, different kinds of pathogenic microbes too reside in the soil which infect the plants through roots. These pathogenic microbes are accountable for considerable destruction to the plants and crops; therefore, they fall in the group of phytopathogens. Native and indigenous microorganism-mediated antagonism to these pathogens usually includes the secretion of various inhibitory substances, such as secondary metabolites, including antimicrobial metabolites, antibiotics, and extracellular enzymes which prevent or limit the spread of infection in plants (Braga et al. 2016; Meena and Yadav 2014). They also help in stimulating the natural defense mechanisms of the plant and facilitate its resistance to various phytopathogens. These beneficial soil microorganisms have been termed as "biopesticides" and have emerged as important alternative tools for biological control (Mishra et al. 2015). Additionally, soil microorganisms also play critical role in improving the physical health of the soil. They stimulate the decomposition of organic matter by turning organic residues into humus, carbon dioxide, and water resulting into release of nutrients which are readily available to the plants. They improve water absorption and buffering capacity of the soil by creating an optimum level of environmental

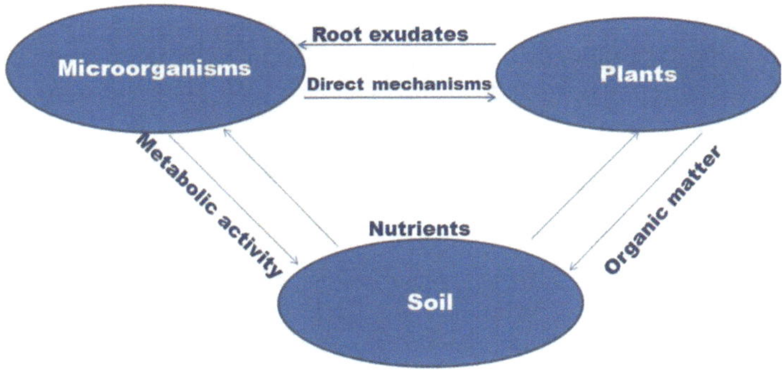

Fig. 5.2 Interactions between soil, microbes, and plant

conditions feasible for root growth like water retention, drainage, and aeration of the soil. Moreover, they are referred to as biological indicators for determining soil health and their effectiveness can be measured through standard analytical procedures (Arias et al. 2005) (Table 5.1).

5.4 Role of Culture-Dependent Techniques in Microbial Diversity Analysis

The identification of microorganisms and their affiliation to a particular taxonomic hierarchy could be a terrible crucial step within the documentation of microbial diversity (Bertilsson et al. 2002) that assists in determining the roles of specific populations. The microbial diversity represents metabolic, structural, morphological, and genetic diversity. Conventionally, microbial diversity was solely assessed through culture-based techniques, of which less than 1% of microbes are culturable as numerous physiological and biochemical tests are required for their full characterization. Moreover, the morphological characteristics of the cell like its shape, elevation, margin, cell wall, motility, flagella, Gram staining reaction, etc. as such might not be adequate for establishing an in-depth classification of microbes (Fakruddin and Mannan 2013; Dhakal et al. 2015). Hence, solely a restricted fraction of microbial wealth has been totally characterized and named. Besides, being very time consuming and laborious, these techniques require prior knowledge of the organisms of interest for their selective and successful culturing from a complex microbial network as well (Boughner and Singh 2016; Meena et al. 2018a). Although problematic, culture-based approaches could be tremendously helpful in understanding the physiological and useful potential of the microbes in gift within the soil. Microbial diversity in soil can be measured primarily through two classic techniques – biochemical-based culture-dependent techniques and molecular-based culture-independent techniques (Fig. 5.3).

5.4.1 Biochemical-Based Culture-Dependent Methods

Culture-dependent methods play a significant role in analyzing the diversity of microbial communities within soil. These methods entail the culturing of microbes in the laboratory by employing diversified culture media, which are designed in such a way that almost nearly all different microbial species might be recovered and afterward be identified through the exploitation of classical techniques which include morphological, physiological, and biochemical characterization (Janssen 2006). Culturing followed by isolation of microorganisms in pure culture is needed to realize in-depth understanding of microbial physiology and their environmental interaction with each other and also to allow access to metabolic pathways encoding genes, which can be distributed all through the genome (Keller and Zengler 2004; Schleifer 2004; Green and Keller 2006; Yadav et al. 2018a). A brief account of the

Table 5.1 Beneficial microbes exhibiting different activities to maintain soil health and plant growth

Microbial group	Genera involved	Source of isolation	Physiological role in plant growth promotion	Microbial transformation(s) in soil	Type of interaction with plant/organic matter	References
Bacteria	*Rhizobium*	Soil and root nodules of nodulating legumes like soya bean, gram, Lucerne, etc.	Nitrogen fixation	Decomposition of organic matter	Symbiotic	Poole et al.(2018)
	Pseudomonas, Bacillus, Serratia	Soil and rhizosphere	Nitrogen mineralization and plant-growth-promoting attributes	–	Non-symbiotic	Li et al. (2017); Soni et al. (2017)
	Nitrosomonas, Nitrobacter	Soil and rhizosphere	Nitrification	Convert soil ammonia into nitrate	Non-symbiotic	Koops and Pommerening-Rose (2001)
	Pseudomonas, Achromobacter	Soil and rhizosphere	Denitrification	Decomposition of organic matter	Non-symbiotic	Kathiravan and Krishnani (2014)
	Thiobacillus, Desulfovibrio, Beggiatoa	Soil and rhizosphere	Sulfur transformation	–	Non-symbiotic	Overmann and Gemerden (2000); Worner and Zimmermann-Timm (2000)
	Gallionella, Leptothrix, Ferribacterium	Soil and rhizosphere	Iron transformation	–	Non-symbiotic	Johnson et al. (2012)
	Klebsiella	Soil, rhizosphere, and leaf nodules of plants like *Psychotria, Pavetta*	Nitrogen fixation	–	Symbiotic	Centifanto and Silaver(1964)

	Azospirillum	Rhizoplane of grasses, cereals, sorghum, and millets	Nitrogen fixation	–	Associative non-symbiotic	Steenhoudt and Vanderleyden (2006)
Actinomycetous	*Frankia*	Soil, rhizosphere, and root nodules of actinorhizal plants other than legumes like *Alnus*, *Myrica*, *Casuarina,* etc.	Nitrogen fixation	Decomposition of organic matter	Endophytic	Van Nguyen and Pawlowski (2017)
Bacteria, fungi, actinomycetes	*Nocardia, Trichoderma, Fomes, Streptomyces*	Soil and rhizosphere	–	Decomposition of organic matter	Saprophytic	Khatoon et al.(2017)
Bacteria and fungi	*Pseudomonas, Bacillus, and Aspergillus*	Soil and rhizosphere	Phosphorus solubilization	Decomposition of organic matter	Non-symbiotic	Alori et al. (2017); Pan et al. (2012); Stainer et al.(1998)
Bacteria and algae	*Azotobacter, Beijerinckia, Anabaena, Nostoc*	Soil, rhizosphere, and algal source includes rhizoplane of higher plants like *Cycas*, *Zamia*, *Heterozamia*, *Gunnera scabra*, and fern *Azolla*	Nitrogen fixation	–	Non -symbiotic	Shridhar (2012)
Ectomycorrhizae	Mostly basidiomycetous fungi like *Boletus, Lactarius, Armillaria*	Rhizoplane of forest trees like *Pinus*	Phosphorus and micronutrient mobilization	Decomposition of organic matter	Symbiotic	Teotia et al.(2017)

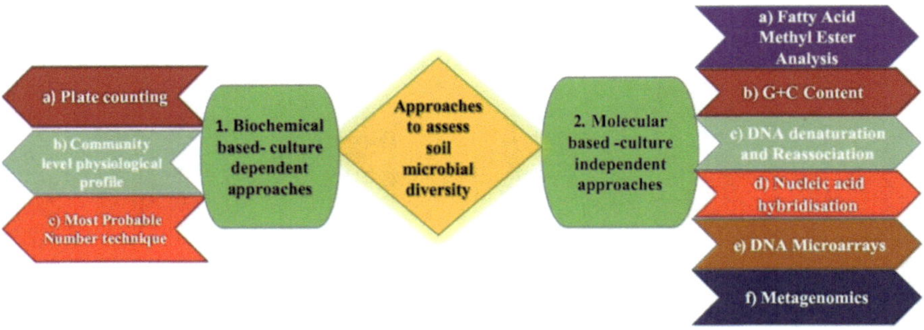

Fig. 5.3 Different approaches for the assessment of soil microbial diversity

various culture-dependent approaches that accustomed quantifying microbial diversity in soil is described beneath the following subheads.

5.4.1.1 Plate Count Techniques

The oldest and most generally used methodology for the evaluation of soil microbial diversity is the plate count technique (pour plate, spread plate, and streak plate). Plate counts are used for the enumeration of colony-forming units following incubation of soil suspension's dilutions on solid agar plates. Besides quick and cheap, these techniques endow with valuable information regarding metabolically active and culturable heterotrophic microbial populations. This is significantly necessary for soil health studies. There are several examples wherever these techniques have disclosed a great deal of microbial diversity related to varied soil quality parameters like suppression of disease and organic matter decomposition (Zhang and Xu 2008; Chandna et al. 2013; Stefani et al. 2015; Jaiswal et al. 2017; Meena et al. 2015d).

In spite of being very popular, the dilution and plating technique has been subject to abundant scrutiny and criticism virtually since its origin. The foremost criticism of this technique is that solely a tiny low fraction of the overall population, as ascertained microscopically, can be grown on laboratory media. It is well documented that by employing plate count technique, solely 1–10 % cells of the overall population as ascertained with direct microscopic counts could be recovered as viable bacteria. For instance, Torsvik et al. (1990a) and Atlas and Bartha (1998) compared the direct microscopic counts of microbes in soil samples and recoverable colony-forming units and reported that less than 0.1% of the microorganisms inhabiting typical agricultural soils can be cultured using current laboratory culture media. This large discrepancy between techniques may be due to the following factors:

1. Environmental samples with low substrate levels may be populated with nutrient-starved or oligotrophic organisms, which are slow growing or non-culturable, and so they don't appear as colonies in a reasonable time frame.
2. Viable organisms could become non-culturable because of extraction and dilution process-mediated stress imposed on them.

3. The competition to form colonies among bacteria might suppress the growth of some microorganisms.
4. Since all microbes have totally different optimal and minimal growth conditions relative to nutrient and electron acceptor requirements, some microbes grow terribly slowly or not in any respect on the plating medium.
5. Inadequate release of the bacteria from soil particles or biofilms during extraction and dilution steps could be ascribed to disparity in recovery of viable bacteria as it allows the clumping of bacteria to give rise to single colony thereby underestimating the actual bacterial load present.

It is not possible to determine which of these factors is responsible for the observed difference between viable and direct counts. As a rule, this distinction in counts is due to viable however non-culturable microbes, because it is believed that the degradation and utilization of dead cells is quite quick. Factors that prohibit their use embrace problem in dislodging bacteria or spores from soil particles or biofilms, appropriate choice of growth medium (Tabacchioni et al. 2000), cultural conditions (temperature, pH, light), failure to grow diversified bacterial and fungal species with current techniques, and the potential for colony distribution and colony–colony impediment (Trevors 1998). Additionally, fast-growing microorganisms including fungi that produce massive spores are favored by plating methods (Dix and Webster 1995). All these shortcomings have great impact on the perceptible diversity of the microbial communities.

5.4.1.2 Community Level Physiological Profile (CLPP)

Garland and Mills (1991) introduced the community level physiological profile (CLPP), additionally called the sole-carbon-source utilization (SCSU) system, which is one of the most frequently used methods that rely on culturing the microbes inhabiting soils for the assessment of their physiological diversity (Garland and Mills 1991; Konopka et al. 1998; Fakruddin and Mannan 2013; Yadav et al. 2017b). This methodology involves three steps. First step involves direct inoculation of environmental samples in microtiter plates containing completely different carbon sources, nutrients, and a redox dye as well followed by incubation, and afterward the activity of heterotrophic microbial communities can be detected by mass spectroscopic technique. Initially, it absolutely was developed as a tool in microbial taxonomy wherein sole carbon substrate utilization profiles (CSUP) of bacteria were used to identify them up to species level. Microbial communities vary in their potential to utilize a variety of carbon sources, which can be evaluated from these profiles. Variations in carbon source utilization patterns form the basis of comparison among totally different soil samples collected from different locations that in turn reflect variations in the physiological functioning of microbial populations.

CLPP has been a preferred choice of researchers that involves the use of BIOLOG system in which 95 completely different carbon sources are used for the metabolic profiling of microorganisms. Now, this technique is extensively offered to evaluate functional microbial diversity in compost ecosystem (Choudhary et al. 2009; Meena et al. 2015e), to determine the impact of assorted environmental factors on catabolic trait mediated biological eminence of explicit soil sites (Haack et al. 1995), and to employ multivariate techniques for comparing metabolic competency among microbial communities (Preston-Mafham et al. 2002). Like Biolog system, API system is also used for the metabolic profiling of microorganisms based on their utilization patterns of different carbon substrates. API system consists of a large number of API strips impregnated with numerous carbon sources that may be accustomed to estimate functional diversity (Torsvik et al. 1990b). The CLPP methodology is often envisaged as a golden opportunity to overcome the drawbacks of each ancient biochemical-based culture-based approach (Preston-Mafham et al. 2002; Frąc et al. 2012; Verma et al. 2015) and new molecular-based culture-independent approaches as these are time consuming and require specialized expertise.

CLPP has become popular because it has many advantages:

(i) Able to differentiate between microbial communities, comparatively simple to use, extremely consistent, and produces a great deal of information regarding metabolic characteristics of microbial communities (Zak et al. 1994).
(ii) It is equipped with an automated measuring apparatus which makes it a simple and fast technique to analyze and offer immense information about vital functional attributes of microbial communities (Broughton and Gross 2000).

Nonetheless, this technique has some disadvantages:

(i) This technique requires specialized expertise for data acquisition and interpretation analysis as it is typically difficult to analyze and interpret the data obtained in CLPP.
(ii) Since microbial communities are a mixture of both fast- and slow-growing organisms, the slow growers might be ignored during this analysis.
(iii) During incubation, growth on secondary metabolites can also occur.
(iv) This technique solely assesses the metabolic diversity of culturable bacteria and soil fungi and is not suitable for slow-growing bacteria as they have the least influence on the microbial metabolic profile.
(v) The metabolic growth response, which involves both cooperative and competitive effects in Biolog EcoPlate wells, may be a significant downside of the CLPP method (Garland 1996, 1997).
(vi) Unable to measure the functional potential of the entire soil microbial community, because solely an awfully restricted range of soil microbial genera grow on CLPP plates (Ros et al. 2008).
(vii) In addition, the carbon sources might not be true representatives of soil microbial communities present in soil (Yao et al. 2000), and therefore, the information obtained in CLPP may not have any practical feasibility.
(viii) BIOLOG sole C-source test plates impregnated with different carbon sources (Campbell et al. 1997) and TTC (triphenyl tetrazolium chloride) are buffered at nearly neutral pH, which is entirely different from the pH of certain acidic or alkaline soils, hence restricting the growth of those microorganisms that are well adapted to grow in these soils.

Several of those factors have conferred disadvantages while determining soil microbial community structure. Still, CLPP is helpful in finding out the functional microbial diversity of soils and may be a valuable tool particularly when employed in conjunction with alternative strategies (Kirk et al. 2004; Buragohain et al. 2017).

Carbohydrate utilization appears to retort most significantly so perturbations take place within the community, as proven by a rise associated with straw (Bossio and Scow 1995) or apronitin (Vahjen et al. 1995) addition to soil or a decline associated with metal stress (Lehman et al. 1997) or by introducing high concentrations of *Corynebacterium* into soil (Vahjen et al. 1995). Additional studies with totally different levels and kinds of stresses that embody synchronous analysis of CLPP and direct measures of sugar utilization like ^{14}C-labeling are required. Insam et al. (1996) suggested that augmented utilization of microbial products that are used as sole carbon sources within the Biolog plates was due to their increased cell turnover and accessibility during later stages of composting. Some specific modifications within microbial communities with dead microbial biomass are required to verify these results; these profiles can then be used as indicators of cell turnover.

This methodology has been used with great success to evaluate prospective metabolic diversity within microbial communities inhabiting contaminated sites (Konopka et al. 1998), plant rhizospheres (Grayston et al. 1998), arctic soils (Derry et al. 1999), herbicide-treated soil (El Fantroussi et al. 1999), or microorganism's inocula (Bej et al. 1991). For example, to assess the diversity of anaerobic microbial communities in an aquifer, Röling et al. (2000) employed Biolog system along with denaturing gradient gel electrophoresis (DGGE) and reported that separation of microbial communities from the polluted aquifer below a landfill site from those of aquifers located upstream or downstream of the landfill became possible because of conjoint application of anaerobic CLPP DGGE.

5.4.1.3 Most Probable Number (MPN) Technique

Another most frequently used culture-dependent approach is the most probable number (MPN) technique that estimates the sizes of microbial populations in liquid media. Like plate count technique, the MPN methods also employ serial dilutions of the sample in suitable media; however, instead of solid media, liquid media are used for incubation. In MPN technique, population size is calculated from the patterns of positive and negative test results obtained across many dilutions and massive replicates per dilution; hence, it provides a statistically significant enumeration of culturable organisms (Roszak and Colwell 1987).

In greater number of cases, higher counts are obtained through MPN technique as compared to plating methods (Reichardt 1978), though exemptions exist (Cassidy et al. 2000). This might be ascribed to the actual fact that colonies could not be formed by some bacteria at the air-solid interface (Eiler et al. 2000). Highly selective media have been used so far in MPN technique, for example, for the culturing of nitrate or sulfate-reducing, sulfur-oxidizing, or methanogenic bacteria (Vester and Ingvorsen 1998) or for the isolation of thermophilic cyanobacteria (Ferris et al. 1996). In liquid cultures, the growth of fast-growing dominant bacteria is generally overgrown by less abundant accompanying species that can merely be eliminated by liquid serial dilutions. Hence, in MPN technique, the highest dilution endows with enrichments or even pure cultures of abundant but fastidious bacteria that might not be detected through traditional enrichment experiments (Ferris et al. 1996). Consequently, MPN methodology can be envisaged as a vital tool for the isolation of novel microorganisms.

In order to evaluate the potential of soil as oil contaminants degrade and to monitor bioremediation, enumeration of soil microorganisms exhibiting hydrocarbon degrading ability provides a classical example of exploiting MPN strategies in soil microbiological analysis (Johnsen 2010; Meena et al. 2016b). Presently, MPN strategies usually use a 96-well microtiter plate setup. As an example, aromatic substrates are typically dissolved in an exceedingly appropriate solvent and added to every well microtiter plate separately to enumerate hydrocarbon degraders (Johnsen et al. 2002). Soil protozoa, particularly heterotrophic flagellates and naked amoebae, are usually tiny and a lot are less amoeboid or pliable. These characteristics create their quantification terribly tough. However, estimation of total protozoan numbers in soil samples become easier through MPN-PCR assay (Fredslund et al. 2001).

MPN technique is more advantageous than DNA-based quantification of gene sequences in two respects: (i) this technique is used to enumerate solely living organisms having a well-established ability to embrace the method of interest, and (ii) the quantification might recover bacteria having completely different metabolic pathways instead of restricting to those whose gene sequences are known. A significant constraint of the MPN methodology is that solely a little range of parallel serial dilutions (usually 3–10) could be processed for every sample. Consequently, the number of isolated obtained from the highest dilutions fall in the order of 10 at the most. The little range of parallels additionally causes the intrinsic statistical ambiguity in cultivation strategies (Grosskopf et al. 1998; Bruns et al. 2002; Yadav et al. 2017a). Hence, the traditional MPN approach, attributable to these limitations, does not appear to be acceptable if the cultivated fraction is more diversified and different kinds of bacteria occur in low numbers.

5.5 Limitations of Culture-Dependent Approaches

1. The main limitation of culture-based techniques is that these are extremely biased toward readily cultivatable organisms, would completely overlook the largest proportion of non-culturable bacteria, i.e., "the great plate count anomaly" (Amann et al. 1995; Torsvik and Ovreas 2002), and hence were not suitable for contemporary studies.
2. The cultivation-dependent approach may underestimate the microbial load in samples because it is likely that at the same time, many species fail to grow at all on the given medium or show so few colony-forming units that they are overlooked in the isolation step.

3. This technique also failed to determine the predominant microorganisms in nature that could not be cultivated by using standard techniques (Amann et al. 1995).
4. Culture-dependent approach do not essentially give in-depth information regarding composition of microbial communities (Orphan et al. 2000), and ecological inferences based on metabolic profiling of culturable bacteria do not seem to reflect the natural populations from which they were isolated.
5. Culture-dependent methods alone are not acceptable to determine the composition of microbial communities (Ward et al. 1990) because genetically distinct isolates cannot be differentiated always on the basis of their morphological characteristics (Chandler et al. 1997a, b).
6. Culture-based approaches selectively allow the growth of faster-growing microorganisms with colony- or biofilm-forming ability (Leadbetter 2003; Ferrari et al. 2004; Kumar et al. 2018).
7. In addition to the current, culture-based techniques may select those organisms having no practical significance in situ (Staley and Konopka 1985).

In spite of these shortcomings, culture-based approach could offer an extremely insightful and terribly easy mean to enumerate and characterize some specific groups if appropriate selective media and cultural growth conditions are provided. As an example, microbial population with a meticulous capability to grow on a specific substrate may be selected. Culturing of microorganism is additionally indispensable (i) to characterize them physiologically, (ii) for trustworthy description of novel species, and (iii) to enumerate those microbial populations having particular growth requirements or having capability to grow on some specific substrates or those that may resist some antibiotics. It is therefore evident that culture-based approaches, besides very helpful for evaluating the physiological potential of microbial communities, do not essentially give conclusive information about their composition (Pearce et al. 2003).

5.6 Role of Culture-Independent Techniques in Microbial Diversity Analysis

It is typically troublesome to fully characterize the soil microbial populations using culture-based approach because these exhibit immense phenotypic and genotypic diversity. Hence, most of these communities are generally unculturable (Dokic et al. 2010; Bhat 2013; Meena et al. 2015c). Additionally, morphological and nutritional criteria-based characterization of the microbial communities is futile to give a normal taxonomic order consistent with the evolutionary relatedness (Agrawal et al. 2015; Varma et al. 2017a). Consequently, focus was turned toward development of molecular-based culture-independent techniques that complement the information provided by culture-based methods (Perez-Trallero et al. 2007) and permit us to give a reliable structural description of microbial communities: (i) without the same old dependency on selective culturing, (ii) to understand these complex communities, (iii) to eliminate the need for culturing beforehand, (iv) to aid in assessing the total microbial diversity present, (v) to allow the characterization of many long-recognized but poorly understood organisms (Head et al. 1998), and (vi) to overcome the shortcomings of cultivation-based strategy that seems to ruthlessly underestimate the actual microbial diversity of the soil (Akkermans et al. 1995, Ovreas and Torsvik 1998). Molecular approaches vary with respect to discriminatory power, reproducibility, ease of use, and ease of interpretation (Lasker 2002; Hall 2007). These approaches can now be considered an indispensable component of the microbiologist's complete research toolbox and, therefore, are at the cutting edge of science. Basically, culture-independent approaches are exploited to infer the composition of microbial communities through:

1. The isolation, extraction, quantification, and identification of those molecules/products that are derived from soil and are specifically belong to a particular microorganism or a group of microorganisms or
2. The use of highly sophisticated fluorescence-based microscopic techniques for the detection of phospholipids, fatty acids, and nucleic acids molecules (Morgan and Winstanley 1997), whereas use of microscopic techniques either for hybridizing fluorescent-labelled nucleic acid probes with total RNA of soil or hybridizing the cells in situ

Advancement in molecular biology tools has made it feasible to investigate the microbial diversity by incorporating the protocols of culture-independent techniques for studying the microbial communities. Presently, a substantial variety of useful approaches are available to study the microbial diversity at molecular level. Mainly, these embrace fatty acid methyl ester (FAME) profile analysis, Guanine plus cytosine (G+C) content, DNA denaturation and reassociation, nucleic acid hybridization, DNA microarrays, and metagenomics.

5.6.1 Fatty Acid Methyl Ester (FAME) Analysis

One of the foremost frequently used methodologies for investigating the soil microbial diversity on the premise of lipid molecules is phospholipid fatty acid (PLFA) technique. This methodology was delineated originally by Bligh and Dyer in the year 1959 and has been used as a culture-independent methodology by omitting the need to culture (Banowetz et al. 2006). Soil microorganisms, plant residues, and organics oil amendments are main sources of soil fatty acids, which are the most abundant class of soil lipids.

The fatty acids play a vital role in monitoring microbial activities that happen in the soil and unravelling the structure of soil microbial communities (Zelles 1999) along with practices pertaining to soil management (Dinel et al. 1998), processes associated with soil organic matter transformations (Jandl et al. 2004), and soil organic matter formation processes (Naafs et al. 2004;Filley et al. 2008; Ram and Meena 2014). The fatty acid profile of soil also provides useful information about soil wettability and its aggregate stability (Jandl et al. 2004) besides phytotoxic properties that may have an effect on seed germination (Edney and Rizvi 1996). Therefore, it can be inferred that the content and distribution of fatty acids in soils can be put forward to acquire imperative information about several vital processes happening in the soil surroundings. Apart from playing significant roles in the field of soil science, fatty acid biomarkers additionally are accustomed to gain information about different sources of organic matter, climate change, and microbial transformations happening in diversity of varied aquatic environments, viz., river, marine and lacustrine sediments, speleothems, and sedimentary rocks (Huang et al. 2008; Verma et al. 2015b).

Phospholipid fatty acid (PLFA) procedure involves the extraction of microbial phospholipids into a single but appropriate solvent phase, which is generally comprised of chloroform, methanol, and phosphate buffer. Extracted lipids afterward are separated into diverse types through solid-phase extraction method followed by transesterification catalyzed by base to yield fatty acid methyl esters (FAMEs) (Findlay 2004). This method is best suited for the fatty acid profiling of living soil microorganisms because phospholipids commence to collapse immediately after the death of the microorganisms (Drenovsky et al. 2004). This method is additionally referred to as the fatty acid methyl ester (FAME) analysis because fatty acid profiling indirectly reflects the growth behavior of bacteria (Fakruddin and Mannan 2013). Fatty acids are considered as signature molecules, which are all pervading in all living organisms. Phospholipids found in cytoplasmic membrane of microorganisms make up a comparatively steady fraction of the cell biomass and are serving as key determinants to differentiate between main taxonomic groups distributed within a microbial community.

Hence, these are used as chemotaxonomic marker as any variation in the fatty acid profile is used to interpret variation in the abundance of microbial populations with respect to reference in the database of pure strains and identified biosynthetic processes (Kamer and Baath 1998; Zelles 1999). Direct extraction of total fatty acid methyl esters or simply PLFAs from soil could be used to assess solely gross changes happening in the structure of microbial communities. It does not permit detection at the species level which, however, is feasible with standard media used for culturing and databases. Several fatty acids isolated from specific microbial groups have been listed by Lechevalier (1989) and Zelles (1999). For instance, biomass of bacteria could be estimated by taking into consideration numerous fatty acids (15,0, a15,0, i15,0, i16,0, i17:0, cy17:0,cy19:0, and 16:1 ω7c), whereas biomass of fungi can be obtained from the 18:2 ω 6c fatty acid (Frostegård and Bååth 1996). The fatty acid content is quite stable and additionally not affected by mutations, plasmids, or injured cells. Although this methodology is quantitative, cheap, sturdy, and highly reproducible, it has some limitations also which are described below:

1. PLFA analysis is laborious and slow; hence, it time consuming and expensive to carry out (Buyer and Sasser 2012).

2. Detection of individual microbial strains or species is typically not possible with this technique; however, overall variations in community composition could be detected instead.
3. Solely fatty acids could not be used to embody particular species because individual species may have various fatty acids and the similar fatty acids may be present in more than one species (Bossio and Scow 1998).
4. Factors like temperature and nutritional requirements may affect the fatty acid composition of the cells. FAME profiles of other species can possibly be confounding (Graham et al. 1995).
5. There are certain fatty acids that aren't perpetually accepted as biomarkers of different microbial taxa of microorganisms and have always been a matter of debate, particularly when analyses of complex systems are involved (Scott et al. 2010; Frostegard et al. 2011). Even so, some specific types of fatty acids are still used as markers for characterizing different microbial populations within a given taxon as per the information available about culturable microflora (Zelles 1999; Scott et al. 2010).

Although, fatty acid biomarkers provide relatively low taxonomic resolution while investigating microbial communities in soil, an inclusive examination of fatty acid profiles obtained from diverse soil lipids could also be envisaged as novel and exciting opportunity for taxonomic studies with respect to biochemical characterization of soil environments (Ferrari et al. 2015).

5.6.2 Guanine Plus Cytosine (G+C) Content (Mol% G+C)

Base composition is the primary property of deoxyribonucleic acid (DNA) used for systemic studies which is usually expressed as mole percentage guanine + cytosine (mol% G+C). Differences in the G+C contents of DNA of different organisms reflect the diversity of microbial populations inhabiting soil (Nusslein and Tiedje 1999). G+C content can be determined by thermal denaturation of DNA, whose value ranges from 25% to 75% in bacteria, although it is constant for a specific organism.

The information inferred from G+C profiles revealed that organisms having similar G+C contents are closely related while those whose G+C contents differ from 3% to 5 % belong to taxonomically related groups (Tiedje et al. 1999). Same base composition does not necessarily confirm the relationship among different organisms; however, differences in base composition could be a worthy proof of missing relationship. This methodology is conferred with some sure benefits: (i) unaffected by PCR biases, (ii) encompasses the entire extracted DNA, (iii) quantitative, and (iv) can reveal the diversity of rare members within microbial communities. Nonetheless, this methodology requires fairly a great amount (up to 50 μg) of DNA (Tiedje et al. 1999). Nusslein and Tiedje (1999) utilized G+C content data along with amplified ribosomal DNA restriction analysis (ARDRA) abundance patterns and rDNA sequence analysis to examine the overall changes taking place microbial communities from a vegetative cover of forest to pasture in a Hawaiian soil. Differences in the microbial diversity were detected with all three methods revealing that plants strongly influence the composition of microbial communities. Authors believed that these methods could be used for examining the microbial communities more systematically as these methods have examined the community at a unique level of resolution.

5.6.3 DNA Denaturation and Reassociation

The kinetics of DNA denaturation and reassociation has been used to resolve the genetic intricacy of microbial populations of a given environmental sample, hence providing a broad-scale investigation of community structure (Torsvik et al. 1990b; Torsvik et al. 1996; Ovreas et al. 2003). While DNA reassociation kinetics reflects the diversity of microbial communities based on diversified sequences present in the given ecosystem, it does not, however, provide information relating to the phylogenetic origin (ancestral history) of the microorganisms present.

The procedure involves the extraction of total community DNA from environmental samples like soil followed by purification and denaturation. Following denaturation, the DNA is permitted to reassociate/reanneal. The rate at which DNA reassociate or hybridize is directly proportional to the concentration of similar/complementary DNA sequences, while the length of dissimilar sequences is reciprocally proportional to its rate of reassociation or hybridization. Moreover, the rate of DNA reassociation or hybridization decreases with an increase in the diversity or intricacy of DNA sequences (Theron and Cloete 2000).

The reassociation reaction rate is mainly controlled by two parameters: (i) DNA concentration (Co) and (ii) incubation time (t), generally defined as half the association value, Cot1/2, i.e., the time required by half of the DNA to reassociate. Cot1/2 value is a vital measure of diversity index under specific conditions as it considers both the concentration and distribution of DNA re-association (Torsvik et al. 1998). On the other hand, degree of similarity of DNA sequences measured through hybridization kinetics can be used to find the similarity between microbial communities that belong to two completely different samples (Griffiths et al. 1999).

5.6.4 Nucleic Acid Hybridization

One of the foremost vital qualitative and quantitative tools in the field of bacterial ecology at molecular level is nucleic acid hybridization using specific probes (Griffiths et al. 1999; Clegg et al. 2000; Theron and Cloete 2000; Yadav et al. 2017). This technique can be accomplished using either on isolated DNA or RNA or is applicable in situ too. Highly specific and well-known sequences are used to design oligonucleotide or polynucleotide probe whose 5V-end is usually stagged with markers (Theron and Cloete 2000). Typically, DNA probes have been used in membrane-based hybridization techniques and, however, may be applied in situ too.

Hence, DNA probes used in membrane hybridization, also referred to as dot blot hybridization, is analogous to traditional Southern hybridization. The in situ technique is additionally referred to as fluorescence in situ hybridization (FISH), wherein fluorescently labelled probes are allowed to hybridize with fixed whole cells.

5.6.4.1 Dot Blot Hybridization

Oligonucleotide samples were traditionally labeled with radioactive isotopes, but nowadays the most frequently used markers are fluorescently labeled markers that include rhodamine or fluorescein derivatives. Quantitative dot blot hybridization technique is used to measure the relative abundance of a particular microbial group wherein the primary step involves lysis of the sample to unleash all nucleic acids and subsequent quantification of apprehensive rRNA sequences with respect to total rRNA using universal primers which are highly specific. Variations in the relative abundance reflect changes in either the microbial population abundance or their activity and, consequently, in the quantity of rRNA content (Theron and Cloete 2000). Hybridization provides vital information about microbes on account of their spatial distribution in environmental samples, though it has several significant limitations listed below:

1. Merely one probe could be used per membrane because probes are labelled with radioactive tags, and hence a separate membrane is required for every probe that makes this technique quite burdensome.
2. Additionally, this technique generates valuable data on relative rRNA abundance. Since cellular rRNA contents vary considerably with growth rate, direct translation of relative rRNA abundance into cell numbers could not work out (Amann et al. 1995).

5.6.4.2 Fluorescent In Situ Hybridization (FISH)

Fluorescent in situ hybridization (FISH), as a culture-independent approach, is one of the most popular DNA hybridization techniques that has been used to envisage uncultured microorganisms from diverse environments (Eiler et al. 2000; Gonzalez-Toril et al. 2003) to examine microbial communities and biofilms as well (Daims et al. 2001; Ferrari et al. 2006; Meena et al. 2014).

Besides, allowing the simultaneous visualization, enumeration, identification, and localization of individual microbial cells in their native environment, it additionally provides comprehensive understanding about the structure and diversity of specific microbial communities along with their spatial distribution and abundance (Amann et al. 1995). For example, cold sulfurous marsh water of Sippemauer Moor (Germany), inhabited by microbial communities comprised of novel archaeal and bacterial species, was explored to detect the presence of strings of pearl-like morphologies using FISH (Rudolph et al. 2001).

FISH depends on the employment of fluorescently labelled oligonucleotide probes that hybridize specifically with target sequences but complementary of the target cell. The most promising candidate used in FISH is 16S rRNA molecule because it is genetically stable, it has both conserved and variable regions in its domain structure, and copy number is also high (Woese 1987). The rRNA targeted region forms the basis for designing oligonucleotide probes at a taxonomic level (Amann et al. 1995); besides, it is extremely useful for constantly designing new probes that target diverse phylogenetic levels (Nercessian et al. 2004; Rusch and Amend 2004) and in assigning a particular phylogenetic group to the targeted microorganism (Amann et al. 1995). Some of the limitations of FISH are given below:

1. One of the most notable limitations of FISH is that either microorganisms themselves or sample debris like soil and mineral particles show autofluorescence which interferes with target cell examination (Moter and Gobel 2000; Bertaux et al. 2007).
2. Low intensity of fluorescent signals from the fluorescently labelled probes is another limitation of FISH which is attributable to inadequate permeabilization of cell walls using standard procedures, rRNA's meager accessibility for the probes, or cell's low ribosomal content because of ablated metabolic activity (Poulsen et al. 1993; Bhatia et al. 1997).
3. The third limitation is the poor sensitivity of hybridization of directly extracted nucleic acids from environmental samples. If sequences are not present in high copy number, such as those from dominant species, probability of detection is low. If sequences do not seem to be in high copy number, like those from dominant species, the likelihood of detection is low.

To overcome the limitations of FISH, several refinements and novel techniques have been reported, viz., exploitation of polynucleotide probes having multiple fluorescent labels and use of peptide nucleic acids (Stender et al. 1999; Zimmermann et al. 2001). For example, these techniques have made it feasible to detect the marine planktons that could have not previously been detectable using traditional oligonucleotide probes (Pernthaler et al. 2003). In addition, increase in signal intensity as compared to conventional FISH resulted from catalyzed reporter deposition (CARD)-FISH (Schönhuber et al. 1997) that allowed, as an example, clear detection of signal over autofluorescent background in some cyanobacteria (Pernthaler et al. 2002). These novel techniques can be well thought out as useful means to detect and identify microbial communities in their environmental niches, though these refinements are not cost-effective as these involve high costs and are time consuming too when compared with conventional FISH (Wagner et al. 2003).

5.6.5 DNA Microarrays

DNA microarrays together with DNA–DNA hybridization have widely been used for the detection and identification of bacterial species (Cho and Tiedje 2001) or for microbial diversity assessment (Greene and Voordouw 2003). DNA microarrays signify a promising mean for the analogous identification of numerous microorganisms with high throughput and furnish a comprehensive view of microbial communities in environmental samples. A single microarray is comprised of thousands of unique and enormously particular DNA sequences that correspond to a single gene; thereby, this approach ought to be treasured in elucidating the diversity of bacteria (DeSantis et al. 2007; Dadhich and Meena 2014). In order to elucidate functional diversity of microbial communities, some specific target genes coding for enzymes like naphthalene dioxygenase, nitrate reductase, nitrogenase, etc. could be exploited in microarray studies. Similarly, environmental DNA "standards" (less than 70% hybridization DNA fragments) that may represent diverse species can also be used in microarray (Greene and Voordouw 2003).

In DNA microarray technique, firstly, the entire extracted environmental DNA is amplified through PCR. The PCR amplicons, which are fluorescently labelled and directly allowed to hybridize with already, attach known molecular probes on the microarrays (Gentry et al. 2006). Following hybridization, positive signals are scored using confocal laser scanning microscopy. Generally, the signal intensity of hybridization on microarrays is directly proportional to targeted organism's abundance. Microarrays have tested beneficial to learn about soil microbial ecology; however, maximum research is of technical nature due to (i) excessive intricacy of the microbial communities and excessive biomass, (ii) higher concentration of those substances that often inhibit the enzymatic reactions, and (iii) the newness of the technology. For enhancing the sensitivity and selectivity of the investigation, optimal conditions are elaborated and suitable probes are developed for the detection of either particular phylogenetic levels or pertinent functional genes.

Compared to conventional membrane-based hybridization methods, the microarray technique allows the rapid evaluation of the samples with replication, involves low cost, is automated, and has low background level (Shalon et al. 1996) which is a sure benefit in investigating the microbial communities. Despite these advantages, it has some shortcomings also which are as described below.

1. The major limitation of microarray technology is cross-hybridization, especially when managing environmental samples.
2. The microarray is no longer helpful for the detection and identification of novel taxa of prokaryotes.
3. Additionally, if the microarray does not have a specific probe for corresponding genus, ecological importance of a genus might be utterly unnoticed.

5.6.6 Metagenomics

Metagenomics, the power of genomic analysis, by definition is a functional and sequence-based investigation of the entire genomes of microbial communities retrieved directly from environmental samples, avoiding the necessity to isolate and culture individual microbial species and which do not require prior knowledge of the microbial populations (Riesenfeld et al. 2004; Chen and Pachter 2005; Zeyaullah et al. 2009). Metagenomics is additionally recognized by other names such as environmental genomics or community genomics or microbial ecogenomics. The word metagenomics is derived from a Greek word "meta" which means "transcendent." The word "meta" is used in two senses. In first sense, "meta" means that this exciting approach may increase the feasibility to know biology at the aggregate level, transcending the every individual organism to focus on the distribution of genes within the microbial communities and how these genes, for performing collective functions, may affect each other's activities. In the second sense, meta additionally acknowledges the necessity to computational strategies that provide in-depth information regarding genetic makeup and activities of those complicated communities that can solely be sampled; however, it can never utterly be characterized. Thus, metagenomics enable the scientists to study all of the genomes in a community as a whole by transcending individual genes and genomes (Schloss and Handelsman 2003; Riesenfeld et al. 2004; Meena et al. 2015a).

Metagenomics approach involves the random sampling of the collective genome (metagenome or microbiome) of microbial communities (Ghazanfar et al. 2010) directly extracted from the environment followed by sequencing (Schloss and Handelsman 2003). Since metagenomics can directly access the collective genome of microbial communities, it gives all- inclusive information about the genetic diversity, composition, and evolution of microbial species along with their interactions within natural environmental niches (Simon and Daniel 2011).

Several environments such as soil, phyllosphere, ocean, acid mine drainage, etc. have been explored through metagenomic approach that have made it possible to get access to evolutionary and functional diversity of uncultivated microflora (Handelsman 2004). Hence, this approach provides insights into the biochemical roles of uncultivated microflora and its interaction with other environmental factors. Environmental metagenomic libraries are the greatest resources of new microbial enzymes and antibiotics that may serve as persuasive candidates in the field of biotechnology, medicine, industry, etc. (Rondon et al. 2000; Riesenfeld et al. 2004). Mainly four steps are involved in the construction of these libraries: (1) total DNA is isolated from an environmental sample, (2) randomly selected DNA fragments are cloned into an appropriate vector through shotgun cloning, and (3) clones are then transformed into a host bacterium, and (4) positive clones are screened. In order to retrieve better information about the metagenome of an environment, small DNA fragments (2–3 kb size) in metagenomic libraries are preferred over larger fragments. For instance, to retrieve the genomes of rare microbial species, at least 1011 genomic clones would be needed (Riesenfeld et al. 2004).

Metagenomic libraries could be screened either through (1) phylogenetic markers or "anchors," such as 16S rRNA and recA, (2) by random sequencing (Tyson et al. 2004; Venter et al. 2004), (3) by conserved genes other than 16S rRNA through multiplex PCR or hybridization (Stein et al. 1996), or (4) by the expression of specific traits, such as antibiotic production or enzymatic activities (Rondon et al. 2000; Lorenz and Schleper 2002; Diaz-Torres et al. 2003; Knietsch et al. 2003; Schloss and Handelsman 2003; Meena et al. 2015b).Various biochemical processes have been discovered in environmental metagenomic libraries. As an example, Rondon et al. (2000) have identified numerous new antibiotics (e.g., turbomycin, terragine);

microbial enzymes, viz., amylases, cellulases, lipases, etc.; and proteins like antiporters in soil metagenomic libraries. Therefore, it is apparent that in the library, volumes are not well ordered. A million of random fragments of DNA extracted from all the coexisting microbes in the sampled community comprises the library instead of containing the genome of one species.

Culture-independent approaches are employed to characterize the diverse microbial communities; some of these are depicted in Table 5.2.

5.6.7 Limitations of Culture-Independent Approaches

1. A great variation in the lysis potency of bacterial or fungal cells might exist between and within microbial communities that are accountable for biasness in molecular-based diversity investigations (Prosser 2002). If the cell extraction strategy is simply too mild and gram negative, not gram-positive bacteria would be lysed. However, if the strategy is incredibly harsh, each gram-negative and gram-positive cell could be lysed but shearing of their DNA could take place (Wintzingerode et al. 1997). Even different fungal cells have different lysis potencies; hence, their spores, mycelia, and mycelia of different ages lyse differently (Prosser 2002).
2. The methodology employed for the extraction of DNA or RNA may lead to bias diversity studies. Shearing of nucleic acids can occur if extraction methods like bead beating are too harsh, which are accountable for the inconveniences that arise throughout sequent PCR detection (Wintzingerode et al. 1997).
3. Different strategies used for the extraction of nucleic acids can lead to completely different yields of product (Wintzingerode et al. 1997).
4. It is pertinent to get rid of inhibitory substances such as humic acids, which may be extracted simultaneously with nucleic acids, when dealing with environmental samples. These substances may interfere with ensuant PCR detection and can cause substantial loss of DNA or RNA that might again lead to bias diversity studies.

5.7 Rhizosphere: As Soil's Microhabitat for Beneficial Microbiota

The term rhizosphere was first defined by Lorentz Hiltner in 1904 (Hartmann et al. 2009; Hiltner 1904) and redefined by Pinton as the precinct of soil in the vicinity of plant roots that comprises the root tissues colonized by microorganisms (Pinton et al. 2001). The rhizosphere consists three discrete zones including endorhizosphere, ectorhizosphere, and rhizoplane (Lynch 1990). Plant roots, soil, and interactions with microbes considerably with variable soil physical and chemical characteristics ultimately result in alteration of microbial communities' dynamics in the rhizosphere zone (Nihorimbere et al. 2011; Novello et al. 2017; Kumar et al. 2017). The rhizosphere effect indicates the influence of plant root exudates on microbes, and the migration of root exudates in the soil directly affects the intensity

Table 5.2 Molecular-based culture-independent approaches employed for the characterization of diverse microbial communities

Sr. No.	Technique employed	Microorganism/phyla identified	Source	References
1.	FAME	*Bacillus, Paenibacillus, Arthrobacter, Micrococcus, Staphylococcus, Kocuria varians, Paenibacillus polymyxa*	Mural paintings of Spain, Austria, and Germany	Heyrman et al. (1999)
		Virgibacillus pantothenticus, Bacillus atrophaeus, Corynebacterium diphtheria	Solar salterns of Goa	Surve et al. (2012)
		Citrobacter freundii and *Pseudomonas aeruginosa*	Soils of Secunderabad	Rajan et al. (2011)
		Pseudomonas aeruginosa and *Pseudomonas putida*	Soil of Houston metropolitan area, USA	Iyer et al. (2011)
2.	G+C content	*Arthrobacter, Brevibacterium linens, Nocardia cellulans, Corynebacterium michiganense,* and *Jensenia canicruria*	Soil	Skyring and Quadling (1970)
		Actinobacteria, Deinococcus, Fusobacteria, Planctomycetes, spirochaetes, and *Proteobacteria*	–	Li and Du (2014)
3.	DNA denaturation and reassociation	α-*Proteobacteria*	Soil	Sandaa et al. (1999)
4.	Nucleic acid hybridization	*Rhizobium leguminosarum*	Soil	Laguerre et al. (1993);
		Pyrococcus and *Thermococcus*	–	Gonzalez and Saiz-Jimenez (2005)
		Simazine-degrading bacteria	Agricultural and natural soil	Martín et al. (2008)
5.	DNA microarrays	Fluorescent *pseudomonads*	–	Cho and Tiedje (2001);
		Methanotrophs	Environmental samples	Bodrossy et al. (2003);
		Methylocystis and *Methylocaldum*	Landfill sites	Stralis-Pavese et al. (2004)
		Rhodococcus, Ralstonia, Comamonas, and *Burkholderia*	Soil microcosms	Rhee et al. (2004)
		Geothrix fermentans, Arthrobacter spp., and *Actinobacteria*	Uranium-contaminated soil	Brodie et al. (2006)

(continued)

Table 5.2 (continued)

Sr. No.	Technique employed	Microorganism/phyla identified	Source	References
		Bacillus bataviensis and Bacillus sp.	Environmental aerosols, Austin, and San Antonio	Brodie et al. (2007)
		Desulfomicrobium spp. and *Desulfonema* and *Desulfomonile*	Tooth pockets and chemocline of the mat	Loy et al. (2002)
		Lactobacillus acidophilus	Skim milk	Azcarate-Peril et al. (2009)
6.	Metagenomics	*Candidatus solibacter usitatus, Bradyrhizobium japonicum, Rhodopseudomonas palustris, Conexibacter woesei, Candidatus koribacter versatilis, Gemmatimonas aurantiaca, Sorangium cellulosum, Mycobacterium tuberculosis, Rhodopirellulabaltica, Myxococcus xanthus*	Forest and vineyard soils in Chile	Castan˜eda and Barbosa (2017)
		Halobacteria, *Balneolaeota, Bacteroidetes, Rhodothermaeota, Nanohaloarchaeota*	Hypersaline soils of the Odiel Salt Marshes in Huelva, Southwestern Spain	Vera-Gargallo et al. (2018)

of microbial activities (Morgan and Whipps 2001). Moreover, plant root exudates are responsible to arbitrate the interactions of plant roots and the microbial communities in the rhizosphere (Chaparro et al. 2013). Plant roots discharge around 5–21% of photosynthetically static carbon in the form of soluble amino acids, organic acids, nucleotides, sugars, vitamins, and many more secondary metabolites (Badri et al. 2013). Root exudates are clustered into two groups: first, low-molecular-weight substances including amino acids, organic acids, phenolic compounds, sugars and, second, high-molecular-weight substances including nucleotides, flavonones, polysaccharides, and proteins. The quality and quantity of root exudates is measured by cultivar, plant species, plant developmental stage, variety, and a wide range of environmental and ecological factors including soil type, pH, and temperature (Badri and Vivanco 2009; Yang et al. 2017; Meena et al. 2017c). These differences influence diversity in microbes in the rhizosphere that possess a significant degree of specificity for distinct plant species. A variety of transport mechanisms are implemented by plants to release substrates into the rhizosphere and export them to microbes (Weston et al. 2012).

Usually, root exudates are released by plant roots through either active (secretions) or passive (diffusates) mechanisms. The low-molecular-weight organic constituents are exuded from plants by a passive channel. This process depends on the polarity of the secreted compounds, membrane permeability along with cytosolic pH, and small polar and uncharged molecules which are transported through direct passive diffusion process (Badri and Vivanco 2009; Meena et al. 2018b). Plant roots release other substrates like secondary metabolites including polysaccharides, proteins, and different membrane-bound proteins (Weston et al. 2012). These transporter proteins include the ATP-binding cassette (ABC) transporters (Sugiyama et al. 2008), the multidrug and toxic compound extrusion (MATE) family (Yazaki 2005), the major facilitator superfamily (Reddy et al. 2012), and the aluminum-activated malate transporter family (Weston et al. 2012; Dhakal et al. 2016). The information of these membrane-bound transport protein functions are not described in the literature as they have been coupled with the transport of a variety of compounds into the rhizosphere. Plant root exudates intervene a multitude of rhizospheric interactions at the species level and at the community level with multitrophic (Fig. 5.4). The rhizospheric microbial community structure alternation depends upon plant genotype, plant developmental stage, exposure to disease-suppressive

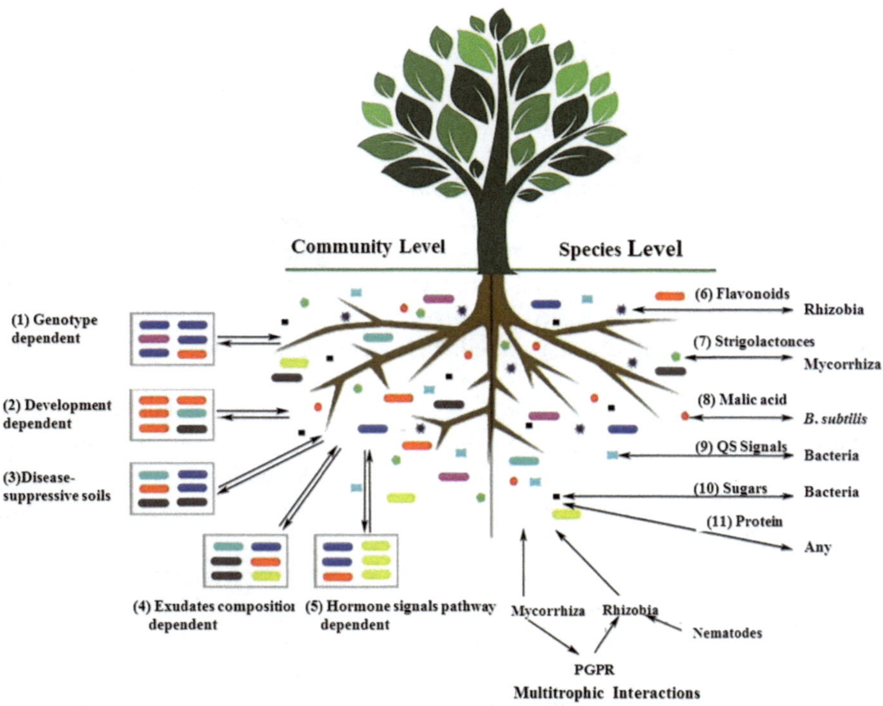

Fig. 5.4 The interactions between plants, rhizosphere microbes, and root exudate compounds mediating rhizospheric interactions and affecting plant microbiome (Huang et al. 2014). Different rods represent different microbial taxonomic groups. Each gray rectangle represents a distinct rhizosphere microbial community; different colored rods represent the 1,000,000 different bacterial genomes (Gans et al. 2005; Sihag et al. 2015). Roesch et al. (2007) investigated that four distinct soils contain *Bacteroidetes*, *Betaproteobacteria*, and qualitative and quantitative distribution of microbes. Circles, pentagons, rectangles, stars, and squares represent a wide range of substances released as root exudates

soils, root exudate composition, and plant hormone signaling (Huang et al. 2014). Specific compounds secreted as root exudates mediate plant–microbe or species level interactions which include flavonoids that act as signaling compounds to stimulate symbiosis between legumes and rhizobia (Abdel-Lateif et al. 2012), strigolactones that initiate mycorrhizal hyphal branching (Akiyama et al. 2005), malic acid that recruits specific plant-growth-promoting rhizobacteria (PGPR) such as *Bacillus subtilis* (Rudrappa et al. 2008; Meena et al. 2017a), termination or initiation of quorum sensing (QS) in bacteria (Gao et al. 2003), and sugars and amino acids that act as chemo-attractants for microbial community (Somers et al. 2004). Other root exudates mediate multitrophic interactions like plant attraction toward nematodes, rhizobacteria, and rhizobia interaction resulting in the increase of nodulation efficiency and rhizobacteria interaction with mycorrhizae increasing colonization efficiency. Zhalnina et al. (2018) also reported consumption of aromatic organic acids exuded by plants including nicotinic, shikimic, salicylic, cinnamic, and indole-3-acetic by rhizosphere bacteria.

5.8 The Rhizosphere Microbiome

Microbial communities show significant performance improving plant's physiology along with its growth and development. Although the importance of the rhizosphere microbiome for plant growth and development is well recognized, a diversity of beneficial rhizo-microbial communities is still unexplored. The knowledge about the existence and behavior of microbes in the rhizosphere microbiome is a critical factor to analyze their significance for plant growth and development. Culture-independent approaches revealed that diversity of soil and rhizosphere microbial communities are extremely unexplored. Next-generation sequencing technologies showed that only a 5% of bacteria have been cultured through the latest molecular methodologies and a significant percentage of the bacterial phyla identified have no cultured representative till date (Mendes et al. 2013; Igiehon and Babalola 2018; Meena et al. 2017b). Following the metagenomic studies with substantial computational analysis, it was predicted that only 1 g of soil can restrain more than *Alphaproteobacteria* as the most abundant bacterial groups.

It has been reported that genes and their functions expressed in the rhizosphere are based on reporter genes. These reporter genes are responsible to distinguish microbial habitat in terms of biological, chemical, and physical stimuli, hence facilitating the evaluation of some specific members of the rhizosphere microbiome (Barret et al. 2011; Datta et al. 2017). A promoter trapping strategy in vivo expression technology (IVET) exploited a broad panel of reporter genes to evaluate various processes in the rhizosphere including reactions of bacteria to nutrients like carbon, nitrogen, phosphorus availability, temperature, and water potential along with metabolic processes (DeAngelis et al. 2005; Herron et al. 2010). Bioreporters were also successfully employed to study bacterial communication and quorum sensing in the rhizosphere and in situ antimicrobial compound production (Ferluga and Venturi 2009; Rochat et al. 2010; Verma et al. 2015a). "Omics" approaches facilitate the detection of gene transcripts and development of proteins or metabolites to offer a comprehensive insight into the genes and functions expressed in the rhizosphere microbiome. Delmotte et al. (2009) first reported bacterial communities in the phyllosphere of *Arabidopsis*, clover plants, and soybean through metaproteogenomic approach. Drigo et al. (2010) provided a conceptual model based on DNA-SIP data in which plants assimilated carbon which was transferred to arbuscular mycorrhizal fungi (AMF), followed by a slower secretion from AMF to stimulate bacterial-fungal populations in the rhizosphere. Collectively, these studies represent a unification of functional and computational approaches which endow these potent tools to infer physiological attributes of microbial communities in the rhizosphere.

5.9 Conclusions and Future Prospects

Current agricultural practices have lessened soil biodiversity, mainly excessive use of chemicals, leading to irremediable unfavorable ecological alterations and ensuring the loss of agricultural productivity and disturbance in soil health. The biological, chemical, and physical characteristics of soil play a key role in maintaining long-term soil health, sustainable agricultural productivity, and negligible environmental impact. Thus, soil health refers to a complete scenario of soil functionality. Although its direct measurement is difficult, the status can be accessed by measuring physical features and SOM content and studying microbial diversity as it is considered as an indicator for soil's fertility assessment. In this regard, considerable progress has been made in the field of molecular biology enabling rapid, reproducible, and reliable techniques for soil microbial communities' analysis. The combination of these modern molecular techniques offers a promising role in determining soil health status. Long-term comparative studies have shown that organic and sustainable practices can facilitate both organic matter accumulation and enhance microbial activity in the soil. Furthermore, the organic carbon lost during conventional agriculture could be recovered through eco-friendly and sustainable soil management practices, imparting contribution in extenuating climate change. The knowledge about novel culture-dependent and molecular culture-independent approaches will draw our attention toward the diverse microbiome of the soil and rhizosphere so that the discovery of new species exhibiting different attributes of plant growth promotion, decompositions of organic wastes, bioremediation, and soil restoration can be explored and promoted in sustaining the ecosystem.

References

Abdel-Lateif K, Bogusz D, Hocher V (2012) The role of flavonoids in the establishment of plant roots endosymbioses with arbuscular mycorrhiza fungi, rhizobia and Frankia bacteria. Plant Signal Behav 7(6):636–641. https://doi.org/10.4161/psb.20039

Agrawal PK, Agrawal S, Shrivastava R (2015) Modern molecular approaches for analyzing microbial diversity from mushroom compost ecosystem. 3 Biotech 5:853–866. https://doi.org/10.1007/s13205-015-0289-2

Akiyama K, Matsuzaki K, Hayashi H (2005) *Plant Sesquiterpenes* induce hyphal branching in arbuscular mycorrhizal fungi. Nature 435(7043):824–827. https://doi.org/10.1038/nature03608
Akkermans ADL, van Elsas JD, de Bruijn FJ (1995) Molecular microbial ecology manual. Kluwer, Dordrecht
Alori ET, Glick BR, Babalola OO (2017) Microbial phosphorus solubilization and its potential for use in sustainable agriculture. Front Microbiol 8:971. https://doi.org/10.3389/fmicb.2017.00971
Amann RI, Ludwig W, Schleifer KH (1995) Phylogenetic identification and in situ detection of individual microbial cells without cultivation. Microbiol Rev 59:143–169
Arias ME, González-Pérez JA, González-Vila FJ, Ball AS (2005) Soil health: a new challenge for microbiologists and chemists. Int Microbiol 8:13–21
Ashoka P, Meena RS, Kumar S, Yadav GS, Layek J (2017) Green nanotechnology is a key for eco-friendly agriculture. J Clean Prod 142:4440–4441
Atlas RM, Bartha R (1998) Microbial ecology: fundamentals and applications. Cummings Publishing Company, Benjamin, pp 281–324
Azcarate-Peril MA, Tallon R, Klaenhammer TR (2009) Temporal gene expression and probiotic attributes of Lactobacillus acidophilus during growth in milk. J Dairy Sci 92:870–886
Badri DV, Vivanco JM (2009) Regulation and function of root exudates. Plant Cell Environ 32(6):666–681. https://doi.org/10.1111/j.1365-3040.2009.01926.x. PMID: 19143988
Badri DV, Zolla G, Bakker MG, Manter DK, Vivanco JM (2013) Potential impact of soil microbiomes on the leaf metabolome and on herbivore feeding behavior. New Phytol 198(1):264–273. https://doi.org/10.1111/nph.12124. PMID: 23347044
Baldock J (2007) Composition and cycling of organic carbon in soil. In: Marschner P, Rengel Z (eds) Nutrient cycling in terrestrial ecosystems. Springer, Berlin, pp 1–36
Banowetz GM, Whittaker GW, Dierksen KP, Azevedo MD, Kennedy AC, Griffith SM, Steiner JJ (2006) Fatty acid methyl ester analysis to identify sources of soil in surface water. J Environ Qual 3:133–140
Barret M, Morrissey JP, O'Gara F (2011) Functional genomics analysis of plant growth-promoting rhizobacterial traits involved in rhizosphere competence. Biol Fertil Soils 47:729–743
Bej AK, Perlin M, Atlas RM (1991) Effect of introducing genetically engineered microorganisms on soil microbial diversity. FEMS Microbiol Ecol 86:169–175
Bertaux J, Gloger U, Schmid M, Hartmann A, Scheu S (2007) Routine fluorescence in situ hybridization in soil. J Microbiol Methods 69:451–460
Bertilsson S, Cavanaugh CM, Polz MF (2002) Sequencing-independent method to generate oligonucleotide probes targeting a variable region in bacterial 16S rRNA by PCR with detachable primers. Appl Environ Microbiol 68:6077–6086
Bhardwaj D, Ansari MW, Sahoo RK, Tuteja N (2014) Biofertilizers function as key player in sustainable agriculture by improving soil fertility, plant tolerance and crop productivity. Microb Cell Factories 13:66. https://doi.org/10.1186/1475-2859-13-66
Bhat AK (2013) Preserving Microbial diversity of Soil ecosystem: a key to sustainable productivity. Int J Curr Microbiol Appl Sci 2:85–101
Bhatia U, Robison K, Gilbert W (1997) Dealing with database explosion: a cautionary note. Science 276:1724–1725
Bligh EG, Dyer WJ (1959) A lipid method of total lipid extraction and purification. Can J Biochem 37:911–917
Bodrossy L, Stralis-Pavese N, Murrell JC, Radajewski S, Weilharter A, Sessitsch A (2003) Development and validation of a diagnostic microbial microarray for methanotrophs. Environ Microbiol 5(7):566–582
Bossio DA, Scow KM (1995) Impact of carbon and flooding on the metabolic diversity of microbial communities in soils. Appl Environ Microbiol 61:4043–4050
Bossio DA, Scow KM (1998) Impacts of carbon and flooding on soil microbial communities: phospholipid fatty acid profiles and substrate utilization patterns. Microb Ecol 35:265–278
Boughner L, Singh P (2016) Microbial ecology: where are we now? Postdoc J 4(11):3–17. https://doi.org/10.14304/surya.jpr.v4n11.2
Brady NC, Weil RR (2014) The nature and properties of soils. Pearson Education Limited, Harlow

Braga RM, Dourado MN, Araujo WL (2016) Microbial interactions: ecology in a molecular perspective. BJM 47:86–98

Brodie EL, De Santis TZ, Joyner DC, Baek SM, Larsen JT, Andersen GL, Hazen TC, Richardson PM, Herman DJ, Tokunaga TK, Wan JM, Firestone MK (2006) Application of a high-density oligonucleotide microarray approach to study bacterial population dynamics during uranium reduction and reoxidation. Appl Environ Microbiol 72(9):6288–6298. https://doi.org/10.1128/AEM.00246-06

Brodie EL, De Santis TZ, Parker JPM, Zubietta IX, Piceno YM, Andersen GL (2007) Urban aerosols harbor diverse and dynamic bacterial populations. Proc Natl Acad Sci 104(1):299–304. https://doi.org/10.1073/pnas.0608255104

Broughton LC, Gross K (2000) Patterns of diversity in plant soil microbial communities along a productivity gradient in a michigan old-field. Oecologia 125:420–427

Bruns A, Cypionka H, Overmann J (2002) Cyclic AMP and acyl homoserine lactones increase the cultivation efficiency of heterotrophic bacteria from the central Baltic Sea. Appl Environ Microbiol 68:3978–3987

Brussard L, de Ruiter PC, Brown GC (2007) Soil biodiversity for agricultural sustainability. Agric Ecosyst Environ 121:233–244

Bulgarelli D, Schlaeppi K, Spaepen S, Van Themaat EVL, SchulzeLefert P (2013) Structure and functions of the bacterial microbiota of plants. Annu Rev Plant Biol 64:807–838. https://doi.org/10.1146/annurev-arplant-050312-120106

Buragohain S, Sharma B, Nath JD, Gogaoi N, Meena RS, Lal R (2017) Impact of ten years of biofertilizer use on soil quality and rice yield on an inceptisol in Assam, India. Soil Res. https://doi.org/10.1071/SR17001

Buyer JS, Sasser M (2012) High throughput phospholipid fatty acid analysis of soils. Appl Soil Ecol 61:127–130

Campbell CD, Grayston SJ, Hirst DJ (1997) Use of rhizosphere carbon sources in sole carbon source tests to discriminate soil microbial communities. J Microbiol Methods 30:33–41

Cassidy MB, Leung KT, Lee H, Trevors JT (2000) A comparison of enumeration methods for culturable Pseudomonas fluorescens cells marked with green fluorescent protein. J Microbiol Methods 40:135–145

Castan˜eda LE, Barbosa O (2017) Metagenomic analysis exploring taxonomic and functional diversity of soil microbial communities in Chilean vineyards and surrounding native forests. PeerJ 5:e3098. https://doi.org/10.7717/peerj.3098

Centifanto YM, Silaver WS (1964) Leaf-nodule symbiosis endophyte of *Psychotria bacteriophila*. J Bacteriol 88(3):776–781

Chandler DP, Brockman FJ, Fredrickson JK (1997a) A Use of 16S rDNA clone libraries to study changes in a microbial community resulting from ex-situ perturbation of a subsurface sediment. FEMS Microbiol Rev 20:217–230

Chandler DP, Li SM, Spadoni CM, Drake GR, Balkwill DL, Fredrickson JK, Brockman FJ (1997b) A molecular comparison of culturable aerobic heterotrophic bacteria and 16S rDNA clones derived from a deep subsurface sediment. FEMS Microbiol Ecol 23:131–144

Chandna P, Nain L, Singh S, Kuhad RC (2013) Assessment of bacterial diversity during composting of agricultural byproducts. BMC Microbiol 13:99–113. https://doi.org/10.1186/1471-2180-13-99

Chaparro JM, Badri DV, Vivanco JM (2013) Rhizosphere microbiome assemblage is affected by plant development. ISME J 8(4):790–803. https://doi.org/10.1038/ismej.2013.196

Chen K, Pachter L (2005) Bioinformatics for whole-genome shotgun sequencing of microbial communities. PLoS Comput Biol 1:24. https://doi.org/10.1371/journal.pcbi.0010024

Cho JC, Tiedje JM (2001) Bacterial species determination from DNA–DNA hybridization by using genome fragments and DNA microarrays. Appl Environ Microbiol 67:3677–3682

Choudhary DK, Agarwal PK, Johri BN (2009) Evaluation of in situ functional activity of casing soils during growth cycle of mushroom (*Agaricus bisporus* (Lange) Imbach) employing community level physiological profiles (CLPPs). Indian J Microbiol 50(1):19–26

Clegg CD, Ritz K, Griffiths BS (2000) %G+C profiling and cross hybridisation of microbial DNA reveals great variation in below-ground community structure in UK upland grasslands. Appl Soil Ecol 14:125–134

Cynthia M, Kallenbach, Frey SD, Grandy AS (2016) Direct evidence for microbial-derived soil organic matter formation and its ecophysiological controls. Nat Commun:1–10. https://doi.org/10.1038/ncomms13630

Dadhich RK, Meena RS (2014) Performance of Indian mustard (*Brassica juncea* L.) in Response to foliar spray of thiourea and thioglycollic acid under different irrigation levels. Indian J Ecol 41(2):376–378

Dadhich RK, Meena RS, Reager ML, Kansotia BC (2015) Response of bio-regulators to yield and quality of Indian mustard (*Brassica juncea* L. Czernj. and Cosson) under different irrigation environments. J Appl Nat Sci 7(1):52–57

Daims H, Nielsen JL, Nielsen PH, Schleifer KH, Wagner M (2001) In -situ characterization of Nitrospira-like nitrite-oxidizing bacteria active in wastewater treatment plants. Appl Environ Microbiol 67:5273–5284

Datta R, Baraniya D, Wang YF, Kelkar A, Moulick A, Meena RS, Yadav GS, Ceccherini MT, Formanek P (2017) Multi-function role as nutrient and scavenger of free radical in soil. Sustain MDPI 9:402. https://doi.org/10.3390/su9081402

Datta R, Kelkar A, Baraniya D, Molaei A, Moulick A, Meena RS, Formanek P (2017a) Enzymatic degradation of lignin in soil: a review. Sustain MDPI 9:1163. https://doi.org/10.3390/su9071163. 1–18

De Santis TZ, Brodie EL, Moberg JP, Zubieta IX, Piceno YM, Andersen GL (2007) High-density universal 16S rRNA microarray analysis reveals broader diversity than typical clone library when sampling the environment. Microb Ecol 53:371–383

DeAngelis KM, Ji PS, Firestone MK, Lindow SE (2005) Two novel bacterial biosensors for detection of nitrate availability in the rhizosphere. Appl Environ Microbiol 71:8537–8547

DeForest JL, Smemo KA, Burke DJ, Elliott HL, Becker JC (2012) Soil microbial responses to elevated phosphorus and pH in acidic temperate deciduous forests. Biogeochemistry 109:189–202

Delmotte N, Knief C, Chaffron S, Innerebner G, Roschitzki B, Schlapbach R, von Mering C, Vorholt JA (2009) Community proteogenomics reveals insights into the physiology of phyllosphere bacteria. Proc Natl Acad Sci U S A 106:16428–16433

Derry AM, Staddon WJ, Kevan PG, Trevors JT (1999) Functional diversity and community structure of micro-organisms in three arctic soils as determined by sole-carbon source-utilization. Biodivers Conserv 8:205–221

Dhakal Y, Meena RS, De N, Verma SK, Singh A (2015) Growth, yield and nutrient content of mungbean (*Vigna radiata* L.) in response to INM in eastern Uttar Pradesh, India. Bangladesh J Bot 44(3):479–482

Dhakal Y, Meena RS, Kumar S (2016) Effect of INM on nodulation, yield, quality and available nutrient status in soil after harvest of green gram. Legum Res 39(4):590–594

Diaz-Torres ML, McNab R, Spratt DA, Villedieu A, Hunt N, Wilson M, Mullany P (2003) Novel tetracycline resistance determinant from the oral metagenome. Antimicrob Agents Chemother 47:1430–1432

Dinel H, Monreal CM, Schnitzer M (1998) Extractable lipids and organic matter status in two soil catenas as influenced by tillage. Geoderma 86:279–293

Dix NJ, Webster J (1995) Fungal ecology. Chapman & Hall, London, pp 332–333

Dokic L, Savic M, Narancic T, Vasiljevic B (2010) Metagenomic analysis of soil microbial communities. Arch Biol Sci 62:559–564

Drenovsky RE, Elliot GN, Graham KJ, Scow KM (2004) Comparison of phospholipid fatty acid (PLFA) and total soil fatty acid methyl esters (TSFAME) for characterizing soil microbial communities. Soil Biol Biochem 36:1793–1800

Drigo B, Pijl AS, Duyts H, Kielak AM, Gamper HA, Houtekamer MJ, Boschker HTS, Bodelier PLE, Whiteley AS, van Veen JA, Kowalchuka GA (2010) Shifting carbon flow from roots into associated microbial communities in response to elevated atmospheric CO_2. Proc Natl Acad Sci U S A 107:10938–10942

Edney NA, Rizvi M (1996) Phytotoxicity of fatty acids present in dairy and hog manure. J Environ Sci Health Part B 31:269–281

Eiler H, Pernthaler J, Grockner FO, Amann R (2000) Culturability and in situ abundance of peragic bacteria from the North Sea. Appl Environ Microbiol 66:3044–3051

Eilers KG, Debenport S, Anderson S, Fierer N (2012) Digging deeper to find unique microbial communities: the strong effect of depth on the structure of bacterial and archaeal communities in soil. Soil Biol Biochem 50:58–65. https://doi.org/10.1016/j.soilbio.2012.03.011

El Fantroussi S, Verschuere L, Verstraete W, Top EM (1999) Effect of phenylurea herbicides on soil microbial communities estimated by analysis of 16S rRNA gene fingerprints and community- level physiological profiles. Appl Environ Microbiol 65:982–988

Fakruddin MD, Mannan KSB (2013) Methods for analyzing diversity of microbial communities in natural environments. Ceylon J Sci (Bio Sci) 42(1):19–13. https://doi.org/10.4038/cjsbs.v42i1.5896

Ferluga S, Venturi V (2009) OryR is a LuxR-family protein involved in interkingdom signaling between pathogenic *Xanthomonas oryzae* pv. *oryzae* and rice. J Bacteriol 191:890–897

Ferrari BC, Oregaard G, Sorensen SJ (2004) Recovery of GFP-labeled bacteria for culturing and molecular analysis after cell sorting using a benchtop flow cytometer. Microb Ecol 48:239–245

Ferrari BC, Tujula N, Stoner K, Kjelleberg S (2006) Catalyzed reporter deposition-fluorescence in situ hybridization allows for enrichment-independent detection of microcolony-forming soil bacteria. Appl Environ Microbiol 72:918–922. https://doi.org/10.1128/AEM.72.1.918–922

Ferrari AE, Ravnskov S, Larsen J, Tonnersen T, Maronna RA, Wall LG (2015) Crop rotation and seasonal effects on fatty acid profiles of neutral and phospholipids extracted from no-till agricultural soils. Soil Use Manag 31:165–175. https://doi.org/10.1111/sum.12165

Ferris MJ, Ruff-Roberts AL, Kopczynski ED, Bateson MM, Ward DM (1996) Enrichment culture and microscopy conceal diverse thermophilic *Synechococcus* populations in a single hot spring microbial mat habitat. Appl Environ Microbiol 62:1045–1050

Filley TR, Boutton TW, Liao JD, Jastrow JD, Gamblin DE (2008) Chemical changes to non aggregated particulate soil organic matter following grassland-to-woodland transition in a subtropical savanna. J Geophys Res 113:G03009. https://doi.org/10.1029/2007JG000564

Findlay RH (2004) Determination of microbial community structure using phospholipid fatty acid profiles. In: Kowalchuk GA et al (eds) Molecular microbial ecology manual, 2nd edn. Kluwer, Dordrecht, pp 983–1004

Frąc M, Oszust K, Lipiec J (2012) Community level physiological profiles (CLPP), characterization and microbial activity of soil amended with dairy sewage sludge. Sensors (Basel) 12(3):3253–3268

Fredslund L, Ekelund F, Jacobsen CS, Johnsen K (2001) Development and application of a most-probable-number–PCR assay to quantify flagellate populations in soil samples. Appl Environ Microbiol 67:1613–1618

Frostegård A, Bååth E (1996) The use of phospholipid fatty acid analysis to estimate bacterial and fungal biomass in soil. Biol Fertil Soils 22:59–65

Frostegard A, Tunlid A, Baath E (2011) Use and misuse of PLFA measurements in soils. Soil Biol Biochem 43:1621–1625

Gans J, Wolinsky M, Dunbar J (2005) Computational improvements reveal great bacterial diversity and high metal toxicity in soil. Science 309:1387–1390

Gao M, Teplitski M, Robinson JB, Bauer WD (2003) Production of substances by *Medicago truncatula* that affect bacterial quorum sensing. Mol Plant-Microbe Interact 16(9):827–834. https://doi.org/10.1094/MPMI.2003.16.9.827

Garland J (1996) Analytical approaches to the characterization of samples of microbial communities using patterns of potential c source utilization. Soil Biol Biochem 28:213–221

Garland J (1997) Analysis and interpretation of community-level physiological profiles in microbial ecology. FEMS Microbiol Ecol 24:289–300

Garland JL, Mills AL (1991) Classification and characterization of heterotrophic microbial communities on the basis of patterns of community-level sole-carbon-source utilization. Appl Environ Microbiol 57:2351–2359

Gentry TJ, Wickham GS, Schadt CW, He Z, Zhou J (2006) Microarray applications in microbial ecology research. Microb Ecol 52:159–175
Ghazanfar S, Azim A, Ghazanfar MA, Anjum MI, Begum I (2010) Metagenomics and its application in soil microbial community studies: biotechnological prospects. J Anim Plant Sci 6:611–622
Gonzalez JM, Saiz-Jimenez C (2005) A simple fluorimetric method for the estimation of DNA–DNA relatedness between closely related microorganisms by thermal denaturation temperatures. Extremophiles 9:75–79. https://doi.org/10.1007/s00792-004-0417-0
Gonzalez-Toril E, Gomez F, Rodriguez N, Fernandez D, Zuluaga J, Marin I, Amils R (2003) Geomicrobiology of the Tinto river, a model of interest for biohydrometallurgy. Hydrometallurgy 71:301–309
Graham JH, Hodge NC, Morton JB (1995) Fatty acid methyl ester profiles for characterization of Glomalean fungi and their endomycorrhizae. Appl Environ Microbiol 61:58–64
Grayston SJ, Wang S, Campbell CD, Edwards AC (1998) Selective influence of plant species on microbial diversity in the rhizosphere. Soil Biol Biochem 30:369–378
Green D, Keller M (2006) Capturing the uncultivated majority. Curr Opin Biotechnol 17:236–240
Greene EA, Voordouw G (2003) Analysis of environmental microbial communities by reverse sample genome probing. J Microbiol Methods 53:211–219
Griffiths BS, Ritz K, Ebblewhite N, Dobson G (1999) Soil microbial community structure: effects of substrate loading rates. Soil Biol Biochem 31:145–153
Grosskopf R, Janssen PH, Liesack W (1998) Diversity and structure of the methanogenic community in anoxic rice paddy soil microcosms as examined by cultivation and direct 16S rRNA gene sequence retrieval. Appl Environ Microbiol 64:960–969
Haack SK, Garchow H, Klug MJ, Forney LJ (1995) Analysis of factors affecting the accuracy, reproducibility, and interpretation of microbial community carbon source utilization patterns. Appl Environ Microbiol 61:1458–1468
Hall N (2007) Advanced sequencing technologies and their wider impact in microbiology. J Exp Biol 210:1518–1525
Han K, Li ZF, Peng R, Zhu LP, Zhou T, Wang LG, Li SG, Zhang XB, Hu W, Wu ZH, Qin N, Li YZ (2013) Extraordinary expansion of a Sorangium cellulosum genome from an alkaline milieu. Sci Rep 3:2101–2107
Handelsman J (2004) Metagenomics: application of genomics to uncultured microorganisms. Microbiol Mol Biol Rev 68:669–685
Hartmann A, Schmid M, van Tuinen D, Berg G (2009) Plant-driven selection of microbes. Plant Soil 321:235–257
Head IM, Saunders JR, Pickup RW (1998) Microbial evolution, diversity and ecology: a decade of ribosomal RNA analysis of uncultivated microorganisms. Microb Ecol 35:1–21
Herron PM, Gage DJ, Cardon ZG (2010) Micro-scale water potential gradients visualized in soil around plant root tips using microbiosensors. Plant Cell Environ 33:199–210
Heyrman J, Mergaert J, Denys R, Swings J (1999) The use of fatty acid methyl ester analysis (FAME) for the identification of heterotrophic bacteria present on three mural paintings showing severe damage by microorganisms. FEMS Microbiol Lett 181:55–62
Hiltner L (1904) U ber neuere Erfahrungen und Probleme auf dem Gebiete der Bodenbakteriologie unter besonderer Berücksichtigung der Gründüngung und Brache. Arb DLG 98:59–78. https://doi.org/10.1007/s12088-009-0021-1
Huang XY, Cui JW, Pu Y, Huang JH, Blyth AJ (2008) Identifying "free" and "bound" lipid fractions in stalagmite samples: an example from Heshang Cave, southern China. Appl Geochem 23:2589–2595
Huang X, Chaparro JM, Reardon KF, Zhang R, Shen Q, Vivanco JM (2014) Rhizosphere interactions: root exudates, microbes, and microbial communities. Botany 92:267–275
Igiehon NO, Babalola OO (2018) Rhizosphere microbiome modulators: contributions of nitrogen fixing bacteria towards. Int J Environ Res Public Health 15:574. https://doi.org/10.3390/ijerph15040574

Insam H, Amor K, Renner M, Crepaz C (1996) Changes in functional abilities of the microbial community during composting of manure. Microb Ecol 31:77–87

Iyer R, Iken B, Tamez T (2011) Isolation, Molecular and Biochemical Identification of Paraoxon Metabolizing *Pseudomonas* Species. J Bioremed Biodegrad 2:132. https://doi.org/10.4172/2155-6199.1000132

Jaiswal AK, Elad Y, Paudel I, Graber ER, Cytryn E, Frenkel O (2017) Linking the below ground microbial composition, diversity and activity to soil borne disease suppression and growth promotion of tomato amended with biochar. Sci Rep 7:44382. https://doi.org/10.1038/srep44382

Jandl G, Leinweber P, Schulten HR, Eusterhues K (2004) The concentrations of fatty acids in organo-mineral particle-size fractions of a Chernozem. Eur J Soil Sci 55:459–469

Janssen PH (2006) Identifying the dominant soil bacterial taxa in libraries of 16S rRNA and 16S rRNA genes. Appl Environ Microbiol 72:1719–1728

Johnsen AR (2010) Introduction to microplate MPN-enumeration of hydrocarbon degraders. In: Timmis KN (ed) Handbook of hydrocarbon and lipid microbiology. Springer, Berlin, pp 4160–4172

Johnsen AR, Bendixen K, Karlson U (2002) Detection of microbial growth on polycyclic aromatic hydrocarbons in microtiter plates by using the respiration indicator WST-1. Appl Environ Microbiol 68:2683–2689

Johnson KW, Carmichael MJ, McDonald W, Rose N, Pitchford J, Windelspecht M, Karatan E, Brauer SL (2012) Increased abundance of Gallionella spp., Leptothrix spp. and total bacteria in response to enhanced Mn and Fe concentrations in a disturbed Southern Appalachian high elevation wetland. Geomicrobiology 29(2):124–138

Kalbitz K, Solinger S, Park JH, Michalzik B, Matzner E (2000) Controls on the dynamics dissolved organic matter in soils: a review. Soil Sci 165(4):277–304

Kamer M, Baath E (1998) Microbial community dynamics during composting of straw material studied using phospholipid fatty acid analysis. FEMS Microbiol Ecol 27:9–20

Kathiravan V, Krishnani KK (2014) *Pseudomonas aeruginosa* and *Achromobacter* sp.: Nitrifying aerobic denitrifiers have a plasmid encoding for denitrifying functional genes. World J Microbiol Biotechnol 30(4):1187–1198

Keller M, Zengler K (2004) Tapping into microbial diversity. Nat Rev Microbiol 2(2):141–150

Khatoon H, Solanki P, Narayan M, Tewari L, Rai JPN (2017) Role of microbes in organic carbon decomposition and maintenance of soil ecosystem. Int J Chem Stud 5(6):1648–1656

Kirk JL, Beaudette LA, Hart M, Moutoglis P, Klironomos JN, Lee H, Trevors JT (2004) Methods of studying soil microbial diversity. J Microbiol Methods 58(2):169–188

Knietsch A, Bowien S, Whited G, Gottschalk G, Daniel R (2003) Identification and characterization of coenzyme B_{12}-dependent glycerol dehydratase- and diol dehydratase-encoding genes from metagenomic DNA libraries derived from enrichment cultures. Appl Environ Microbiol 69:3048–3060

Konopka A, Oliver JL, Turco RF (1998) The use of carbon source utilization patterns in environmental and ecological microbiology. Microb Ecol 35:103–115

Koops HP, Pommerening-Rose A (2001) Distribution and ecophysiology of the nitrifying bacteria emphasizing cultured species. FEMS Microbiol Ecol 37(1):1–9

Kumar S, Meena RS, Pandey A, Seema (2017) Soil acidity management and an economics response of lime and sulfur on sesame in an alley cropping system. Int J Current Microb Appl Sci 6(3):2566–2573

Kumar S, Meena RS, Yadav GS, Pandey A (2017a) Response of sesame (*Sesamum indicum* L.) to sulphur and lime application under soil acidity. Int J Plant Soil Sci 14(4):1–9

Kumar S, Meena RS, Bohra JS (2018) Interactive effect of sowing dates and nutrient sources on dry matter accumulation of Indian mustard (*Brassica juncea* L.). J Oilseed Brassica 9(1):72–76

Laguerre G, Bardin M, Amarger N (1993) Isolation from soil of symbiotic and non-symbiotic Rhizobium leguminosarum by DNA hybridization. Can J Microbiol 39(12):1142–1149. https://doi.org/10.1139/m93-172

Land M, Hauser L, Jun S-R et al (2015) Insights from 20 years of bacterial genome sequencing. Funct Integr Genomics 15:141–161

Lasker BA (2002) Evaluation of performance of four genotypic methods for studying the genetic epidemiology of *Aspergillus fumigatus* isolates. J Clin Microbiol 40:2886–2892
Leadbetter JR (2003) Cultivation of recalcitrant microbes: cells are alive, well and revealing their secrets in the 21st century laboratory. Curr Opin Microbiol 6:274–281
Lechevalier MP (1989) Lipids in bacterial taxonomy. In: O'Leary WM (ed) Practical handbook of microbiology. CRC Press, Boca Raton, pp 57–67
Li XQ, Du D (2014) Variation, Evolution, and Correlation Analysis of C+G Content and Genome or Chromosome Size in Different Kingdoms and Phyla. PLoS One 9(2):e88339. https://doi.org/10.1371/journal.pone.0088339
Li HB, Singh RK, Singh P, Song QQ, Xing YX, Yang LT, Li YR (2017) Genetic diversity of nitrogen-fixing and plant growth promoting pseudomonas species isolated from sugarcane rhizosphere. Front Microbiol 8:1268. https://doi.org/10.3389/fmicb.2017.01268
Liebeke M, Brözel VS, Hecker M, Lalk M (2009) Chemical characterization of soil extract as growth media for the ecophysiological study of bacteria. Appl Microbiol Biotechnol 83:161–173. https://doi.org/10.1007/s00253-009-1965-0
Lima-Perim JE et al (2016) Linking the composition of bacterial and archaeal communities to characteristics of soil and flora composition in the Atlantic Rainforest. PLoS One 11(1):e0146566
Lochhead AG, Chase FE (1943) Qualitative studies of soil micro-organisms: Nutritional requirements of the predominant bacterial flora. Soil Sci 55:185
Lorenz P, Schleper C (2002) Metagenome—a challenging source of enzyme discovery. J Mol Catal B Enzym 19:13–19
Loy A, Lehner A, Lee N, Adamczyk J, Meier H, Ernst J, Schleifer KH, Wagner M (2002) Oligonucleotide microarray for 16S rRNA gene-based detection of all recognized lineages of sulfate-reducing prokaryotes in the environment. Appl Environ Microbiol 68(10):5064–5081. https://doi.org/10.1128/AEM.68.10.5064–5081.2002
Lynch JM (1990) The rhizosphere. Wiley, New York
Madsen EL (2005) Identifying microorganisms responsible for ecologically significant biogeochemical processes. Nat Rev Microbiol 3:439–446
Manlay RJ, Feller C, Swift MJ (2007) Historical evolution of soil organic matter concepts and their relationships with the fertility and sustainability of cropping systems. Agric Ecosyst Environ 119:217–233
Martín M, Gibello A, Lobo C, Nande M, Garbi C, Fajardo C, Barra-Caracciolo A, Grenni P, Martínez-Iñigo MJ (2008) Application of fluorescence in situ hybridization technique to detect simazine-degrading bacteria in soil samples. Chemosphere 71(4):703–710
Meena H, Meena RS (2017) Assessment of sowing environments and bio-regulators as adaptation choice for clusterbean productivity in response to current climatic scenario. Bangladesh J Bot 46(1):241–244
Meena RS, Yadav RS (2014) Phonological performance of groundnut varieties under sowing environments in hyper arid zone of Rajasthan, India. J Appl Nat Sci 6(2):344–348
Meena RS, Yadav RS (2015) Yield and profitability of groundnut *(Arachis hypogaea* L) as influenced by sowing dates and nutrient levels with different varieties. Legum Res 38(6):791–797
Meena RS, Yadav RS, Meena VS (2014) Response of groundnut (*Arachis hypogaea* L.) varieties to sowing dates and NP fertilizers under Western Dry Zone of India. Bangladesh J Bot 43(2):169–173
Meena RS, Dhakal Y, Bohra JS, Singh SP, Singh MK, Sanodiya P (2015a) Influence of bioinorganic combinations on yield, quality and economics of mungbean. Am J Exp Agric 8(3):159–166
Meena RS, Meena VS, Meena SK, Verma JP (2015b) The needs of healthy soils for a healthy world. J Clean Prod 102:560–561
Meena RS, Meena VS, Meena SK, Verma JP (2015c) Towards the plant stress mitigate the agricultural productivity: a book review. J Clean Prod 102:552–553
Meena RS, Yadav RS, Meena H, Kumar S, Meena YK, Singh A (2015d) Towards the current need to enhance legume productivity and soil sustainability worldwide: a book review. J Clean Prod 104:513–515

Meena RS, Yadav RS, Reager ML, De N, Meena VS, Verma JP, Verma SK, Kansotia BC (2015e) Temperature use efficiency and yield of groundnut varieties in response to sowing dates and fertility levels in Western Dry Zone of India. Am J Exp Agric 7(3):170–177

Meena H, Meena RS, Singh B, Kumar S (2016a) Response of bio-regulators to morphology and yield of clusterbean [*Cyamopsis tetragonoloba* (L.) Taub.] under different sowing environments. J Appl Nat Sci 8(2):715–718

Meena RS, Bohra JS, Singh SP, Meena VS, Verma JP, Verma SK, Shiiag SK (2016b) Towards the prime response of manure to enhance nutrient use efficiency and soil sustainability a current need: a book review. J Clean Prod 112:1258–1260

Meena RS, Gogaoi N, Kumar S (2017a) Alarming issues on agricultural crop production and environmental stresses. J Clean Prod 142:3357–3359

Meena RS, Kumar S, Pandey A (2017b) Response of sulfur and lime levels on productivity, nutrient content and uptake of sesame under guava (*Psidium guajava* L.) based agri-horti system in an acidic soil of eastern Uttar Pradesh, India. J Crop Weed 13(2):222–227

Meena RS, Meena PD, Yadav GS, Yadav SS (2017c) Phosphate solubilizing microorganisms, principles and application of microphos technology. J Clean Prod 145:157–158

Meena H, Meena RS, Lal R, Singh GS, Mitran T, Layek J, Patil SB, Kumar S, Verma T (2018a) Response of sowing dates and bio regulators on yield of clusterbean under current climate in alley cropping system in eastern U.P. Indian Legum Res 41(4):563–571

Meena RS, Kumar V, Yadav GS, Mitran T (2018b) Response and interaction of *Bradyrhizobium japonicum* and *Arbuscular mycorrhizal* fungi in the soybean rhizosphere: a review. Plant Growth Regul 84:207–223

Mendes R, Garbeva P, Raaijmakers JM (2013) The rhizosphere microbiome: significance of plant beneficial, plant pathogenic, and human pathogenic microorganisms. FEMS Microbiol Rev 37:634–663

Mishra J, Tewari S, Singh S, Arora NK (2015) Biopesticides: where we stand? In: Arora NK (ed) Plant microbes symbiosis: applied facets. Springer, New Delhi, p 37. https://doi.org/10.1007/978-81-322-2068-82

Morgan JAW, Whipps JM (2001) Methodological approaches to the study of rhizosphere carbon flow and microbial population dynamics. In: Pinton A, Varanini Z, Nannipieri P (eds) The rhizosphere. Biochemistry and organic substances at the soil-plant interface. Marcel Dekker, Inc., New York, pp 373–409

Morgan JA, Winstanley E (1997) Microbial biomarkers. In: VanElsas JD, Trevors JT, Wellington EM (eds) Modern soil microbiology. Marcel Dekker, Inc., New York, pp 331–352

Moter A, Gobel UB (2000) Fluorescence in situ hybridization (FISH) for direct visualization of microorganisms. J Microbiol Methods 41:85–112

Muller DB, Vogel C, Bai Y, Vorholt JA (2016) The plant microbiota: systems-level insights and perspectives. In: Bonini NM (ed) Annual review of genetics, vol 50. Annual Reviews, Palo Alto, pp 211–234

Naafs DFW, van Bergen PF, Boogert SJ, de Leeuw JW (2004) Solvent extractable lipids in an acid andic forest soil: variations with depth and season. Soil Biol Biochem 36:297–308

Nercessian O, Prokofeva M, Lebedinski A, Haridon LS, Cary C, Prieur D, Jeanthon C (2004) Design of 16S rRNA-targeted oligonucleotide probes for detecting cultured and uncultured archaeal lineages in high-temperature environments. Environ Microbiol 6:170–182

Nielsen MN, Winding A (2002) Microorganisms as indicators of soil health. NERI Technical Report No. 388. National Environmental Research Institute, Ministry of the Environment, Denmark. http://www.dmu.dk

Nihorimbere V, Ongena M, Smargiassi M, Thonart P (2011) Beneficial effect of the rhizosphere microbial community for plant growth and health. Biotechnol Agron Soc 15:327–337

Nouri E, Breuillin-Sessoms F, Feller U, Reinhardt D, Dutilh BE (2014) Phosphorus and nitrogen regulate arbuscular mycorrhizal symbiosis in *Petunia hybrid*. PLoS One 9(3):e90841. https://doi.org/10.1371/journal.pone.0090841

Novello G, GamaleroE BE, Boatti L, Mignone F, MassaN CP, Lingua G, Berta G (2017) The rhizosphere bacterial microbiota of *Vitis vinifera* cv. pinot noir in an integrated pest management vineyard. Front Microbiol. https://doi.org/10.3389/fmicb.2017.01528
Nusslein K, Tiedje JM (1999) Soil bacterial community shift correlated with change from forest to pasture vegetation in a tropical soil. Appl Environ Microbiol 65:3622–3626
Orphan VJ, Taylor LT, Hafenbrad lD, DeLong EF (2000) Culture-dependent and culture-independent characterization of microbial assemblages associated with high-temperature petroleum reservoirs. Appl Environ Microbiol 66:700–711
Overmann J, Gemerden HV (2000) Microbial interactions involving sulfur bacteria: implications for the ecology and evolution of bacterial communities. FEMS Microbiol Rev 24:591–599
Ovreas L, Torsvik V (1998) Microbial diversity and community structure in two different agricultural soil communities. Microb Ecol 36:303–315
Ovreas L, Daae FL, Heldal M, Torsvik V, Rodriguez-Valera F (2003) Characterization of microbial diversity in hypersaline environments by melting profiles and reassociation kinetics in combination with terminal restriction fragment length polymorphism (T-RFLP). Microb Ecol 46:291–301
Pan I, Dam B, Sen SK (2012) Composting of common organic wastes using microbial inoculants. 3 Biotech 2(2):127–134. https://doi.org/10.1007/s13205-011-0033-5
Pearce DA, van der Gast CJ, Lawley B, Ellis-Evans JC (2003) Bacterioplankton community diversity in a maritime Antarctic lake, determined by culture-dependent and culture-independent techniques. FEMS Microbiol Ecol 45:59–70
Perez-Trallero E, Montes M, Orden B, Tamayo E, Garcia-Arenzana JM, Marimon JM (2007) Phenotypic and genotypic characterization of *Streptococcus pyogenes* isolates displaying the MLSB phenotype of macrolide resistance in Spain, 1999 to 2005. Antimicrob Agents Chemother 51:1228–1233
Pernthaler A, Pernthaler J, Amann R (2002) Fluorescence *in situ* hybridization and catalyzed reporter deposition for the identification of marine bacteria. Appl Environ Microbiol 68:3094–3101
Pernthaler J, Pernthaler A, Amann R (2003) Automated enumeration of groups of marine picoplankton after fluorescence *in situ* hybridization. Appl Environ Microbiol 69:2631–2637
Pham VHT, Kim J (2016) Improvement for isolation of soil bacteria by using common culture media. J Pure Appl Microbiol 10(1):49–59
Pinton R, Varanini Z, Nannipieri P (eds) (2001) The rhizosphere: biochemistry and organic substances at the soil–plant interface. Marcel Dekker, New York
Poole P, Ramachandran V, Terpolilli J (2018) Rhizobia: from saprophytes to endosymbionts. Nat Rev Microbiol 16:291–303
Poulsen LK, Ballard G, Stahl DA (1993) Use of rRNA fluorescence in situ hybridization for measuring the activity of single cells in young and established biofilms. Appl Environ Microbiol 59:1354–1360
Preston-Mafham J, Boddy L, Randerson P (2002) Analysis of microbial community functional diversity using sole-carbon-source utilization profiles – a critique. FEMS Microbiol Ecol 42:1–14
Prosser JI (2002) Molecular and functional diversity in soil microorganisms. Plant Soil 244:9–17
Rajan A, Aruna N, Kaur S (2011) Comparative study of FAME and sequence analysis for identification of Bacteria. Biotechnol Bioinf Bioeng 1(3):319–323
Ram K, Meena RS (2014) Evaluation of pearl millet and mungbean intercropping systems in Arid Region of Rajasthan (India). Bangladesh J Bot 43(3):367–370
Rashida MI, Mujawar LH, Shahzade T, Almeelbi T, Ismail IMI, Oves M (2016) Bacteria and fungi can contribute to nutrients bioavailability and aggregate formation in degraded soils. Microbiol Res 183:26–41
Raynaud X, Nunan N (2014) Spatial ecology of bacteria at the microscale in soil. PLoS One 9:e87217

Reddy VS, Shlykov MA, Castillo R, Sun EI, Saier MH (2012) The major facilitator superfamily (MFS) revisited. FEBS J 279(11):2022–2035. https://doi.org/10.1111/j.1742-4658.2012.08588.x. PMID:22458847

Reichardt WT (1978) Einfuhrung in die Methoden der Gewassermikrobiologie. Gustav-Fischer Verlag, Stuttgart

Rhee SK, Liu X, Wu L, Chong SC, Wan X, Zhou J (2004) Detection of genes involved in biodegradation and biotransformation in microbial communities by using 50-mer oligonucleotide microarrays. Appl Environ Microbiol 70(7):4303–4317. https://doi.org/10.1128/AEM.70.7.4303-4317.2004

Riesenfeld CS, Schloss PD, Handelsman J (2004) Metagenomics: genomic analysis of microbial communities. Annu Rev Genet 38:525–552

Rochat L, Pechy-Tarr M, Baehler E, Maurhofer M, Keel C (2010) Combination of fluorescent reporters for simultaneous monitoring of root colonization and antifungal gene expression by a biocontrol Pseudomonad on cereals with flow cytometry. Mol Plant-Microbe Interact 23:949–961

Roesch LFW, Fulthorpe RR, Riva A, Casella G, Hadwin AKM, Kent AD, Daroub SH, Camargo FAO, Farmerie WG, Triplett EW (2007) Pyrosequencing enumerates and contrasts soil microbial diversity. ISME J 1:283–290

Röling WFM, van Breukelen BM, Braster M, Goeltom MT, Groen J, van Verseveld HW (2000) Analysis of microbial communities in a landfill leachate polluted aquifer using a new method for anaerobic physiological profiling and 16S rDNA based fingerprinting. Microb Ecol 40:177–188

Rondon MR, August PR, Bettermann AD, Brady SF, Grossman TH, Liles MR, Loiacono KA, Lynch BA, MacNeil IA, Minor C, Tiong CL, Gilman M, Osburne MS, Clardy J, Handelsman J, Goodman RM (2000) Cloning the soil metagenome: a strategy for accessing the genetic and functional diversity of uncultured microorganisms. Appl Environ Microbiol 66:2541–2547

Ros M, Goberna M, Pascual JA, Klammer S, Insam H (2008) 16S rDNA analysis reveals low microbial diversity in community level physiological profile assays. J Microbiol Methods 72:221–226

Roszak DB, Colwell RR (1987) Survival strategies of bacteria in the natural environment. Microbiol Rev 51:365–379

Rudolph C, Wanner G, Huber R (2001) Natural communities of novel Archaea and Bacteria growing in cold sulfurous springs with a string-of-pearls-like morphology. Appl Environ Microbiol 67:2336–2344

Rudrappa T, Czymmek KJ, Pare PW, Bais HP (2008) Root-secreted malic acid recruit's beneficial soil bacteria. Plant Physiol 148(3):1547–1556. https://doi.org/10.1104/pp.108.127613

Rusch A, Amend JP (2004) Order-specific 16S rRNA-targeted oligonucleotide probes for (hyper) thermophilic archaea and bacteria. Extremophiles 8:357–366

Sandaa RA, Torsvik V, Enger Ò, Daae FL, Castberg T, Hahn D (1999) Analysis of bacterial communities in heavy metal-contaminated soils at different levels of resolution. FEMS Microbiol Ecol 30:237–251

Schleifer K (2004) Microbial diversity: facts, problems and prospects. Syst Appl Microbiol 27:3–9

Schloss PD, Handelsman J (2003) Biotechnological prospects from metagenomics. Curr Opin Bioechnol 14:303–310

Schmidt H, Eickhorst T (2014) Detection and quantification of native microbial populations on soil-grown rice roots by catalyzed reporter deposition-fluorescence in- situ hybridization. FEMS Microbiol Ecol 87:390–402

Schönhuber W, Fuchs B, Juretschko S, Amann R (1997) Improved sensitivity of whole-cell hybridization by the combination of horseradish peroxidase-labeled oligonucleotides and tyramide signal amplification. Appl Environ Microbiol 63:3268–3273

Scott SA, Davey MP, Dennis JS, Horst I, Howe CJ, Lea Smith DJ, Smith AG (2010) Biodiesel from algae: challenges and prospects. Curr Opin Biotechnol 21:277–286

Shalon D, Smith SJ, Brown PO (1996) A DNA microarray system for analyzing complex DNA samples using Two-color fluorescent probe hybridization. Genome Res 6:639–645. https://doi.org/10.1101/gr.6.7.639

Sharma PD (2005) Terrestrial environments. In: Environmental microbiology. Alpha Science International, Harrow, pp 27–51

Shridhar B (2012) Review: nitrogen fixing microorganisms. Int J Microbiol Res 3:46–52

Sihag SK, Singh MK, Meena RS, Naga S, Bahadur SR, Gaurav, Yadav RS (2015) Influences of spacing on growth and yield potential of dry direct seeded rice (*Oryza sativa* L.) cultivars. Ecoscan 9(1-2):517–519

Simon C, Daniel R (2011) Metagenomic analyses: past and future trends. Appl Environ Microbiol 77:1153–1161

Skyring GW, Quadling C (1970) Soil bacteria: a principal component analysis and guanine–cytosine contents of some Arthrobacter–coryneform soil isolates and of some named cultures. Can J Microbiol 16(2):95–106. https://doi.org/10.1139/m70-017

Somers E, Vanderleyden J, Srinivasan M (2004) Rhizosphere bacterial signalling: a love parade beneath our feet. Crit Rev Microbiol 30(4):205–240. https://doi.org/10.1080/10408410490468786

Soni R, Kumar A, Kanwar SS, Pabbi S (2017) Efficacy of liquid formulation of versatile rhizobacteria isolated from soils of Northern Western Himalayas on *Solanum lycopersicum*. IJTK 16(4):660–668

Stainer A, Levi-Minzi R, Riffaldi R (1998) Maturity evaluation of organic wastes. Biocycle 29:54–56

Staley JT, Konopka A (1985) Measurement of in situ activities of non-photosynthetic organisms in aquatic and terrestrial habitats. Annu Rev Microbiol 39:321–346

Steenhoudt O, Vanderleyden J (2006) *Azospirillum*, a free-living nitrogen-fixing bacterium closely associated with grasses: genetic, biochemical and ecological aspects. FEMS Microbiol Rev 24(4):487–506

Stefani FOP, Bell TH, Marchand C, de la Providencia IE, El Yassimi A, St-Arnaud M, Hijri M (2015) Culture-dependent and -independent methods capture different microbial community fractions in hydrocarbon-contaminated soils. PLoS One 10(6):e0128272. https://doi.org/10.1371/journal.pone.0128272

Stein JL, Marsh TL, Wu KY, Shizuya H, De Long EF (1996) Characterization of uncultivated prokaryotes: isolation and analysis of a 40-kilobase-pair genome fragment front a planktonic marine archaeon. J Bacteriol 178:591–599

Stender H, Lund K, Petersen KH, Rasmussen OF, Hongmanee P, Miorner H, Godtfredsen SE (1999) Fluorescence *in situ* hybridization assay using peptide nucleic acid probes for differentiation between tuberculous and nontuberculous *mycobacterium* species in smears of *mycobacterium* cultures. J Clin Microbiol 37:2760–2765

Stralis-Pavese N, Sessitsch A, Weilharter A, Reichenauer T, Riesing J, Csontos J, Murrell JC, Bodrossy L (2004) Optimization of diagnostic microarray for application in analysing landfill methanotroph communities under different plant covers. Environ Microbiol 6(4):347–363

Sugiyama A, Shitan N, Yazaki K (2008) Signaling from soybean roots to Rhizobium: an ATP-binding cassette-type transporter mediates genistein secretion. Plant Signal Behav 3(1):38–40. https://doi.org/10.4161/psb.3.1.4819. PMID:19704765

Surve VV, Patil MU, Dharmadhikari SM (2012) FAME and 16srDNA sequence analysis of halophilic bacteria from solar salterns of Goa: a comparative study. Int J Sci Res Publ 2(8):1–8

Tabacchioni S, Chiarini L, Bevivino A, Cantale C, Dalmastri C (2000) Bias caused by using different isolation media for assessing the genetic diversity of a natural microbial population. Microb Ecol 40:169–176

Taylor CB (1951) Nature of the factor in soil-extract responsible for bacterial growth-stimulation. Nature 168:115–116

Teotia P, Kumar M, Prasad R, Kumar V, Tuteja N, Varma A (2017) Mobilization of micronutrients by mycorrhizal fungi. In: Varma A et al (eds) Mycorrhiza – function, diversity, state of the art. https://doi.org/10.1007/978-3-319-53064-2_2

Theron J, Cloete TE (2000) Molecular techniques for determining microbial diversity and community structure in natural environments. Crit Rev Microbiol 26:37–57
Tiedje JM, Asuming-Brempong S, Nusslein K, Marsh TL, Flynn SJ (1999) Opening the black box of soil microbial diversity. Appl Soil Ecol 13:109–122. https://doi.org/10.1016/S0929-1393(99)00026-8
Torsvik V, Ovreas L (2002) Microbial diversity and function in soil: from genes to ecosystems. Curr Opin Microbiol 5:240–245
Torsvik V, Salte K, Soerheim R, Goksoeyr J (1990a) Comparison of phenotypic diversity and DNA heterogeneity in a population of soil bacteria. Appl Environ Microbiol 56:776–781
Torsvik V, Goksoyr J, Daae FL (1990b) High diversity in DNA of soil bacteria. Appl Environ Microbiol 56:782–787
Torsvik V, Sorheim R, Goksoyr J (1996) Total bacterial diversity in soil and sediment communities—a review. J Ind Microbiol 17:170–178
Torsvik V, Daae FL, Sandaa R, Ovreas L (1998) Review article: novel techniques for analysing microbial diversity in natural and perturbed environments. J Biotechnol 64:53–62
Trevors JT (1998) Bacterial biodiversity in soil with an emphasis on chemically-contaminated soils. Water Air Soil Pollut 101:45–67
Tyson GW, Chapman J, Hugenholtz P, Allen EE, Ram RJ, Richardson PM, Solovyev VV, Rubin EM, Rokhsar DS, Banfield JS (2004) Community structure and metabolism through reconstruction of microbial genomes from the environment. Nature 428:37–43
Vahjen W, Munch JC, Tebbe CC (1995) Carbon source utilization of soil extracted microorganisms supplemented with genetically engineered and non-engineered *Corynebacterium glutamicum* and a recombinant peptide at the community level. FEMS Microbiol Ecol 18:317–328
Van Nguyen T, Pawlowski K (2017) *Frankia* and actinorhizal plants: symbiotic nitrogen fixation. In: Mehnaz S (ed) Rhizotrophs: plant growth promotion to bioremediation, Microorganisms for sustainability, vol 2. Springer, Singapore
Varma D, Meena RS, Kumar S (2017) Response of mungbean to fertility and lime levels under soil acidity in an alley cropping system in Vindhyan Region, India. Int J Chem Stu 5(2):384–389
Varma D, Meena RS, Kumar S, Kumar E (2017a) Response of mungbean to NPK and lime under the conditions of Vindhyan Region of Uttar Pradesh. Legum Res 40(3):542–545
Venter JC, Remington K, Heidelberg JF, Halpern AL, Rusch D, Eisen JA, Wu D, Paulsen I, Nelson KE, Nelson W, Fouts DE, Levy S, Knap AH, Lomas MW, Nealson K, White O, Peterson J, Hoffman J, Parsons J, Tillson HB, Pfannkoch C, Rogers YH, Smith OH (2004) Environmental genome shotgun sequencing of the Sargasso Sea. Science 304:66–74
Vera-Gargallo B, Navarro-Sampedro L, Carballo M, Ventosa A (2018) Metagenome Sequencing of Prokaryotic Microbiota from Two Hypersaline Soils of the Odiel Salt Marshes in Huelva, Southwestern Spain. Genome Announc 6(9):e00140–e00118. https://doi.org/10.1128/genomeA.00140-18
Verma JP, Jaiswal DK, Meena VS, Meena RS (2015) Current need of organic farming for enhancing sustainable agriculture. J Clean Prod 102:545–547
Verma JP, Meena VS, Kumar A, Meena RS (2015a) Issues and challenges about sustainable agriculture production for management of natural resources to sustain soil fertility and health: a book review. J Clean Prod 107:793–794
Verma SK, Singh SB, Prasad SK, Meena RN, Meena RS (2015b) Influence of irrigation regimes and weed management practices on water use and nutrient uptake in wheat (*Triticum aestivum* L. Emend. Fiori and Paol.). Bangladesh J Bot 44(3):437–442
Vester F, Ingvorsen K (1998) Improved most-probable-number method to detect sulfate-reducing bacteria with natural media and a radiotracer. Appl Environ Microbiol 64(5):1700–1707
Wagner M, Horn M, Daims H (2003) Fluorescence *in situ* hybridisation for the identification and characterization of prokaryotes. Curr Opin Microbiol 6:302–309
Ward DM, Weller R, Bateson MM (1990) 16S ribosomal RNA sequences reveal numerous uncultured microorganisms in a natural community. Nature 345:63–65

Weston LA, Ryan PR, Watt M (2012) Mechanisms for cellular transport and release of allelochemicals from plant roots into the rhizosphere. J Exp Bot 63:3445–3454. https://doi.org/10.1093/jxb/ers054

Wintzingerode FV, Göbel-Ulf B, Stackebrandt E (1997) Determination of microbial diversity in environmental samples: pitfalls of PCR-based rRNA analysis. FEMS Microbiol Rev 21:213–229. https://doi.org/10.1111/j.1574-6976.1997.tb00351.x

Woese CR (1987) Bacterial evolution. Microbiol Rev 51:221–271

Worner U, Zimmermann-Timm H (2000) *Beggiatoa leptomitiformis* – a filamentous sulfur-oxidizing bacterium colonizing laboratory-made aggregates. Limnol Ecol Manag Inland Waters 30(2):215–221

Yadav GS, Babu S, Meena RS, Debnath C, Saha P, Debbaram C, Datta M (2017) Effects of godawariphosgold and single supper phosphate on groundnut (*Arachis hypogaea*) productivity, phosphorus uptake, phosphorus use efficiency and economics. Indian J Agric Sci 87(9):1165–1169

Yadav GS, Lal R, Meena RS, Babu S, Das A, Bhomik SN, Datta M, Layak J, Saha P (2017a) Conservation tillage and nutrient management effects on productivity and soil carbon sequestration under double cropping of rice in North Eastern Region of India. Ecol Indic. http://www.sciencedirect.com/science/article/pii/S1470160X17305617

Yadav GS, Lal R, Meena RS, Datta M, Babu S, Das A, Layek J, Saha P (2017b) Energy budgeting for designing sustainable and environmentally clean/safer cropping systems for rainfed rice fallow lands in India. J Clean Prod 158:29–37

Yadav GS, Das A, Lal R, Babu S, Meena RS, Saha P, Singh R, Datta M (2018a) Energy budget and carbon footprint in a no-till and mulch based rice–mustard cropping system. J Clean Prod 191:144–157

Yadav GS, Das A, Lal R, Babu S, Meena RS, Patil SB, Saha P, Datta M (2018b) Conservation tillage and mulching effects on the adaptive capacity of direct-seeded upland rice (*Oryza sativa* L.) to alleviate weed and moisture stresses in the North Eastern Himalayan Region of India. Arch Agron Soil Sci. https://doi.org/10.1080/03650340.2018.1423555

Yang Y, Wang N, Guo X, Zhang Y, Ye B (2017) Comparative analysis of bacterial community structure in the rhizosphere of maize by high-throughput pyrosequencing. PLoS One 12(5):e0178425. https://doi.org/10.1371/journal.pone.0178425

Yao H, He Z, Wilson MJ, Campbell CD (2000) Microbial biomass and community structure in a sequence of soils with increasing fertility and changing land use. Microb Ecol 40:223–237

Yazaki K (2005) Transporters of secondary metabolites. Curr Opin Plant Biol 8(3):301–307. https://doi.org/10.1016/j.pbi.2005.03.011. PMID:15860427

Zak JC, Willig MR, Moorhead DL, Wildman HG (1994) Functional diversity of microbial communities: a quantitative approach. Soil Biol Biochem 26:1101–1108

Zelles L (1999) Fatty acid patterns of phospholipids and lipopolysaccharides in the characterisation of microbial communities in soil: A review. Biol Fertil Soils 29:111–129

Zeyaullah M, Kamli MR, Islam B, Atif M, Benkhayal FA, Nehal M, Rizvi MA, Ali A (2009) Metagenomics- an advanced approach for noncultivable micro-organisms. Biotechnol Mol Biol Rev 4:49–54

Zhalnina K, Louie KB, Hao Z, Mansoori N, da Rocha UN, Shi S, Cho H, Karaoz U, Loqué D, Benjamin P, Bowen, Firestone MK, Northen TR, Brodie EL (2018) Dynamic root exudate chemistry and microbial substrate preferences drive patterns in rhizosphere microbial community assembly. Nat Microbiol 3:470–480. https://doi.org/10.1038/s41564-018-0129-3

Zhang L, Xu Z (2008) Assessing bacterial diversity in soil. J Soils Sediments 8:379–388. https://doi.org/10.1007/s11368-008-0043-z

Zimmermann J, Ludwig W, Schleifer KH (2001) DNA polynucleotide probes generated from representatives of the genus *Acinetobacter* and their application in fluorescence *in situ* hybridization of environmental samples. Syst Appl Microbiol 24:238–244

Conservation Agriculture Practices to Improve the Soil Water Management and Soil Carbon Storage in Mediterranean Rainfed Agro-Ecosystems

6

Iván Francisco García-Tejero, Rosa Carbonell, Rafaela Ordoñez, Francisco Perea Torres, and Víctor Hugo Durán Zuazo

Abstract

Water is the most limiting natural resource in agro-ecosystems of arid and semi-arid environments. In this regard, many areas of Southern Europe and other countries with similar climatic characteristic are being affected by the climate change, and many efforts oriented to achieve the sustainability of agriculture are being developed. Water limitations are forcing to implement different strategies to improve the water usage in agriculture, with significant constraints for the case of rainfed systems, where the improvement of water management must be focused to those soil management systems able to increase the soil water holding capacity. Implementing conservation agriculture (CA) practices such as minimum tillage (MT), direct drilling (DD), or the use of cover crops in perennial systems allows to improve the soil water retention and the its disposal for the crop during the maximum evapotranspiration period. Additionally, CA not only implies management strategies focused to climate change adaptation but also mitigation, encouraging the role of soil as a carbon sink. This work summarizes the advantages of these strategies, focusing on the effects of DD versus conventional tillage (CT) in rainfed systems for annual crops in terms of soil water conservation and carbon storage. On overall, DD practices allow to retain more water in the soil profile, as in the first centimetres as in the deeper zones of soil

I. F. García-Tejero · F. P. Torres
Centro IFAPA "Las Torres-Tomejil", Seville, Spain

R. Carbonell · R. Ordoñez
Centro IFAPA "Alameda del Obispo", Córdoba, Spain

V. H. Durán Zuazo (✉)
Centro IFAPA "Camino de Purchil", Granada, Spain
e-mail: victorh.duran@juntadeandalucia.es

R. S. Meena (ed.), *Soil Health Restoration and Management*,
https://doi.org/10.1007/978-981-13-8570-4_6

profile, with a faster soil water depletion in CT plots Moreover, significant higher CO_2 emissions are promoted in CT systems in comparison to DD, in which the soil acts as a potential carbon sink. In conclusion, CA practices can act with a double role: as mitigation and as adaptation strategies to climate change.

Keywords
Direct drill · Conventional tillage · Soil water dynamic · Soil physical properties

Abbreviations

AOC	Active organic carbon
CA	Conservation agriculture
DD	Direct drill
FC	Field capacity
GHG	Greenhouse emission gasses
MCP	Multi-sensor capacitance probes
MT	Minimum tillage
OM	Organic matter
PWP	Permanent wilting point
SOC	Soil organic carbon
SWAP	Soil water atmosphere plant
ρ_b	Bulk density

6.1 Introduction

Water scarcity and rainfall irregularity are the main characteristics of climate in many regions of Southern Europe, with the driest period coinciding with those months in which the evapotranspiration is higher (La Jeunesse et al. 2016; Lorite et al. 2018; Meena and Meena 2017). Moreover, according to the last forecast published by the Intergovernmental Panel on Climate Change (IPCC 2014), water depletions are taking place in many areas of Southern Europe, and it is expected that, in the following years, this will occur with a higher significance. Consequently, the last predictive models for climate change predict higher depletions of soil water reserves, together with an increasing of reference evapotranspiration. Thus, soil water extraction by crops will significantly increase (≅30% on average, depending of the crop and management system), promoting an imbalance between the crop water demand and the available resources (García-Tejero et al. 2014; Datta et al. 2017b; Meena et al. 2016b).

Improving the water conservation and productivity turns out to be even more necessary under the current and expected conditions, with many efforts oriented to

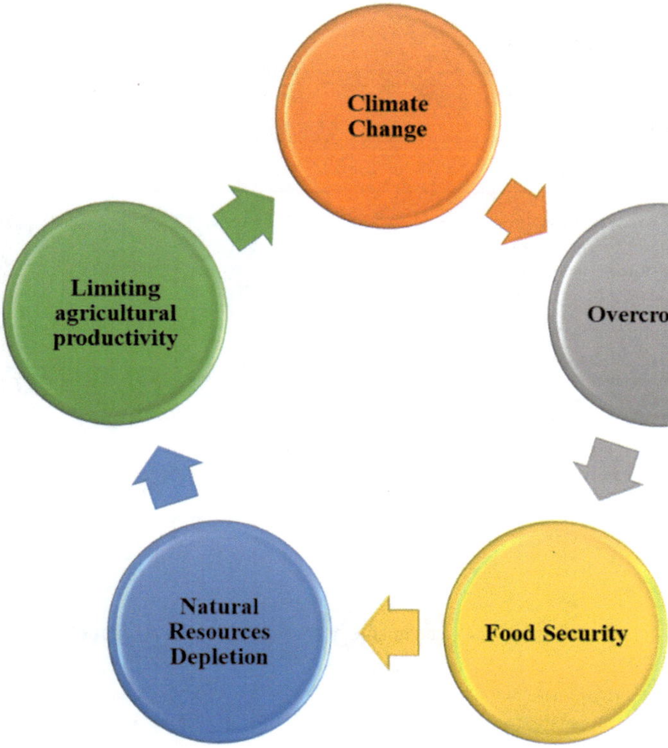

Fig. 6.1 Main limiting factors to achieve a new paradigm of sustainable agriculture

reach a maximum agricultural profitability together with natural resources conservation within the wide context of "sustainable agriculture" (García-Tejero et al. 2011; Yadav et al. 2018b). However, this fact does not result easy, within the current context of climate change, overpopulation (actually there are 7,5 billion of people, being expected to reach 9,5 in the following 30 years), the food scarcity, the overexploitation of natural resources, and the threshold falling of land productivity (as result of the last 70 years of intensive agriculture and natural resources degradation) (Iglesias and Garrote 2018) (Fig. 6.1).

Taking into account this current scenario, the different stakeholders involved in the agricultural sector should focus their efforts by, on one hand, recovering the quality of natural resources involved in the farming practices throughout different sustainable practices and, on the other hand, promoting a new paradigm of sustainable agriculture, mitigating the potential effects of this activity in the climate change, and adapting the strategies to the new limitations, always with the aim of achieving an equilibrium between agriculture, environment, and food security (García-Tejero et al. 2014; Meena et al. 2015a; Dhakal et al. 2016) (Fig. 6.2).

It is not easy to implement new strategies of agricultural production, reaching similar production values and keeping (or even improving) the natural resources

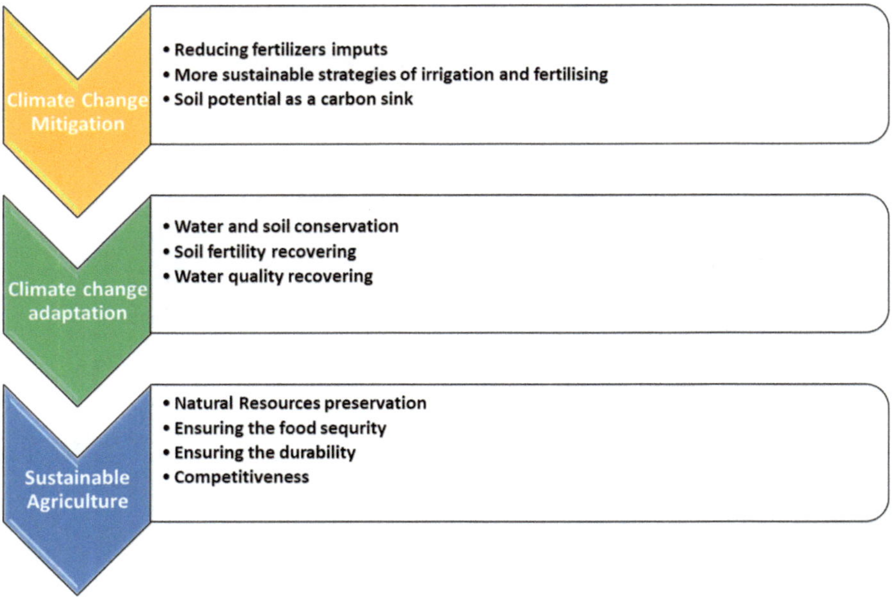

Fig. 6.2 Global actions to achieve a sustainable agriculture system under climate change conditions

quality, especially, when these are scarcer and their quality has been progressively ameliorated. These strategies should be focused in two complementary ways: reducing those actions in order to promote a climate change mitigation and others related to climate change adaptation. In this regard, conservation agriculture (CA) practices encompass those techniques focused on reducing or eliminating the soil tillage, in rainfed systems for annual crops, and keeping a natural cover in permanent crops with different soil management strategies such as direct drilling (DD) and minimum tillage (MT). Moreover, CA represents a wider concept than DD, focusing on three main strategies: minimizing the soil disturbance, crop rotation, and permanent organic soil covers (Kassam et al. 2012; Ashoka et al. 2017; Meena et al. 2015c) (Fig. 6.3).

CA techniques in Mediterranean countries as well as worldwide were initially introduced mainly by economic factors (Soane and Ball 1998), with the aim of reducing the crop production cost, especially those related to the energy consumes. Actually, around 125.000 ha around the world are cultivated under CA practices, representing 10% of global arable surface (Kertész and Madarász 2014; Meena et al. 2015d; Sihag et al. 2015). During the twenty-first century, the adoption of these practices has significantly increased, around 7.000 ha year^{-1} (Kassam et al. 2014), with the most rapid expansion in North and South America, Australia, and some countries of Africa (Friedrich et al. 2012). Likewise, the adoption of CA by European farmers is still meagre compared to other regions of the world (Derpsch 2005; Ram and Meena 2014). For the case of Spain, CA and DD in particular began

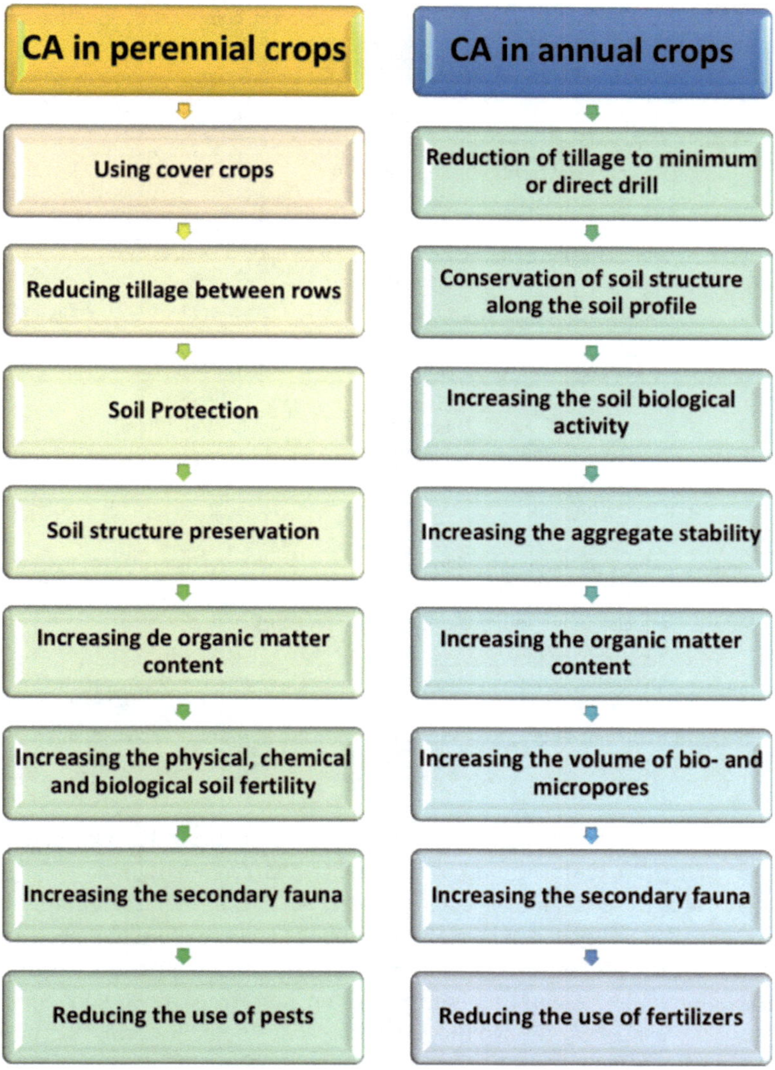

Fig. 6.3 Activities and effects of conservation agriculture (CA) practices in perennial and annual crops systems

in the early 1980s to improve soil water retention and reduce erosion (Fernández 1997), with rapid implementation mainly in rain-fed systems (Bescana et al. 2006; Kumar et al. 2017b).

The positive impact of CA techniques (Table 6.1) includes the major conservation and improvement of soil organic matter with slow mineralization that directly improves physical soil properties (Pelegrin et al. 1990; Ball et al. 1994; Suwardji and Eberbach 1998; Lars et al. 2003; Varma et al. 2017b) in different types of soils

Table 6.1 Positive impacts of CA practices in the different components of soil-water-plant-atmosphere system

Actions	Effects	Level	References
No tillage	Higher water conservation	Soil and Water	Vercauteren (2013)
	Soil physical fertility	Soil and Water	Pelegrin et al. (1990), Ball et al. (1994); Lars et al. (2003); Hatfield et al. (2001); López-Bellido et al. (2007)
Land cover crops	Increasing the SOC	Soil, Plant, Atmosphere	Kirkby et al. (2013)
	Carbon sequestration	Soil, Atmosphere	Dhruw et al. (2009), Wani and Qaisar (2014)
	Increasing biomass carbon	Soil, Plant	Macedo et al. (2006, 2008); Meena et al. (2015a)
Crop rotations	Increasing productivity and sustainability	Soil, Plant	Keeler et al. (2009)
	Controlling soil erosion	Soil, Water	Ratnadass et al. (2013); Lithourgidis et al. (2011)
Intercropping	Increasing biological diversity	Plant	Lithourgidis et al. (2011), Kumar et al. (2016)
	Increasing the soil biological fertility	Soil, Plant	Fustec et al. (2010)
	Increasing water conservation	Water	Ghanbari et al. (2010)
	Reducing soil erosion	Soil and plant	Hauggaard-Nielsen et al. (2007)

and crops in many environments (Christian and Bacon 1990; Richard et al. 1995; Lowther et al. 1996; Ordoñez et al. 2007; Meena and Yadav 2015).

The advantages of CA are especially noticeable in rain-fed agriculture systems, where water availability depends exclusively on the amount of rainfall and its distribution, mainly in those regions such as Southern Europe, with an adverse climatology in relation to the rainfall irregularity, and the duration of the dry periods. These conditions are especially unfavourable for crop development during the period in which the soil water removal by the crop is highest (Perea et al. 2006; Datta et al. 2017a; Meena et al. 2017a). Considering that in rainfed systems it is not possible (or very difficult) to increase the soil water content (by means of irrigation), the only way of improving the water management is throughout physical soil properties management. In this agreement, Hatfield et al. (2001) and López-Bellido et al. (2007) argued that modifying the physical soil properties influence the soil water extraction by the crop or by means of the evaporation process thus finally affecting the soil water dynamic along the profile. Consequently, CA practices in rainfed systems, especially under these particular weather conditions, could be the first step to advance in the soil water conservation and agricultural sustainability (Ordoñez et al. 2007; Kumar et al. 2017a).

6.2 Improvements of Soil Physical Properties Linked to No-Tillage Systems

Water is the dynamic element of the complex system "soil-water-plant-atmosphere (SWAP)", where each component is perfectly integrated and their conditions mainly depend on the relationships between them. Water is responsible for material and energy fluxes that occur in the SWAP system and for communication and interconnection of the processes involved in the functioning of this system. In this sense, the principle that governs the water movement in the soil is the spontaneous flux of water, from areas with high water potential to those where this potential energy is lower. This water potential gradient is the force that causes the flow in the soil; and between this, the plant and atmosphere (Fig. 6.4).

Thus, those practices able to ease the water uptake by the crops should be considered.

The soil water balance is governed by the processes of infiltration, redistribution, drainage, evaporation, and crop water extraction, which are closely related, determining the *inputs* and *outputs* of SWAP system (Fig. 6.5).

Soil management determines the physical properties and the OM content, being especially relevant the soil characteristics in the first centimetres of the top layer (0–5 cm), which acts as the interface between the gas phase (atmosphere) and solid phase (soil) of the complex SWAP. In this sense, all these actions, movements, and managements developed in the first centimetres of soil will determine the following changes in the deeper zones of soil profile (Shaver et al. 2002; Yadav et al. 2018a; Meena et al. 2017b).

Bulk density, total and effective porosity, aggregate stability, sorptivity, and saturated hydraulic conductivity are the most affected physical properties for tillage system; these properties are very related to the presence and type of soil organic matter and determine water holding capacity of soil and the energy status of the water retained in the soil.

Bulk density (ρ^b) characterizes the density of the solid soil matrix and considers the relationships between volume and mass of the solid soil phases. It is defined as the mass of dry soil per volume of this. Tillage initially decreases the ρ_b of soil surface, but at long term, the passing of machinery induces a progressive soil

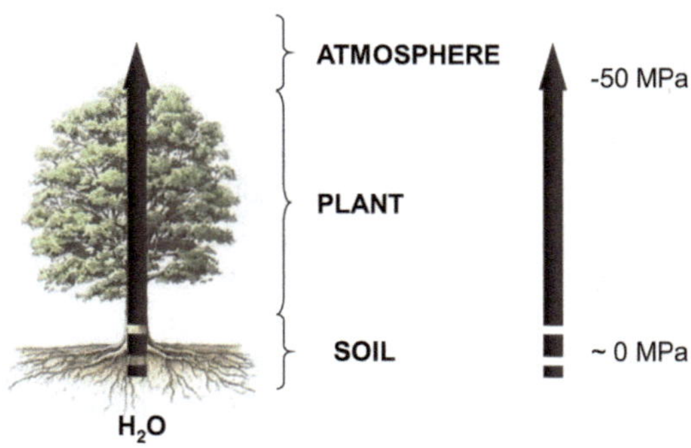

Fig. 6.4 Water extraction in the soil-plant-atmosphere system

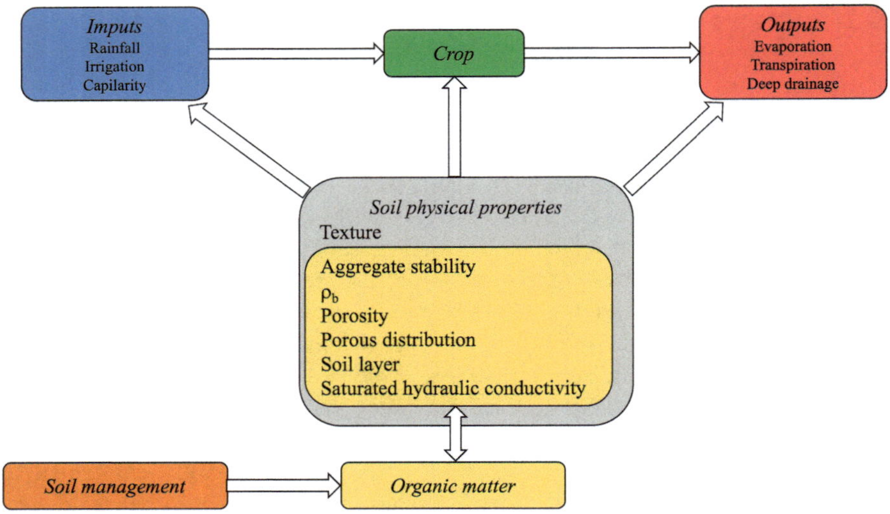

Fig. 6.5 Simplified diagram of soil water balance (SWB) and the relationships between tillage, organic matter, soil physical properties, and their effects in the SWB

compaction, increasing ρ_b along the soil profile and promoting the formation of an arable soil layer. By the contrast, Dao (1996) showed that those soils under NT conditions were able to keep a lower bulk density than those subjected to traditional soil management methods, especially in the first 5 cm of soil. Hammel (1989) found that direct drill promoted an increase of bulk density in comparison to tillage systems in the surface 0.3 m of soil. However, along the time, differences were nonexistent when averaged over a depth of 0–0.5 m. This property is very related to the total and effective porosity.

Porosity links the relative pore space of a soil with its total volume. Its value generally ranges from 0.3 to 0.6, and it is very influenced by size particle distribution, soil organic matter content, soil water content, or tillage system. However, this value does not offer information about the kind of pores and its distribution.

Obviously, bulk density and total porosity are highly related. Any practices able to decrease the bulk density will increase total soil porosity (Shaver et al. 2002).

CA practices promote the increasing of soil porosity, especially in the first layers of soil surface, although the effective pores could descend in number, stability, and continuity, compared with conventional tillage (CT) systems (Roseberg and McCoy (1992). These effective pores are more related to the saturated hydraulic conductivity and the number of free pores when a rain event takes place (Ahuja et al. 1984, 1989).

When we talk about the soil porosity, it is important to distinguish two basic types: (i) a textural porosity (due to the size distribution of soil particles) and (ii) structural porosity, which comprised the micropores, cracks, bio-pores, and structural macropores. While the former is difficult to change, since the texture is an inherent physical property of the medium and very stable over time, the latter is greatly affected by management (Dexter 2004).

This structural porosity and pores distribution are highly related to the aggregate stability (soil structure) (Hillel 1998). In this sense, there are two main factors implicated in aggregate stability: clay and organic matter contents. Clay minerals allow the flocculation process, a requisite for the soil aggregation. Secondly, organic polymers (organic matter) promoted the cementation of these structures. Considering that the clay content cannot be naturally modified (at least for a long time period), all those practices able to increase the organic matter should be considered. In this sense, incorporating the crop residues or pruning or avoiding those practices which induce a faster organic matter mineralization should be considered. Keeping the soil aggregate stability is a major factor to maintaining a porous matrix and an adequate pores size distribution. Several works show that land cover and a greater amount of organic matter in soils without tillage avoid degradative processes such as loss of structure and soil compaction (Raper et al. 2000). These processes are favoured by the loss of organic matter associated with intensive tillage practices (Ordoñez et al. 2007; Meena et al. 2016a; Verma et al. 2015a).

The progressive loss of porosity is a result of soil compaction. This leads to a decrease of soil hydraulic conductivity, infiltration capacity, and, therefore, the water storage potential of soil profile. Tillage promotes a progressive and continuous soil compaction and the formation of ploughing layers (Fig. 6.6), as it has been previously discussed by other authors such as García-Tejero et al. (2007b) and Varma et al. (2017a).

In this sense, soil acts as a dynamic system of water storage, which returns to the atmosphere by means of the evaporation and transpiration processes. The composition of soil mineral and organic phases and their structure will be decisive factors in the storage, preservation, and availability of water for the crop. The amount of water and its energy status in the soil are two important factors affected by soil physical properties. In this sense, texture and structure will be key determinants of the behaviour of the characteristic curve of soil moisture, which relates the soil water content and the energy status (Hillel 1998). These curves can be determined throughout pedotransfer functions considering several parameters such as texture, organic matter content, bulk density, or soil structure (Schaap et al. 2001; Dadhich and Meena 2014; Meena et al. 2018b). Although texture is an inherent property of the medium and difficult to alter (at least at short–medium term), the structure can easily be altered throughout the soil management. Maintaining a good soil structure will improve the soil water properties, highly dependent on organic matter content and conservation of the pore space.

During the infiltration process, water tends to fill all the medium pores. Initially, it is distributed by the microporous space of the soil matrix, adhering to the particles of this and being held with great force. Once soil is saturated, the infiltrated water tends to occupy the larger-pore spaces, where the force in the water is held down considerably, thus being more available for crop (Field Capacity, FC). As the water fills all the pores of larger size (saturated soil), water will distribute so that it tends to occupy the deeper zones of the soil profile (infiltration). Similarly, when soil is saturated and begins to dry, the water leaves the larger-pore spaces, where the retention force between water and soil is lower. As it is dried, the required force to extract

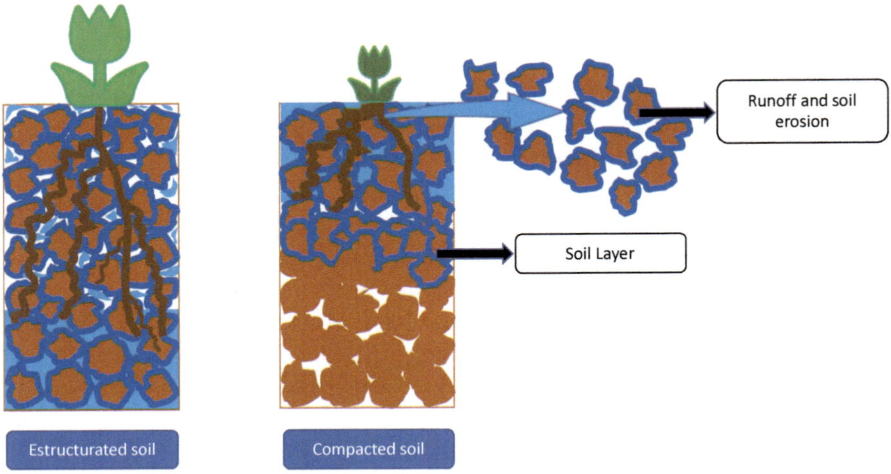

Fig. 6.6 Effects of soil compaction evolution along the soil profile

the water from the soil increases, till it becomes impossible for the extraction for the crop (permanent wilting point). Said this, the water movement in soil and their availability for the crop depends not only on the stored quantity but also a set of physical properties inherent in the medium that determine the force with which it is retained. These properties are basically the texture of the medium and the structuring of their elementary particles (Dexter 1988; Yadav et al. 2017c). The preservation of a proper soil structure along the profile largely determines the soil capability of water storage. Thus, the presence of a soil layer is promoted by (i) soil compaction by the agricultural machinery to pass along,(ii) soil degradation and rinsing out of fine soil particles to deeper zones, (iii) and soil crusting making it difficult for the soil to store water in deeper zones (Fig. 6.6). As a result, the soil water content is below its natural potential, and the crop development is affected.

Several works have shown the advantages of CA practices relating to the soil water dynamic. Muriel et al. (2005) concluded that CA practices improve the soil water retention capacity, especially in the first 30 ft deep. These authors reported that the discharge rate was slower in a soil under DD, which impacted positively on the development of spring-summer crops, where production-limiting factor is undoubtedly the availability water. Jiménez et al. (2005) found that for annual crops, in soils subjected to DD or MT, there was more available water at crop disposition. In this sense, for annual crops under rainfed systems and soils subjected to DD or MT, more water was accessible to the crop, especially during the maximum evapotranspirative demand period, being this water retained in the deeper zones of soil profile (García-Tejero et al. 2007b; Kumar et al. 2018; Meena et al. 2017c).

The characteristic curve of soil moisture retention gives the relationship between soil water content and the applied pressure, mainly determined by the size distribution of soil pores. The study of this function is a good indicator of physical quality of soil. It gives us an idea of their pore distribution, which in turn is related to soil structure, degree of compaction, hydraulic conductivity, and other relevant physics and chemical properties. Dexter (2004) studied the relationships between the slope of the function in the inflection point of the retention curve. In this sense, changes in the curve from convex to concave, and the slope value is related to some of the most remarkable soil physical properties. This point is reached when the soil begins to lose the water trapped in the structural micropores. The area between the saturation and inflexion points includes the structural pores, being the area defined below the inflection point corresponding with the textural porosity (Richard et al. 2001; Meena and Yadav 2014). Structural microporosity determines the slope value, with the range of porosity more influenced by the majority of physical properties and soil management (Dexter 1988). Soils with a lower slope in the retention curve have a greater ability to retain and conserve water. Also, different management practices may cause variations in the behaviour of the curve, mainly due to changes in the slope of discharging. García-Tejero et al. (2009) founded significant differences in these moisture retention curves, related to changes in the slope value of the inflexion point (Fig. 6.7).

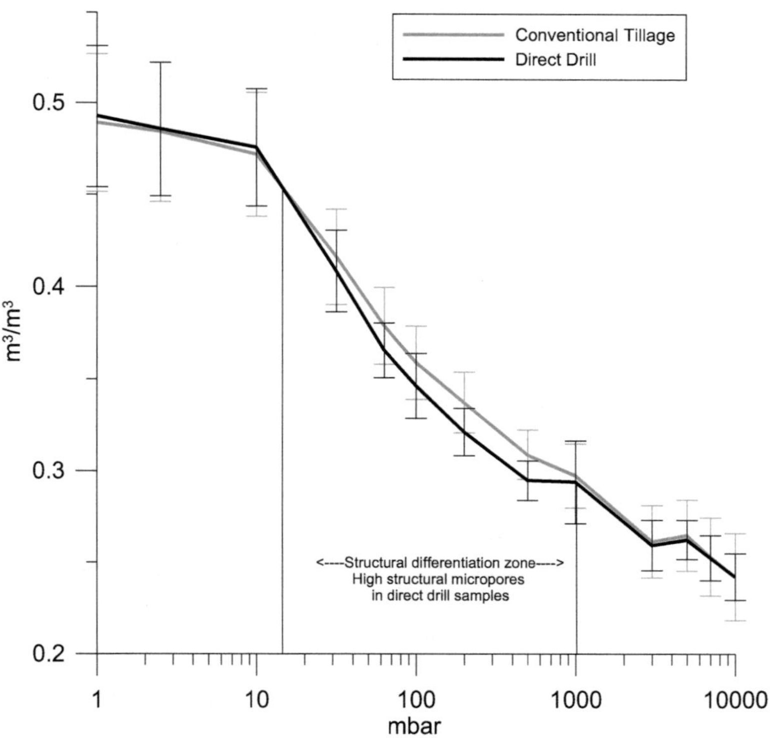

Fig. 6.7 Soil moisture retention curve for different soil managements

6.3 Effects of Direct Drill in the Dynamic of Soil Water Content in Rainfed Systems of Semi-arid Environments

When water is the most limiting factor for agricultural practices in rainfed systems, it is crucial to introduce those soil management strategies able to improve the soil water retention, and after this, the disposal of this natural resource when the crop water demand is maximum (Ordoñez et al. 2007; Buragohain et al. 2017; Meena et al. 2015b). In this sense, in annual summer crops, the greatest water losses from the profile during the first months after sowing take place through direct evaporation from the soil (López-Bellido et al. 2007; Meena et al. 2015e), whereas, as the crop is growing, transpiration rate progressively increases, and hence, water storage is the main limiting factor, determining the survival and final production (Allen et al. 1998).

Considering the different processes described in the previous sections of this chapter, CA practices in rainfed areas under the climatological characteristics of Southern Europe allow improvement of soil physics and chemical conditions, which are directly related to the soil water holding capacity (Ordoñez et al. 2007; Dhakal et al. 2015).

With the aim of defining the effects of different soil management systems in the spatial and temporal dynamic of soil water content, this section shows the main results of a long-term experience, developed during four consecutive years in a clayey soil, with a classical crop rotation of wheat-sunflower-pea field, focusing the efforts in determining the advantages of each system (DD and CT) in terms of soil water conservation.

6.3.1 Location

The experience was developed in an experimental orchard located in Carmona, Seville (SW Spain) (37° 24′ N, 5° 35′ W), which belongs to the Andalusian Institute of Training and Agricultural Research (IFAPA). The soil is a Chromic Haploxerert (USDA 1999), with a textural composition of 0.69% of sand, 35.53% of silt, and 57.53% of clay and a variable bulk density of ≅ 1.31 g cm^{-3} at FC, depending on the soil water content because of the presence of smectites (expandable clays). The organic matter content was below to 1.5%, and the soil water content at FC (−0.033 MPa) and PWP (−1.5 MPa) were 0.39 0.24 m^3 m^{-3}, respectively.

The Ap culturing horizon has approximately 25 cm of depth, and the underlying Bw horizon reached as much as 65 cm.

The climate of the study area is typically semi-arid Mediterranean, with an annual rainfall of 495 mm and a potential evapotranspiration of 1600 mm, with a large inter- and intra-annual variability and average winter and summer temperatures of 10° and 35 ° C, respectively (Perea 2000).

More information about the experimental site characteristic can be found in the following references: Perea (2000) and Ordoñez et al. (2007).

Fig. 6.8 Enviroscan probes: sensors, circuit, and data logger

6.3.2 Experimental Plots and Field Measurements

The experimental work was developed using different plots that since 1982 has been continuously subjected to the same soil management system. Specifically, these plots from the establishment have been subjected to DD and CT, with the aim of studying the long-term effects of soil management on soil physical and hydraulic properties. Four plots per management system were monitored, each one of them with 2700 m^2 and a randomized complete block design. During the studied period (2003–2007), the crop rotation was as follows: 2003–2004, sunflower; 2004–2005, wheat; 2005–2006, sunflower; and 2006–2007, pea.

Two probes for measuring the soil water content were installed in each monitoring experimental plot of CT and DD; each probe has five multisensor capacitance probes (MCP) (Enviroscan, Sentek PTY Ltd., Kent Town, Australia, 1999) installed at 10, 20, 30, 60, and 90 cm in soil depth.

Each sensor consists of two metallic rings of 5.1 cm in diameter and 7.5 cm in length, separated by one of plastic, which acts as a condenser. These sensors are connected to an oscillator and a set of circuits that modify the oscillation frequency. This frequency depends on the variation of the surrounding soil capacitance, which is influenced directly by the soil water content. The measurements were made at a frequency of 4 h. This information was stored in a data logger, and later unloaded. The whole system was fed across a photovoltaic cell connected to the system (Fig. 6.8).

6.3.3 Soil Water Content vs. Soil Management System

Table 6.2 shows the data of the annual water balance and crop yield in the different soil management systems during the 4-year monitoring period.

According to the general weather conditions registered during the studied seasons, 2003–2004 and 2006–2007 could be considered as two periods of low water

Table 6.2 General weather conditions during the monitoring period

	2003–2004		2004–2005		2005–2006		2006–2007	
ETo (mm)	1.620		1.741		1.581		1.504	
Rainfall (mm)	580		228		434		560	
AWB (mm)	−1.040		−1.513		−1.147		−944	
Crop	Sunflower		Wheat		Sunflower		Legume	
Seed time (DOY)	74		341		80		311	
Harvest (DOY)	242		178		244			
	DD	CT	DD	CT	DD	CT	DD	CT
Annual yield (kg ha^{-1})	1.017	1.756	2.192	2.469	945	847	–	–
Historical yield (kg ha^{-1})	1.030	1.130	3.563	3.759	1.030	1.130	1.441	1.259

DOY Day of the year, *DD* Direct drilling, *CT* Conventional tillage

deficit, with similar conditions to those registered in wet seasons. By contrast, 2004–2005 registered very severe weather conditions, with the highest deficit.

During 2003–2004, the difference between rainfall and ET_0 was of −1.040 mm, involving an important water deficit for crops. In this season, the soil water was observed that in the first 30 cm of soil profile, the available water content was higher in those plots under DD than in those under CT. These differences were even more clear in the first 10 cm of soil depth, diminishing gradually at greater depths (Fig. 6.9). Something similar was detected in the deeper zones. The most significant differences were detected during the discharge period (low rainfall and high ET_0), with a lower rate of water depletion process in DD in comparison to CT treatment (Fig. 6.9).

Moreover, the soil water dynamic in the deeper zones was noticeable. As it can be observed in Figure 6.10, a higher amount of water was putted at disposition of crop during the dry period in those plots under DD in comparison to CT. Consequently, the plots under DD were able to retain more water, putting it at crop disposition during the periods with the highest values of evapotranspiration.

Similar results were observed by Muriel et al. (2005) and Jiménez et al. (2005), who reported that techniques such as CA improved the soil water conservation as in the first centimetres in the deeper zones of soil profile, being this water available for crop extraction during the period of maximum evapotranspirative demand.

The following season (2004–2005) was the driest in terms of annual rainfall and evapotranspiration rate (Table 6.2). The collected rainfalls were not able to recover the soil water contents reached during the previous season, this being 40% below to those values registered in 2003–2004 (Fig. 6.10). As a consequence, the water retained in the soil profile was not able to reach the deeper zones (60–90 cm). By contrast, the water scarcity promoted that the rainfall water was retained in the first centimetres of soil profile. In this sense, there are no significant differences between the soil water contents registered in both soil management systems considered in this experience.

During 2005–2006 (Fig. 6.11), accumulated rainfall and ET_0 were more similar to those registered in the first studied season (434 mm of rainfall and 1581 mm of ET_0, Table 6.1). In this sense, rainfall was almost twice than was detected the previous season, although somewhat lower than those registered in the first studied

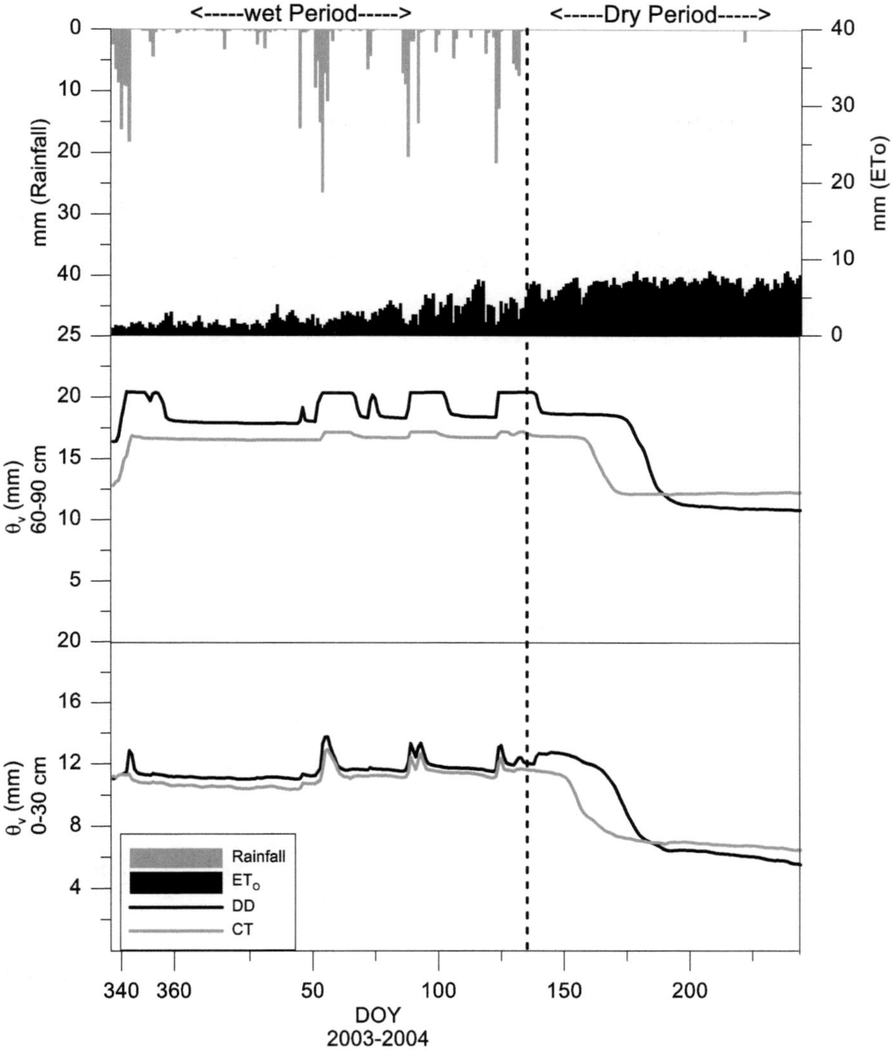

Fig. 6.9 Soil water balance from 0 to 30 and from 60 to 90 cm of soil depth during 2003–2004

seasons. Likewise, these rainfalls allowed partially restoring the soil water content in both systems (Fig. 6.11).

So, the most significant differences between soil management systems were observed in the soil contents of the deeper zones. By contrast, in the first centimetres of soil profile, no significant differences were observed between DD and CT, although during the discharge period the soil water extraction in the experimental plots under CT was faster than was observed in DD. The water uptake in DD plots was significantly higher than those registered under CT, relying on with more available water for crop during the extraction period. The higher contents of soil water in

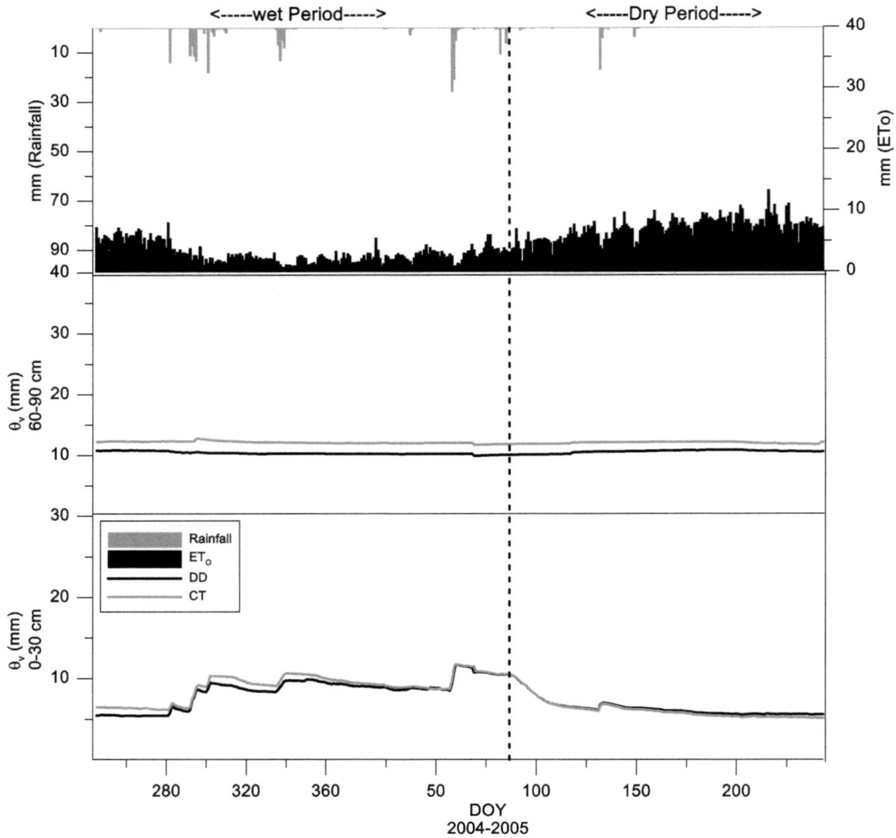

Fig. 6.10 Soil water balance from 0 to 30 and from 60 to 90 cm of soil depth during 2004–2005

the deeper zones would be related to the continuity of the natural soil structure, with a higher soil infiltration in those plots under DD, minimizing the erosion and runoff processes, as it has been discussed in the previous sections of this chapter.

Considering the rainfall events during this season, initially (autumn and beginning of winter), rainfalls were retained in the first centimetres of soil profile. Once the first soil layers were saturated, the excess of water was drained to deeper zones, being this water kept in the following layers of soil (spring rainfall events). Similar results to this finding were reported by Moreno et al. (2005) or Ordoñez et al. (2007), among others.

Finally, during 2006–2007 (Fig. 6.12), the weather conditions were like those registered in the first studied season. This fact promoted that, again, the soil water content in the first centimetres were higher in DD plots, as it can be seen in Figure 6.12. Moreover, it was remarkable that the higher provision of water took place in the experimental plots under DD during the discharge period, showing that CA practices in rainfed systems provide more water for the crop, especially during the maximum evapotranspirative demand period.

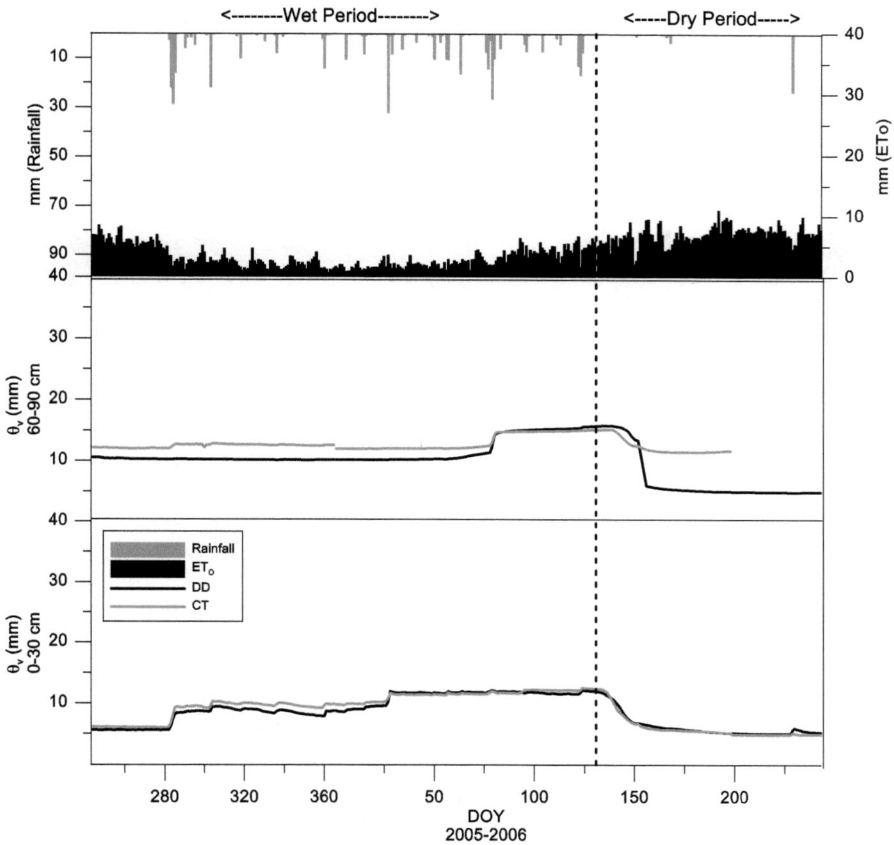

Fig. 6.11 Soil water balance from 0 to 30 and from 60 to 90 cm of soil depth during 2005–2006

Unlike the observed first season, the most significant differences between DD and CT were not observed in the first centimetres of soil profile, these being more evident in the deeper zones, although, on overall, the plots subjected to DD practices were able to retain 30% more of water than those under CT.

As a final point, according to the effects of soil management on yield, there were not significant differences between treatments (Table 6.1). In this sense, the crop yield for each season were on average, in line with those produced in the last 25 years, except in the 2006–2007 season, in which the crop was not harvested, being incorporated into the soil as an organic input in both treatments.

These results allow us to conclude that DD offers the ability to maintain sustainable crop yield levels, with the advantages that this practice entails, as both soil properties are concerned, being the substantial savings for the farmer mainly in terms of fuel.

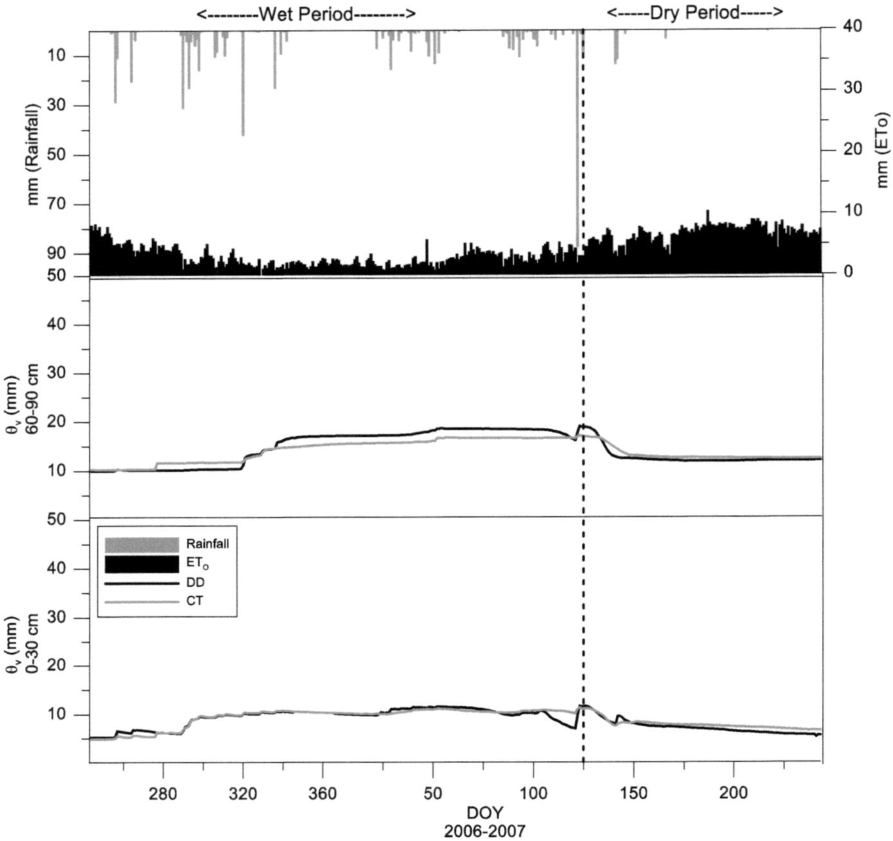

Fig. 6.12 Soil water balance from 0 to 30 and from 60 to 90 cm of soil depth during 2006–2007

6.4 Soil Carbon Storage in Rainfed Systems Under Conservation Agriculture Practices

Currently, for many developing countries, the main issue related to agricultural practices is the food security, the economic development, and the climate change adaptation. However, a wide number of these countries are increasing their interest in those actions aimed to mitigate the advance of climate change by means of reduction of the greenhouse gas emissions from agricultural activities (Wilkes et al. 2013; Yadav et al. 2017b).

The agriculture and land-use changes are responsible for the emission of an important fraction of the greenhouse gasses (GHG), such as carbon dioxide (CO_2), methane (CH_4), or nitrous oxide (N_2O). Soil management contributes with 25% of the total anthropogenic emissions: 10–14% directly from the production processes and the remaining (12–17%) as a response to changes in the soil conservation (as deforestation) (Paustian et al. 2016; Dadhich et al. 2015).

In this regard, soil can act as source or sink of carbon; thus, there is a direct relationship between the different strategies of soil management and climate change mitigation. In this line, soil is the largest reservoir of organic carbon (SOC) in the earth, with three times more than the atmosphere (Ciais 2013; Meena et al. 2018a).

In areas with Mediterranean climate, the SOC is limited by the low biomass production, and hence the opportunity of increasing the soil capability as a carbon sink is lower than other areas. The introduction of CA, such as the DD and MT practices, the use of cover crops, and the crop rotation, has significant environmental benefits (Kassam et al. 2012; Verma et al. 2015c). Among them, the increasing of SOC pointed out these strategies as a proper alternative to mitigate the climate change, and reducing the CO_2atmospheric, these are retained as SOC (Márquez-García et al. 2013; Yadav et al. 2017a).

During the photosynthesis process, CO_2 is fixed by plants and incorporated to the vegetal tissues. When plants die, the carbon fixed to leaves, roots, branches, and stems is broken down in the soil, being transformed in OM, increasing the levels of this parameter and, hence, the soil fertility. CA introduces important changes in the SOC dynamic, encouraging the soil carbon sequestration. This process is related to the reduction of OM aeration, causing a less soil disturbance and minimizing its oxidation (OM is less accessible to microorganisms), and as result, SOC increases (Fig. 6.13).

With the aim of evaluating the potential advantages of CA in relation to climate change mitigation, and the role of agricultural soil as a carbon sink, this section shows the main results related to a long-term experience to assess the soil capability of increasing the atmospheric carbon sequestration and minimizing GHG emissions (concretely CO_2) resulting from agricultural activity, when two different soil management systems are applied: DD and CT.

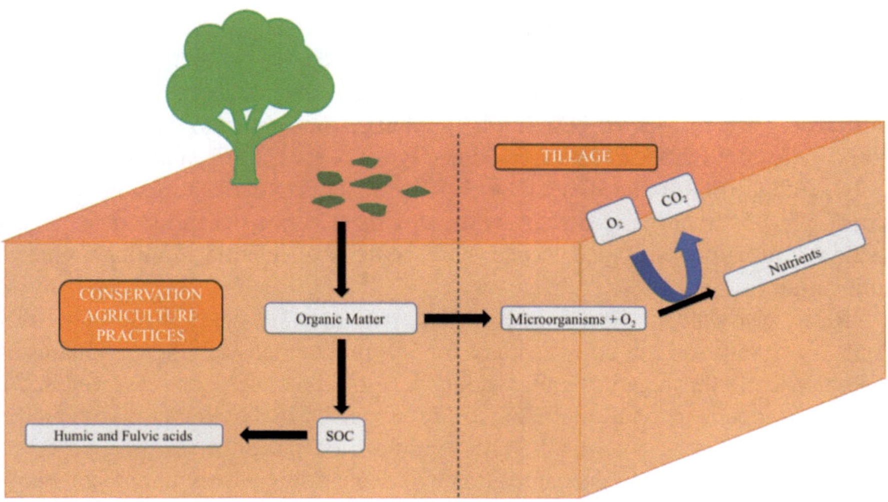

Fig. 6.13 Schematic representation of carbon sequestration or carbon emissions, depending on the soil management system

6.4.1 Location, Sampling, and Data Analysis

The trial was conducted during three consecutive seasons (2006–2007, 2007–2008, and 2008–2009) in the same plots previously described in Sect. 6.3.1, taking measurements in plots under DD and CT systems. In order to evaluate the temporal evolution of OM content and SOC contents, there were defined ten points per experimental plot by means of GPS, collecting one sample per point at different depths (0–5, 5–10, and 10–20 cm). Subsequently, once a month, soil samples were translated to the soil laboratory where these were analysed. The total organic carbon (TOC) was determined using the Walkley-Black methodology, proposed by Nelson and Sommer (1982). SOC and active organic carbon (AOC) were determined using the methodology proposed by Cambardella and Elliot (1992).

Relating to the CO_2 flow measurements, these were monthly collected, simultaneous with the soil sampling previously described. Measurements were developed by using a portable infrared absolute and differential EGM-4 gas analyser (PP Systems CO., Amesbury, MA, USA) connected to a soil respiration camera. Additional information can be found in Carbonell-Bojollo et al. (2011, 2015). Finally, and coinciding with punctual actions in which a soil alteration (tillage) was done, CO_2 fluxes were monitored in both management systems at 2, 4, 6, 24, and 48 h after this.

6.4.2 Effects of Direct Drill on the Soil Organic Carbon Retention and CO_2 Emissions

Table 6.3 shows the contents of SOC and AOC registered at different depths in the two studied soil management systems.

Significant differences in terms of SOC were detected in the different depths considered. This fact was especially evident during the first season, with 6 t ha^{-1} of SOC more in DD in comparison to CT. Regarding the AOC contents, significant differences were observed between both management systems, this being a very good indicator of the soil organic fraction (Oyonarte et al. 2007; Meena et al. 2014).

Overall the SOC in DD plots was 14–20% higher than that detected in CT plots, for 0–5 and 5–10 cm, respectively, evidencing that CA practices encourage the soil action as a carbon sink and, hence, have benefits in relation to climate change mitigation.

Relating to the CO_2 emissions from soil, two factors encourage the CO_2 emissions from soil: soil humidity and aeration (O_2 presence in the first centimetres of soil profile). In this regard, significant higher respiration rates were observed in CT plots in comparison to DD plots (Fig. 6.14), especially in those months with higher rainfall (from February to April). These rainfalls increased the soil water content, which facilitates the OM decomposition by soil microorganisms and carbon emission as CO_2 (Carbonell-Bojollo et al. 2015; Verma et al. 2015b).

Moreover, not only rainfall events promote the carbon release from soil. Another important question is related to the relationship between tillage and CO_2 emissions.

Table 6.3 Soil organic carbon and active organic carbon at different depths during the studied seasons in plots under direct drill and conventional tillage

	0–5 cm		5–10 cm		10–20 cm		0–20	
	CT	DD	CT	DD	CT	DD	CT	DD
SOC (t ha^{-1})								
Onset	7.4b	7.3a***	8.0b	9.4a***	8.2a	8.4a	18.1	21.4
2006–2007	7.5b	7.2a***	8.0b	9.3a***	7.8b	8.3a***	17.7	23.8
2007–2008	8.1b	9.1a***	7.8b	9.8a***	8.9a	8.0b*	19.2	20.5
2008–2009	8.6a	9.7a*	7.1b	8.8a***	7.0b	8.6a**	16.4	20.6
AOC (t ha^{-1})								
Onset	0.44 a	0.55 a	0.44 a	0.59 a	0.45 a	0.54 a	1.0	1.3
2006–2007	0.36 a	0.56 b**	0.36 b	0.55 a**	0.37 b	0.50 a*	0.8	1.2
2007–2008	0.27a	0.34 a	0.29 a	0.38 a	0.32 a	0.41 a	0.7	0.9
2008–2009	0.22 b	0.93 a***	0.17 b	0.79 a***	0.19 b	0.59 a***	0.6	1.8

CT Conventional tillage, *DD* Direct drill, *SOC* Soil organic carbon, *AOC* Active organic carbon. Different letters within a column show significant differences between seasons ($p < 0.05$). *, **, and *** show significant differences between soil management systems at 95, 99, and 99.9% of confidence level

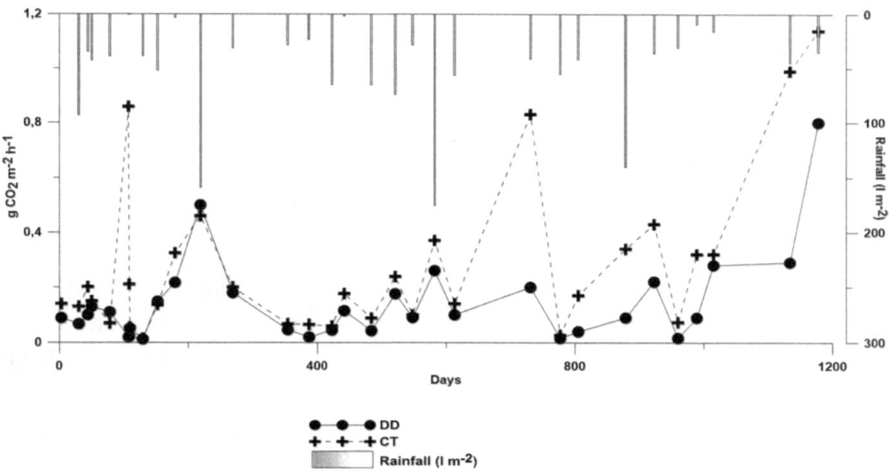

Fig. 6.14 CO_2 emissions in direct drill (DD) and conventional tillage (CT) plots under rainfall events

In this sense, according to Almaraz et al. (2009) or Oorts et al. (2007), conventional agriculture tends to increase CO_2 fluxes to atmosphere, especially the following days after any eventual soil modification, for instance, after tillage. Table 6.3 summarizes the accumulated daily emission in both soil management systems in those days in which some tillage was applied. Maximum differences between both systems ranged between 39% and 90%, depending on the maximum soil temperature and soil water content in the moment of readings.

Table 6.4 Maximum daily emissions of CO_2 after tilling, maximum difference between them, and moment after tilling in which it is detected in both management systems

Season	Maximum daily emission (kg ha^{-1})		Maximum difference between systems	T_{max}	Accumulated rainfall in the month previous to the tillage	θ_v (m^3 m^{-3})
	CT	DD				
2006–2007	38.5	8.4	87% (4 h)	21.2	127.8	20.5
	20.3	8.5	74% (4 h)	17.7	38.8	10.1
2007–2008	6.3	3.8	39% (0 h)	34.2	11.0	2.9
	13.7	9.1	63% (2 h)	16.0	66.0	11.4
2008–2009	22.0	6	73% (0 h)	18.7	95.2	18.3
	30.0	8	90% (4 h)	31.3	44.6	10.6

CT Conventional tillage, *DD* Direct drill, T_{max} Maximum soil temperature detected during measurements, θ_v Soil water content

Considering a whole period of 24 h, during the first studied season, plots under CT showed CO_2 emissions between 30.1 and 11.8 kg ha^{-1} more than those plots under DD, whereas for the second and third season, these were of 2.5 and 4.6 kg ha^{-1} and 116 and 22 kg ha^{-1}.

The direct relationship found between CO_2 emissions and soil water content was noticeable in CT plots [CO_2 (kg ha^{-1}) = $1.40\theta_v + 4.56$; r = 0.77], not being so clear for the case of DD plots. Additionally, the lowest value in CT coincided with a day in which the soil temperature limited the soil microorganism's activity (Table 6.4).

6.5 Conclusions

In rainfed systems, water use efficiency is directly influenced by soil management and the chemical and physical fertility of soil. In this sense, CA practices allow to maintain an optimum range of soil fertility, improving the soil organic matter content, hence easing the structural stability of soils. Conventional tillage promotes the increasing organic matter decomposition, which is directly related to land degradation, the loss of soil structure, and the increasing of CO_2 emissions. Linked to this, soil aggregate degradation implies to diminish the porosity and the increasing of bulk density, these properties being directly related to the soil capability of rainwater harvesting. In this sense, DD allows increasing the soil water retention not only in the first layers of soil profile but also in the deeper zones. This fact is directly related to the higher continuity of soil profile, avoiding the compaction of the deeper soil layers. Moreover, this higher continuity along the soil profile promotes a higher capability of storing water to be making available for the crops during the drier periods.

Additionally, CA not only fosters the soil water conservation in semi-arid environments but also facilitates the increase of OM and carbon sequestration, acting as a carbon sink and reducing the CO_2 emissions to atmosphere. This fact becomes the CA in a set of strategies with a double paper in relation to climate change: on one hand, acting as an adaptation strategy and, on the other hand, as a mitigation strategy against climate changes.

In conclusion, CA practices comprise a bundle of benefits for rainfed systems, all of them involved in the land and water conservation; these are especially beneficial in Mediterranean agro-ecosystems or areas with low soil fertility.

References

Ahuja LR, Naney RE, Green RE, Nielsen DR (1984) Macroporosity to characterize spatial variability of hydraulic conductivity and effects of land management. Soil Sci Soc Am J 48:699–702

Ahuja LR, Cassel DK, Bruce RR, Barnes BB (1989) Evaluation of spatial distribution of hydraulic conductivity using effective porosity data. Soil Sci 148:404–411

Allen RG, Pereira LS, Raes D, Smith M (1998) Crop evapotranspiration —guidelines for computing crop water requirements, FAO Irrigation and drainage paper 56. Food and Agriculture Organization, Rome

Almaraz JJ, Zhou XM, Mabood F, Madramootoo C, Rochette P, Ma BL, Smith DL (2009) Greenhouse gas fluxes associated with soybean production under two tillage systems in southwestern Quebec. Soil Tillage Res 104(1):134–139

Ashoka P, Meena RS, Kumar S, Yadav GS, Layek J (2017) Green nanotechnology is a key for eco-friendly agriculture. J Clean Prod 142:4440–4441

Ball BC, Lang RW, Robertson EAG, Franklin MF (1994) Crop performance and soil conditions on imperfectly drained loams after 20–25 years of conventional tillage or direct drilling. Soil Tillage Res 31:97–118

Bescana P, Imaz MJ, Virto I, Enrique A, Hoogmoed WB (2006) Soil water retention as affected by tillage and residue management in semiarid Spain. Soil Tillage Res 87:19–27

Buragohain S, Sharma B, Nath JD, Gogaoi N, Meena RS, Lal R (2017) Impact of ten years of bio-fertilizer use on soil quality and rice yield on an inceptisol in Assam, India. Soil Res. https://doi.org/10.1071/SR17001

Cambardella CA, Elliot ET (1992) Particulate soil organic matter change across a grassland cultivation sequence. Soil Sci Soc Am J 56:777–783

Carbonell-Bojollo R, González-Sánchez E, Veróz-González O, Ordóñez-Fernández R (2011) Soil management systems and short-term CO_2 emissions in a clayey soil in southern Spain. Sci Total Environ 409:2929–2935

Carbonell-Bojollo R, González-Sánchez E, Repullo-Ruibérriz de Torres M, Ordóñez-Fernández R, Domínguez-Jiménez J, Basch G (2015) Soil organic fractions under conventional and no-till management in a long-term study in southern Spain. Soil Res 53:113–124

Christian DG, Bacon EGT (1990) A long-term comparison of ploughing, tine cultivation and direct drilling on the growth and yield of winter cereals and oilseed rape on clayey and silty soils. Soil Tillage Res 18:311–331

Ciais P(2013) The physical science basis. Contribution of working group to the fifth Assessment report of the Intergovernmental panel on climate change (eds Stocker TF et al.) pp 465–470

Dadhich RK, Meena RS (2014) Performance of Indian mustard (*Brassica juncea* L.) in response to foliar spray of thiourea and thioglycollic acid under different irrigation levels. Indian J Ecol 41(2):376–378

Dadhich RK, Meena RS, Reager ML, Kansotia BC (2015) Response of bio-regulators to yield and quality of Indian mustard (*Brassica juncea* L. Czernj. and Cosson) under different irrigation environments. J Appl Nat Sci 7(1):52–57

Dao TH (1996) Tillage system and crop residue effects on surface compaction of a Paleustoll. Agron J 88:141–148

Datta R, Baraniya D, Wang YF, Kelkar A, Moulick A, Meena RS, Yadav GS, Ceccherini MT, Formanek P (2017a) Multi-function role as nutrient and scavenger of free radical in soil. Sustain MDPI:402. https://doi.org/10.3390/su9081402

Datta R, Kelkar A, Baraniya D, Molaei A, Moulick A, Meena RS, Formanek P (2017b) Enzymatic degradation of lignin in soil: a review. Sustain MDPI 1163(9):1–18. https://doi.org/10.3390/su9071163

Derpsch R (2005) The extent of conservation agriculture adoption worldwide: implications and impact. In: Proceedings 3rd World Congress on Conservation Agriculture, Nairobi, p 15. pp (CD-Rom)

Dexter AR (1988) Advances in characterization of soil structure. Soil Tillage Res 11:199–238

Dexter AR (2004) Soil physical quality. Part I. theory, effects of soil texture, density, and organic matter on root growth. Geoderma 120:201–214

Dhakal Y, Meena RS, De N, Verma SK, Singh A (2015) Growth, yield and nutrient content of mungbean (*Vigna radiata* L.) in response to INM in eastern Uttar Pradesh, India. Bangladesh J Bot 44(3):479–482

Dhakal Y, Meena RS, Kumar S (2016) Effect of INM on nodulation, yield, quality and available nutrient status in soil after harvest of green gram. Legum res 39(4):590–594

Dhruw SK, Singh LJ, Singh AK (2009) Storage and sequestration of carbon by leguminous and non-leguminous trees on red lateritic soil of Chhattisgarh. Indian Forester 135(4):531–538

Fernández C (1997) Historia y evolución de los sistemas de laboreo. El laboreo de Conservación. Cap 1. In: Torres LG, y Fernández PG (eds) Agricultura de Conservación: Fundamentos Agroquímicos, Medioambientales y Económicos. Asociación española del laboreo de conservación/Suelos Vivos, Córdoba, España

Friedrich T, Derpsch R, Kassam A (2012) Global overview of the spread of conservation agriculture Field and actuarial science reports. Available in: http://aciar.gov.au/files/node/13993/

Fustec J, Lesuffleur F, Mahieu S, Cliquet JB (2010) Nitrogen rhizodeposition of legumes: a review. Agron Sustain Dev 30:57–66

García-Tejero I, Jiménez JA, Martínez G, Vanderlinden K, Muriel JL, Perea F (2007b) Conservación y disponibilidad del agua en el suelo en función del tipo de laboreo. Vida Rural, vol 251, pp 29–31

García-Tejero IF, Espejo AJ, Martínez G, Vanderlinden K, Durán ZVH, Muriel JL (2009) Efectos del laboreo en la curva de retención hídrica de un suelo bajo diferentes sistemas de manejo. Vida Rural 297:32–37

García-Tejero IF, Durán ZVH, Rodríguez CR, Muriel FJL (2011) Water and sustainable agriculture, Springerbriefs in Agriculture, 94 p

García-Tejero IF, Durán ZVH, Muriel FJL (2014) Towards sustainable irrigated Mediterranean agriculture: implications for water conservation in semi-arid environments. Water Int 39:635–648

Ghanbari A, Dahmardeh M, Siahsar BA, Ramroudi M (2010) Effect of maize (L.) – cowpea (Vigna unguiculata L.) intercropping on light distribution, soil temperature and soil moisture in arid environment. J Food Agric Environ 8:102–108

Hammel JE (1989) Long term tillage and crop rotation effects on bulk density and soil impedance in northern Idaho. Soil Sci Soc Am J 53:1515–1519

Hatfield JL, Sauer TJ, Prueger JH (2001) Managing soils to achieve greater water use efficiency: a review. Agron J 93:271–280

Hauggaard-Nielsen H, Jornsgaard B, Kinane J, Jensen ES (2007) Grain legume–cereal intercropping: the practical application of diversity, competition and facilitation in arable and organic cropping systems. Renew Agric Food Syst 23:3–12

Hillel D (1998) Chapter 5: Soil structure and aggregation. In: Environmental soil Physics. Academic, London, pp 101–126

Iglesias A, Garrote L (2018) Local and corrective actions for adaptation to use less water for agriculture in the Mediterranean region. In: Garcia Tejero IF, DuránZuazo VH (eds) Water scarcity and sustainable agriculture in semiarid environment: tools, strategies and challenges for woody crops. Academic, Oxford, pp 73–84

IPCC (2014) Climate change: impacts, adaptations and vulnerabilities. Part B. Regional aspects (Barros V, Field CB, eds). Cambridge University Press, 688 p

Jiménez JA, García-Tejero I, Vanderlinden K, Perea F, Muriel JL (2005) Balance de agua en suelos arcillosos bajo laboreo convencional y siembra directa. In: Actas Congreso Internacional sobra Agricultura de Conservación; Córdoba, Spain. Asociación Española Agricultura de Conservación/Suelos Vivos, Federación Europea Agricultura de Conservación, Diputación de Córdoba. pp. 397–402

Kassam A, Derpsch R, Friedrich T (2014) Global achievements in soil and water conservation: the case of conservation agriculture. Int Soil Water Cons Res 2(1):5–13

Kassam A, Friedrich T, Derpsch R, Lahmar R, Mrabet R, Basch G, González- Sánchez EJ, Serraj R (2012) Conservation agriculture in the dry Mediterranean climate. Field Crop Res 132:7–17

Keeler BL, Hobbie SE, Kellogg LE (2009) Effects of long-term nitrogen addition on microbial enzyme activity in eight forested and grassland sites: implications for litter and soil organic matter decomposition. Ecosystems 12:1–15

Kertész A, Madarász B (2014) Conservation agriculture in Europe. Int Soil Water Conserv Res 2:91–96

Kirkby CA, Richardson EA, Wade JL, Batten DG, Blanchard C, Kirkegaard AJ (2013) Carbon-nutrient stoichiometry to increase soil carbon sequestration. Soil Biol Biochem 60:77–86

Kumar S, Sheoran S, Kumar SK, Kumar P, Meena RS (2016) Drought: a challenge for Indian farmers in context to climate change and variability. Progress Res Int J 11:6243–6246

Kumar S, Meena RS, Pandey A, Seema (2017a) Soil acidity management and an economics response of lime and sulfur on sesame in an alley cropping system. Int J Curr Microb Appl Sci 6(3):2566–2573

Kumar S, Meena RS, Yadav GS, Pandey A (2017b) Response of sesame (*Sesamum indicum* L.) to sulphur and lime application under soil acidity. Int J Plant Soil Sci 14(4):1–9

Kumar S, Meena RS, Bohra JS (2018) Interactive effect of sowing dates and nutrient sources on dry matter accumulation of Indian mustard (*Brassica juncea* L.). J Oilseed Brassica 9(1):72–76

La Jeunesse I, Cirelli C, Aubin D, Larrue C, Soddu A (2016) Is climate change a threat for water uses in the Mediterranean region? Results from a survey at local scale. Sci Total Environ Part B 543:981–996

Lars JM, Per S, Karl JR, Kari T (2003) Spatial and temporal effects of direct drilling on soil structure in the seedling environment. Soil Tillage Res 71:163–173

Lithourgidis AS, Dordas CA, Damalas CA, Vlachostergios DN (2011) Annual intercrops: an alternative pathway for sustainable agriculture. Aust J Crop Sci 5:396–410

López-Bellido RJ, López BL, Benítez EJ, BFJ L (2007) Tillage system, preceding crop, and nitrogen fertilizer in wheat crop I. Soil water content. Agron J 99:59–65

Lorite IJ, Ruiz RM, Gabaldón LC, Cruz BM, Porras R, Santos C (2018) Water management and climate change in semiarid environments. In: García-Tejero IF, Durán-Zuazo VH (eds) Water scarcity and sustainable agriculture in semiarid environment: tools, strategies and challenges for woody crops. Academic Press-Elsevier, Amsterdam, pp 3–40

Lowther WL, Horrell RF, Fraser WJ, Trainor KD, Johnstone PD (1996) Effectiveness of a strip seeder direct drill for pasture establishment. Soil Tillage Res 38:161–174

Macedo MO, Campello EFC, Andrade AG, de FSM (2006) Establishment of legume trees on heaps of blast furnace slag. Floresta Ambiente 13:20–25

Macedo MO, Resende AS, Garcia PC, Boddey RM, Jantalia CP, Urquiaga S, Campello EFC, Franco AA (2008) Changes in soil C and N stocks and nutrient dynamics 13 years after recovery of degraded land using leguminous nitrogen-fixing trees. For Ecol Manag 255:1516–1524

Márquez-García F, González SEJ, Castro GS, Ordóñez-Fernández R (2013) Improvement of soil carbon sink by cover crops I olive orchards under semiarid conditions. Influence of the type of soil and weed. Span J Agric Res 11:335–346

Meena H, Meena RS (2017) Assessment of sowing environments and bio-regulators as adaptation choice for clusterbean productivity in response to current climatic scenario. Bangladesh J Botany 46(1):241–244

Meena RS, Yadav RS (2014) Phonological performance of groundnut varieties under sowing environments in hyper arid zone of Rajasthan, India. J App Nat Sci 6(2):344–348
Meena RS, Yadav RS (2015) Yield and profitability of groundnut *(Arachis hypogaea* L) as influenced by sowing dates and nutrient levels with different varieties. Legum Res 38(6):791–797
Meena RS, Yadav RS, Meena VS (2014) Response of groundnut (*Arachis hypogaea* L.) varieties to sowing dates and NP fertilizers under Western dry zone of India. Bangladesh J Bot 43(2):169–173
Meena RS, Dhakal Y, Bohra JS, Singh SP, Singh MK, Sanodiya P (2015a) Influence of bioinorganic combinations on yield, quality and economics of Mungbean. Am J Exp Agric 8(3):159–166
Meena RS, Meena VS, Meena SK, Verma JP (2015b) The needs of healthy soils for a healthy world. J Clean Prod 102:560–561
Meena RS, Meena VS, Meena SK, Verma JP (2015c) Towards the plant stress mitigate the agricultural productivity: a book review. J Clean Prod 102:552–553
Meena RS, Yadav RS, Meena H, Kumar S, Meena YK, Singh A (2015d) Towards the current need to enhance legume productivity and soil sustainability worldwide: a book review. J Clean Prod 104:513–515
Meena RS, Yadav RS, Reager ML, De N, Meena VS, Verma JP, Verma SK, Kansotia BC (2015e) Temperature use efficiency and yield of groundnut varieties in response to sowing dates and fertility levels in Western dry zone of India. Am J Exp Agric 7(3):170–177
Meena H, Meena RS, Singh B, Kumar S (2016a) Response of bio-regulators to morphology and yield of clusterbean [*Cyamopsis tetragonoloba* (L.) Taub.] under different sowing environments. J App Nat Sci 8(2):715–718
Meena RS, Bohra JS, Singh SP, Meena VS, Verma JP, Verma SK, Shiiag SK (2016b) Towards the prime response of manure to enhance nutrient use efficiency and soil sustainability a current need: a book review. J Clean Prod 112:1258–1260
Meena RS, Gogaoi N, Kumar S (2017a) Alarming issues on agricultural crop production and environmental stresses. J Clean Prod 142:3357–3359
Meena RS, Kumar S, Pandey A (2017b) Response of sulfur and lime levels on productivity, nutrient content and uptake of sesame under guava (*Psidium guajava* L.) based agri-horti system in an acidic soil of eastern Uttar Pradesh, India. J Crop Weed 13(2):222–227
Meena RS, Meena PD, Yadav GS, Yadav SS (2017c) Phosphate solubilizing microorganisms, principles and application of microphos technology. J Clean Prod 145:157–158
Meena H, Meena RS, Lal R, Singh GS, Mitran T, Layek J, Patil SB, Kumar S, Verma T (2018a) Response of sowing dates and bio regulators on yield of clusterbean under current climate in alley cropping system in eastern U.P. Indian Leg Res 41(4):563–571
Meena RS, Kumar V, Yadav GS, Mitran T (2018b) Response and interaction of *Bradyrhizobium japonicum* and *Arbuscular mycorrhizal* fungi in the soybean rhizosphere: A review. Plant Growth Regul 84:207–223
Moreno F, Murillo JM, Madejón E, Girón IF, Pelegrín F (2005) Laboreo de conservación: efectos a largo plazo sobre la calidad del suelo. In: Libro de Actas del Congreso Internacional de Agricultura de Conservación. AEAC-SV, FEAC, Diputación de Córdoba (eds). , 515–520 pp
Muriel JL, Vanderlinden K, Perea F, Jiménez JA, García-Tejero I, Pérez JJ (2005) Régimen hídrico en suelos arcillosos de campiña sometidos a distintos sistemas de manejo. In: Actas Congreso Internacional sobre. Agricultura de Conservación, Córdoba, España, pp 537–542
Nelson DW, Sommer LE (1982) Total carbon, organic carbon and organic matter. In: Page RH, Keeny DR (eds) Methods of soil analysis, II. Chemical and microbiological properties, 2nd edn. Soil Science Society of America, Madison
Oorts K, Merckx R, Grehan E, Labreuche E, Nicolardot B (2007) Determinants of annual fluxes of CO_2 and N_2O in long-term no-tillage and conventional tillage systems in northern France. Soil Tillage Res 95:133–148
Ordoñez R, González FP, Giráldez JV, Perea F (2007) Soil properties and crop yields alter 21 years of direct drilling trials in southern Spain. Soil Tillage Res 94:47–54
Oyonarte C, Mingorance M, Durante P, Piñeiro G, Barahona E (2007) Indicators of change in the organic matter in arid soils. Sci Total Environ 378:133–137

Paustian K, Lehmann J, Ogle S, Reasy D, Robertson P, Smith P (2016) Climate-smart soils. Nature. https://doi.org/10.1038/nature17174

Pelegrin F, Moreno F, MartínAJ CM (1990) The influence of tillage methods on soil physical properties and water balance for a typical crop rotation in SW Spain. Soil Tillage Res 16:345–358

Perea F (2000) Agronomía del laboreo de conservación en los vertisuelos de la campiña andaluza. Unpublished PhD dissertation. Department of Agronomy, University of Cordoba, Cordoba, Spain

Perea F, Jiménez JA, García-Tejero I, Vanderlinden K, Muriel JL (2006) Caracterización hidroclimática en vertisuelos de la campiña de Carmona. CAREL4:389–1407

Ram K, Meena RS (2014) Evaluation of pearl millet and mungbean intercropping systems in arid region of Rajasthan (India). Bangladesh J Bot 43(3):367–370

Raper RL, Reeves DW, Burmester CH, Schwab EB (2000) Tillage depth, tillage timing and cover crop effects on cotton yield, soil strength and tillage energy requirements. Appl Eng Agric 16:379–385

Ratnadass A, Blanchard E, Lecompte P (2013) Ecological interactions within the biodiversity of cultivated systems. In: Hainzelain (ed) Cultivating biodiversity to transform agriculture. Quae, Cirad, pp 141–179

Richard G, Boiffin J, Duval Y (1995) Direct drilling of sugar beet (*Beta vulgaris* L.) into a cover crop: effects on soil physical conditions and crop establishment. Soil Tillage Res 34:169–185

Richard G, Cousin I, Sillon JF, Bruand A, Guérif J (2001) Effect of compaction on the porosity of a silty soil: influence on unsaturated hydraulic properties. Eur J Soil Sci 52:49–58

Roseberg RJ, McCoy EL (1992) Tillage and traffic-induced changes in macroporosity and macropore community: air permeability assessment. Soil Sci Soc Am J 56:1261–1267

Schaap MG, Leij FJ, van Geuchten MT (2001) ROSETTA: a computer program for estimating soil hydraulic parameters with hierarchical pedotransfer functions. J Hydrol 251:163–176

Shaver TM, Peterson GA, Ahuja LR, Westfall DG, Sherrod LA, Dunn G (2002) Surface soil physical properties after twelve years of dryland no-till management. Soil Sci Soc Am J 66:1296–1303

Sihag SK, Singh MK, Meena RS, Naga S, Bahadur SR, Gaurav YRS (2015) Influences of spacing on growth and yield potential of dry direct seeded rice (*Oryza sativa* L) cultivars. The Ecoscan 9(1–2):517–519

Soane BD, Ball BC (1998) Review of management and conduct of long-term tillage studies with special reference to a 25-yr experiment on barley in Scotland. Soil Tillage Res 45:17–37

Suwardji P, Eberbach PL (1998) Seasonal changes of physical properties of an OxicPaleustalf (Red Kandosol) after 16 years of direct drilling or conventional cultivation. Soil Tillage Res 49:65–77

USDA (1999) Soil taxonomy. A basic system of soil classification for making and interpreting soil surveys, 2nd ed. Agricultural Handbook 436. US Government Printing Office, Washington (DC)

Varma D, Meena RS, Kumar S (2017a) Response of mungbean to fertility and lime levels under soil acidity in an alley cropping system in Vindhyan region, India. Int J Chem Stu 5(2):384–389

Varma D, Meena RS, Kumar S, Kumar E (2017b) Response of mungbean to NPK and lime under the conditions of Vindhyan region of Uttar Pradesh. Legum Res 40(3):542–545

Vercauteren M (2013) Impact of conservation agriculture (CA) on water conservation and yield in Nanyuki, Kenia. Master's dissertation of Science in bioscience engineering: Agriculture. Gent University, 62 pp. Available at: https://lib.ugent.be/fulltxt/RUG01/002/063/597/RUG01-002063597_2013_0001_AC.pdf

Verma JP, Jaiswal DK, Meena VS, Meena RS (2015a) Current need of organic farming for enhancing sustainable agriculture. J Clean Prod 102:545–547

Verma JP, Meena VS, Kumar A, Meena RS (2015b) Issues and challenges about sustainable agriculture production for management of natural resources to sustain soil fertility and health: a book review. J Clean Prod 107:793–794

Verma SK, Singh SB, Prasad SK, Meena RN, Meena RS (2015c) Influence of irrigation regimes and weed management practices on water use and nutrient uptake in wheat (*Triticum aestivum* L. Emend. Fiori and Paol.). Bangladesh J Bot 44(3):437–442

Wani NR, Qaisar KN (2014) Carbon percent in different components of tree species and soil organic carbon pool under these tree species in Kashmir valley. Curr World Environ 9(1):174–181

Wilkes A, Tennigkeit T, Solymosi K (2013) National integrated mitigation planning in agriculture: a review paper. FAO, Rome

Yadav GS, Babu S, Meena RS, Debnath C, Saha P, Debbaram C, Datta M (2017a) Effects of godawariphosgold and single supper phosphate on groundnut (*Arachis hypogaea*) productivity, phosphorus uptake, phosphorus use efficiency and economics. Indian J Agric Sci 87(9):1165–1169

Yadav GS, Lal R, Meena RS, Babu S, Das A, Bhomik SN, Datta M, Layak J, Saha P (2017b) Conservation tillage and nutrient management effects on productivity and soil carbon sequestration under double cropping of rice in north eastern region of India. Ecol Indic. http://www.sciencedirect.com/science/article/pii/S1470160X17305617

Yadav GS, Lal R, Meena RS, Datta M, Babu S, Das LJ, Saha P (2017c) Energy budgeting for designing sustainable and environmentally clean/safer cropping systems for rainfed rice fallow lands in India. J Clean Prod 158:29–37

Yadav GS, Das A, Lal R, Babu S, Meena RS, Saha P, Singh R, Datta M (2018a) Energy budget and carbon footprint in a no-till and mulch based rice–mustard cropping system. J Clean Prod 191:144–157

Yadav GS, Das A, Lal R, Babu S, Meena RS, Patil SB, Saha P, Datta M (2018b) Conservation tillage and mulching effects on the adaptive capacity of direct-seeded upland rice (*Oryza sativa* L.) to alleviate weed and moisture stresses in the north eastern Himalayan region of India. Arch Agron Soil Sci. https://doi.org/10.1080/03650340.2018.1423555

Terraced Subtropical Farming: Sustainable Strategies for Soil Conservation

7

Víctor Hugo Durán Zuazo, Carmen Rocío Rodríguez Pleguezuelo, Belén Cárceles Rodríguez, Baltasar Gálvez Ruiz, Saray Gutiérrez Gordillo, Pedro Cermeño Sacristán, Simón Cuadros Tavira, and Iván Francisco García-Tejero

Abstract

Terracing is a soil conservation strategy applied worldwide to prevent erosion and runoff on sloping lands. Orchard terraces can considerably reduce soil loss due to water erosion if they are well planned, correctly constructed and properly maintained. Terraces have to be combined with additional soil conservation practices, of which the most important is the maintenance of a soil cover, especially during the rainy period. On the coastal strip of the provinces of Granada and Malaga (south-eastern Spain), irrigated subtropical fruit species have been introduced and cultivated on terraces with a considerable importance as the only European producer region. The subtropical farming in this zone also has strong socio-economic impact. In the present chapter, land-use changes were analysed in a selected representative watershed over 29 years. According to the findings, formerly, 97.5% of the watershed was devoted to traditional Mediterranean crops; however, after this period, due to abandonment, this area was reduced to 17.6% and increased in subtropical fruit crops (26.6%), shrubland (29.8%) and abandoned cropland (24.6%). The main driving force in land-use change has been intensive irrigation on terraces planted with subtropical crops, which are economically more profitable than traditional rainfed crops, almond and olive, which have been replaced or abandoned. The intensification of subtropical farming in terraces provokes environmental effects, especially those regarding soil and water resources, which need to be minimized. The results support the recommendation of using plant covers on the taluses of subtropical crop terraces in

V. H. Durán Zuazo (✉) · C. R. Rodríguez Pleguezuelo · B. C. Rodríguez · B. G. Ruiz
Centro IFAPA "Camino de Purchil", Granada, Spain
e-mail: victorh.duran@juntadeandalucia.es

S. G. Gordillo · P. C. Sacristán · I. F. García-Tejero
Centro IFAPA "Las Torres-Tomejil", Seville, Spain

S. C. Tavira
Dpto. Ingeniería Forestal, Universidad de Córdoba, Córdoba, Spain

R. S. Meena (ed.), *Soil Health Restoration and Management*,
https://doi.org/10.1007/978-981-13-8570-4_7

order to control soil erosion and improve the soil quality in the taluses of orchard terraces. In this sense, compared to bare soil, thyme and native spontaneous vegetation plant covers reduced the runoff with 94% and 93% and declined erosion with 71% and 79%, respectively. That is to avoid the collapse of the structure and make more feasible the subtropical fruit cultivation in the study area. Thus, it is possible to mitigate the impact of subtropical farming on terraces by adopting sustainable measures for soil and water conservation.

Keywords

Aromatic and medicinal plants · Orchard terraces · Plant covers · Subtropical crops · Soil erosion

Abbreviations

AMP	Aromatic-medicinal plant
B_S	Bare soil
K	Potassium
KINEROS	Kinematic runoff and erosion model
L_A	Lavender
LUC	Land-use change
N	Nitrogen
P	Phosphorus
R_O	Rosemary
SOC	Soil organic carbon
SOM	Soil organic matter
S_A	Satureja
S_V	Native spontaneous vegetation
SWAT	Soil and water assessment tool
T_H	Thyme
TN	Total nitrogen
USLE	Universal soil loss equation
WEPP	Water erosion prediction project

7.1 Introduction

Globally, soil erosion has been accelerated for the last centuries due to the intensification and increase of certain human activities such as the expansion of intensive agriculture, deforestation urbanization, etc. (Jefferson et al. 2013; Ashoka et al. 2017). The global soil erosion has been estimated in 22 ± 6 Gt $year^{-1}$ of which 10–20% finish in downstream river channels (Dotterweich 2013; Meena and Meena

2017). Soil erosion and sediment distribution imply several on-site problems such as loss soil fertility, land degradation and decrease of productivity. In the same way, these phenomena can lead to off-site problems: high frequency of floods, freshwater deterioration and degradation of aquatic habitats.

In this context, crop terraces are one of the most characteristic agricultural features in mountainous areas in many parts of the world (Ore and Bruins 2012; Cao et al. 2013; Tarolli et al. 2014; Yadav et al. 2018b), at the same time as being among the most visible prints of humans in the landscape (Arnáez et al. 2015). The main objectives for constructing terraces are to encourage water and soil retention (Cots-Folch et al. 2006; Meena et al. 2015c), promote water–soil infiltration (Yuan et al. 2003), diminish runoff velocity and thus mitigate erosion (Durán et al. 2005; Li et al. 2014; Meena et al. 2016a), allow machinery and humans to work and establish irrigation systems. Nowadays terracing is promoted since it is the best management practice for soil and water conservation. This ancient farming technique provides a good basis for increasing the usefulness of steep slopes throughout the world (Ruecker et al. 1998; Quine et al. 1999; Drechsler and Settele 2001; Kasai et al. 2001; Buragohain et al. 2017).

Terraces have been used traditionally in the Mediterranean Basin, as demonstrated by many archaeological studies (Ferro et al. 2017; Porat et al. 2018; Dadhich and Meena 2014). The exact historical origin of terraces remains unclear, although Barker et al. (2007) contended that terraces originated in Southeast Asia and spread to the Mediterranean Basin (Galletti et al. 2013). In Spain, terraces probably date to the Bronze Age (Asins 2006). During the eighteenth and nineteenth centuries, many mountainous areas of Spain were intensively terraced, with the increase of the rural population (Arnáez et al. 2011; Meena et al. 2018a). However, since the 1950s, because of the rural emigration (demographic shift to urban areas), the landscape changed in many rural areas of Spain such as the abandonment of agricultural activities and surges in other economic activities, such as tourism.

However, many mountainous areas in Spain have been terraced during the last decades, which has been one of the main driving forces for agriculture development. Specifically, along the coasts of the provinces of Granada and Málaga (Andalusia, southern Spain), considerable economic investment has been made to design wide and high terraces with a regular layout for commercial purposes.

In this coast of Granada, the great fertility of meadow soils and the easy access to water have permit that this area has been characterized by an important agriculture tradition throughout centuries. Among the most traditionally rooted crops in this area, we can mention the sugar cane (*Saccharum officinarum* L.), which was a great economic incentive, leaving in the landscape some bastions such as mills. The crisis of the sugar cane promoted the total abandonment of this crop and the associated industry. However, during the middle of the twentieth century, subtropical fruit crops were introduced, reaching high profitability (Fig. 7.1), since Spain is the only producer country of these fruits in Europe. In agreement with cultivated area as well as to the economic importance, the leading is avocado (*Persea americana* Mill.), with 15,182 ha in Spain, from which 13,274 ha are located in Andalusia, followed by mango (*Mangifera indica* L.), 4433 ha in Spain and 3317 in Andalusia (ESYRCE 2017).

Fig. 7.1 Avocado (**a**) and mango (**b**) orchard terraces for subtropical farming in south-eastern Spain

Today more than 50% of Spanish avocado fruits are exported to other countries in the European market, where it is well accepted, due to the relative nearness and good quality. In the same line, mango is the third worldwide tropical fruit in terms of production, after banana and tropical pineapple, and the fifth in relation to all types of fruits. According to the Observatory of Prices and Markets (OPM-CAPDR 2015), 19,341 t of mango were harvested in Andalusia during 2014, with an increasing rate of 15% over the two previous seasons. Export volume rates as well as the value of total trade was augmented in the last decade, being the value considerably risen in the last few years, even doubling in the case of mango.

At a global scale, the agricultural intensification has considerably boosted yield through agrochemicals, but has led to a decline in soil quality and functionality by nutrient diffuse losses (Saikia et al. 2013; Srivastava et al. 2016; Yadav et al. 2017b).

Preventive and corrective measures to palliate this problem need to be implemented urgently, especially in those areas with high risk of soil erosion and runoff generation. Soil degradation on cultivated areas represent a severe threat to soil resources, especially in the Mediterranean, due to their peculiar features (climatic, edaphic and topographical conditions) (Guerra et al. 2016). Terraced and cultivated lands are not an exception; subtropical fruits deserve a particular attention because, aside representing one of the most important crops in terms of income and employment in the area, they also have proven to be the form of agricultural use that needs to be more sustainable by preserving the soil (Durán et al. 2011; Meena et al. 2015b). In this context, the importance of plant covers in controlling water erosion is crucial. That is, vegetation influences erosion mainly by intercepting rainfall and protecting the soil surface against the impact of rainfall drops and by intercepting runoff (Durán and Rodríguez 2008; Varma et al. 2017a).

Nitrogen (N) as nitrate (NO_3-N) leaching is currently a major source of concern because of its direct impact on drinking water, its relationship with eutrophication of coastal sea water and its contribution to the pollution of the atmosphere with ammonia or nitrogen oxides as was pointed out by Bouwman et al. (2013). Additionally, phosphorus (P) is often applied in amounts that exceed the uptake capability of agricultural plants. The P losses from agricultural fields to surface waters are erosive surface runoff and subsurface transfer, especially risky when excessive loading of fertilizers is incorporated to sandy soils with limited phosphate sorption (Mabilde et al. 2017; Datta et al. 2017a).

On the other hand, important potassium (K) concentrations in surface runoff and subsurface water are thought to have less critical impacts on groundwater quality and on the overall environment. In fact, K leaching is in direct relation to the amount of exchangeable K in the soil, which reflects the level of K input and resulting K surpluses (Kayser et al. 2012; Meena et al. 2016b).

In this context, the subtropical crops installed on the terraces may often be a potential source of pollution, encouraged by the prevalent climatic conditions (as mentioned above, scarce but often of torrential rainfall) and to the excessive agricultural inputs (fertilizers, pesticides, herbicides, etc.).

Given the importance of these emerging crops for the economy of the southern provinces of Granada and Malaga, and the complex relationships of soil–plant–water in these agroecosystems, the primary objective of this chapter is to characterize the chief environmental problems originated by subtropical farming on steeply sloping terrain along the Granada coast. The most important sustainable measures for soil and water conservation in orchard terraces are proposed from an agro-environmental perspective. In the first section, we analyse the land-use changes due to establishment of new irrigated subtropical fruit trees in typical Mediterranean watershed, and in the second section we discuss the impact of aromatic plant covers and native spontaneous vegetation on soil erosion, runoff and probable pollution risk by NPK in the terraces. Finally, we discuss the environmental benefits of plant cover to improve and preserve the soil organic carbon in the taluses of orchard terraces.

7.2 Land-Use Changes Based on Introducing New Subtropical Fruits

Land-use change (LUC) is vital in the ongoing global change phenomena being directly related to human health, environmental refugees, food security, biodiversity, water and soil quality, runoff and sedimentation rates, etc. (Durán et al. 2011, 2014; Rutten et al. 2014; Eitelberg et al. 2016; Carpio et al. 2016; Qi et al. 2018; Dhakal et al. 2015; Yadav et al. 2017c). Additionally, LUCs have altered the ecosystems of the Mediterranean area being dependent to the vagaries and complexities of political, social, economic and religious and cultural components. The industrialization and pressure from tourism have led to major socio-economic changes in rural areas, based on the abandonment of subsistence terraced hillside land in favour of cash-crop cultivation on better soils in the plains, providing far higher net output (Puigdefábregas and Mendizabal 1998).

Traditional elements of the Mediterranean Basin (i.e. hedgerows, irrigation ditches, rough pastures, ponds and terraces) provide habitats for organisms and thus maintain biodiversity (Plexida et al. 2018; Yadav et al. 2018a). In particular, as stated before terracing, an agricultural measure for harvesting water and controlling soil erosion has an ancient history of converting landscapes into stepped agroecosystems in the Mediterranean area.

In the Mediterranean region, and in the traditional rainfed farming in Andalusia (southern Spain), soils on sloping land and cultivated for many years have been continuously degraded by water erosion. Currently, terracing continues, sometimes with heavy financial investment, resulting in pronounced alterations in the soil profile. The new structures commonly found in the study area are a reverse-sloped bench terrace type with a toe drain measuring 160–170 m long, with a platform of 2–3 m wide and the talus 3–5 m high (Fig. 7.1).

The economy of the Granada coast has been based on tourism, concretely, in the town of Almuñécar (south-eastern Spain), and according to IECA (2015), 55.5% was involved in tourism and services, and 7.2% of the active population was employed in agriculture and fishing. In the 1980s, intensive irrigated farming was established based in tropical and subtropical crops in newly constructed machinery-made terraces. These structures are being used to cultivate avocado (*Persea americana* Mill.), mango (*Mangifera indica* L.), loquat (*Eriobotrya japonica* L.), cherimoya (*Annona cherimola* Mill.), lychee (*lychee chinensis* Sonn.) and other crops. The new machinery-made terraces have completely changed the landscape of this zone, since old terraces were cut by hand and built of stone. The terrace construction employs heavy machinery that need a significant economic investment (3300 € ha^{-1}).

7.2.1 Pilot Watershed

We study the LUCs in the agricultural landscape for 29 years (1978–2007), concretely in a pilot watershed located in Almuñécar, which typifies adjacent watersheds in the coast of Granada. The small agricultural watershed was located approximately

1.7 km north of the city of Almuñécar. The watershed area is 343 ha and ranges in altitude from 80 to 720 m. The topography is mountainous with an average slope exceeding 50% with subtropical to semi-hot within the Mediterranean subtropical climatic category according to Elías and Ruiz (1977). The closeness to the sea and to the Penibaetic mountain system in the north reduces the northern winds that results in a unique microclimate in Europe suitable for subtropical farming.

The average annual rainfall depth in the monitoring area is 449.0 mm. Throughout the summer, the pilot watershed had only base flow, peaking in the months of heaviest rainfall (December–January). And despite scattered light rains, runoff is lowest during July and August, because the watershed storage becomes depleted and the rainfall has to satisfy the evaporation, transpiration and soil-storage demands before generating runoff.

Details of LUC dynamics in the watershed over the period of 29 years were analysed from diverse site information such as land-use maps (1,50,000 for 1978 and 1,10,000 for 2007), aerial photographs and field observations.

7.2.2 Land-Use Dynamics

In the 1970s on marginal land in many semiarid environments, as stated by Faulkner et al. (2003), the rainfed crops such as olive, almond and vineyard expanded rapidly. In this line, during 1978 the dominant land-use types of the watershed area were almond plantations, fallow cereal-growing land and vineyard with 64.2, 24.7 and 6.7%, respectively. In contrast, during 2007 the areas for almond and vineyard were lowered to 17 and 0.6%, respectively. Olive plantations and fallow cereal-growing land had disappeared altogether as shown in Table 7.1. After 29 years, the abandonment of olive plantations in the watershed was due to the more profitable new irrigated crops and due to several factors affecting olive in general – that is, the

Table 7.1 Land-use changes in the pilot watershed for 29 years

	1978		2007	
Land-use type	(ha)	(%)	(ha)	(%)
Almond (*Prunus dulcis* Mill.)	220.6	64.2	58.3	17
Fallow land and legume-cereal mixture	85	24.7	–	0.0
Vineyard (*Vitis vinifera* L.)	22.9	6.7	2.0	0.6
Olive (*Olea europaea* L.)	6.5	1.9	–	0.0
Shrubland	6.7	1.9	102.5	29.8
Avocado (*Persea americana* L.)	–	–	66.1	19.2
Mango (*Mangifera indica* L.)	–	–	13.5	3.9
Loquat (*Eriobotrya japonica* L.)	–	–	8.2	2.4
Cherimoya (*Annona cherimola* Mill.)	–	–	3.9	1.1
Abandoned cropland	–	–	84.5	24.6
Greenhouse	–	–	2.4	0.7
Others	2.0	0.6	2.4	0.6
Total	343.7	100	343.7	100

competition with other regions having a comparative advantage that influences their economic sustainability, which is the case of Jaen and Cordoba provinces.

In addition, in 1978, an important part of the watershed was devoted to fallow land in rotation with a legume–cereal mixture (24.7%). This was part of the traditional and predominant farming in which the use of synthetic inputs not being indispensable to improve soil fertility but leaving the land free from cultivation for some time. This type of agricultural system can also be contemplated to be a traditional Spanish rural activity with food production for local use during the 1970s.

The LUC is arguably the most pervasive socio-economic force driving changes and degradation of ecosystems. Such disturbance of the land affects important ecosystem processes and services, which can have wide-ranging and long-term impacts as was reported by many authors (Table 7.2). Crop production and irrigation water diversions have brought many consequences especially in areas where water resources are scarce (García and Durán 2018; Meena et al. 2015d; Kumar et al. 2017b).

On the other hand, native vegetation in general provides many ecosystem services, supporting biodiversity, providing critical habitat for wildlife, remove carbon dioxide from the atmosphere, intercept precipitation, slow down surface runoff and reduce soil erosion and flooding (Yapp et al. 2010; Wang et al. 2018; Sihag et al. 2015). Also, it has long been recognized that LUC can cause soil and water degradation (Table 7.3). Runoff and sediments from agricultural lands are leading sources of water pollution both in inland and coastal waters. The impact of LUC on soil quality has been reported by many studies as is shown in Table 7.4. In addition, LUC provides many economic and social benefits, but often comes at a substantial cost to the environment, which is the case of highly profitable subtropical fruits. Therefore, the establishment of subtropical crops on terraced hillsides requires planned sustainable agricultural measures based on the conservation of soil and water. This watershed, as in many other agricultural systems in the Mediterranean area, the priority should be given to the adoption and implementation of environmental planning strategies for sustainable land use.

7.2.2.1 Mediterranean Subtropical Farming

According to results, new woody crops established within the pilot watershed amounting to 26.6% of the area. Traditional dryland cultivation of almonds and fallow areas was changed to irrigation and the whole olive area was turned into subtropical plantations (91.7 ha). In this sense, in recent years, these new woody irrigated crops were planted mainly on new machinery-made terraces. The most extended subtropical crop was avocado, grown on 66.1 ha, amounting to 19.2% of the watershed area (Table 7.1). In this context, the cultivation of avocado on the coast of Granada started in the early 1960s and expanded during the 1980s, and about 85% of this fruit is exported to the European market, since this fruit is preferred to that coming from overseas, which loses quality due to the transport conditions and time.

The main difficulty of avocado cultivation on terraces, as well as for the rest of the crops in the study area, is the cost of energy needed for pumping irrigation water to high levels, and the price of water about 0.30–0.40 € m^{-3}. Another obstacle is the large size of the trees due to the high application rates of N fertilizers on terraces,

Table 7.2 Land-use change (LUC) based in abandonment of agricultural activities in Mediterranean areas

Authors	Objective	Main findings
Rodrigo et al. (2018)	105 rainfall simulations were carried out in agriculture lands of the Mediterranean belt (vineyards, almond, orange and olive orchards) and in paired abandoned lands to assess the impact of land abandonment on soil and water losses	After abandonment, soil detachment decreased drastically in the olive and orange orchards, while vineyards did not show any difference, and almond registered higher erosion rates after the abandonment. Terraced oranges and olives recovered a dense vegetation cover after the abandonment, while the sloping terrain of almond and vineyards enhanced the development of crusts and rills and a negligible vegetation cover resulted in high erosion rates. The contrasted responses to land abandonment in the Mediterranean agricultural lands suggest that it should be programmed and managed with soil erosion control strategies for some years to avoid land degradation
Romero et al. (2017)	Evaluation of ecosystem evolution in abandoned fields in Valencia, Murcia and Andalusia and the application of different methodological approaches	In Valencia, the main responses were the recovery of vegetation after land abandonment and an increase in organic matter and soil infiltration capacity. In Murcia, except for some terraced areas on marls where erosion processes following abandonment were important, land abandonment resulted in vegetation recovery, improved soil properties and reduced surface wash and soil losses. In Andalusia, along climatological gradients, it showed the relationship between vegetation patterns and soil moisture and the control that climate exerts on hydrological and erosive behaviour. This study in the Mediterranean demonstrated that abandonment can result in recovery of the geo-ecosystem as vegetation and soil quality improvements. The marls areas in Murcia were the exception with low soil quality and low vegetation cover, and as a consequence showed evidence of high erosion rates after abandonment
Hernández et al. (2016)	The main objective of this study was to define the historical impact of land-use and land-cover change (LUCC) on soil quality (SQ) and soil quality index (SQI) in a Mediterranean landscape	The results indicate that most LUCC dynamics were observed in the Espinal woodland, where agriculture and grazing activities have been developed historically. The SQI showed significant SQ deterioration when native forests (SQI = 0.82) degrade and are transformed by anthropic interventions like annual crops (SQI = 0.27), perennial crops (SQI = 0.34) or grassland (SQI = 0.36). However, a slight improvement in SQ compared to the other land uses was found in the dense espinal (SQI = 0.46).

(continued)

Table 7.2 (continued)

Authors	Objective	Main findings
Almagro et al. (2013)	Explore the links between heterogeneity in soil respiration (Rs), aboveground plant biomass and belowground properties in three representative land-use types in a dry Mediterranean ecosystem	Conversion of native woodland to agricultural land and subsequent land abandonment leads to profound changes in the relationships between Rs, aboveground biomass and belowground properties in this dry Mediterranean ecosystem. While above- and belowground vegetation are the primary controls on spatial variability in labile soil C pools and Rs in the open forest and olive grove sites, a complete lack of influence of current vegetation patterns on soil C pools and respiration rates in the abandoned agricultural field was observed
Vaccari et al. (2012)	Assess the carbon flux measurements from an abandoned Mediterranean ecosystem and integration into a modeling framework, with soil sampling and historical information, evaluating the past and predicted SOC evolution in relation to LUC	Results showed that the abandoned ecosystem was able to rapidly recover the SOC depleted during agricultural exploitation, increasing SOC content by about 30% above the 1990s. The model simulations predicted an increasing SOC sequestration until the middle of this century under unmanaged conditions. However, re-cultivation of these soils may potentially cause a SOC loss ranging from 4.1 to 5.8 kg C m^{-2} in 90 years, if appropriate soil conservation management practices are not implemented.
Laliberté et al. (2010)	Explore how LUC impacts functional redundancy and response diversity in plant communities	The land-use intensification significantly reduced both functional redundancy and response diversity, although specific relationships varied considerably among the different land-use gradients. These results indicate that intensified management of ecosystems for resource extraction can increase their vulnerability to future disturbances

which reduces the size of fruits in relation to tree canopy, slows down the manual harvest and leads to a high risk of N pollution of water bodies (Durán et al. 2004; Varma et al. 2017b).

A new subtropical crop, cherimoya, not planted in the watershed before 1978, covered 3.9 ha by 2007 (Table 7.1); Spain is the first cherimoya producer in the world, followed by traditional producers such as Peru and Chile (Vanhove and Van Damme 2013; Kumar et al. 2018). This crop began to be cultivated during the nineteenth century, most cherimoya orchards on the coast of Granada are planted in flat areas, and 90% of this fruit is consumed in Spain and the remaining 10% is exported to European markets (Durán et al. 2006a; Meena and Yadav 2015).

Table 7.3 Land-use change (LUC) studies and its effects on soil erosion in Mediterranean areas

Authors	Objective	Main findings
Cerdà et al. (2018)	Eleven monitoring years in two paired plots (abandoned vs. control) with four subplots to determine how soil and water losses evolved after abandonment within an agricultural area	The results showed that during the two years after abandonment there was an increase in sediment yield followed by a decrease. Once the field was abandoned, a sudden increase in runoff (× 2.1 times) and sediment concentration (× 1.2 times) was found due to the lack of vegetation and tillage. After one year, the sediment concentration and, after two years, the runoff rates were lower in the abandoned than in the tilled plots. This short transition period ended in contrasting responses between the control and abandoned plot as the impact of abandonment resulted in 21 times less sediment yield after nine years of abandonment. Agriculture land abandonment resulted in lower erosion rates over the long term but showed an increase in soil and water losses over the short term (two years)
Lizaga et al. (2018)	Assess how LUCs after widespread land abandonment affect soil redistribution	In cropland soils, mean erosion (62.6 Mg ha^{-1} $year^{-1}$) and deposition rates (55.2 Mg ha^{-1} $year^{-1}$) were significantly higher than in the other land uses. The lowest mean erosion rates (2.4 Mg ha^{-1} $year^{-1}$) were found in natural forests and the lowest mean deposition (2.6 Mg ha^{-1} $year^{-1}$) in pine afforestation evidencing the soil stabilization achieved in the last decades due to revegetation. A sediment budget with the interpolated rates results in a specific sediment yield of 4.15 Mg ha^{-1} $year^{-1}$. These results outline the impact of LUCs on soil redistribution in fragile mountain agroecosystems
Navas et al. (2017)	Fallout Caesium-137 (^{137}Cs) has been used to study soil redistribution in abandoned fields located in the Central Spanish Pyrenees	Considering all the fields and all geomorphic positions within the fields, the greatest ^{137}Cs losses and gains were found in the fields with the longest duration of abandonment, indicating more intense soil redistribution. Irrespective of the timing of abandonment, the ranges of ^{137}Cs inventories in the fields were found to be proportional to the water erosion index. The ^{137}Cs technique demonstrated that patterns of sediment redistribution were closely related to the topographic and physiographic characteristics of the slopes
Ohana et al. (2015)	Estimate the effects of 20 years of land-cover changes (LCCs) on rainfall-runoff relations in an extreme rainfall event using the KINEROS model	An increase in runoff volume and peak discharge between the time periods as a result of LCCs. A strong relationship was detected between vegetation cover and the runoff volume. The LCCs with most pronounced effects on runoff volumes were related to urbanization and vegetation removal

(continued)

Table 7.3 (continued)

Authors	Objective	Main findings
Serpa et al. (2015)	The impacts of climate and LUCs on stream flow and sediment export were evaluated for a humid and a dry Mediterranean catchment, using the SWAT model	Compared to the baseline scenario (1971–2000), climate change scenarios (A1B y B1) showed a decrease in annual rainfall for both catchments (humid: −12%; dry: −8%), together with increases in rainfall during winter. LUCs were derived from a socio-economic storyline in which traditional agriculture is replaced by more profitable land uses (i.e. corn and commercial forestry at the humid site; sunflower at the dry site). Climate change projections showed a decrease in streamflow for both catchments, whereas sediment export decreased only for the humid. LUCs resulted in an increase in streamflow, but the erosive response differed between catchments. The combination of climate and LUC scenarios led to a reduction in streamflow for both catchments, suggesting a domain of the climatic response. As for sediments, contrasting results were observed for the humid (A1B: −29%; B1: −22%) and dry catchment (A1B: +222%; B1: +5%), which is mainly due to differences in the present-day and forecasted vegetation types. The results highlight the importance of climate-induced LUC impacts, which could be similar to or more severe than the direct impacts of climate change alone
Nadal et al. (2013)	Analyse runoff and sediment yield from land under various traditional and current land uses in Mediterranean mountain areas	The results indicate that (i) cereal cultivation on steep slopes represents a major problem for soil conservation; (ii) farmland abandonment has led to a marked reduction in runoff and sediment yield as a consequence of rapid plant recolonization, particularly by dense shrubs; and (iii) the natural transformation of abandoned fields into grazing meadows has reduced runoff and sediment yield. Land use trends in the Mediterranean mountains are mainly characterized by generalized farmland abandonment and a decrease in livestock pressure. From a hydrological and geomorphological point of view, the main consequences have been a reduction in overland flow from the hillslopes, and a reduction in sediment sources, with differences up to one order of magnitude in sediment yield from dense shrub cover and grazing meadow areas compared with areas under shifting agriculture
Alatorre et al. (2012)	The soil erosion and sediment delivery WATEM/SEDEM model was applied to a small catchment in the Central Spanish Pyrenees	The results demonstrated a reduction of about one order of magnitude in gross erosion (from 3180 to 350 Mg year^{-1}) and sediment delivery (from 11.2 to 1.2 Mg year^{-1} ha^{-1}) during the last decades as a result of the abandonment of traditional land uses (mostly agriculture) and subsequent vegetation recolonization

(continued)

Table 7.3 (continued)

Authors	Objective	Main findings
Maetens et al. (2012)	Study effects of land use on annual soil loss (SL), annual runoff (R) and annual runoff coefficient (RC)	Construction sites have the highest mean annual RC (57%) and SL (325 Mg ha^{-1} $year^{-1}$). Bare soil, vineyards and tree crops have high mean annual RC (5–10%) and SL (10–20 Mg ha^{-1} $year^{-1}$). Cropland and fallow show similar annual RC (8.0 and 7.3%), but lower SL (6.5 and 5.8 Mg ha^{-1} $year^{-1}$). Plots with (semi-)natural vegetation cover show lowest annual RC (<5%) and SL (<1 Mg ha^{-1} $year^{-1}$). Plot length and slope gradient correlations with R and SL depend on land-use type and are not concurrent for R and SL. Most land-use types show positive correlations between annual R and SL. Plots in cold climates have higher RC than plots in temperate and pan-Mediterranean climates. Annual SL in the pan-Mediterranean is less than in temperate zones, due to stony or clayey soils having a low erodibility. Annual RC in the pan-Mediterranean was higher than in temperate zones. Annual R increases strongly with increasing precipitation (P) above 500 mm $year^{-1}$, while annual SL was found to stabilize at $P > 500$ mm $year^{-1}$. For shrubland, annual SL was found to decrease for $P > 250$–500 mm $year^{-1}$, which is attributed to an accompanying increase in vegetation cover
Nadeu et al. (2012)	Examine the effect of erosion processes and LUC on the stock, type, and stability of organic carbon in two medium-sized subcatchment in south-eastern Spain	LUC influenced the dominating erosion processes and, thus, the source of eroding sediments. Carbon isotopes used as tracers revealed that, in one of the subcatchments, the deposited sediments were derived from deep soil through non-selective erosion processes and channel incision. In the other subcatchment, topsoil material was predominantly eroded. Assessments of the overall carbon budget at the catchment scale should consider which sources are responsible for the main part of sediments and how these may vary over time due to LUCs
Durán et al. (2012)	Analyse the impact of LUCs on runoff and sediment yield in a semiarid watershed in south-eastern Spain (Lanjarón), and its implications for water quality	The runoff was highly variable, ranging between 53.4 and 154.7 mm $year^{-1}$, with an average of 97.6 mm $year^{-1}$. In contrast, sediment yields were more regular, averaging 1.8 Mg ha^{-1} $year^{-1}$. Formerly, 32% of the watershed was forested and runoff was more regular, despite the typical Mediterranean rainfall cycle; however, due to forest area reduction to 17% and the increase in abandoned farmland area (18%) in recent decades, the runoff variability has increased. Greater amounts of solutes (32.7 Mg ha^{-1} $year^{-1}$) were exported, so that this water is considered as poor for irrigation use. In addition, the nutrient concentrations of the water discharged were lower than threshold limits cited in water-quality standards for agricultural use and for potable water, with the exception of K (65.9 mg L^{-1}), which may degrade surface waters as well as irrigated soils. Thus, hydrological and erosive processes depended on the watershed features, but also on prior conditions in combination with the characteristics of rainfall

(continued)

Table 7.3 (continued)

Authors	Objective	Main findings
Lana-Renault et al. (2011)	The stream flow response of two catchments was compared: one catchment was extensively used for agriculture in the past, and the other is covered by dense natural forest	Differences in soil depth and permeability, together with differences in vegetation cover, may explain the contrasting dominant runoff generation processes operating in each catchment and consequently the differences between their hydrograph characteristics
Boix-Fayos et al. (2009)	The effect of LUCs on the SOC stock and the soil C transported by water erosion and buried in depositional wedges behind check-dams in a Mediterranean catchment in south-eastern Spain.	Changes in land use patterns in the catchment between 1956 and 1997 (57% decrease in areas dedicated to agriculture and 1.5-fold increase of the total forest cover) induced an accumulation rate of total SOC of 10.73 g m^{-2} $year^{-1}$. Mineral associated organic carbon was the main soil carbon pool (70%). Particulate organic carbon was highest in the shrubland soils (33%). The average sediments/soil enrichment ratio at the subcatchment scale (8–125 ha) was 0.59–0.43 g kg^{-1}. Eroded soil C accounted for between 2 and 78% of the soil C stock in the first 5 cm of depth in the subcatchments. The C erosion rate varied between 0.008 and 0.2 t ha^{-1} $year^{-1}$. Observed changes in land use (decrease in agricultural areas) reduced soil C erosion, although sediments from non-agricultural sources are richer in organic C. At catchment scale from the 4% of the soil C stock mobilized by water erosion, 77% is buried in the sediment wedges behind check-dams. Soil C replacement due to increased vegetation cover between 1974 and 1997 represented a 36% of the original SOC stock. All together represent an erosion-induced sink of soil organic C of 40% compared to the original levels of 23 years before

Similarly, loquat, another crop not planted before 1978, occupies 8.2 ha in the pilot watershed in 2007. The easy adaptation of loquat to the Mediterranean climate has led to its rapid expansion throughout the Mediterranean Basin. China is the world's largest producer of loquat with more than 400,000 t in 120,000 ha (Lin 2007). The trees are usually planted in a single row on terraces with platforms of 3–4 m wide that hinder the mechanization of loquat plantations as it does for other subtropical species within the watershed.

Finally, mango orchards not planted before 1978 with area devoted to this crop in the watershed of 13.5 ha in 2007 (Table 7.1). In the last decades, a strong increasing trend of mango cultivation in this area has been reported (ESYRCE 2017). Within the watershed, as in the adjacent watersheds, the most extensively produced commercial cultivars are Osteen and Keitt (Durán et al. 2003; Rodríguez et al. 2016; Datta et al. 2017b).

Table 7.4 Impact of land-use change (LUC) on soil quality in Mediterranean areas

Authors	Objective	Main findings
Lozano et al. (2017)	CarboSOIL model to predict changes in SOC stocks in a semi-natural area of southern Spain for three-time horizons (2040, 2070 and 2100), considering two circulation models (BCM2 and ECHAM5) and three IPCC scenarios (A1b, A2, B2). The effects of potential LUCs from natural vegetation (Mediterranean evergreen oak woodland) to agricultural land (olives and cereal) on SOC stocks were also evaluated	Predicted values of SOC contents correlated well those measured (R^2 ranging from 0.71 at 0–25 cm to 0.97 at 50–75 cm) showing the efficiency of the model. Results showed substantial differences among time horizons, climate and land use scenarios and soil depth with larger decreases of SOC stocks in the long term (2100-time horizon) and particularly in olive groves. The combination of climate and land use scenarios (in particular, conversion from current 'dehesa' to olive groves) resulted in yet higher losses of SOC stocks, e.g. −30, −15 and −33% in the 0–25, 25–50 and 50–75 cm sections, respectively. This study shows the importance of SOC stocks assessment under both climate and land use scenarios at different soil sections and point towards possible directions for appropriate land use management in Mediterranean semi-natural areas
Novara et al. (2017)	Estimate the influence of different soil regions (areas characterized by a typical climate and parent material association) and bioclimates (zones with homogeneous climatic regions and thermotype indices) on SOC dynamics after agricultural land abandonment	Results showed that abandonment of cropland soils increased SOC stock by 9.03 Mg C ha^{-1} on average, ranging from 5.4 to 26.7 Mg C ha^{-1} in relation to the soil region and bioclimate. Considering the 14,337 ha of abandoned lands in Sicily (Italy), the CO_2 emission as a whole was reduced by 887,745 Mg CO_2. It could be concluded that abandoned agricultural fields represents a valid opportunity to mitigate agriculture sector emissions in the study area

(continued)

Table 7.4 (continued)

Authors	Objective	Main findings
Serpa et al. (2017)	The impacts of climate and LUCs on total nitrogen (TN), phosphorus (TP) and copper (Cu) exports from a humid Mediterranean catchment were evaluated using the SWAT model	SWAT reproduced reasonably well TN, TP and Cu exports in providing an adequate baseline scenario as well as a suitable model parameterization for assessing the impacts of climate and LUCs under the A1B and B1 emission scenarios for the end of the twenty-first century. Climate changes itself led to a decline in annual TN and TP exports under both emission scenarios mostly due to a decrease in runoff and erosion induced by a reduction in rainfall, but it hardly affected Cu exports largely due to its strong immobilization in soils. LUCs per se resulted in an increase in streamflow, but the changes in water quality varied markedly according to the scenarios. A substantial decrease in TN, TP and Cu exports was observed under scenario A1B, due to a reduction in vineyard areas. Under scenario B1, TP exports decreased much less while TN exports hardly changed, reflecting differences in the preferential transport pathways of these compounds. Cu exports also remained the same, as no changes occurred in the vineyard areas. The combination of climate and LUC revealed additive impacts on the exports of all three contaminants, emphasizing the importance of integrated approaches to define adaptive land management practices that can ensure the future sustainability of Mediterranean water resources
Lozano et al. (2016)	Determine the effects of LUC [Mediterranean evergreen oak woodland (MEOW-dehesa) to olive grove and cereal] on SOC and N concentrations and stocks, and on the soil quality in The Cardeña-Montoro Nature Reserve (southern Spain)	The LUC from MEOW-dehesa to olive groves increased the SOC concentration (MEOW-dehesa: 19.4 g kg^{-1}; 7.2 g kg^{-1}; olive grove: 41.8 g kg^{-1}; 15.4 g kg^{-1} in the surface and subsurface layers respectively) and stock (22.3% higher in olives than in MEOW-dehesa and 17% higher in olives than in cereal), but did not affect the N concentration and decreased N stock in the surface layer. The conversion from MEOW-dehesa to cereal did not affect SOC stock and concentration; it only decreased the N concentration and stock in the surface layer (N: 1.6 g kg^{-1} and 1.1 g kg^{-1} in MEOW-dehesa and cereal respectively; N stock 4.2 Mg ha^{-1} and 2.9 Mg ha^{-1} in MEOW-dehesa and cereal, respectively). After the LUC to olive stratification ratios increased, however, in cereal decreased. Additionally, it is important to point out that all the stratification ratios calculated were >2

(continued)

Table 7.4 (continued)

Authors	Objective	Main findings
Nadal et al. (2016)	Assess the effects of land abandonment, LUC and afforestation practices on SOC dynamics. Five different land covers [bare soil, permanent pastureland, secondary succession, *Pinus sylvestris* (PS) and *Pinus nigra* (PN) afforestation] were analysed	The findings showed that (i) SOC were higher in the PN sites in the topsoil (10 cm), (ii) when all the profiles were considered no significant differences were observed between pastureland and PN, (iii) SOC accumulation under secondary succession is a slow process and (iv) pastureland should also be considered due to the relative importance in SOC stocks. The first step of SOC stabilization after afforestation is the formation of macro-aggregates promoted by large inputs of SOC, with a high contribution of labile organic matter. However, our respiration experiments did not show evidence of SOC stabilization. Finally, SOC mineralization was higher in the top layers and values decreased with depth
Trigalet et al. (2016)	Study the gradual changes in SOC pools during secondary succession	The significant increase in SOC attached to sand particles or occluded in aggregates (S+A) over time indicated aggregation processes along the vegetation recovery stages. Carbon concentrations within S+A or rSOC fractions did not change over time. In contrast, SOC bound to silt and clay (s+c) fraction significantly increased after three decades from 7.9 to 20.4 g C kg^{-1} at the final stage of the chronosequence. In addition to the frequently reported increases in particulate organic matter (POM), carbon was also allocated to the more stable SOC fractions. Taking active carbon (DOC + POM; high turnover rate) and intermediate carbon (s+c, S+A; low turnover rate) as indicators for C turnover, the proportion of active carbon increased from 17% to 39% along the chronosequence. Consequently, these results showed that the proportion of labile carbon increases along the secondary succession while, at the same time, the slow cycling carbon fractions sequester carbon (from 0.6 to 1.12 kg C m^{-2})
Muñoz et al. (2015)	Propose a methodology to assess the impact of land use and land cover change dynamics on SOC stocks at different depths	Land use and land cover changes between 1956 and 2007 implied soil organic C losses in all soil groups, resulting in a total loss of 16.8 Tg (~ 0.33 Tg $year^{-1}$). Afforestation increased SOC mostly in the topsoil, and forest contributed to sequestration of 8.62 Mg ha^{-1} of SOC (25.4%). Deforestation processes implied important C losses, particularly in Cambisols, Luvisols and Vertisols

(continued)

Table 7.4 (continued)

Authors	Objective	Main findings
Chrenková et al. (2014)	Compare the effect of LUC from agriculture to forestry on soil aggregation. Soils from agricultural land (dry crops with conventional tillage management using ploughing), and nearby forest, where agriculture was abandoned between 20 and 50 years ago were compared. Selected physico-chemical properties [total content of aggregates (TCA), aggregate stability (AS), soil organic matter (SOM), mean weight diameter (MWD) of aggregates and water repellency (WR) among others] were studied	As expected, in all soils, the AS was significantly lower in agricultural than in forest sites. In forest sites the AS varied between 46 and 82% while in agricultural sites it ranged between 14 and 45%. The results showed strong positive correlation of AS with SOM. The lowest values of AS in forest areas were found in wettable sandy soils. The relative increase of AS in forest soils compared to agricultural land use ranged from 39 to 79%. In particular, the lower SOM and absence of WR are responsible for the lower values of AS in agricultural land use due to the long-term conventional tillage. Soil WR plays a key role in soil aggregation and its stability, especially for sandy soils. The MWD was in most cases higher in forest than in agricultural sites, except for two wettable sandy soils with the lowest initial AS. Here was demonstrated that the prolonged use of conventional tillage using ploughing has a strong influence on AS in all types of soils studied. The recovery after decades of land use abandonment not only is dependent on how is the increase of SOM, but also is controlled by some soil properties such as the texture and the presence of WR. In some cases the presence of hydrophobic substances can play a key role in the formation and stabilization of soil aggregates, contributing to maintaining good levels of SOM and avoiding higher levels of soil degradation
Fernández et al. (2014)	Examine the effects of LUC from natural forest (NF) to olive groves (OG) in a toposequence (summit-SM, back slope-BS and toe slope-TS) and to determine SOC, N and the stratification ratio (SR) of SOC in a Mediterranean area in a Calcisols–Regosols–Vertisols sequence	Total SOC was low for both NF and OG soils. In the case of NF, there was a pattern of decrease from the highest SM to the lowest topographical position TS; whereas for OG, the highest SOC was found in TS (from 50.2 to 1.1 g kg^{-1}). SOC was more than 40% higher in NF than in OG in SM and BS. The total SOC stock in NF soils remained evenly distributed in the three topographical positions. However, the trend for the OG soils was an increase of SOC stocks along the hillside (SM: 68.9–4.8 Mg ha^{-1}; BS: 82.3–12.3 Mg ha^{-1} and TS: 158.7–15.4 Mg ha^{-1}). N followed a pattern of decrease with depth in the OG soils, but not in NF soils. As for N stock, it was higher in NF than in OG, with the exception of NF-TS and OG-TS. SR values in NF had a general trend of increasing from higher to lower topographical positions, despite the fact that SOC decreased from higher to lower topographical positions. This indicates that soil quality is higher in lower topographical positions even though there is less SOC. SR values in OG soils had a similar trend as in NF soils (but with lower values). Therefore, soil quality was higher in lower topographical positions for both land uses

(continued)

Table 7.4 (continued)

Authors	Objective	Main findings
Parras et al. (2013)	Determine the long-term effects of LUC [arable crops (AC) by olive grove (OG) and vineyards (V)] and to determine SOC, total nitrogen (TN), C:N ratio and their stratification in soil	This LUC principally affected the horizon thickness, texture, bulk density, pH, organic matter, organic carbon, TN and C:N ratio. The LUC had a negative impact in the soil, affecting the SOC and TN stocks. The conversion from AC to V and OG involved the loss of the SOC stock (52.7 and 64.9 % to V and OG respectively) and the loss of the TN stock (42.6 and 38.1% to V and OG, respectively). With respect to the stratification ratios (SRs), the effects were opposite; 46 years after LUC increased the SRs (in V and OG) of SOC, TN and C:N ratio
Durán et al. (2013)	Evaluation the impact of different land-cover types on soil erosion and runoff in agroforestry landscape, characterizing trends and relationships in soil properties under the semiarid Mediterranean climate of south-eastern Spain	Forest dominated by *Pinus* stands exhibited lowest amounts of erosion and runoff, contrasting with abandoned farmland. Olive had higher erosion than did almond, cereals, or grasslands, but with the highest runoff rate under almond groves. The erosion and runoff response to shrubland showed an intermediate situation between forest and farmland–grassland uses. Under forest and shrubland, better soil properties were determined, especially higher SOC and total N, and lower soil-bulk density. Erosion was highly dependent on runoff, bulk density, soil organic C and the degree of soil surface covered
Marzaioli et al. (2010)	The impact of different land use types on soil quality was evaluated by measuring several soil properties that are sensitive to stress or disturbance	Data indicated a clear difference in soil quality among the studied areas: low soil quality in almost all permanent crops; intermediate soil quality in shrublands, grazing lands, coniferous forest and middle-hill olive grove (the only crop with an herb layer on the soil surface); high soil quality in mixed forests. Results suggested that the permanent crop management had generally a strong negative impact on soil quality, while the moderate grazing activity and the crop management that leaves an herb cover on the soil had a lower negative impact. Nevertheless, the abandonment of cultivated lands, with consequent development of shrublands, produced an improvement of soil quality suggesting a good recovery capacity in the studied soil
Muñoz et al. (2011)	Improve and test methodologies to assess land cover change (LCC) dynamics and temporal and spatial variability in C stored in vegetation at a wide scale	The study area has supported important changes during the studied period with significant consequences for vegetation C stocks, mainly due to afforestation and intensification of agriculture, resulting in a total vegetation C stock of 156.08 Tg in 2007, with an increase of 17.24 Tg since 1956. This study demonstrates the importance of LCC for C sequestration in vegetation from Mediterranean areas

(continued)

Table 7.4 (continued)

Authors	Objective	Main findings
Padilla et al. (2010)	Measure implications of LUC on C sequestration services, changes in land use and associated carbon sequestration potential throughout the twentieth century in a rural area at the foothills of the Sierra Nevada range (SE Spain) were explored	They found that forest systems replaced dry land farming and pastures from the middle of the century onwards as a result of agricultural abandonment and afforestation programmes. The area has always acted as a carbon sink with sequestration rates ranging from 28,961 in 1921 to 60,635 t CO_2 year^{-1} in 1995, mirroring LUC. Conversion from pastures to woodland, for example, accounted for an increase in C sequestration above 30,000 t CO_2 year^{-1} by the end of the century. However, intensive deforestation would imply a decrease of approximately 66% of the bulk CO_2 fixed. Thus, woodland conservation was essential to maintain the ecosystem services that underlie C sequestration in the study area
Zucca et al. (2010)	Evaluate the impacts of LUC (forest by pasture) on soil properties	The comparison of forest and pasture soils showed a significant soil loss (in terms of soil thickness, −22%) and a clear decrease in organic carbon storage (−64% on average). An increase in bulk density (+44%) and a change in microporosity and its vertical distribution were also observed, respectively by field measurements and micromorphological quantification on digital images
Khresat et al. (2008)	This study was carried out to evaluate the effects of deforestation on physical and chemical properties of soils under native forest in the Mediterranean region of north-western Jordan	Findings showed that many forests were changed into cultivated lands at a rate more than the reforestation. Subsequently, adverse effects on the studied physical and chemical properties were observed. The most affected properties were particle size distribution, bulk density of surface soil and subsoil. Organic matter and cation-exchange capacity decreased in cultivated soil as compared to the forest soil. Cultivated soils were found to exhibit a significantly lower status in physical and chemical soil properties as compared to forest soils. This general decline in the soil physical and chemical properties, in turn, contributed to soil erosion, reduction of soil fertility and land degradation

7.3 Environmental Consequences of Subtropical Farming on Terraces

Terraces have been widely utilized in agriculture, being one of the most common agricultural features in mountainous regions of Spain (Lasanta et al. 2013), Italy (Tarolli et al. 2014), Israel (Ore and Bruins 2012), etc. These are structures differing in size and shape which consist in a flat section, normally grazed or cultivated, and a vertical or almost vertical wall usually done by stone, grass, scrub or trees. This wall can have from a few centimetres to several metres.

According to Durán et al. (2005), terracing of sloping land for subtropical crops can deteriorate soil properties, especially reducing soil organic matter and changing the distribution and stability of soil aggregates, above all in bare soil areas (taluses). The taluses of new terraces that are completely unprotected by vegetation urgently need the implementation of plant covers to reduce soil erosion and runoff, and to maintain and conserve the terrace structure. Moreover, as stated by Rodríguez et al. (2009), plant covers in this type of Mediterranean environment encourages atmospheric carbon fixation and recycles plant nutrients by incorporating to the natural cycles.

On the other hand, the intensification of irrigated subtropical farming has led to the application of chemical products to maximize fruit production. Particularly, this type of agriculture relies on the use of chemical fertilizers, herbicides, fungicides, insecticides, plant-growth regulators, etc. In this sense, avocado is the crop with the highest fertilizer requirements, while loquat and mango require higher volume of water due to the higher number of trees per hectare (400–600 trees) (Durán et al. 2003). Therefore, subtropical farming in the watershed has augmented water consumption, coinciding with the summer and with the highest water demand for tourism. Local farmers often apply higher nutrient amounts of NPK than required by the crops, and such excesses represent high environmental pollution risk, requiring a detailed assessment of nutrient rates. Consequently, high fertilizer application in subtropical intensive agriculture is often one of the main sources of nutrient leaching to the environment, associated with a reduced quality of groundwater and surface water (Braden and Shortle 2013; Yadav et al. 2017a).

Another problem of the intensification of agriculture based on subtropical farming in the area is the overexploitation of the Río Verde aquifer (Calvache and Pulido 1991; Meena and Yadav 2014). Due to the scarcity of fresh surface-water resources during drought years the water supply is covered by the exploitation of aquifers, which sometimes leads to marine intrusion processes which increase groundwater salinity (Calvache and Pulido 1996), affecting the irrigation wells and plantations that use this water for irrigation (Durán et al. 2016; Dadhich et al. 2015). Severe climatic events combining water scarcity and droughts are predicted to increase in intensity, frequency and geographic extent as a result climate change. To grow crops successfully in this adverse scenario, farmers will need to adjust to less available water (García and Durán 2018; Meena et al. 2015a). The application of efficient water management strategies to such deficit irrigation will be a key element to increase agricultural water productivity in areas affected by water scarcity.

The LUCs are provoked by a number of natural and human forces. Whereas natural phenomena such as climate change are felt only over a long period of time, the impact of human activities is immediate and often radical and detrimental (Liu et al. 2018; Verma et al. 2015a). Concretely, the main driving force which significantly affects land use is highly profitable subtropical farming, which has reshaped the landscape.

This new activity has exerted heavy pressure on the environment due to the intense use of soil and water resources. Therefore, it is essential to encourage and optimize the equilibrium between water demand and water availability by research on water requirements of the existing subtropical crops in the area, and for soils, it is needed to protect against erosion by better land-use planning (Durán et al. 2011).

After the abandonment of traditional (stone) terraces in 1978 occupied by almonds and olives, new orchard terraces were built for subtropical crops, exacerbating soil degradation (water erosion, soil-nutrient losses, carbon losses, etc.). The establishment of subtropical crops on terraced hillsides requires planned sustainable agricultural measures based on the analysis of water and nutrient balances to avoid water waste and to preserve groundwater from pollution by fertilizers. Thus, the subtropical farming production in this fragile high-altitude ecosystem requires urgent implementation of agro-environmental strategies to mitigate its impact.

7.4 Plant Covers in Terraces for Erosion and Runoff Prevention

Soil is a limited and exhaustible resource which is the basis for human societies and for the production of goods, services and resources (Ochoa et al. 2016; Meena et al. 2017a). In this context, water erosion is an environmental problem that ranks as one of the most important problems worldwide with long-lasting effects exerting both agro-ecosystem services and soil functions (Costantini et al. 2018; Kumar et al. 2017a). Verheijen et al. (2009) proposed a definition of tolerable soil erosion as 'any actual erosion rate at which a deterioration or loss of one or more soil functions does not occur'. In this sense, for Europe, it was estimated that the upper limit of tolerable soil erosion is equal to soil formation (1.4 Mg ha^{-1} $year^{-1}$), while the lower limit to be 0.3 Mg ha^{-1} $year^{-1}$. Unfortunately, soil erosion rates in the present for tilled, cultivated lands in Europe range from 3 to 40 times greater than the upper limit. This situation is aggravated in the Mediterranean basin, characterized by unpredictable rainfall fluctuations from year to year.

In the coast of Granada, where agriculture is based on subtropical farming, the detached soil particles from the taluses of terraces accumulates on the platform of the terrace below, hindering fruit harvesting and plantation maintenance. Local farmers usually remove vegetation from the taluses because most of these plants are weeds. However, the importance of vegetation in controlling erosion and runoff is widely accepted (Durán and Rodríguez 2008; Atucha et al. 2013; Vallebona et al. 2016; Meena et al. 2014). The extreme rain events in this area could provoke the collapse of terrace structure (Fig. 7.2). In addition, since topsoil is usually relatively rich in nutrients, eroded soil typically contains about three times more nutrients than the soil remaining on the eroded land.

On the other hand, aromatic and medicinal plants (AMPs) have traditionally been utilized in the study area. Their multiple uses are well known whether fresh, frozen or dry or for producing extracted essential oils (Gresta et al. 2008; Barata et al. 2016; Dhakal et al. 2016). In Spain, according to ESYRCE (2017), the area devoted to AMPs was more than 13,406 ha from which 93% of the surface under rainfed conditions and 5,133 ha in Andalusia (southern Spain). Therefore, the maintenance and cultivation of these AMPs constitute one of the major economic activities for local farmers in the Mediterranean mountains.

Fig. 7.2 Rills and gullies in the taluses of avocado (**a**), mango and loquat (**b**) orchard terraces after a heavy rainfall event

7.4.1 Aromatic and Medicinal Plant Covers in the Taluses

In this section, we present the main findings of an experiment designed to test the response of runoff, soil erosion and nutrient losses to plant covers during two hydrological years. To overcome with this, were installed 12 closed erosion and runoff plots of 4 m × 4 m (16 m^2) on the taluses of the terraces (Fig. 7.3). Four aromatic-medicinal plants were tested: thyme, *Thymus mastichina* L. (T_H); lavender, *Lavandula dentata* L. (L_A); satureja, *Satureja obovata* Lag. (S_A); and rosemary, *Rosmarinus officinalis* L. (R_O); each replicated twice. The planting grid of AMPs was 40 × 40 cm, with about 81 plants per erosion and runoff plot. Also, two plots were left with native spontaneous vegetation (S_V) growing in the monitoring area. Finally, two plots were left with bare soil (B_S) as a control without any plant cover.

Fig. 7.3 Closed erosion and runoff plots used for the experiment

Sediment concentration in runoff was determined. In surface runoff, samples were determined for NO_3^-, NH_4^+, $H_2PO_4^{3-}$ and K concentrations (APHA, AWWA, WPCF 1995), and each sediment sample was analysed for N, P and K content (MAPA 1994).

In these plots according to the findings for the yearly data, the control B_S had significantly higher rates of soil erosion and runoff (26.4 Mg ha^{-1} and 55.7 mm year^{-1}, respectively) than in the rest of plots with plant covers. The plant cover plots measured the following rates in runoff for R_O, S_A, T_H, L_A and S_V of 41.7, 38.2, 16.9, 16.1 and 12.4 mm year^{-1}, respectively, whereas annual soil erosion for S_A, R_O, T_H, S_V and L_A of 18.0, 13.4, 5.5, 4.4 and 3.2 Mg ha^{-1} year^{-1}, respectively. This means that S_V reduced runoff and soil erosion with respect to B_S by 78 and 83%, respectively. On the other hand, L_A and T_H were also highly effective plant covers in reducing runoff (71 and 88%, respectively) and soil erosion with respect to B_S (70 and 79%, respectively). These results for soil erosion on bare soil were much higher than those reported by Bautista (1999) for erosion plots of 0–8 Mg ha^{-1} year^{-1}, Castillo et al. (1997) and Romero et al. (1999) between 0.012 and 1.84 Mg ha^{-1} year^{-1}, and by Durán et al. (2005) for bare soil in the same area of 9.1 Mg ha^{-1} year^{-1}. Similarly, in vineyards, Ramos et al. (2015) reported soil erosion ranged from less than 1.0 Mg ha^{-1} year^{-1}, in the driest year, to 13.9 Mg ha^{-1} year^{-1}, in the wettest.

Table 7.5 presents the analysis of variance concerning the response of average runoff and soil erosion to tested plant covers. The lowest soil erosion rates were found under T_H and S_V (0.14 and 0.17 Mg ha^{-1}, respectively), the values of which differed significantly from those of the other plant covers used. B_S plot recorded the highest erosion rates (2.36 Mg ha^{-1}). In terms of runoff, significantly lower rates for S_V and T_H were found in comparison with B_S (0.7, 0.9 and 3.3 mm, respectively).

Table 7.5 Mean soil erosion and runoff after each storm event for each plant cover

Plant cover	Soil erosion (Mg ha^{-1})	Runoff (mm)
Thymus mastichina (T_H)	0.14 a ± 0.1	0.7 a ± 0.6
Spontaneous vegetation (S_V)	0.17 a ± 0.12	0.96 a ± 0.79
Lavandula dentata (L_A)	0.52 ab ± 0.4	1.81 ab ± 0.9
Rosmarinus officinalis (R_O)	0.79 b ± 0.5	2.09 ab ± 1.3
Satureja obovata (S_A)	1.43 bc ± 0.7	2.46 ab ± 2.0
Bare soil (B_S)	2.36 c ± 0.9	3.32 b ± 1.6

Different letters within the same columns are statistically different at the level 0.01 (LSD). (± standard deviation)

However, the rest of the plant covers (L_A, S_A and R_O) did not significantly differ from each other (Table 7.5). The trend for runoff and for soil loss was higher during the second study year, when the highest rainfall was recorded. Compared to bare soil, T_H and S_V reduced the runoff with 94 and 93% and reduced erosion with 71 and 79%, respectively. The least effective for soil erosion among the plant covers was S_A, which reduced soil loss by only 39%, and the least effective regarding runoff was R_O, which reduced it by only 26%, with respect to B_S. In general, plant covers softened the mechanical impact of the raindrops on the soil surface of the taluses, diminishing the superficial runoff and thereby aiding soil conservation in the taluses of orchard terraces.

7.4.1.1 Plant–Nutrient Concentration in Runoff

The average NO_3 concentration in the runoff samples during the monitoring period ranged from 6.1 to 24.3 mg L^{-1} for L_A and S_V, respectively, with the highest variability for this latter (Fig. 7.4a). In this sense, the reduction of plant–nutrient fluxes by AMPs followed the pattern: $S_V > B_S > R_O > T_H > S_A > L_A$, reducing S_V in 75% the NO_3 transportation in relation to the L_A covers. Most of the dissolved nitrogen in the runoff came presumably from N fertilizers applied by fertigation in terraces rather than from the soil.

In most cases, with exception of L_A and S_A plots, the average NO_3 exceeded the 10 mg L^{-1}, that is, the limit for drinking water (US EPA 1976). Moreover, NO_3 concentrations in all plots were below the 50 mg L^{-1}, maximum limit set by European directive for drinking water. Additionally, the NO_3 in S_V (43.6 mg L^{-1}) not exceeded the recommended limit (45 mg L^{-1}) for drinking water by the BIS (1991) and the WHO (1996).

The average K concentrations in runoff from the plots ranged from 5.8 to 24.1 mg L^{-1} for S_A and S_V, respectively (Fig. 7.4a). These important K concentrations were presumably due to the application of KNO_3, which is usually applied in terraced mango plantations (Durán et al. 2006a, b; Varma et al. 2017b). All plots with exception for S_A, the K concentration were more than 12 mg L^{-1} [the limit for drinking water established by the European Community (EEC 2000)].

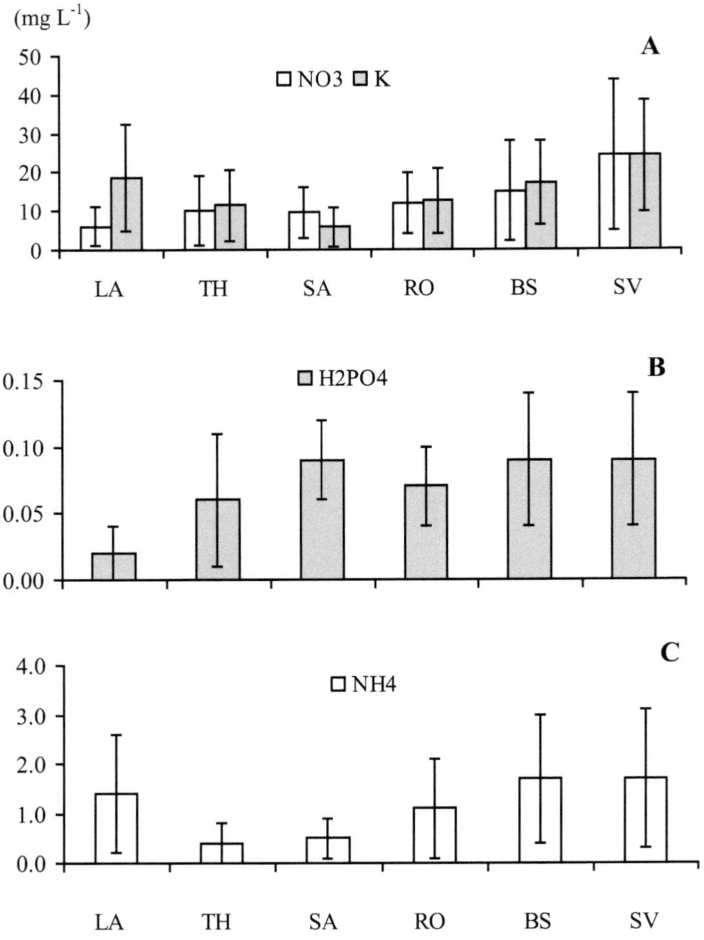

Fig. 7.4 Average nitrate and potassium (**a**), phosphate (**b**) and ammonium (**c**) concentrations of surface runoff collected from the taluses of orchard terraces. S_V Spontaneous vegetation, T_M *Thymus mastichina*, L_A *Lavandula dentata*, S_A *Satureja obovata*, R_O *Rosmarinus officinalis*, *BS* bare soil. Vertical bars are standard deviation

Phosphate ($H_2PO_4^{3-}$) concentrations in runoff from plots ranged from 0.02 to 0.1 mg L^{-1} for L_A and B_S (Fig. 7.4b). Similarly to B_S, the highest concentrations (0.09 mg L^{-1}) were found from S_V and S_A plots. In this regard, the average $H_2PO_4^{3-}$ concentrations consistently exceeded the limit concentration associated with eutrophication of surface water of 0.01 mg P L^{-1} in agreement with Vollenweider (1968). In relation to minimize the soluble P mobility, Balogh and Walker (1992) reported that low $H_2PO_4^{3-}$ concentrations in runoff is because it is actively taken up by plants and readily absorbed and/or precipitated with Fe, Al and Ca in soils.

The ammonium (NH_4^+) concentration ranged from 0.4 to 1.7 mg L^{-1} for T_H and B_S/S_V plots, respectively (Fig. 7.4c). Therefore, $NH_4^+$was fourfold higher (76%) for bare soil and spontaneous vegetation than for thyme cover crop. In this context, the average NH_4^+ in runoff in most tested plant covers for monitored erosive events exceeded 1.5 mg L^{-1}, this concentration being standard for public water supplies (WHO 1996).

Concentrations of NH_4^+ were significantly much lower than NO_3^-; this dominance is probably due to its high solubility and lower affinity for the adsorption sites in the soil as pointed out by Southwick et al. (1995).

In general, the AMP covers were effective in retaining soil and plant–nutrient elements on the talus with the steep slope. Thus, controlling runoff and sediments regulated the nutrient flow at the same time as releasing nutrients from litter in a biological cycle that is absent from bare soil in taluses.

7.4.1.2 Plant–Nutrient Transportation in Sediments

Table 7.6 shows the results for the $N\text{-}NO_3$, $N\text{-}NH_4$, $P\text{-}PO_4$ and K losses per area by runoff and N, P and K losses per area in sediments. The greatest total $N\text{-}NO_3$ losses per area were recorded in the B_S plot, while the lowest were measured in T_H and L_A. For $P\text{-}PO_4$ and K, the highest losses were again found in B_S. The B_S plot had the highest rate of nutrient losses in terms of runoff per area while the lowest were fixed in T_H, S_V and L_A, except for K, for which the lowest losses rates were in the S_A plots. These $N\text{-}NO_3$ annual losses from bare soil were higher than those recorded by Durán et al. (2005) for similar conditions (probably due to the more aggressive rainfall events registered during the monitoring period), and much lower than those reported by Ramos and Martínez (2006). However, nitrate transportation in this study was consistent with those found in vineyards by Ramos et al. (2015) with 270 mg m^{-2} $year^{-1}$.

$P\text{-}PO_4$ transportation recorded ranged from 0.012 to 0.040 kg ha^{-1} $year^{-1}$ for T_H and S_V, respectively, which were similar to those found by Francia et al. (2006), who recorded rates between 0.07 and 0.29 kg ha^{-1} $year^{-1}$ in olive orchards but lower than those registered (0.5 Kg ha^{-1} $year^{-1}$) by Ramos et al. (2015). These latter authors found a reduction in soil losses of 57% by introducing filter strips and drainage terraces.

Finally, K transportation by runoff ranged from 114.3 to 289.2 mg m^{-2} $year^{-1}$ for S_A and B_S, and from 21.1 to 100.5 mg m^{-2} $year^{-1}$ for L_A and B_S, respectively in sediments (Table 7.6). These K-loss rates were lower than those reported by Francia et al. (2006) (47.0–333.8 mg m^{-2} $year^{-1}$). Plant–nutrient transportation in agricultural lands may be affected by many factors such as soil-management systems, rainfall characteristics, solubility of nutrients and the hydrological characteristics (Schroeder et al. 2004; Verma et al. 2015b).

Bare soil plots produced the highest nutrient losses, which were diminished using plant covers, most effectively by L_A, S_V and T_H, except for K transportation, which showed the greatest decrease in S_A and T_H plots. Therefore, when altering the natural ecosystem by the construction of terraces for subtropical farming, plant covers with potential utilization (aromatic, medicinal or melliferous) may be used in the taluses of the terraces in order to prevent soil erosion and eventual structure collapse.

Table 7.6 Annual nutrient losses with runoff and eroded soil under the different plant covers

Plant cover	In runoff				In eroded soil		
	NO_3-N	NH_4-N	PO_4-P	K	N	P	K
	(mg m^{-2} yr^{-1})						
R_O	366.2 ± 28.4	17.6 ± 4.1	3.1 ± 0.9	227.6 ± 28.7	528.0 ± 25.4	15.8 ± 4.8	52.9 ± 14.8
S_A	312.8 ± 15.9	40.0 ± 6.7	2.5 ± 0.7	114.3 ± 33.8	796.0 ± 42.1	19.9 ± 5.8	63.7 ± 45.1
T_H	148.2 ± 25.4	7.4 ± 5.5	1.2 ± 0.8	179.4 ± 19.8	218.8 ± 28.9	7.8 ± 2.5	44.1 ± 11.7
L_A	80.6 ± 8.7	19.2 ± 3.8	1.9 ± 1.1	214.8 ± 45.7	78.9 ± 9.4	3.8 ± 1.5	21.1 ± 17.3
S_V	302.5 ± 17.8	11.7 ± 2.6	1.9 ± 0.7	216.2 ± 25.4	187.9 ± 12.8	12.1 ± 4.9	99.3 ± 31.8
B_S	367.8 ± 34.7	44.1 ± 4.5	4.0 ± 1.2	289.2 ± 31.4	1025.4 ± 65.7	34.8 ± 10.8	100.5 ± 34.7

R_O *Rosmarinus officinalis*, S_A *Satureja obovata*, T_H *Thymus mastichina*, L_A *Lavandula dentata*, S_V Spontaneous vegetation, B_S Bare soil. ± standard deviation

7.4.2 Soil Organic Carbon and Aromatic and Medicinal Plant Covers

The soil organic carbon (SOC) content was augmented by effect of AMP covers, this being higher after 3 years than bare soil plot (Fig. 7.5), that is, B_S plot decreased the SOC content in 15% after the 3-year monitoring period, in contrast with R_O and S_V plots that increased the SOC in 8% for the same period. The loss of SOC from bare soil plot was expected because the soil aggregates easily broken down and consequently the finer particles could be transported by erosion. In this sense, it is widely accepted that soil erosion is the major factor in land degradation, reducing the productivity and organic matter content (Labrière et al. 2015; Wang et al. 2018; Meena et al. 2015e), overall the soil quality. The mean SOC content increased consistently due to the effect of plant covers, by a mean value of 6%. In short, all plant covers encouraged the sequestering of carbon, which boosts the available organic matter and consequently the soil quality and health by influencing aeration and water retention as well as serving as a major repository and reserve source of plant nutrients. Thus, protection of the soil surface against organic carbon losses by sediment is feasible by using AMP covers, this being one of the most effective conservation practices in this type of environment.

In general, the positive effect of plant covers in reducing soil erosion and runoff is widely accepted, especially those concerning with the improvement and conservation the soil quality. Table 7.7 shows some advantages and disadvantages of plant covers.

Numerous factors have been reported to markedly influence the effect of terracing on soil and water conservation, including the terrace structure, cultivation, field management, climate, soil texture, land use and topography (Arnáez et al. 2007; Ehigiator and Anyata 2011; Liu et al. 2011; Park et al. 2014; Meena et al. 2017b). For example, in central Palestine, soil erosion was 3–20 times greater at non-terraced sites than at terraced sites (Hammad et al. 2004, 2006). In Tanzania, terracing with proper field management reduced soil loss by 96.3% (Wickama et al. 2014). In

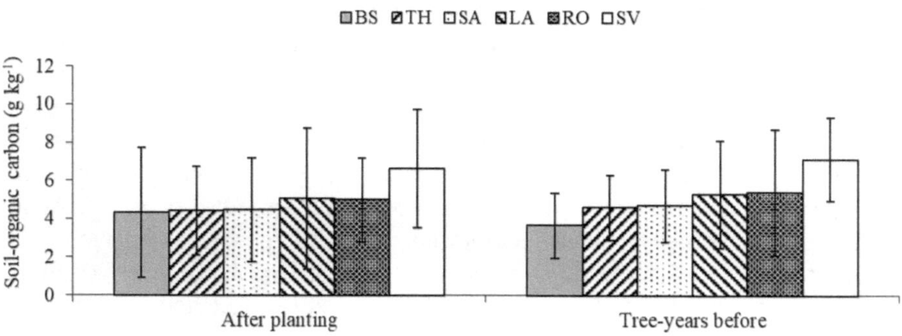

Fig. 7.5 Average soil organic carbon content under different plant covers during the 3-year monitoring period. S_V Spontaneous vegetation, T_M *Thymus mastichina*, L_A *Lavandula dentata*, S_A *Satureja obovata*, R_O *Rosmarinus officinalis*, B_S bare soil. Vertical bars are standard deviation

Table 7.7 Advantages and disadvantages in of plant covers

Advantages	Disadvantages
Soil erosion and runoff control and increase residue cover	Time-consuming activity for its management
Improve rainfall interception/increase soil water infiltration	Additional costs for planting, harvesting or killing
Augmented soil organic carbon	Reduce or increase soil moisture effects
Improve soil physico-chemical properties/reduced soil compaction	Difficult to incorporate cover crops with tillage
Economic value (i.e. AMPs)	Second additional activity
Recycle nutrients	May increase disease risks
Improve weed control and biodiversity	May increase insect pests
Improvement of soil–water content	Need water for its development
Atmospheric carbon sequestration	Additional time for harvesting
Wildlife habitat and landscape aesthetics	Allelopathic effects

AMPs aromatic and medicinal plants

Jiangxi Province, the red-soil hilly area of South China, the efficiency of soil and water conservation by terraces both can be up to 99.9% (Zuo and Li 2004; Chen et al. 2017; Verma et al. 2015c).

In a Mediterranean context, and in a regional scale, Bevan and Conolly (2011) concluded that in the terraced landscape of the island of Antikythera in Greece, only a small part of the total erosion came from the terraced areas (less than 15%). Lal (1982) compared terraced and non-terraced catchments respecting to soil erosion and runoff and these were 15.3 and 2.7 times higher, respectively, in non-terraced. Additionally, erosion rates on terraces may vary also with the used crop. In this line, Zhang et al. (2008, 2014) or Van Dijk (2002) demonstrated that erosion amount was higher if wheat, maize, peanuts, ginger or cassava.

In a catchment scale, hydrological connectivity is a key element in the behaviour of flooding and erosion during a storm. That is, terraces change the path of runoff and, therefore, the transport of sediment (Cammeraat 2004). Bellin et al. (2009) concluded that the terrace maintenance slowed runoff for events with a return period shorter than 8–10 years. Consequently, due to the reduction in hydrological connectivity in terraced areas, sediment production is limited to fields near the channel and almost all sediments are retained on plots or in drainage channel at the base of the terraces (Van Dijk et al. 2005; Ram and Meena 2014). In contrary, Nunes et al. (2016) studied hydrological and erosion processes in the terraced landscape system in Portugal, which is characterized by wet season irrigation to keep soils saturated and avoid frost on winter pastures. They concluded that these structures promoted saturation, runoff generation and sediment yield but in a way not significant for soil conservation.

Moreover, the use of plant covers, or stone walls, may protect the terrace risers. Van Dijk and Bruijnzeel (2004) calculated the total soil loss from terrace risers with little or no plant coverage and obtained that it can reach 200 t ha^{-1} $year^{-1}$ in humid climates. This rate could decrease up to 69% if the risers have dense vegetation cover.

However, the abandonment of these structures has proven to be an important driver for increasing runoff and erosion production. For example, in Mediterranean areas, the runoff coefficient on abandoned terraces is two times greater than cultivated terraces, and the erosion of abandoned terraces is directly related to the amount of plant cover, soil characteristics, environmental conditions and the time since abandonment (Arnáez et al. 2015; Meena et al. 2018b). Also, Calsamiglia et al. (2017) studied the effects of the abandonment and deterioration of these structures on the catchment by using a geomorphometric index of connectivity in a small Mediterranean catchment with terraces. Among other conclusion, they found that the deterioration of these structures increased connectivity index in 73%, in which the failure of them augmented runoff and thus accelerating the hydraulic processes, causing their collapse.

Thus, an important part of the studies related with terraces and soil and water conservation point that soil loss from cultivated slopes is much higher when terraces are not built, since terraces increase infiltration rates, decrease sediment transport and change hydro-morphological connectivity in watersheds. Table 7.8 shows some selected studies in relation to terraced lands and the effects on soil and water conservation.

7.5 Conclusion

In the pilot watershed studied, the main driving force affecting the land-use types was agriculture, mainly based on subtropical farming; this activity exerts heavy pressures on the environment because of the intense use of soil and water resources. After the abandonment of traditional stone terraces occupied by almonds, vineyards and olives, new orchard terraces were built for subtropical crops, enhancing soil and water degradation. Thus, given the increasing trend in the cultivation of subtropical crops, especially mango and avocado on terraces in the coming years, in the future, will be crucial the adoption and implementation of environmental planning strategies for sustainable land use.

The switch to cultivating subtropical crops in the terraced mountains of the coast of Granada province (south-eastern Spain) has left taluses without vegetation and thus has promoted permanent soil erosion, deteriorating soil physical properties, plant–nutrient depletion and loss of soil fertility while increasing the risk of landslides and endangering the stability of these structures. The implementation of aromatic plant covers on the taluses of subtropical orchard terraces substantially reduces soil erosion and runoff.

On the other hand, the presence of plant nutrients is due to the use of fertilizers for subtropical fruit production, which emerged to the soil surface in the talus as well as the terrace by water evaporation. These excesses, accumulating on the surface, were susceptible to dissolution by rains and transport towards lower levels. Therefore, the conservative use of fertilizer is important in these zones in order to avoid heavy agricultural pollution.

Similarly, these nutrient losses can be reduced by using plant covers in comparison to the bare soil treatment, especially with thyme, lavender and spontaneous plant covers. The cultivation of AMPs could represent extra income for farmers and

Table 7.8 Soil degradation studies on terraced lands

Authors	Objective	Main findings
Brandolini et al. (2018)	Investigate the relationships between landslide magnitude and land use conditions of agricultural terraced slopes	The results revealed that abandoned terraced slopes have been affected by a higher amount of mobilized debris volumes than still-cultivated terraces. Furthermore, terraces abandoned for a short time (less than 25–30 years) resulted in the most hazardous land use class, showing erosion rates that were approximately 2 and 3 times higher than terraced slopes abandoned a long time ago (more than 25–30 years) and still-cultivated terraces, respectively
Camera et al. (2018)	Quantify the effectiveness of terrace maintenance on protecting cultivated land against soil erosion	Terrace maintenance could reduce soil erosion by a factor of 3.8. The sediment traps were found to be an effective method for understanding and quantifying soil erosion in terraced mountain environments
Gathuga et al. (2018)	Use SWAT model to evaluate the impacts of structural conservation measures on water and sediment yield from a catchment in central Kenya	Model simulation results indicated that terraces and grassed waterways would significantly impact water and sediment yield at the catchment outlet. Terraces were found to provide the greatest reduction in sediment yield, by 81% from the baseline scenario, while grassed waterways reduced sediment yield by 54%. Terraces indicated a reduction in surface runoff by 30% from the base annual average value of 202 mm
Londero et al. (2018)	Study the hydrology of agricultural hillslopes under no-till management, with and without terraces in southern Brazil	The results show higher soil and water losses in the catchment without terraces. Terracing reduced peak flow rates and the total surface runoff. The no-till system without terraces was unable to adequately control surface runoff and soil erosion. Surface runoff and sediment yield were higher under no-till without terraces than under no-till with terraces
Chen et al. (2017)	The aim was to evaluate the roles of terracing on water erosion control in China	The results confirmed that terracing significantly and positively affected water erosion control. In terms of different terrace structures, bench terraces were better with respect to runoff and sediment reductions. In addition, a significant positive correlation between slope gradient (3°–15° and 16°–35°) and the effect of terracing on water erosion control was observed with the greatest decreases in water erosion occurred at slopes of 26°–35° and 11°–15°
Khelifa et al. (2017)	A terraced agricultural catchment was used to analyse and parameterize the effects of bench terraces on water and sediment yield using SWAT model	SWAT indicated that the local terraces, established on approximately 50% of the watershed area, reduced surface runoff by around 19% and sediment yield by around 22%

(continued)

Table 7.8 (continued)

Authors	Objective	Main findings
Martínez et al. (2017)	Study the impact of lithology in the first stages of soil erosion by water in terraces	The results showed that the highest soil loss and runoff occurred on marl substrata. Marls also showed the shallowest infiltration front and lowest infiltration rate. Limestones and, especially, metamorphic areas, showed a lower degree of soil erosion, higher infiltration rates and deeper infiltration fronts
Ramos (2016)	Erosion processes under different climate change scenarios were evaluated as well as the effects of implementing drainage terraces in vineyards	The results confirmed the relevance of extreme events on annual soil losses. The WEPP model gave satisfactory results in predicting runoff and soil losses, although the soil losses recorded after some extreme events were under-predicted. The construction of drainage terraces, perpendicular to the maximum slope, 3 m wide and 30 m between terraces, may lead to an average decrease in soil losses of about 45%
Ramos et al. (2015)	The effects of introducing drainage terraces and vegetative filter strips in a small basin with vineyards. Soil and nutrient losses were simulated in scenarios with and without soil conservation measures, using SWAT model	The simulation confirmed the effectiveness of introducing drainage terraces and filter strips to reduce soil and nutrient losses in a water basin dominated by small vineyards. Drainage terraces with a separation of 24 m, which is equivalent to eight rows of vines, would be able to reduce soil losses by about 50%, on average, while using 3 m-wide filter strips could reduce soil losses by an estimated 42%, on average. If both measures were combined, the total reduction in soil losses could be about 57%, on average. Nutrient losses were significantly reduced by the introduction of both measures
Schmidt and Zemadim (2015)	The effectiveness of the simulated conservation practices (terraces, bunds and residue management) is evaluated using SWAT model considering investment decisions on different terrain types	Results suggest that a mixed strategy of terracing on steep slopes and residue management on flat and middle slopes dramatically decrease surface runoff and erosion. A landscape-wide investment of terraces and bunds throughout the watershed landscape provides the greatest reduction in surface flow and erosion. However, the type and amount of investment in sustainable land management activities have different implications with respect to labour input and may be cost-prohibitive in the medium term
Li et al. (2014)	Erosion losses were investigated in a field experiment based on erosion plots on a 21% slope in southern China	The planting of grass on the riser is the most important factor affecting runoff and sediment yield, the next most important factor was the construction of a bund built on the edge of the terrace. Although the terrace bed slope does not significantly affect the runoff, it does benefit the soil. A terrace combined with vegetation measures is effective in controlling runoff and soil erosion. Among various terraces, the level terrace is apparently the most beneficial for soil conservation purposes

(continued)

Table 7.8 (continued)

Authors	Objective	Main findings
Zhang et al. (2014)	This study assessed the pattern of soil redistribution and soil particle size redistribution for two types of terraces: with and without embankments	Tillage erosion dominated the soil redistribution, with smaller contributions made by water erosion in the embankment terraces. The establishment of embankments at the lower end of the terrace obstructed the formation and development of overland water flow, thereby creating a line of null downslope soil transport leading to tillage-induced soil accumulation on the upslope side of the embankment with little granulometric sorting
Sánchez et al. (2013)	Current water erosion was evaluated in rangelands by establishing a runoff lot and using the Universal Soil Loss Equation	Results indicated that soils displayed a severe erosion of 16.67 Mg ha^{-1} $year^{-1}$. The agroforestry system with *Leucaena lecocephala* (Lam.) de Wit and *Moringa oleifera* (Lam.) in terraces recorded soil erosion rate of 2.17 Mg ha^{-1} $year^{-1}$, therefore avoiding the loss of 14.5 Mg ha^{-1} $year^{-1}$. Thus, agroforestry technology in terraces was an efficient method for the control of water erosion by reducing the 87% of soil loss
Sharda et al. (2013)	The performance of an intercrop-based conservation bench terrace (CBT) system for controlling runoff and soil erosion and improving crop productivity in comparison to the conventional system of farming on sloping borders was evaluated in a sub-humid climate	Analysis of runoff and soil loss data revealed that the CBT system reduced runoff and soil loss by about 80 and 88%, respectively, compared to the conventional system. The results of storm analysis suggest that CBT system was effective in reducing runoff and soil loss from storms of all sizes as compared to the conventional system. The CBT system responded mainly to large storms due to the storage effects in the impoundment (recipient) area. The C factor of USLE was estimated as 0.31 and 0.28 for intercrop-based conventional and CBT systems, respectively. Analysis of crop yield data revealed that the CBT system was about 18% more productive than the conventional system in terms of maize equivalent yields, primarily due to better in situ rainwater conservation
Mupenzi et al. (2012)	Assess the system of radical terraces	The radical terraces have a positive impact on environment where it contributed to limiting water flow, which was the main cause of soil degradation. The results of the study showed that the radical terraces contributed to increase in the farm productivity, fight against erosion and contributed to poverty reduction

(continued)

Table 7.8 (continued)

Authors	Objective	Main findings
Bazzoffi and Gardin (2011)	Comparative assessment of soil erosion by USLE with five increasing intensity of terrace degradation: conserved partially damaged, very damaged, partially removed, removed, each of which corresponding to different values of the indexes of verification for Good Agricultural and Environmental Condition (GAEC)	The results of the analysis showed a significant increase in erosion in relationship with increasing degradation of terraces, concluding that the GAEC standard retain terraces in relation to the primary objective in reducing erosion. In both areas, forest on totally degraded terraces was able to decrease erosion below the tolerance threshold (11.2 Mg ha^{-1} $year^{-1}$), in the same manner as conserved terraces do for other soil uses. At Lamole (Tuscany), the natural vegetation in evolution on completely degraded terraces was able to decrease erosion below the tolerance threshold. On the contrary, at Costaviola (Calabria) with the same soil use and level of terrace degradation, soil erosion remained above the tolerance threshold. Terraces, although degraded, continue to play a role in the protection of soil against erosion in abandoned areas
Lesschen et al. (2009)	Model runoff and erosion for a semiarid catchment using a multi-scale approach based on hydrological connectivity was used	The results showed that the spatial distribution of vegetation patches and agricultural terraces largely determined hydrological connectivity at the catchment scale. Runoff and sediment yield for the scenario without agricultural terraces were, respectively, a factor four and nine higher compared to the current situation
Yang et al. (2009)	The SWAT model to assess the efficacy of flow diversion terraces (FDT) systems on maintaining surface water quality at the watershed level was evaluated	The results indicated that FDT implemented contributed to the reduction of sediment yield by 4 Mg ha^{-1} $year^{-1}$ on average, which represented a reduction of 56%. Also found that the FDT system as a soil conservation beneficial management practices not only reduced the sediment loading at a watershed level, but also reduced water yield during the summer growing seasons. FDT systems contributed to the reduction of water yield by 158 mm $year^{-1}$, which represented a reduction of 20%

(continued)

Table 7.8 (continued)

Authors	Objective	Main findings
Ni and Zhang (2007)	Two toposequences (a long slope and terraced field series) were selected, to determine the effects of soil redistribution rates and topographic changes on P, K and $CaCO_3$ contents, and examine the contribution of water and tillage erosion to the variation and distribution pattern in these chemical properties within different landscapes	For the long slope, soil loss occurred at upper slope positions and soil accumulation was present at lower slope positions. However, terrace banks create a line of zero downslope transport of soil. Tillage erosion as the dominant soil redistribution process was found to be a major contributor of the variation in soil chemical properties in the terraced landscape, while water erosion plays an important role in the variation in soil chemical properties in the long slope landscape. The distribution of extractable K is closely linked to soil redistribution. No relationship between extractable P concentrations and ^{137}Cs inventories on the terraced fields was found. The $CaCO_3$ contents were strongly and significantly negatively correlated with extractable P irrespective of landscape types
Hammad et al. (2006)	Study the effects of rainfall characteristics and terracing conservation practice on erosion and runoff	The terraced conservation system reduced the negative effect of intense rainfall, resulting in a lower amount of runoff and erosion than in the non-terraced system. SOC, Mg, Ca and K showed significant differences between terraced and non-terraced treatments, which could be attributed to the differences in erosion between both treatments
Sang et al. (2006)	Study the impact of bench terracing on soil erosion and the use of weeds and plant residues to control soil and nutrient losses	Rills are the primary form of erosion on bare bench terraces. In addition, some rills develop into gullies that can run from the upper terrace down to the lower terrace. Rills and gullies are rarely found on weed-covered terraces. The results of an erosion experiment using model slope plots and artificial rainfall systems showed that plots covered with weeds and plant residues had much less surface runoff and soil and nutrient losses than bare soil
Zhang et al. (2006)	A toposequence of terraced fields as well as a long slope was selected to determine effects of soil redistribution rates and processes on SOC stocks and dynamics	The data reflect a difference in erosion process; along the long slope transect, water erosion is the dominant process, while in the terraced landscape soil distribution is mainly the result of tillage erosion. The data support the theory that water erosion may cause a loss of SOC due to selective removal of the most labile fraction of SOC, while on the other hand tillage erosion only transports the soil over short distances with less effect on the total SOC stock

(continued)

Table 7.8 (continued)

Authors	Objective	Main findings
Durán et al. (2005)	Evaluate soil loss and runoff in taluses of orchard terraces	This study highlighted the severity of erosion of 9.1 Mg ha^{-1} $year^{-1}$ and runoff of 100 mm $year^{-1}$in taluses of orchard terraces of south-east Spain, reflecting the urgency of planning strategies to protect these structures against chronic destruction
Gebremichael et al. (2005)	Examine the factors that control the effectiveness of bunds installed on cropland in reducing soil erosion	Stone bunds have led to a 68% reduction in annual soil loss due to water erosion. This reduction was due to the accumulation of sediment behind the stone bunds, which occurs faster in the early years after construction and decreases as the depression behind the bunds becomes filled with sediment. New stone bunds are particularly effective in trapping sediment in transport, but regular maintenance and increase in height of the bunds is necessary to maintain their effectiveness. The average USLE P factor for stone bunds in the study area is estimated to be 0.32
Van Dijk and Bruijnzeel (2004)	Report measurements of sediment yield from terrace risers and beds and from terrace units	The results demonstrate that soil loss from the terraces occurs in two stages: rainfall-driven transport by splash and shallow overland flow (wash) from the terrace riser and bed to a central drain is followed by a combination of onward wash transport of fine sediment and entrainment by runoff of coarser sediment deposited in the drain. A model (TEST: Terrace Erosion and Sediment Transport) was developed, describing these processes as a function of vegetation and soil surface cover and the presence of a layer of deposited sediment
Li and Lindstrom (2001)	Quantify soil quality parameters in terraced and steep hillslopes and determine relationships between soil redistribution from tillage erosion and water erosion	Water erosion was the primary cause for the overall decline in soil quality on the steep cultivated hillslope while tillage erosion had a comparable contribution to overall level in soil quality on the terraced hillslope. Soil movement by tillage controlled the spatial patterns in SOM, N and P on both terraced and steep cultivated hillslopes

an environmentally friendly measure to bolster the stability of the taluses of the orchard terraces and help minimize the risk of pollution from agricultural runoff. Therefore, a nutrient balance is essential in planning fertilizer application, optimizing efficiency and minimizing damage from intensive farming on terraces.

Here we show that it is extremely important to take into account the driven forces in order to make more sustainable the agricultural activities in this study area (Fig. 7.6). Concretely, the use of plant covers on steep slopes (214% in the present study) is advisable for erosion control, for maintaining and increasing SOC and for

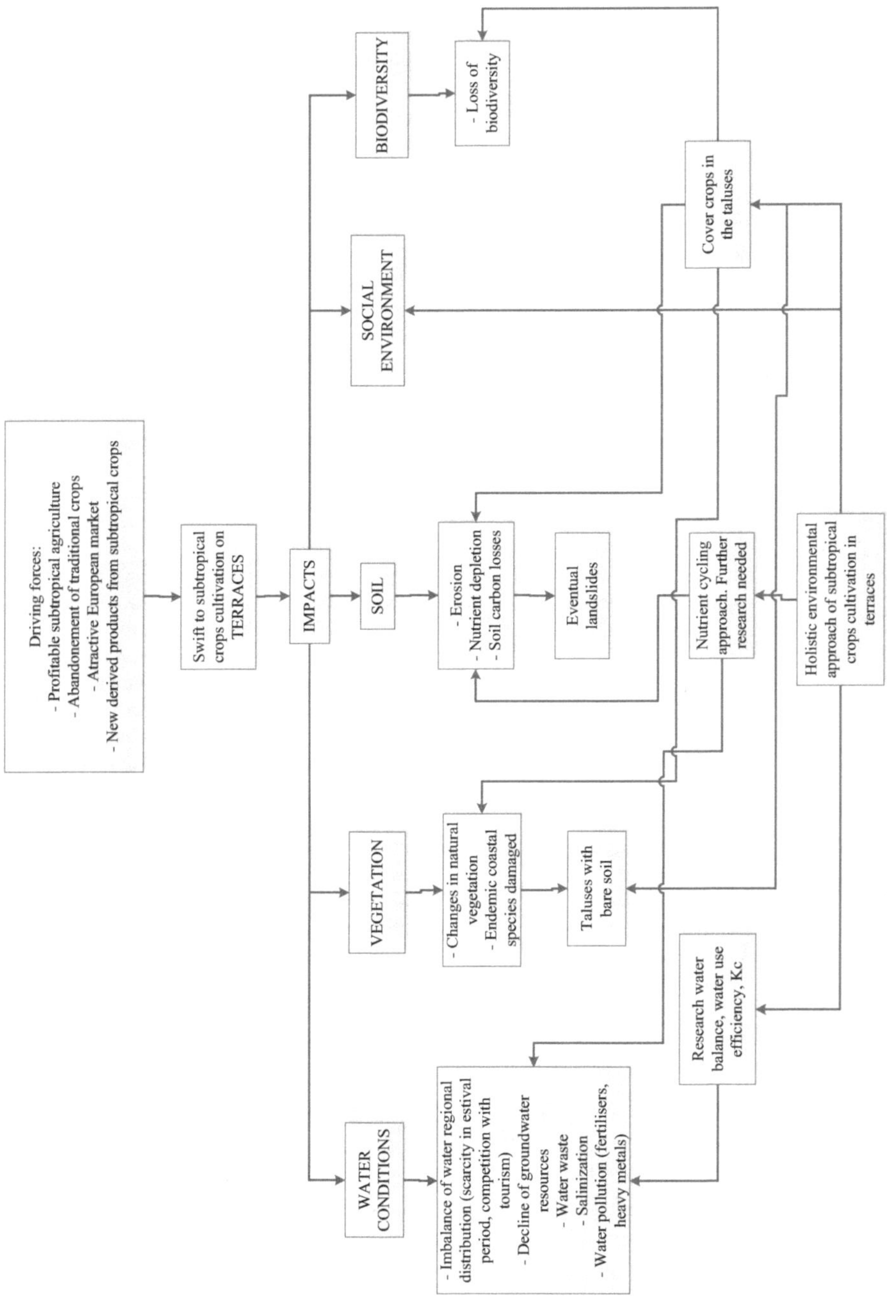

Fig. 7.6 Main driven forces shift to subtropical farming in the Mediterranean environment

recycling nutrients and consequently for an efficient regulation of the flow of nutrients by runoff and sediments. Vegetation makes a nutrient pool available while moderating surface runoff and sediment movement, both of which are major nutrient carriers.

Acknowledgements Part of this publication was sponsored by the research project, 'Impact of deficit irrigation on productivity of subtropical fruit crops: tools for sustainable water stress management' (PP.AVA.AVA201601.8), and cofinanced by the European Regional Development Fund (ERDF) within the Operational Programme Andalusia 2014–2020 'Andalucía is moving with Europe'.

References

Alatorre LC, Beguería S, Lana RN, Navas A, García RJM (2012) Soil erosion and sediment delivery in a mountain catchment under scenarios of land use change using a spatially distributed numerical model. Hydrol Earth Syst Sci 16:1321–1334

Almagro M, Querejeta JI, Boix FC, Martínez MM (2013) Links between vegetation patterns, soil C and N pools and respiration rate under three different land uses in a dry Mediterranean ecosystem. J Soils Sediments 13:641–653

APHA, AWWA, WPCF (ed) (1995) Standard methods for the examination of water and wastewater, 17th edn. APHA, AWWA, WPCF, Washington, DC

Arnáez J, Lasanta T, Ruiz FP, Ortigosa L (2007) Factors affecting runoff and erosion under simulated rainfall in Mediterranean vineyards. Soil Tillage Res 93:324–334

Arnáez J, Lasanta T, Errea MP, Ortigosa L (2011) Land abandonment, landscape evolution, and soil erosion in a Spanish Mediterranean mountain region: the case of Camero Viejo. Land Degrad Dev 22:537–550

Arnáez J, Lana RN, Lasanta T, Ruiz FP, Castroviejo J (2015) Effects of farming terraces on hydrological and geomorphological processes. A review. Catena 128:122–134

Ashoka P, Meena RS, Kumar S, Yadav GS, Layek J (2017) Green nanotechnology is a key for eco-friendly agriculture. J Clean Prod 142:4440–4441

Asins S (2006) Linking historical Mediterranean terraces with catchment, harvesting and distribution structures. In: Morel JP et al (eds) The archeology of crop and gardens. Ediplugia, Bari, pp 21–40

Atucha A, Merwin IA, Brown MG, Gardiazabal F, Mena F, Adriazola C, Lehmann J (2013) Soil erosion, runoff and nutrient losses in an avocado (*Perseaamericana* Mill) hillside orchard under different groundcover management systems. Plant Soil 368:393–406

Balogh JC, Walker WJ (1992) Golf course management and construction. Lewis Publishers, Chelsea, MI

Barata AM, Rocha F, Lopes V, Carvalho AM (2016) Conservation and sustainable uses of medicinal and aromatic plants genetic resources on the worldwide for human welfare. Ind Crop Prod 15:8–11

Barker G, Gilbertson D, Mattingly D (2007) Archaeology and desertification—the Wadi Faynan landscape survey, Southern Jordan. Council for British Research in the Levant and Oxbow Books, Oxford

Bautista S (1999) Regeneración post-incendio de un pinar (*Pinushalepensis* Miller) en ambiente semiárido. Erosión del suelo y medidas de conservación a corto plazo. Ph.D. University of Alicante, Alicante. Spain

Bazzoffi P, Gardin L (2011) Effectiveness of the GAEC standard of cross compliance retain terraces on soil erosion control. Ital J Agron 6(s1):e6 43–51

Bellin N, van Wesemael B, Meerkerk A, Barbera GG (2009) Abandonment of soil and water conservation structures in Mediterranean ecosystems: a case study from south east Spain. Catena 76:114–121

Bevan A, Conolly J (2011) Terraced fields and Mediterranean landscape structure: an analytical case study from Antikythera, Greece. Ecol Model 222:1303–1314

BIS (1991) Bureau of Indian Standards. Drinking water specification. IS: 10500:1991. New Delhi. India

Boix-Fayos C, de Vente J, Albaladejo J, Martínez MM (2009) Soil carbon erosion and stock as affected by land use changes at the catchment scale in Mediterranean ecosystems. Agric Ecosyst Environ 133:75–85

Bouwman AF, Beusen AHW, Griffioen J, Van Groenigen JW, Hefting MM, Oenema O, Van Puijenbroek PJTM, Seitzinger S, Slomp CP, Stehfest E (2013) Global trends and uncertainties in terrestrial denitrification and N2O emissions. Philos Trans R Soc B-Biol Sci 368:1–11

Braden JB, Shortle S (2013) Agricultural sources of water pollution. Encycl Energy, Nat Res Environ Econ 3:81–85

Brandolini P, Cevasco A, Capolongo D, Pepe G, Lovergine F, Del Monte M (2018) Response of terraced slopes to a very intense rainfall event and relationships with land abandonment: a case study from Cinque Terre (Italy). Land Degrad Dev 29:630–642

Buragohain S, Sharma B, Nath JD, Gogaoi N, Meena RS, Lal R (2017) Impact of ten years of biofertilizer use on soil quality and rice yield on an inceptisol in Assam, India. Soil Res. https://doi.org/10.1071/SR17001

Calsamiglia A, Fortesa J, García CJ, Lucas BME, Calvo CA, Estrany L (2017) Spatial patterns of sediment connectivity in terraced lands: anthropogeni controls of catchmentsensivity. Land Degrad Dev 29:1198–1210

Calvache ML, Pulido BP (1991) Saltwaterintrusion into small coastal aquifer (rioverde, Almuñécar SE Spain). J Hydrol 129:195–213

Calvache ML, Pulido BP (1996) Modelización de medidas de corrección de la intrusión marina en los acuíferos de Río Vélez, Río Verde y Castell de Ferro (provincias de Málaga y Granada). Estud Geol 52:269–277

Camera C, Djuma H, Bruggeman A, Zoumides C, Eliades M, Charalambous K, Abate D, Faka M (2018) Quantifying the effectiveness of mountain terraces on soil erosion protection with sediment traps and dry-stone wall laser scans. Catena 171:251–264

Cammeraat LH (2004) Scale dependent thresholds in hydrological and erosion response of a semi-arid catchment in southeast Spain. Agric Ecosyst Environ 104:317–332

Cao Y, Wu Y, Zhang Y, Tian J (2013) Landscape pattern and sustainability of a 1300- year-old agricultural landscape in subtropical mountain areas, Southwestern China. Int J Sustain Dev World Ecol 20:349–357

Carpio AJ, Oteros J, Tortosa TS, Guerrero CJ (2016) Land use and biodiversity patterns of the herpetofauna: the role of olive groves. Acta Oecol 70:103–111

Castillo V, Martínez MM, Albaladejo J (1997) Runoff and soil loss response to vegetation removal in a semiarid environment. Soil Sci Am J 61:1116–1121

Cerdà A, Rodrigo CJ, Novara A, Brevik EC, Vaezi AR, Pulido M, Giménez MA, Keesstra SD (2018) Long-term impact of rainfed agricultural land abandonment on soil erosion in the Western Mediterranean basin. Prog Phys Geogr 42(2):202–219

Chen D, Wei W, Chen L (2017) Effects of terracing practices on water erosion control in China: a meta-analysis. Earth Sci Rev 173:109–121

Chrenková K, Mataix SJ, Dlapa P, Arcenegui V (2014) Long-term changes in soil aggregation comparing forest and agricultural land use in different Mediterranean soil types. Geoderma 235–236:290–299

Costantini EAC, Castaldini M, Diago MP, Giffard B, Zombardo A (2018) Effects of soil erosion on agro-ecosystem services and soil functions: a multidisciplinary study in nineteen organically farmed European and Turkish vineyards. J Environ Manag 223:614–624

Cots-Folch R, Martínez CJA, Ramos MC (2006) Land terracing for new vineyard plantations in the north-eastern Spanish Mediterranean region: landscape effects of the EU Council Regulation policy for vineyards restructuring. Agric Ecosyst Environ 115:88–96

Dadhich RK, Meena RS (2014) Performance of Indian mustard (*Brassica juncea* L.) in Response to foliar spray of thiourea and thioglycollic acid under different irrigation levels. Indian J Ecol 41(2):376–378

Dadhich RK, Meena RS, Reager ML, Kansotia BC (2015) Response of bio-regulators to yield and quality of Indian mustard (*Brassica juncea* L. Czernj. and Cosson) under different irrigation environments. J App Nat Sci 7(1):52–57

Datta R, Baraniya D, Wang YF, Kelkar A, Moulick A, Meena RS, Yadav GS, Ceccherini MT, Formanek P (2017a) Multi-function role as nutrient and scavenger off reeradical in soil. Sustain MDPI 9:402. https://doi.org/10.3390/su9081402

Datta R, Kelkar A, Baraniya D, Molaei A, Moulick A, Meena RS, Formanek P (2017b) Enzymatic degradation of lignin in soil: a review. Sustain MDPI 9(7):1163. https://doi.org/10.3390/su9071163

Dhakal Y, Meena RS, De N, Verma SK, Singh A (2015) Growth, yield and nutrient content of mungbean (*Vigna radiata* L.) in response to INM in eastern Uttar Pradesh, India. Bangladesh J Bot 44(3):479–482

Dhakal Y, Meena RS, Kumar S (2016) Effect of INM on nodulation, yield, quality and available nutrient status in soil after harvest of green gram. Legum Res 39(4):590–594

Dotterweich M (2013) The history of human-induced soil erosion: geomorphic legacies, early descriptions and research, and the development of soil conservation—a global synopsis. Geomorphology 201:1–34

Drechsler M, Settele J (2001) Predator-prey interactions in rice ecosystems: effects of guild composition, trophic relationships, and land use changes – a model study exemplified for Philippine rice terraces. Ecol Model 137:135–159

Durán ZVH, Rodríguez PCR (2008) Soil-erosion and runoff prevention by plant covers, a review. Agron Sustain Dev 28:65–86

Durán ZVH, Martínez RA, Aguilar RJ, Franco TD (2003) El cultivo del mango (*Mangifera indica* L.) en la costa granadina. Granada, España. 141 p

Durán ZVH, Martínez RA, Aguilar RJ (2004) Nutrient losses by runoff and sediment from the taluses of orchard terraces. Water Air Soil Pollut 153:355–373

Durán ZVH, Aguilar JR, Martínez RA, Franco TD (2005) Impact of erosion in the taluses of subtropical orchard terraces. Agric Ecosyst Environ 107:199–210

Durán ZVH, Rodríguez PCR, Franco TD, Martín PFJ (2006a) El cultivo del chirimoyo (*AnnonacherimoliaMill*). Granada, Spain, 105 p

Durán ZVH, Francia MJR, Rodríguez PCR, Martínez RA, Cárceles RB (2006b) Soil erosion and runoff prevention by plant covers in a mountainous area (SE Spain): implications for sustainable agriculture. Environmentalist 26:309–319

Durán ZVH, Rodríguez PCR, Flanagan DC, García TI, Muriel FJL (2011) Sustainable land use and agricultural soil. In: Lichtfouse E (ed) Alternative farming systems, biotechnology, drought stress and ecological fertilisation, sustainable agriculture reviews, vol 6. Springer, Dordrecht, pp 107–192

Durán ZVH, Francia MJR, García TI, Rodríguez PCR, Martínez RA, Cuadros TS (2012) Runoff and sediment yield from a small watershed in south-eastern Spain (Lanjarón): implications for water quality. Hydrol Sci J 57:1610–1625

Durán ZVH, Francia MJR, García TI, Martínez RA (2013) Implications of land-cover types for soil erosion in a semiarid mountain slopes: towards sustainable land use in problematic landscapes. Acta Ecol Sin 33:272–281

Durán ZVH, Rodríguez PCR, Cuadros TS, Francia MJR (2014) Linking soil organic carbon stocks to land-use types in a Mediterranean agroforestry landscape. J Agric Sci Technol 16:667–679

Durán ZVH, Rodríguez PCR, Flanagan DC, García TI, Muriel FJL (2016) Fruit yield of mango irrigated with saline waters. In: Young EP (ed) Mango: production, properties and health benefits. Nova Science Publishers, Hauppauge, pp 109–122

EEC (2000) Water framework directive, European Community Directive; 2000/60
Ehigiator OA, Anyata BU (2011) Effects of land clearing techniques and tillage systems on runoff and soil erosion in a tropical rain forest in Nigeria. J Environ Manag 92:2875–2880
Eitelberg AD, van Vliet J, Doelman CJ, Stehfest E, Verburg HP (2016) Demand for biodiversity protection and carbon storage as drivers of global land change scenarios. Glob Environ Chang 40:101–111
Elías F, Ruiz L (1977) Agroclimatología de España. Cuaderno INIA, No 7, Madrid, Spain
ESYRCE (2017) Ministerio de Agricultura, Alimentación y Medio Ambiente. Available at: https://www.mapama.gob.es/es/estadistica/temas/estadisticas-agrarias/boletin2017sm_tcm30-455983.pdf. Accessed 17 Aug 2018
Faulkner H, Ruiz J, Zukowskyj P, Downward S (2003) Erosion risk associated with rapid and extensive agriculture clearances on sipersive materials in southeast Spain. Environ Sci Pol 6:115–127
Fernández RML, Lozano GB, Parras AL (2014) Topography and land use change effects on the soil organic carbon stock of forest soils in Mediterranean natural areas. Agric Ecosyst Environ 195:1–9
Ferro VC, Lang C, Kaal J, Stump D (2017) When is a terrace not a terrace? The importance of understanding landscape evolution in studies of terraced agriculture. J Environ Manag 202:500–513
Francia MJR, Durán ZVH, Martínez RA (2006) Environmental impact from mountainous olive orchards under different soil-management systems (SE Spain). Sci Total Environ 358:46–60
Galletti CS, Ridder E, Falconer SE, Fall PF (2013) Maxentmodeling of ancient and modern agricultural terraces in the Troodos foothills, Cyprus. Appl Geogr 39:46–56
García TIF, Durán ZVH (2018) Water scarcity and sustainable agriculture in semiarid environment: tools, strategies and challenges for woody crops. Academic Press-Elsevier, Amsterdam, Netherlands, 624 p
Gathuga JN, Sang JK, Maina CW (2018) Modelling the impacts of structural conservation measures on sediment and water yield in Thika-Chania catchment, Kenya. Int Soil Water Conserv Res 6(2):165–174
Gebremichael D, Nyssen J, Poesen J, Deckers J, Haile M, Govers G, Moeyersons J (2005) Effectiveness of stone bunds in controlling soil erosion on cropland in the Tigray Highlands, northern Ethiopia. Soil Use Manag 21:287–297
Gresta F, Lombardo GM, Siracusa L, Ruberto G (2008) Saffron, an alternative crop for sustainable agricultural systems. A review. Agron Sustain Dev 28:95–112
Guerra CA, Maes J, Geijzendorffer I, Metzger MJ (2016) An assessment of soil erosion prevention by vegetation in Mediterranean Europe: current trends of ecosystem service provision. Ecol Indic 60:213–222
Hammad AHA, Haugen LE, Borresen T (2004) Effects of stonewalled terracing techniques on soil-water conservation and wheat production under Mediterranean conditions. Environ Manag 34:701–710
Hammad AHA, Børresen T, Haugen LE (2006) Effects of rain characteristics and terracing on runoff and erosion under the Mediterranean. Soil Tillage Res 87:39–47
Hernández A, Arellano EC, Morales MD, Miranda MD (2016) Understanding the effect of three decades of land use change on soil quality and biomass productivity in a Mediterranean landscape in Chile. Catena 140:195–204
IECA (2015) Instituto de Estadística y Cartografía de Andalucía, Sistemas de indicadores. Available at http://juntadeandalucia.es/organismos/agriculturapescaydesarrolloruralconsejeria/sobre-consejeria/estadisticas/paginas/agrarias-anuario.html
Jefferson AJ, Wegmann KW, Chin A (2013) Geomorphology of the Anthropocene: understanding the surficial legacy of past and present human activities. Anthropocene 2:1–3
Kasai M, Marutani T, Reid LM, Trustrum NA (2001) Estimation of temporally averaged sediment delivery ratio using aggradational terraces in headwater catchments of the Waipaoa river, North Island, New Zealand. Earth Surf Process Landf 26:1–16

Kayser M, Benke M, Isselstein J (2012) Potassium leaching following silage maize on a productive Sandy soil. Plant Soil Environ 58:545–550

Khelifa WB, Hermassi T, Strohmeier S, Zucca C, Ziadet F, Boufarona M, Habaieb H (2017) Parameterization of the effect of bench terraces on runoff and sediment yield by SWAT modelling in a small semi-arid watershed in Northern Tunisia. Land Degrad Dev 28:1568–1578

Khresat S, Al-Bakri J, Al-Tahhan R (2008) Impacts of land use/cover change on soil properties in the Mediterranean region of Northwestern Jordan. Land Degrad Dev 19:397–407

Kumar S, Meena RS, Pandey A, Seema (2017a) Soil acidity management and an economics response of lime and sulfur on sesame in an alley cropping system. Int J Curr Microb App Sci 6(3):2566–2573

Kumar S, Meena RS, Yadav GS, Pandey A (2017b) Response of sesame (*Sesamum indicum* L.) to sulphur and lime application under soil acidity. Int J Plant Soil Sci 14(4):1–9

Kumar S, Meena RS, Bohra JS (2018) Interactive effect of sowing dates and nutrient sources on dry matter accumulation of Indian mustard (*Brassica juncea* L.). J Oilseed Brassica 9(1):72–76

Labrière N, Locatelli B, Laumonier Y, Freycon V, Bernoux M (2015) Soil erosion in the humid tropics: asystematic quantitative review. Agric Ecosyst Environ 203:127–139

Lal R (1982) Effect of slope length and terracing on runoff and erosion on a tropical soil. In: Walling D (ed) Recent development in the explanation and prediction erosion and sediment yield. IAHS Publication, Wallingford, UK

Laliberté E, Wells JA, DeClerck F, Metcalfe DJ, Catterall CP, Queiroz C, Aubin I, Bonser SP, Ding Y, Fraterrigo JM, McNamara S, Morgan JW, Sánchez MD, Vesk PA, Mayfield MM (2010) Land-use intensification reduces functional redundancy and response diversity in plant communities. Ecol Lett 13:76–86

Lana-Renault N, Latron J, Karssenberg D, Serrano MP, Regüés D, Bierkens MFP (2011) Differences in stream flow in relation to changes in land cover: a comparative study in two sub-Mediterranean mountain catchments. J Hydrol 411:366–378

Lasanta T, Arnáez J, Ruiz FP, Lana-Renault N (2013) Agricultural terraces in the Spanish mountains: an abandoned landscape and a potential resource. Bol AGE 63:487–491

Lesschen JP, Schoorl JM, Cammeraat LH (2009) Modelling runoff and erosion for a semi-arid catchment using a multi-scale approach based on hydrological connectivity. Geomorphology 109:174–183

Li Y, Lindstrom MJ (2001) Evaluating soil quality–soil redistribution relationship on terraces and steep hillslope. Soil Sci Soc Am J 65:1500–1508

Li XH, Yang J, Zhao CY, Wang B (2014) Runoff and sediment from orchard terraces in Southeastern China. Land Degrad Dev 25:184–192

Lin SQ (2007) World loquat production and research with special reference to China. Acta Hortic 750:37–44

Liu SL, Wang C, Zhang XL, Yang JJ, Qiu Y, Wang J (2011) Soil and water conservation effect of different terrace configurations in land consolidation Project. J Soil Water Conserv 25:59–62

Liu R, Xiao L, Liu Z, Dai J (2018) Quantifying the relative impacts of climateand human activities on vegetation changes at the regional scale. Ecol Indic 93:91–99

Lizaga I, Quijano L, Gaspar L, Navas A (2018) Estimating soil redistribution patterns with ^{137}Cs measurements in a Mediterranean mountain catchment affected by land abandonment. Land Degrad Dev 29:105–117

Londero AL, Minella JPG, Deuschle D, Schneider FJA, Boeni M, Merten GH (2018) Impact of broad-based terraces on water and sediment losses in no-till (paired zero-order) catchments in southern Brazil. J Soils Sediments 18:1159–1175

Lozano GB, Parras AL, Cantudo PM (2016) Land use change effects on stratification and storage of soil carbon and nitrogen: application to a Mediterranean nature reserve. Agric Ecosyst Environ 231:105–113

Lozano GB, Muñoz RM, Parras AL (2017) Climate and land use changes effects on soil organic carbon stocks in a Mediterranean semi-natural area. Sci Total Environ 579:1249–1259

Mabilde L, De Neve S, Sleutel S (2017) Regional analysis of groundwater phosphate concentrations under acidic sandy soils: edaphic factors and water table strongly mediate thesoil P-groundwater P relation. J Environ Manag 203:429–438

Maetens W, Vanmaercke M, Poesen J, Jankauskas B, Jankauskiene G, Ionita I (2012) Effects of land use on annual runoff and soil loss in Europe and the Mediterranean: a meta-analysis of plot data. Prog Phys Geogr 36(5):599–653

MAPA (1994) Métodos Oficiales de Análisis. Tomo III Secretaría General Técnica del Ministerio de Agricultura, Pesca y Alimentación, Madrid, Spain

Martínez HC, Rodrigo CJ, Romero DA (2017) Impact of lithology and soil properties on abandoned dryland terraces during the early stages of soil erosion by water in south-east Spain. Hydrol Proc 31(17):3095–3109

Marzaioli R, D'Ascoli R, De Pascale RA, Rutigliano FA (2010) Soil quality in a Mediterranean area of Southern Italy as related to different land use types. Appl Soil Ecol 44:205–212

Meena H, Meena RS (2017) Assessment of sowing environments and bio-regulators as adaptation choice for clusterbean productivity in response to current climatic scenario. Bangladesh J Bot 46(1):241–244

Meena RS, Yadav RS (2014) Phonological performance of groundnut varieties under sowing environments in hyper arid zone of Rajasthan, India. J App Nat Sci 6(2):344–348

Meena RS, Yadav RS (2015) Yield and profitability of groundnut *(Arachis hypogaea* L) as influenced by sowing dates and nutrient levels with different varieties. Legum Res 38(6):791–797

Meena RS, Yadav RS, Meena VS (2014) Response of groundnut (*Arachis hypogaea* L.) varieties to sowing dates and NP fertilizers under Western Dry Zone of India. Bangladesh J Bot 43(2):169–173

Meena RS, Dhakal Y, Bohra JS, Singh SP, Singh MK, Sanodiya P (2015a) Influence of Bioinorganic Combinations on Yield, Quality and Economics of Mungbean. Am J Exp Agric 8(3):159–166

Meena RS, Meena VS, Meena SK, Verma JP (2015b) The needs of healthy soils for a healthy world. J Clean Prod 102:560–561

Meena RS, Meena VS, Meena SK, Verma JP (2015c) Towards the plant stress mitigate the agricultural productivity: a book review. J Clean Prod 102:552–553

Meena RS, Yadav RS, Meena H, Kumar S, Meena YK, Singh A (2015d) Towards the current need to enhance legume productivity and soil sustainability worldwide: a book review. J Clean Prod 104:513–515

Meena RS, Yadav RS, Reager ML, De N, Meena VS, Verma JP, Verma SK, Kansotia BC (2015e) Temperature use efficiency and yield of groundnut varieties in response to sowing dates and fertility levels in Western Dry Zone of India. American J Exp Agri 7(3):170–177

Meena H, Meena RS, Singh B, Kumar S (2016a) Response of bio-regulators to morphology and yield of clusterbean [*Cyamopsis tetragonoloba* (L.) Taub.] under different sowing environments. J App and Nat Sci 8(2):715–718

Meena RS, Bohra JS, Singh SP, Meena VS, Verma JP, Verma SK, Shiiag SK (2016b) Towards the prime response of manure to enhance nutrient use efficiency and soil sustainability a current need: a book review. J Clean Prod 112:1258–1260

Meena RS, Gogaoi N, Kumar S (2017a) Alarming issues on agricultural crop production and environmental stresses. J Clean Prod 142:3357–3359

Meena RS, Meena PD, Yadav GS, Yadav SS (2017b) Phosphate solubilizing microorganisms, principles and application of microphos technology. J Clean Prod 145:157–158

Meena H, Meena RS, Lal R, Singh GS, Mitran T, Layek J, Patil SB, Kumar S, Verma T (2018a) Response of sowing dates and bio regulators on yield of clusterbean under current climate in alley cropping system in eastern U.P., Indian. Leg Res 41(4):563–571

Meena RS, Kumar V, Yadav GS, Mitran T (2018b) Response and interaction of *Bradyrhizobium japonicum* and *Arbuscular mycorrhizal* fungi in the soybean rhizosphere: a review. Plant Growth Regul 84:207–223

Muñoz RM, De la Rosa D, Zavala LM, Jordán A, Anaya RM (2011) Changes in land cover and vegetation carbon stocks in Andalusia, Southern Spain (1956–2007). Sci Total Environ 409:2796–2806

Muñoz RM, Jordán A, Zavala LM, De la Rosa D, Abd-Elmabod SK, Anaya RM (2015) Impact of land use and land cover changes on organic carbon stocks in Mediterranean soils (1956-2007). Land Degrad Dev 26(2):168–179

Mupenzi JL, Li L, Ge J, Habumugisha JD, Habiyaremye G, Ngamije J, Baragahoranye I (2012) Radical Terraces in Rwanda. East Afr J Sci Tech 1(1):53–58

Nadal RE, Lasanta T, García RJM (2013) Runoff and sediment yield from land under various uses in a Mediterranean mountain area: long-term results from an experimental station. Earth Surf Process Landf 38:346–355

Nadal RE, Cammeraat E, Pérez CE, Lasanta T (2016) How do soil organic carbon stocks change after cropland abandonment in Mediterranean humid mountain areas? Sci Total Environ 566–567:741–752

Nadeu E, Berhe AA, de Vente J, Boix-Fayos C (2012) Erosion, deposition and replacement of soil organic carbon in Mediterranean catchments: a geomorphological, isotopic and land use change approach. Biogeosciences 9:1099–1111

Navas A, Quine TA, Walling DE, Gaspar L, Quijano L, Lizaga I (2017) Relating intensity of soil redistribution to land use changes in abandoned Pyrenean fields using fallout caesium-137. Land Degrad Dev 28:2017–2029

Ni SJ, Zhang JH (2007) Variation of chemical properties as affected by soil erosion on hillslopes and terraces. Eur J Soil Sci 58:1285–1292

Novara A, Gristina L, Sala G, Galati A, Crescimanno M, Cerdà A, Badalamenti E, La Mantia T (2017) Agricultural land abandonment in Mediterranean environment provides ecosystem services via soil carbon sequestration. Sci Total Environ 576:420–429

Nunes JP, Bernard JL, Rodríguez BML, Marisa SJ, de Oliveira ACC, Jacob JJ (2016) Hydrological and erosion processes in terraced fields: observations from a humid mediterranean region in Northern Portugal. Land Degrad Dev 29:596–606

Ochoa PA, Fries A, Mejía D, Burneo JI, Ruiz SJD, Cerdà A (2016) Effects of climate, land cover and topography on soil erosion risk in a semiarid basin on the Andes. Catena 140:31–42

Ohana NL, Karnieli A, Egozi R, Givati A, Peeters A (2015) Modeling the effects of land-cover change on rainfall-runoff relationships in a semiarid, eastern Mediterranean watershed. Adv Meteorol vol. 2015., Article ID 838070, 16 pages

OPM-CAPDR(2015)ObservatoriodePreciosyMercados.Frutalessubtropicales(2015)Consejeríade Agricultura, Pesca y Desarrollo Rural. http://www.juntadeandalucia.es/agriculturaypesca/observatorio/servlet/FrontController?action=Subsector&table=3940&ec=subsector&subsector=34 [Accessed 12 March 2018]

Ore G, Bruins HJ (2012) Design features of ancient agricultural terrace walls in the Negev desert: human-made geodiversity. Land Degrad Dev 23:409–418

Padilla FM, Vidal B, Sánchez J, Pugnaire FI (2010) Land-use changes and carbon sequestration through the twentieth century in a Mediterranean mountain ecosystem: implications for land management. J Environ Manag 91:2688–2695

Park JY, Yu YS, Hwang SJ, Kim C, Kim SJ (2014) SWAT modeling of best management practices for Chungju dam watershed in South Korea under future climate change scenarios. Paddy Water Environ 12:S65–S75

Parras AL, Martín CM, Lozano GB (2013) Impacts of land use change in soil carbon and nitrogen in a Mediterranean agricultural area (Southern Spain). Solid Earth 4:167–177

Plexida S, Solomou A, Poirazidis K, Sfougaris A (2018) Factors affecting biodiversity in agrosylvopastoral ecosystems with in the Mediterranean Basin: a systematic review. J Environ Manag 151:125–133

Porat N, López IG, Lensky N, Elinson R, Avni Y, Elgart SY, Faershtein G, Gadot Y (2018) Using portable OSL reader to obtain a time scale for soil accumulation and erosion in archaeological terraces, the Judean Highlands, Israel. Quat Geochronol. https://doi.org/10.1016/j.quageo.2018.04.001

Puigdefábregas J, Mendizabal T (1998) Perspectives on desertification: Western Mediterranean. J Arid Environ 39:209–224

Qi X, Fu Y, Wang YR, Ng NC, He Y (2018) Improving the sustainability of agricultural land use: an integrated framework for the conflict between food security and environmental deterioration. Appl Geogr 90:214–223
Quine TA, Walling DE, Zhang X (1999) Tillage erosion, water erosion and soil quality on cultivated terraces near Xifeng in the Loess Plateau, China. Land Degrad Dev 10:251–274
Ram K, Meena RS (2014) Evaluation of pearl millet and mungbean intercropping systems in Arid Region of Rajasthan (India). Bangladesh J Bot 43(3):367–370
Ramos MC (2016) Soil losses in rainfed Mediterranean vineyards under climate change scenarios. The effects of drainage terraces. AIMS Agric Food 1(2):124–143
Ramos MC, Martínez CJA (2006) Nutrient loses by runoff in vineyards of the Mediterranean Alt Penedès region (NE Spain). Agric Ecosyst Environ 113:356–363
Ramos MC, Benito C, Martínez CJA (2015) Simulating soil conservation measures to control soil and nutrient losses in a small vineyard dominated basin. Agric Ecosyst Environ 213:194–208
Rodrigo JC, Martínez HC, Iserloh T, Cerdà A (2018) Contrasted impact of land abandonment on soil erosion in Mediterranean agriculture fields. Pedosphere 28(4):617–631
Rodríguez PCR, Durán ZVH, Martín PFJ, Franco TD (2009) Litter decomposition and nitrogen release in a sloping Mediterranean subtropical agroecosystem on the coast of Granada (SE, Spain): effects of floristic and topographic alterations on slope. Agric Ecosyst Environ 134:79–88
Rodríguez PCR, Franco TD, Francia MJR, Durán ZVH (2016) Producción de frutos de mango Osteen y Keitt con distintos portainjertos. Vida Rural 417:21–26
Romero DA, Cammeraat LH, Vacca A, Kosmas C (1999) Soil erosion at three experimental sites in the Mediterranean. Earth Surf Process Landf 24:1243–1256
Romero DA, Ruiz SJD, Robledano AF, Brevik EC, Cerdà A (2017) Ecosystem responses to land abandonment in Western Mediterranean Mountains. Catena 149:824–835
Ruecker G, Schad P, Alcubilla MM, Ferrer C (1998) Natural regeneration of degraded soils and site changes on abandoned agricultural terraces in Mediterranean Spain. Land Degrad Dev 9:179–188
Rutten M, van Dijk Wilbert M, van RooijHenk H (2014) Land use dynamics, climate change, and food security in Vietnam: a global-to-local modeling approach. World Dev 59:29–46
Saikia SP, Bora D, Goswami A, Mudoi KD, Gogoi A (2013) A review on the role of Azospirillum in the yield improvement of non-leguminous crops. Afr J Microbiol Res 6:1085–1102
Sánchez E, Ortega BV, Domínguez HF, Ortega EM, Can-Chulim A, Sarmiento BD (2013) Soil erosion control using agroforestry terraces in San Pedro Mixtepec, Oaxaca, Mexico. Int J Agric Sci 3(6):423–439
Sang AJ, Mihara M, Horaguchi Y, Yamaji E (2006) Soil erosion and participatory remediation strategy for bench terraces in northern Thailand. Catena 65:258–264
Schmidt E, Zemadim B (2015) Expanding sustainable land management in Ethiopia: scenarios for improved agricultural water management in the Blue Nile. Agric Water Manag 158:166–178
Schroeder PD, Radcliffe DE, Cabrera ML (2004) Rainfall timing and poultry litter application rate effects on phosphorous loss in surface runoff. J Environ Qual 33:2201–2209
Serpa D, Nunes JP, Santos J, Sampaio E, Jacinto R, Veiga S, Lima JC, Moreira M, Corte RJ, Keizer JJ, Abrantes N (2015) Impacts of climate and land use changes on the hydrological and erosion processes of two contrasting Mediterranean catchments. Sci Total Environ 538:64–77
Serpa D, Nunes JP, Keizer JJ, Abrantes N (2017) Impacts of climate and land use changes on the water quality of a small Mediterranean catchment with intensive viticulture. Environ Pollut 224:454–465
Sharda VN, Sena DR, Shrimali SS, Khola OPS (2013) Effects of an intercrop-based conservation bench terrace system on resource conservation and crop yields in a sub-humid climate in India. Trans ASABE 56(4):1411–1425
Sihag SK, Singh MK, Meena RS, Naga S, Bahadur SR, Gaurav, Yadav RS (2015) Influences of spacing on growth and yield potential of dry direct seeded rice (*Oryza sativa* L.) cultivars. Ecoscan 9(1-2):517–519

Southwick LM, Willis GH, Jonhson DC, Selim HM (1995) Leaching of nitrate, atrazine, and metribuzin from sugar cane in Southern Louisiana. J Environ Qual 24:684–690

Srivastava P, Singh R, Tripathi S, Singh RA (2016) An urgent need for sustainable thinking in agriculture-An Indian scenario. Ecol Indic 67:611–622

Tarolli P, Preti F, Romano N (2014) Terraced landscape: from an old best practice to a potential hazard for soil degradation due to land abandonment. Anthropocene 6:10–15

Trigalet S, Gabarrón GMA, Van Oost K, van Wesemael B (2016) Changes in soil organic carbon pools along a chronosequence of land abandonment in southern Spain. Geoderma 268:14–21

U.S. EPA (1976) Quality criteria for water. US Environmental Protection Agency, United States Government Printing Office, Washington, DC

Vaccari FP, Lugato E, Gioli B, D'Acqui L, Genesio L, Toscano P, Matese A, Miglietta F (2012) Land use change and soil organic carbon dynamics in Mediterranean agro-ecosystems: the case study of Pianosa Island. Geoderma 175-176:29–36

Vallebona C, Mantino A, Bonari E (2016) Exploring the potential of perennial crops in reducing soil erosion: a GIS-based scenarioanalysis in southern Tuscany, Italy. Appl Geogr 66:119–131

Van Dijk AIJM (2002) Water and Sediment Dynamics in Bench Terraced Agricultural Steeplands in West Java, Indonesia. PhD thesis. Vrije Universiteit Amsterdam

Van Dijk AIJM, Bruijnzeel LA (2004) Runoff and soil loss from bench terraces. 2. An event - based erosion process model. Eur J Soil Sci 55:317–334

Van Dijk AIJM, Bruijnzeel LA, Vertessy RA, Ruijter J (2005) Runoff and sediment generation on bench-terraced hillsides: measurements and up-scaling of a field-based model. Hydrol Proced 19:1667–1685

Vanhove W, Van Damme O (2013) Value chains of cherimoya (*Annona cherimola* Mill.) in a centre of diversity and its on-farm conservation implications. Trop Conserv Sci 6:158–180

Varma D, Meena RS, Kumar S (2017a) Response of mungbean to fertility and lime levels under soil acidity in an alley cropping system in Vindhyan Region, India. Int J Chem Stu 5(2):384–389

Varma D, Meena RS, Kumar S, Kumar E (2017b) Response of mungbean to NPK and lime under the conditions of Vindhyan Region of Uttar Pradesh. Legum Res 40(3):542–545

Verheijen FGA, Jones RJA, Rickson RJ, Smith CJ (2009) Tolerable versus actual soil erosion rates in Europe. Earth Sci Rev 94:23–38

Verma JP, Jaiswal DK, Meena VS, Meena RS (2015a) Current need of organic farming for enhancing sustainable agriculture. J Clean Prod 102:545–547

Verma JP, Meena VS, Kumar A, Meena RS (2015b) Issues and challenges about sustainable agriculture production for management of natural resources to sustain soil fertility and health: a book review. J Clean Prod 107:793–794

Verma SK, Singh SB, Prasad SK, Meena RN, Meena RS (2015c) Influence of irrigation regimes and weed management practices on water use and nutrient uptake in wheat (*Triticum aestivum* L. Emend. Fiori and Paol.). Bangladesh J Bot 44(3):437–442

Vollenweider RA (1968) Scientific fundamentals of the eutrophication of lakes and flowing waters with particular reference to nitrogen and phosphorus. OECD Report DAS/CSI.6827, Paris, France

Wang Y, Ran L, Fang N, Shi Z (2018) Aggregate stability and associated organic carbon and nitrogen as affected by soil erosion and vegetation rehabilitation on the Loess Plateau. Catena 167:257–265

WHO (ed) (1996) World Health Organization, Guidelines for drinking-water quality. Health criteria and other supporting information, vol 1, 2nd edn. WHO, Geneva, Switzerland

Wickama J, Okoba B, Sterk G (2014) Effectiveness of sustainable land management measures in West Usambara highlands, Tanzania. Catena 118:91–102

Yadav GS, Babu S, Meena RS, Debnath C, Saha P, Debbaram C, Datta M (2017a) Effects of godawariphosgold and single supper phosphate on groundnut (*Arachis hypogaea*) productivity, phosphorus uptake, phosphorus use efficiency and economics. Indian J Agric Sci 87(9):1165–1169

Yadav GS, Lal R, Meena RS, Babu S, Das A, Bhomik SN, Datta M, Layak J, Saha P (2017b) Conservation tillage and nutrient management effects on productivity and soil carbon sequestration under double cropping of rice in North Eastern Region of India. Ecol Ind http://www.sciencedirect.com/science/article/pii/S1470160X17305617

Yadav GS, Lal R, Meena RS, Datta M, Babu S, Das, Layek J, Saha P (2017c) Energy budgeting for designing sustainable and environmentally clean/safer cropping systems for rainfed rice fallow lands in India. J Clean Prod 158:29–37

Yadav GS, Das A, Lal R, Babu S, Meena RS, Saha P, Singh R, Datta M (2018a) Energy budget and carbon footprint in a no-till and mulch based rice–mustard cropping system. J Clean Prod 191:144–157

Yadav GS, Das A, Lal R, Babu S, Meena RS, Patil SB, Saha P, Datta M (2018b) Conservation tillage and mulching effects on the adaptive capacity of direct-seeded upland rice (*Oryza sativa* L.) to alleviate weed and moisture stresses in the North Eastern Himalayan Region of India. Arch Agron Soil Sci. https://doi.org/10.1080/03650340.2018.1423555

Yang Q, Meng FR, Zhao Z, Chowb TL, Benoy G (2009) Assessing the impacts of flow diversion terraces on stream water and sediment yields at a watershed level using SWAT model. Agric Ecosyst Environ 132:23–31

Yapp G, Walker J, Thackway R (2010) Linking vegetation type and condition to ecosystem goods and services. Ecol Complex 3:292–301

Yuan T, Fengmin L, Puhai L (2003) Economic analysis of rainwater harvesting and irrigation methods, with an example from China. Agric Water Manag 60:217–226

Zhang J, Quine TA, Ni S, Ge F (2006) Stocks and dynamics of SOC in relation to soil redistribution by water and tillage erosion. Glob Chang Biol 12:1834–1841

Zhang JH, Su ZA, Liu GC (2008) Effects of terracing and agroforestry on soil and water loss in hilly areas of the Sichuan Basin, China. J Mountain Sci 5:241–248

Zhang JH, Wang Y, Zhang ZH (2014) Effect of terrace forms on water and tillage erosion on a hilly landscape in the Yangtze River Basin, China. Geomorphology 216:114–124

Zucca C, Canu A, Previtali F (2010) Soil degradation by land use change in an agropastoral area in Sardinia (Italy). Catena 83:46–54

Polyculture Management: A Crucial System for Sustainable Agriculture Development

8

Katarzyna Adamczewska-Sowińska and Józef Sowiński

Abstract

Polyculture is a system for the cultivation of a few crops together, in the same space and at the same time. These methods of crop production have been known and used for thousands of years. Since the 1970s, the system of intensive agriculture has dominated, and the use of environmentally friendly methods for food and feed production has been limited, as has the use of the polyculture system (PS). This paper presents different methods of PS, and special attention is paid to the importance of methods for sustainable agriculture, with a focus on soil protection and the effect of polyculture on soil fertilities. A special issue presented here are living mulches and companion crops (CC) methods in agriculture and horticulture production. Soil surface cover is an important practice for the slowdown of degradation processes to increase soil fertilities. Polyculture and plant cover (companion crop or living mulches) have many environmental benefits: protection of soil against water and wind erosion, stabilization of soil temperature, reservoir of water in the soil profile, effect on soil fertilities, biological activity, and physical soil characteristics. Living mulches or CC are an element of biological control and compete with weeds and reduce pest attacks and disease infection.

Plant-plant interaction provides important information helpful for species selection for different polyculture systems. Various crop interactions are presented, and crop selections both recommended and not recommended for PS are

K. Adamczewska-Sowińska
Horticulture Department, Wroclaw University of Environmental and Life Sciences, Wroclaw, Poland
e-mail: katarzyna.adamczewska-sowinska@upwr.edu.pl

J. Sowiński (✉)
Institute of Agroecology and Plant Production, Wroclaw University of Environmental and Life Sciences, Wroclaw, Poland
e-mail: jozef.sowinski@upwr.edu.pl

R. S. Meena (ed.), *Soil Health Restoration and Management*,
https://doi.org/10.1007/978-981-13-8570-4_8

characterized. Special attention is paid to the role of allelochemicals for species selection.

The polyculture system based on commonly known methods of legume and non-legume crop cultivation. The importance of nitrogen fixation phenomena and ways of nitrogen transport from legume crops to non-legume crops is presented.

Better understanding of the polyculture system benefits and popularization of those crop production methods was the main aim of that chapter. More popular should be agriculture system which has more ecological and environmental impact on both crop-crop and crop-environment.

Keywords
Polyculture system · Aboveground and belowground competition · Environment protection · Nitrogen fixation

Abbreviation

B	Cloddiness index
CC	Companion crops
ΔMWD	Water stability index
LM	Living mulch
MC	Mixed cropping
MWDa	Mean weighed diameter of aggregate (dry method)
MWDg	Mean weighed diameter of aggregate (wet method)
N	Nitrogen
$N\text{-}NO_3^-$	Nitrate nitrogen
PS	Polyculture system
S	Misting index
W	Structure index
Wod	Waterproof

8.1 Introduction

A practical cropping system on a farm depends on many factors: available natural and industrial resources, available technologies and technics, and also connections between the farm and other enterprises or business entities with which the farm cooperates. Different methods are used in the various cropping systems. Crops are grown together or separately (at short or long intervals) in the same field as crop rotation or monoculture, which is called a conventional cropping system (Bullock 1992). This method of crop cultivation dominates because of its high productivity, income level, and suitability to combined management practice (Blanco-Canqui, Lal 2010). At present, the use of a monoculture system or crop rotation is the basis of cultivation systems in agriculture around the world, providing the proper use of environmental resources but sometimes also causing a negative technological impact on the environment (Orr et al. 2012). Crop rotation is a temporal and spatial arrangement of crops which is yearly planned for each farm by each farmer. A properly designed crop rotation system fully utilizes interactions between subsequent crops and between crops and the environment. The origin of crop rotation dates back to Charles Townshend, an English politician who designed a method for improving farming production during the agrarian revolution. At the beginning of the seventeenth century, the four-field crop rotation concept was introduced from Holland to England. Townshend popularized crop rotation which consisted of turnips (*Brassica rapa* var. *rapa*) as a root crop, wheat (*Triticum aestivum* L.), barley (*Hordeum vulgare* L.) with undersown clover, and clover (*Trifolium* spp.). That system, called Norfolk Crop Rotation (until today), was used instead of the low yielding 3-year strip rotation with fallow field (Bruns 2012).

Another option is growing two or more crops in the same field at the same time in a single year. This system includes two groups of cropping: sequential cropping – growing two or more crops one after the other, in the same field. Currently, this is a common practice in those regions with long growing seasons and favorable weather conditions (tropical and sub-tropical zones). Another option is when the farmer grows two or more crops simultaneously using a different arrangement (at the same time in a single space) system. This system is called polyculture, mixed cropping (MC), or intercropping (Gliessman 1986). Crop polycultures are a practical application of the main ecological principles and also regulate interaction with natural biomes: biodiversity, plant interactions (competition/stimulation), and other natural forms of regulation (Vandermeer et al. 1998; Głąb et al. 2017). Diversity of cultivated crops increases the potential for an ecological balance among cultivated species but mostly in the soil. Polyculture offers more effective protection against soil degradation and the proliferation of pests and pathogenic bacteria and fungi than monocultures. Diverse plant associations (cultivation of two or more crop species) maximize the usage of resources in a complementary way and ensure good and stable yield more effectively than sole crops (Corre-Hellou et al. 2006). Polyculture is not a commonly used cropping system in developed countries but has a significant influence on food security in the developing world.

Cultivation of a few crops together, on the same area and at the same time, is traditional and as common as the well-known intercropping system (Pleasant 2006). Intercropping probably existed early in agriculture's evolution and was an important stage of the crop domestication process, which has been proposed as a second period of domestication (Plucknett and Smith 1986). Based on the practice of the Iroquois, it is suggested that inhabitants of Latin and South America have cultivated maize (*Zea mays* L.), pumpkin (*Cucurbita* sp.), and bean (*Phaseolus* sp.) (Photo 8.1) – the so-called three sisters or Dioheka (Pleasant 2006, Hart 2008) – for over 3000 years. This system is described as a symbiotic plant association of North and Central America and has no comparable system elsewhere (Pleasant 2016). On the opposite side of the world, intercropping also has a long tradition. In China, cultivation of cereals with mixtures of other species has a 1000-year-old tradition, and it is still widespread in present Chinese agriculture covering around 28–34 million ha

Photo 8.1 Three sisters planting system. (Sowinski J.)

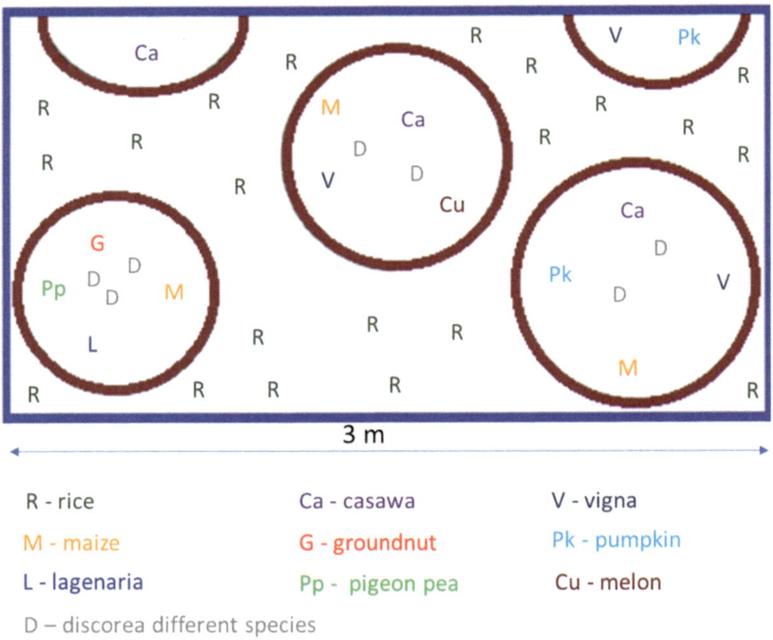

Fig. 8.1 Polyculture mound system in Nigeria. (Adopted: Okigbo and Greenland 1976)

(Li 2001; Li et al. 2007). Also in Africa and India, polyculture has a multi-century history, and this system is still used (Mahapatra 2011; Garrity et al. 2012).

This system of production is popular throughout the tropical zone because it mostly minimizes the risks of crop failure. For example, in Africa in the 1970s, cowpea was the most popular legume crop and was grown in combination (on prevailing areas – 94%) with other crops. In Nigeria, MC systems dominated, and farmers cultivated up to 13 species on the same area (Fig. 8.1) (Okigbo and

Greenland 1976). At some places of tropical zones, farmers cultivate sometimes 30 or more crops on the same land (Dewar 2007).

8.2 Polyculture System for Sustainable Agriculture

Polyculture is an element of sustainable agriculture system which more effectively uses both the area occupied by crops and labor resources than does monoculture (Baldy and Stigter 1997); it also better utilizes environmental resources and provides a higher, more stable yield in variable environmental conditions lowering yield failure risk. This system also has a positive economic effect on the magnitude of inputs and outputs and the stabilization of the food supply chain (Beets 1982).

The main advantages of polyculture are (Knörzer et al. 2009):

- Effect of field diversity, increased yield quantity and provision of greater stability, reduced risk of crop yield failure.
- More efficient and complementary use of resources (biological, e.g., nitrogen from legumes, N-fixing plants as well as soil resources, and social, e.g., human labor) and maximization of land use and provision of better environmental indicators.
- Reduction in the amount of artificial fertilizer and pesticide used.
- Intercrops positively suppress weed competition and reduce crop sensitivity to insects and disease outbreaks.
- Reduced yield damage and impurities from weeds, diseases, and pests.
- Allows more than one harvest per year.
- Improves soil structure, physical and chemical properties (mostly if crops with different root systems are used – taproot vs fibrous).
- Polyculture system covers land better and improves soil water and wind erosion control.
- Better human nutrition.

The main drawbacks of polyculture are (Knörzer et al. 2009):

- Mechanization processes are limited, i.e., difficulties for some practices in sowing technology (different sowing rate, depth, row spacing, and time), fertilizer and pesticide application (rate, selection or application date).
- Difficulties may occur during harvest and crop separation.
- Not commonly recommend cash or staple crops.
- Difficulties for weed control and herbicide selection (sometimes not recommended herbicides for sown crops in polyculture).
- Management demand higher compared to sole crop production, and better education of farmers is required.

Polyculture as a term has different meanings and specific patterns which are named differently. Most similar to natural phytocoenosis (like natural association) is the MC system (Fig. 8.2, Photo 8.2 and 8.3). Therefore, MC is definitely different

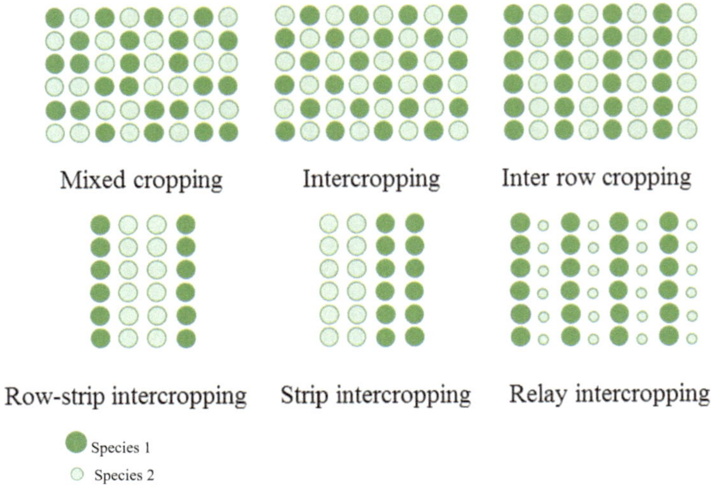

Fig. 8.2 Different polyculture system patterns

Photo 8.2 Examples of different polyculture systems. (Sowinski J., Adamczewska-Sowinska K.)

from other polyculture systems (Lithourgidis et al. 2011; Liu and Song 2012; Yang et al. 2014) (Table 8.1).

Polyculture as a multi-cropping system term includes other different cultivation patterns: intercropping, companion planting, alley cropping, relay intercropping, living mulches (LM), and others. All these cultivation methods allow the use of land, nutrient energy, and other products in a more balanced way compared to

Photo 8.3 Companion crops system. Initial growth (**a**) and full mixture vegetation (**b**) of Persian clover (*Trifolium resupinatum* L.) with annual ryegrass (*Lolium multiflorum* L.). (Sowinski J.)

Table 8.1 Mixed cropping and intercropping comparison

Basis for comparison	Mixed cropping	Intercropping
Model	No any specific sequences of sowing seeds. Easy for planning. System most similar to natural plant diversity with different stands of crops	Follows a definite pattern of sowing seeds. Different models of sowing arranged. Model of cropping system needs specific technic regime
Seeds	Before sowing seeds are mixed and sowed together, use the same seed drill	Seeds not mixed before sowing. Special seed drill or sowing technic should be used
Fertilizer and pesticide	Special selection and the same fertilizer and pesticide is applied to all sown crops at the same field	Specific fertilizer and pesticide could be applied to each crop separately. Model of intercrops has the effect on pesticide and fertilizer application technic
Aim	Minimize the risk of crop failure and stable yield on different environment conditions	Increase the productivity of the crop
Competition	Competition between the crops exist	Competition between the crops does not exist
Harvest	At the same time for the same use	Could be used separately for different purposes

monoculture, but it is not popular at present as an agricultural management system. In some regions, crops are based on mixed system production. Xu et al. (2013) revealed that polyculture accounted for 75% of wheat production areas in China.

Another form of PS is relay intercropping. This system is used for special groups of species such as forage legumes (small seed species) which can be grown together with a companion crop, usually spring cereals (barley and oats – *Avena sativa* L.) or annual grasses, and fast-growing legumes (Sowiński 2014). Small seed forage legumes develop very slowly, directly after sowing. The undersown legumes in CC at the beginning of vegetation do not compete for environment resources (water, light, nutrients). They start to grow more vigorously as the CC are harvested. Companion crops reduce weed competition and weed biomass production at CC harvest term. Common vetch and mixtures of spring barley with common vetch could be recommended as biological methods for weed control or significant suppression of weed growth.

8.3 Living Mulches: Special Type of Polyculture System

Another specific polyculture system, mostly used in horticultural production and based on simultaneous crop cultivation on the same space, is called the living mulches LM system. This method of crop production has mainly been used with perennial crops in orchards, but also this method is also becoming more popular in agriculture as well as vegetable field production. Living mulches are different from other polyculture patterns, and the system relies on a minimum two-crop cultivation, in which one is sown and constitutes the harvested crop (main crop), while the second is sown under the main crop as a cover growing crop for soil protection as green manure for sustainable agriculture production (Baumann et al. 2000, Adamczewska-Sowińska and Kołota 2002). Selected species could be cultivated as cover and main crops in different patterns depending on species characteristics and methods of sowing or planting (Fig. 8.3).

Living mulches constitute a special production system, in which conditions are temporarily created, where the species sown in inter-rows develop properly and

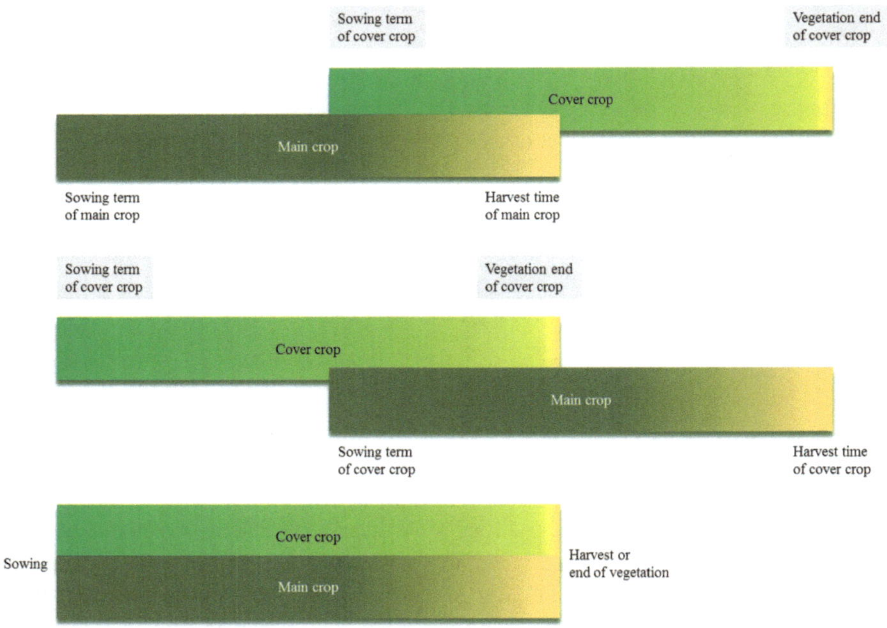

Fig. 8.3 Cover crop management system (example of main crop and cover plant vegetation season). Possibilities of sowing and harvest time

interact beneficially with each other. The species of plants used as LM should be appropriately selected and present the least potential competition relative to the cultivated plants. The species used as LM should be characterized by a rapid initial rate of growth; the crop canopy may be low, but it should contribute well to soil cover, with little demand for nutrients and medium biomass productivity. It is also advisable to decrease the competitiveness in the initial period of growth by delaying LM sowing. Brainard and Bellinder (2004) report that species used as LM often reduce the growth of cultivated plants without showing any effect on weed species or reduce the growth of both cultivated plants and weeds. Interspecies competitiveness for the habitat conditions may result from premature sowing of LM species (Akanvou et al. 2002, Carof et al. 2007).

8.3.1 Environmental Benefits of Living Mulches

Living mulches are a good source of organic matter. After the end of the growth of cultivated plants and the fulfillment of their protective function as living mulches, they are incorporated as a green fertilizer. Liang et al. (2002) show that these methods have been known from around 500 BC in China, where production of mung bean and sesame was encouraged to enrich soil fertility.

Nutritional value (of living mulches) is comparable to the fertilization value of feces from silkworms or well-decomposed farmyard manure. Since the Qin dynasty, attention has been paid to the reduction of the progress of soil degradation and the implementation in the area of less plateau technologies of production, which were aimed at protecting soil against excessive water and wind erosion and consequent depletion of soil fertility. Covering the soil surface with plant material or other covers has many benefits including soil temperature stabilization, accumulation and preservation of water in the soil profile, improvement of the physical properties of the soil, increasing the biological activity, reducing the occurrence of weeds, slowing down the process of soil salinity, and increasing the crop productivity (Liang et al. 2002). Primarily, the LM system took into account cereal cultivation with trees (agroforestry system), and later this system included many species: hemp, soybean, mung bean, rise, cotton, and cereal crops with green fertilizer plants (Knörzer et al. 2009). In Europe, the first reports date from the sixteenth century, where the effectiveness of buckwheat and pea as an LM for increasing soil fertility was discussed.

The importance of mulches definitely increased when soil conservation practices were introduced into US agriculture. "Dust Bowl" problems in the 1930s meant that the US agriculture administration recognized wind soil erosion problems and implemented the practice of new tillage methods with cover biomass, cover crops, and LM (Hartwig and Ammon 2002).

Currently, LM systems in the United States are used for different forms of sustainable agriculture, mostly soil protection purposes, including (Lu et al. 2000):

- Water and wind erosion control.
- Nitrogen recovery.
- Fall cover.

- Summer biomass production.
- Nitrogen production (biologically – by bacteria – fixed N uptake by legumes).
- Weed suppression.
- Short-period cover crops.
- Heat tolerance.
- Fumigation effect.

Vegetable production with LM is a rather difficult issue and cannot be carried out in all conditions. Many problems have been noted with this system in terms of production; for example, vegetable yields decrease because of mutual competitiveness for water, nutrients, and light. The competitiveness of living mulches can be reduced in many ways (Table 8.2) but mainly by selection of appropriate crops.

8.3.2 Species Selection for Living Mulches and Their Characteristic

Living mulches are recommended in orchard and perennial vegetable production, e.g., rhubarb and asparagus. Cover crops selected for this type of production should also be perennial but not exclusively so. The living mulch system is also recommend for annual agriculture and vegetable crops, for this system may be selected species with long vegetation periods and late harvesting dates (cabbage, leek, maize and sweet maize, tomato, pepper, eggplant, and celery). The living mulch system has also not been widely investigated in cereal production as a bi-cropping system (Sowiński 2004; Thorsted et al. 2006a; Hiltbrunner and Liedgens 2008).

Table 8.2 The methods of competitiveness reduction of living mulches relative to cultivated plant (Adamczewska-Sowinska et al. 2009)

Indirect action	Direct action
Selection of vegetable species and living mulches	Inhibition of the growth of living mulches by mowing, crumbling, and shallow mixing with soil Mowing of aboveground parts causes reduction of the growth of root system resulting in competitiveness decreasing
Selection of sowing and planting dates	
Optimal fertilization and irrigation applied directly to cultivated plants	
Applying fast-growing cultivars of vegetables with initial vigor growth	Spraying the biomass of living mulches with sublethal dose of herbicide. Application of sublethal dose causes that mulch from dead biomass remains in inter-row of cultivated plants
Enhancing the distance between rows of plants and living mulches	

For LM practice, the following species are mostly recommended (recommendation for temperate climate conditions). Sowing terms are different compared to main (trade) crops and depend on many environment factors:

- Species from the Fabaceae family:
  - Clover (various species mostly white clover, *Trifolium repens* L., and subterranean clover – *Trifolium subterraneum* L.).
  - Hairy vetch – *Vicia villosa* Roth.
  - Bird's-foot trefoil – *Lotus corniculatus* L.
  - Serradella – *Ornithopus sativus* Brot.
- Grass species from the Poaceae family:
  - Perennial ryegrass – *Lolium perenne* L.
  - Kentucky bluegrass – *Poa pratensis* L.
  - Red fescue – *Festuca rubra* L.
  - Cereals: rye – *Secale cereale* L., wheat, barley, oats, sorghum, *Sorghum bicolor* (L.) Moench.
- Others species:
  - Winter rapeseed – *Brassica napus* L.
  - White mustard – *Sinapis alba* L.
  - Tansy phacelia – *Phacelia tanacetifolia* Benth.
  - French marigold – *Tagetes patula* L.

Plants from the Fabaceae family qualify to be used in intercropping with species with high demands for nitrogen such as maize or most vegetables. Among these, white clover is more often used in crop production as a living mulch.

Because of its morphological traits, white clover offers suitable coverage for the soil surface and provides good protection for the main crops against competition from weeds. Simultaneously, this species does not compete for water because its taproot system reaches deeper soil layers than do cultivated crops. White clover uptakes water from a depth of up to 100 cm, whereas grasses such as red fescue or bluegrass operate only up to 20 cm (Leary and De Frank 2000).

White clover is sensitive to unfavorable conditions of water stress during the initial growth period, especially during germination and plant emergence. A deficit of water and high temperature during these periods lead to a decrease in white clover biomass yield. The seeds of white clover (and other plants from the legumes) need high amounts of water to germinate; about 150–300% of water is soaked up compared to their mass. During vegetation, this species is characterized by high transpiration coefficients, and the highest water uptake is observed in the period of rapid biomass growth. White clover uses more water during its growth than grass; therefore, it is susceptible to periods of water deficiency. This crop grows well if soil moisture is not limited (Hofmann et al. 2007). Water deficit in the soil reduces the growth of the root system, and the decay results in a decrease in the rate of growth.

The grass traits which qualify them for use as an LM are their low canopy type and rapid tillering. Clump and loose grass are recommended. Perennial ryegrass is a loose-clump grass with a strong developed bunch root system that in 80% of cases is located in surface soil. Few roots reach 75–150 cm. Ryegrass belongs to those species adapted to moderate sea climates; therefore, high humidity of air and soil and a moderate temperature favor its development. Ryegrass creates a greater root system mass compared to white clover, regardless of the humidity conditions (Lucero et al. 1999).

The most interesting species that can be used as an accompanying cash crop and that fulfil all the functions of LM and phytosanitary plants belong to the genus marigold (*Tagetes patula* L., *Tagetes erecta* L.). The roots of these plants exude sulfur compounds, thiophenes with their derivate – α-terthienyl – with its destructive force on nematodes. Also, it has not been proved that the inhibition of the development of soil bacteria close to the rhizosphere of marigolds and organic matter left in the soil stimulates the development of naturally occurring microorganisms which compete with nematodes. The phytosanitary effect of marigolds occurs after prolonged living mulch cultivation – a minimum 3–4 months (Photo 8.4).

Many studies confirm the significant influence of weather conditions on biomass yield, of both underground and aboveground parts of cover plants. The biomass yield decreases under the influence of long periods of drought as well as lower than average and excessive rainfall. Such conditions impair soil surface cover by the plants, which do not then fulfil their protective role. Our own research has shown that years with unfavorable weather conditions resulted in decreasing biomass yield for white clover compared to better years (with favorable weather conditions), i.e., 48–55%, whereas for ryegrass the yield reduction amounted to 20–47% (Adamczewska-Sowińska 2004).

It is very important to select an appropriate sowing date and the date of occurrence of LM in production. The degree of competition toward the cultivated plant is dependent on the abovementioned factors and also the length of protective activity of LM crop on the soil and yield of produced biomass. The most appropriate date for sowing the seeds of living mulches, for most vegetable species, is the second part of their growing season, after having passed the critical stage of sensitivity due to the presence of competitive species (Figs. 8.4, 8.5, and 8.6).

Another solution is sowing or planting crops that constitute main yield into growing living mulches. Both species grow together simultaneously, and then the

Photo 8.4 French marigold long-term growth with celery *Apium graveolens* L. var. *dulce* (Mill.) Pers. as living mulch. (Adamczewska-Sowinska K.)

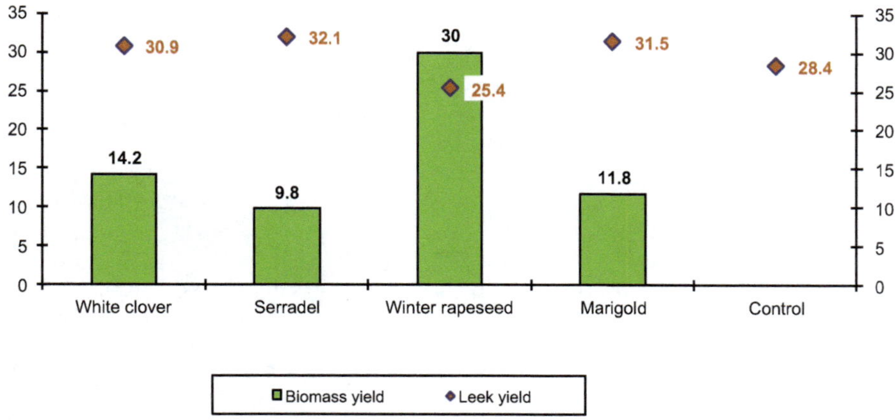

Fig. 8.4 The marketable yield of leek depending on biomass of living mulches (t ha^{-1})

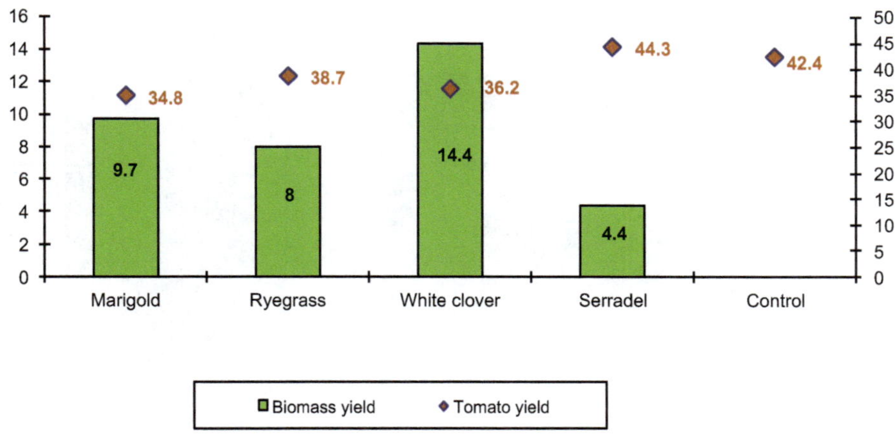

Fig. 8.5 The marketable yield of tomato depending on biomass of living mulches (t ha^{-1})

development of the companion crop is inhibited by mowing or application of herbicides. Living mulches species with short-growing seasons or sown in the previous season and wintering in the field, flower and end vegetation earlier and so its becomes less competitive. Subclover is well adapted to natural conditions. This clover has a prostrate habit, a height reaching max. 10–15 cm with a long creeping stem offering good soil surface coverage. For living mulch, crop emergence and fast soil cover are very important for erosion control and competition with weeds. Subclover covers the soil more quickly than white clover (27 days for subterranean clover versus 36 days for white clover to cover 50% of soil surface) (Ramseier https://www.hafl.bfh.ch/fileadmin/docs/Forschung). Additionally, subterranean clover is a self-regenerating crop and is used for land revegetation and as a green manure in its area of origin (Photo 8.5). In natural conditions, this annual crop dies

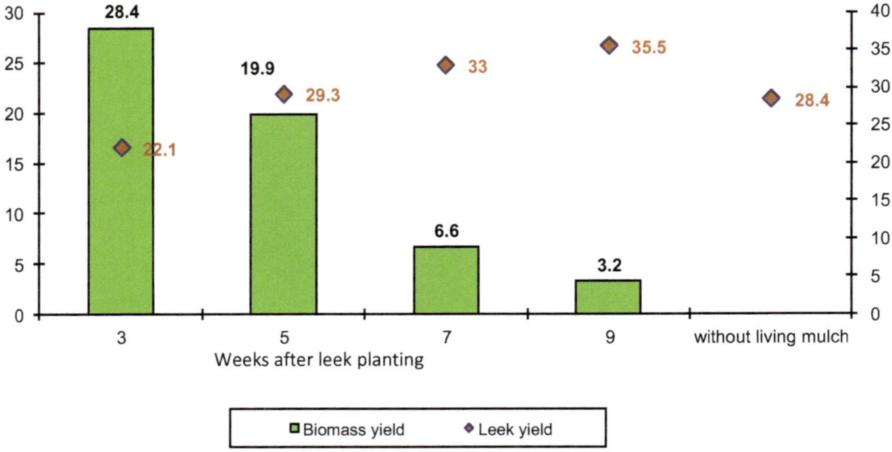

Fig. 8.6 Effect of living mulches sowing term on biomass yield and marketable yield of leek (t ha^{-1})

Photo 8.5 Subclover recommended as living mulches. (Adamczewska-Sowinska K.)

after flowering and seed set and does not compete for environmental resources. Dried subclover biomass covers the soil surface as organic mulch.

Living mulches are seldom used for cereal production which is mainly because of lower yield compared to standard methods of production. Yield reduction is caused by competition from LM for space and environmental resources (light, water, and nutrients) (Thorsted et al. 2006b). To increase main crop yields, Hiltbrunner and Liedgens (2008) suggest limitation of available resources via the reduction of cover crop competition. Dynamic growth of perennial cover crops such as white clover competes with cereals in initial stages of growth – directly after sowing. Thorsted et al. (2006b) and Hiltbrunner et al. (2007) suggest the use of selected genotypes of cover and main crop varieties or use of technologies which increase main crop competition.

Sowiński (2005) conducted a field experiment lasting a few years focusing on winter wheat monoculture cultivation with LM with white clover. As mentioned above, the indispensable element in this system of production is the reduction of the

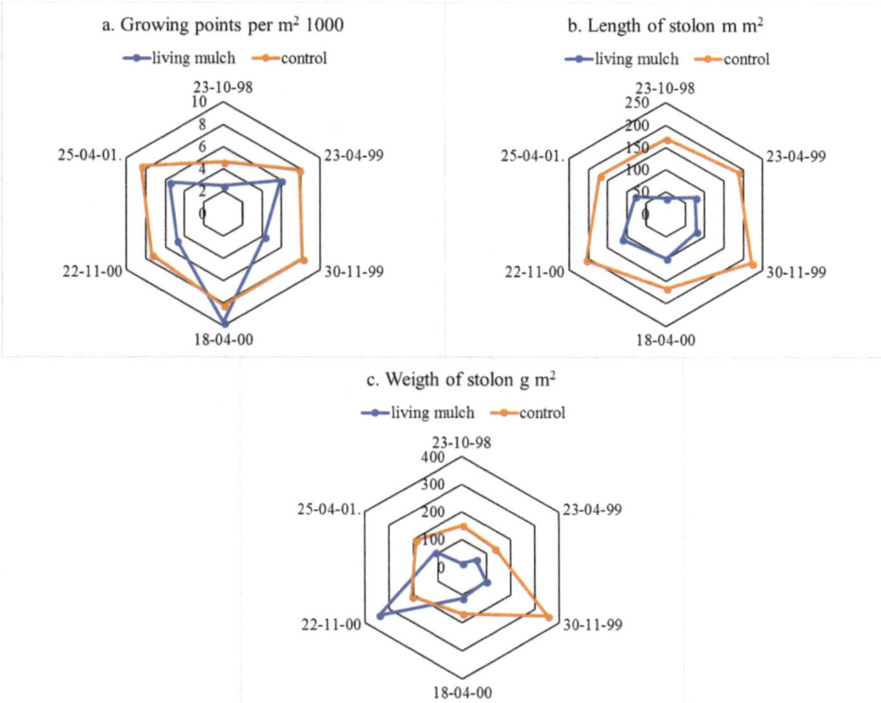

Fig. 8.7 The effects of sowing method on white clover competition parameters

competitiveness of LM in initial growing stages of winter wheat growth. This can be obtained by grazing with ruminants, mowing and biomass harvesting, or herbicide application for suppression of clover growth. In that research, sublethal doses of herbicide inhibited growth of LM compared to white clover grown in control conditions. Parameters used to evaluate the clover competition characteristics are the number of growth points per area units as well as the length and mass of stems lying on the soil (covering) surface (Fig. 8.7a–c). On average, the number of growth points was mostly higher for white clover pure stands: at about 37% compared to clover cultivated as a living mulch. The length of stolons at all dates of measurements was higher in clover pure stands. On average, the length of stolons on non-suppressed white clover was 133% higher and 62% heavier compared to the LM system, when lower doses of herbicide were applied.

8.4 Plant-Plant and Plant-Environment Interactions

Plant-plant interference has a crucial role in regulating the phytocoenosis composition and type of ecosystems (Brooker 2006). Polyculture component crops (the individual crops that constitute polyculture) make more efficient use of environment resources than when the same species are grown separately. Crops cultivated in mixed form have a synergetic effect on plant association and increased diversity and

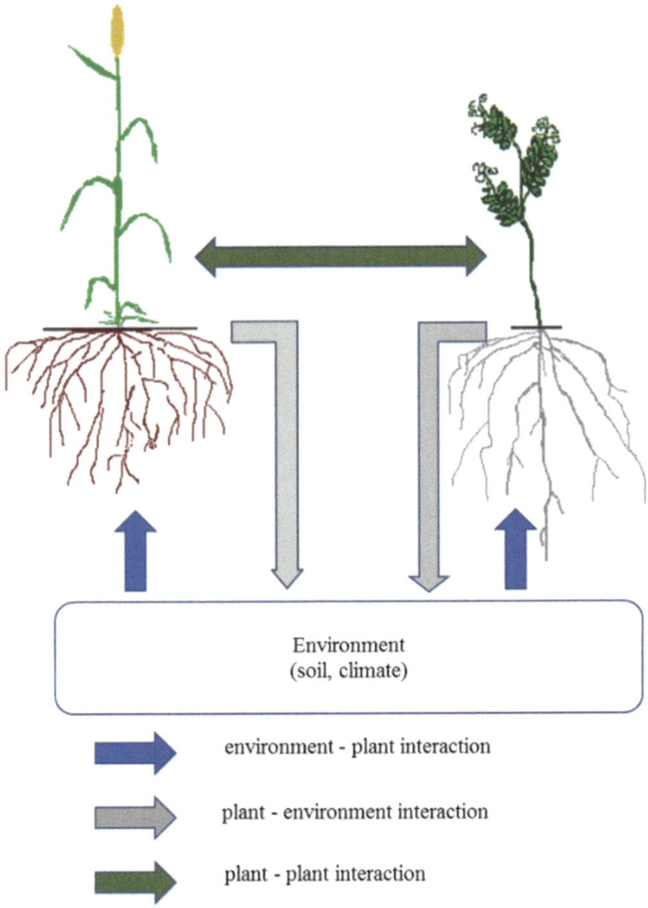

Fig. 8.8 Plant-plant and plant-environment interaction

create niches for other organisms and may influence crop productivity. A beneficial effect of polyculture cultivation may come from the different methods of environmental resource capture and interaction between crops (Fig. 8.8). Crops have different root systems and aboveground part habits. Differences in root system and aboveground part morphology (positively and negatively) influence plant-plant interference. Different plant architectures (habitus) – branchy or close, stem height, physiology, growth dynamic, and development cycle of cultivated crops in polyculture – lead to wider access to environmental resources and their more effective use than when these crops are cultivated as monocultures (Sobkowicz and Podgórska-Lesiak 2007).

Crops cultivated together in mixtures or intercropping systems interact in very different ways (Vandermeer 1989). The predominant interference is **competition** which is the process between two plants (two populations) in which at least one

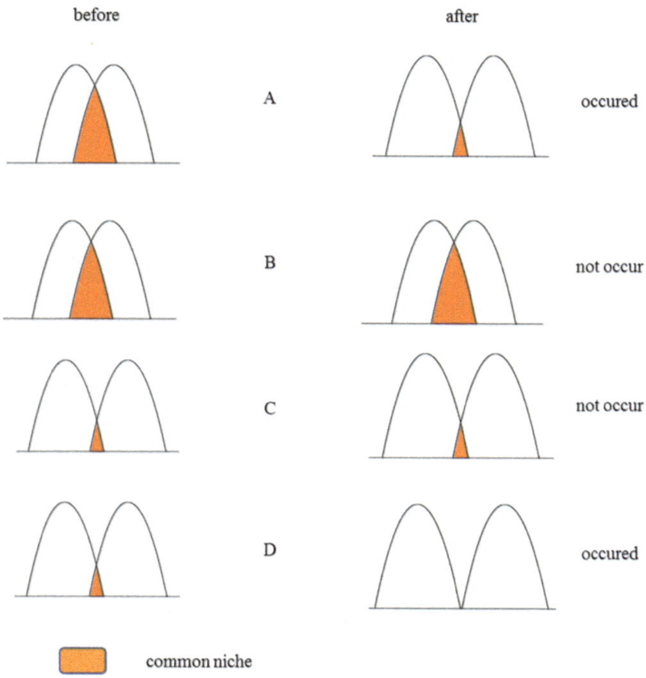

Fig. 8.9 Interaction between two species. Fig. 8.9A, B, high potential competition from the beginning of growth; Fig 8.9C, D – low potential competition from the beginning of growth. (Adopted: Vandermeer 1989)

exerts a negative effect on the other (Fig. 8.9). Crops which characterize potentially high competition at initial growing stages "example A" are not recommended for polyculture cultivation. High initial competition potential creates a low common niche at the end of growing season and a lower production effect. Crops compete in common niches for different environmental resources, i.e., growth spaces, light (sun energy and radiation), water, and nutrients.

Competition occurs for resources on aboveground parts (light, space), as well as for underground resources – i.e., competition between root systems (space, water, nutrients) (Fig. 8.10).

Competition between plants belonging to the same species (or the same group of species) is called intraspecific competition. This type of competition most often takes place on sole stands and in strip intercropping and row-strip intercropping. Intraspecific competition may decrease population size and influence their spreading and densities. Competition between plants belonging to different species is called interspecific competition. This occurs mostly on PS such as mixed cropping, intercropping as well as inter-row intercropping (Table 8.3). This has an effect on population size and spread and additionally may influence diversity and may reinforce selective environmental factors.

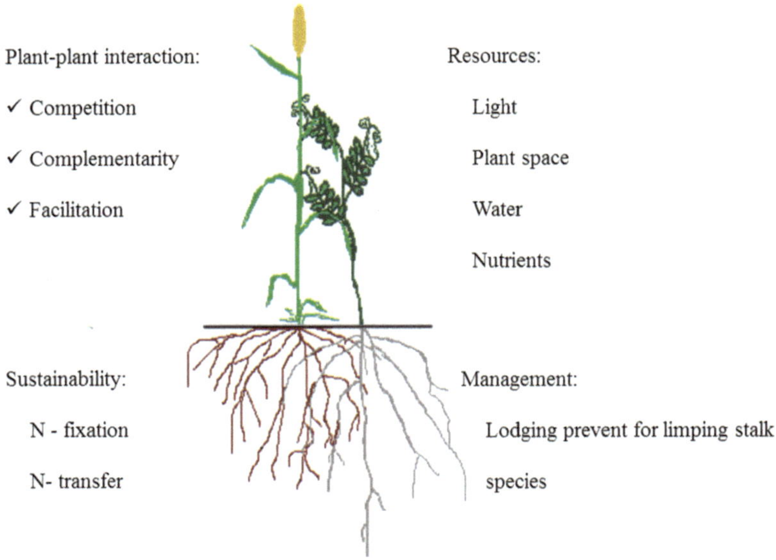

Fig. 8.10 Complexes of interaction between plant-plant interaction (example for cereals intercropped with legumes)

Table 8.3 Recommendation for mixture cultivation. Examples of good and bad companion species

Crops	Crops recommend for mixture cultivation	Crops not recommend for mixture cultivation
Maize	Sunflower, soybeans, peas, and other legumes, pumpkin or other cucurbits, amaranth	Cabbage, tomato, celery
Other cereals	Lupin, pea, vetch, faba beans, soybean	Rapeseed, maize, potato, sugar beet
Potatoes	Beans, cabbage, marigold, dill	Pumpkin and other cucurbits, sunflower, tomatoes
Peas	Maize, beans, cucumber, radish, carrot, turnip, fennel, dill, cabbage, lettuce, celery, spinach, red beet, marigold	Onion, garlic
Tomatoes	Carrot, parsley, celery, onions, basil, oregano, dill, pepper, lettuce, leek, spinach, beans, marigold	Cabbage, potatoes
Peppers	Tomatoes, carrot, onion, basil	Beans, cabbage crops
Beans	Maize, sunflower, cabbage, cucumber, lavender, potato, lettuce, red beet, early whole red beet, celery, spinach, tomato, dill, savory	Onion, garlic
Cabbage	Bean, cucumber, fennel, dill, marigold, endive, lettuce, lamb lettuce, early whole red beet plant, carrot, red beet, radish, tomato, celery	Potato, onion
Cucumber	Bean pea, fennel, dill, calendula, cabbage, lettuce, red beet, celery, spinach, maize, onion	Potato, radish, tomato
Carrots	Pea, cabbage, lettuce, leek, radish, tomato, celery, spinach, onion, fennel, calendula	

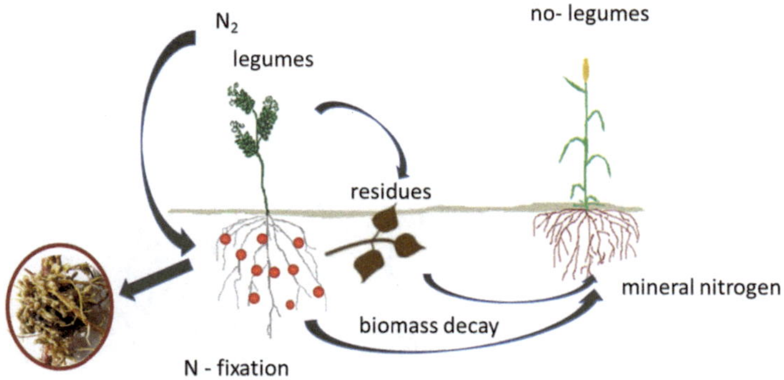

Fig. 8.11 Plant-plant complementarities interaction – example

Another interference between crops is complementarity in which a minimum of two species require proximity for positive interaction (Fig. 8.11). A good example of complementarity is nitrogen transfer between crops cultivated on non-legume–legume mixtures. Non-legume and legume crops acquire soil and fertilizer nitrogen (compete) until legume crops start to fix atmospheric nitrogen. From the onset of nitrogen fixation, competition between non-legumes and legumes species for nutrient sources decreases (Jensen 1996). The best examples of complementarity in polyculture are maize, beans, and pumpkins. Pumpkin growth is very fast, and ground coverage at initial stages is reduced by competition from weeds and water losses. Upright growth of maize and beans preserves humidity in the canopy during the growing season and allows the absorbance of light on the top and bottom parts of the canopy (Brooker et al. 2015).

Niche complementarity at PS leads to resource sharing and also facilitates the growth and yield of crops. Some groups of crops support the growth of others, e.g., legumes facilitate cereal growth through symbiosis with bacterial biologically fixed nitrogen which is released by roots to the soil. Biologically fixed nitrogen or nitrogen obtained from decaying dropped legume tissue has a positive effect on cereal growth (Sobkowicz and Podgórska-Lesiak 2007; Brooker et al. 2015). In this system, cereal may also facilitate limp stalk growth of some legumes (lens, vetch, or pea). Straight, stiff cereal stems prevent legumes from lodging. Legume cultivation occurs on sole stand lodges. Flattened legume canopies preserve the humidity that stimulates disease development and leads to partial or complete decay of pods. Flattened legume canopies are difficult to harvest and seeds are contaminated with soil particles.

Facilitative interactions may be associated with environmental factors, e.g., soil fertilities and reactions, attraction of crops for beneficial organisms, etc.

For PS, interaction between root systems of different crops has a key role for plant survival. Roots uptake nutrients and water from the same (sole cropping system) or different soil layer (polyculture). Plants grown together in mixed cropping, intercropping, and inter-row intercropping compete in the same soil volume, and their growth depends on the availability of the same set of resources (Depuydt 2014). At the beginning of the vegetation season, crops compete only for water and

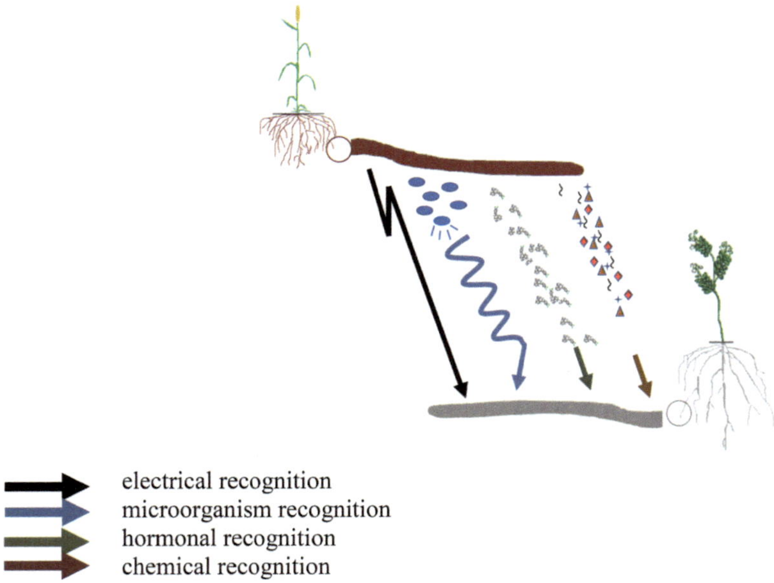

Fig. 8.12 Root-root interaction mechanism. (Modification from Depuydt 2014)

nutrients. Later, during growing development and leaf canopy formation, crops also compete for access to light. Initial growth stages are most important for crops, and crops recognize the surrounding area using different root signal methods: electrical, hormonal, and soluble metabolites from root exudates and from microorganisms associated with crops (Falik et al. 2003; Steenhoudt and Vanderleyden 2000; Prithiviraj et al. 2007; Biedrzycki et al. 2010; Lesuffleur et al. 2007, Semchenko et al. 2014) (Fig. 8.12). Competition on belowground plant parts is important in practice, and this is also a research area of interest (Depuydt 2014). For PS, belowground species interactions are very important and decide on crop selection and the final effect of polyculture cultivation.

Among other effects of chemicals, secondary plant metabolites are the most well-known interference mechanism (Głąb et al. 2017; Farooq et al. 2011). Bioactive substances are released as root exudation and affect plant growth from initial stages (Kumar et al. 2009, Lesuffleur and Cliquet 2010). Plants have a number of chemical interaction mechanisms, e.g., volatilization, foliar leaching, or decomposition of residues and leaf litter (Fig. 8.13). Roots of crops release many chemical substances into the soil, and these influence the growth and development of other plants. Depending on the type of substance and species, some of these chemicals are typically detrimental (inhibitors) and some beneficial (stimulator) for intercropped species (Ghafarbi et al. 2012). Bioactive substances – allelochemicals – cause allelopathy, i.e., they lead to a direct or indirect effect on one species by another through chemical signals.

A large number of allelochemicals have been discovered in recent years (Badri and Vivanco 2009, Bais et al. 2006, Belz 2007, Doré et al. 2004). The structures and properties of these substances may be classified as hydrophobic or hydrophilic

(Czarnota et al. 2003). Damour et al. (2015) more than 20 different groups of substances have been identified which have an influence on germination, physiology processes, growth and development, and crop survival.

Li et al. (2010) outlined a more specific structure:

- Water-soluble organic compounds.
- Lactones.
- Long-chain fatty acids.
- Quinones.
- Phenolics.
- Cinnamic acid and its derivatives.
- Coumarins.
- Flavonoids.
- Tannins.
- Steroids and terpenoids.

Recently many research groups have been strenuously working toward an understanding of the functions of plant physiology processes and the actions of allelochemical mechanisms. Unfortunately, little is known in this area, and difficulties come not only from the large numbers of substances but also from the complexity of their biotic and abiotic mechanisms (Belz 2007).

Plant's active substances are produced and accumulated in different plant parts: leaves, roots, stems, rhizomes, flowers, fruits, and seeds (Głąb et al. 2017) (Fig. 8.13). The mechanisms of allelochemicals actions on plant growth are manifold and rely on volatilizing compounds, leaching from aboveground plant parts – mostly leaves – and belowground exudation or decay of plant residues (Ben-Hammouda et al. 2001; Bonanomi et al. 2006; Kumar et al. 2009). Once transferred into the soil, these compounds may cause one of two types of allelopathic effect, detrimental (inhibitory) or beneficial (stimulatory) (Ghafarbi et al. 2012).

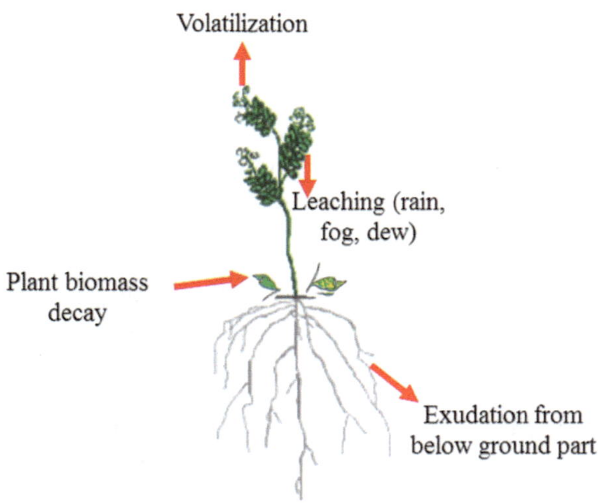

Fig. 8.13 Allelopathic response of plant tissue

8.5 Effect of Polyculture on Soil Properties

8.5.1 Importance for Maintenance of Soil Fertility

Management of soil fertility may optimize the efficiency of applied nutrients, crop productivity, and quality of yield. Soil fertility can be maintained by incorporation into the soil of fertilizer (organic and/or mineral) and also organic matter residue. Biomass complex processes (decay and mineralization) have a positive effect on biological activities, on soil structure and health, and also on soil humus content. Humus is soluble in soil pores as a residual form of organic matter has a high potential for improving soil fertility, mostly on light sandy soil. It is estimated that organic matter contains up to 95% of soil nitrogen and up to 40% of soil phosphorus (Rangarajan 2012).

A positive effect of polyculture (diverse plant communities) on soil fertility has been observed in some research (Balvanera et al. 2006; Zak et al. 2003). More diverse biomass has a positive effect on microbial communities and mineralization processes. This type of organic matter "feeds" microorganisms more intensively than monoculture biomass. Microbes release some nutrients when organic matter decays, and nutrients may also be released when microbes die and decompose. Deep rooting or taproot system of crops such as legumes or others often exceed the top horizon (Tables 8.4 and 8.5) and improve soil structure (Jensen et al. 2011). Biomass decay processes not only release nutrients but also positively affect soil porosities through free pores mostly following taproots (Fig. 8.14). Soil pores are the main factor of soil structure improvement which influence drainage, water retention, crop nutrition, soil compaction, and water infiltration rate (Fig. 8.15). Pores are the route for water flow and the growth of roots into soil. Pores also provide places for air, which is very important for crop vegetation and decide about porosity – i.e., soil volume which is not solid material.

The white clover mulch system for winter wheat cultivation influences the nitrogen availability and efficiency of biological nitrogen resources, and nitrogen utilization efficiency decreases by 36% compared to the wheat traditional production system (Sowiński and Wojciechowski 2018). For plant growth, the most limiting compounds apart from nitrogen is phosphorus (Vance et al. 2000). The polyculture system increases the capacity of crops for phosphorus uptake. Some crops such as white lupin and wheat grown together use acid (especially citric acid) for utilization of light acid leachable phosphorus (lupin) and water-leachable soil phosphorus pools

Table 8.4 Groups of vegetable crops depending on root system

Spread of the main part of root system		
Vertical	Horizontal	Species
Up 30 cm	Up 40 cm	Radish, spinach, onion, cucumber
Up 60 cm	Up 40 cm	Lettuce
	45–75 cm	Bean, cauliflower, peas
> 75 cm	Up 40 cm	Horseradish
	45–75 cm	Red beet, carrots, cabbage, asparagus
	250 cm	Rhubarb

Table 8.5 Vertical deep and horizontal spread of root system for some crops (Rangarajan 2012)

Crops	Deep (cm)	Spatial – diameter (cm)
Fibrous root system		
Oat	150	25
Barley	140	25
Rye	150	25
Wheat	15	12
Potato	90	40
Maize	175	100
Sorghum	175	65
Taproot system		
Pea	140	60
Soybean	200	50
Turnip	150	75
Lucerne	300	13

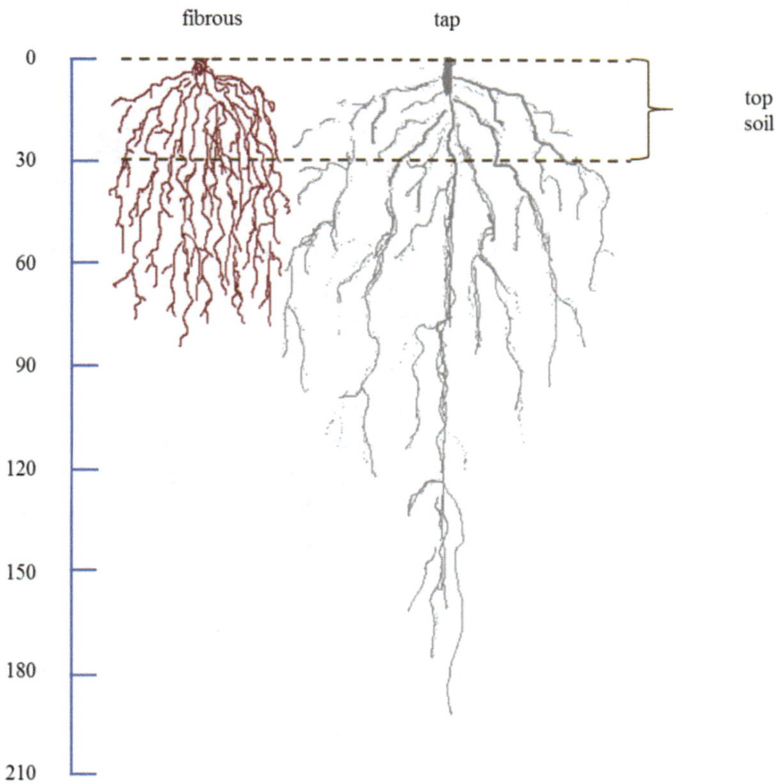

Fig. 8.14 Main type of root system

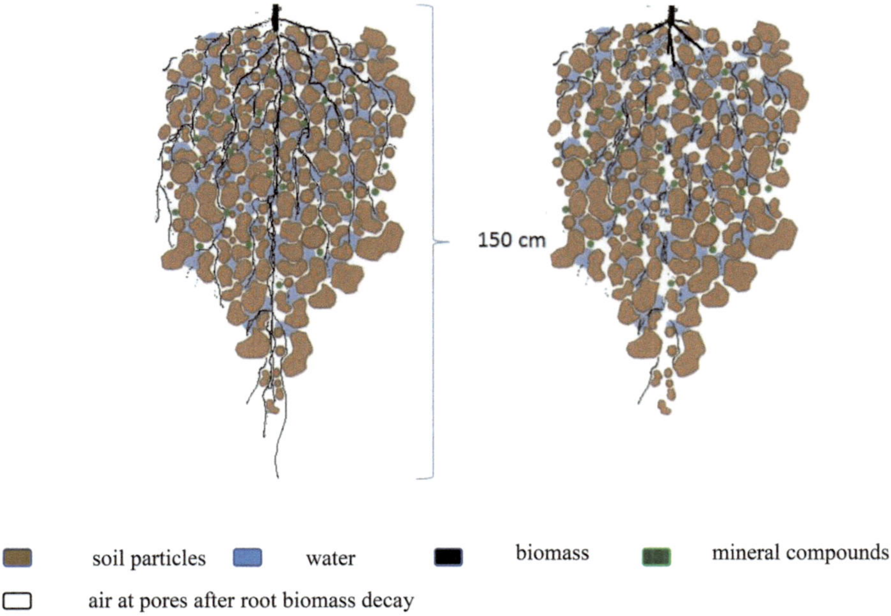

Fig. 8.15 Main soil components. (**a**) Before root biomass decay, (**b**) air-free spaces after root biomass decay. (Sowiński T)

(wheat) (Cu et al. 2005). The complementarities of polycultures for phosphorus uptake also determine vertical and horizontal soil exploration by root systems (volume of the rhizosphere) (Hinsinger et al. 2011). For PS, the volume capacity of the rhizosphere is much bigger than that with sole cultivation which improves the potential for phosphorus availability and uptake efficiency. The intercropping system and plant-root interactions have a key role in the mobilization of unavailable nutrients such as phosphorus (Latati et al. 2016).

Potassium is uptaken by crops in large amounts and, after incorporation of plant residues, is quickly released into the environment (Deguchi et al. 2010). Polyculture systems influence potassium uptake dynamics and reduce losses by leaching (Askegaard and Eriksen 2008). White clover sown as LM for maize competes for potassium and does not contribute positively to the maize potassium uptake. Cover crops or LM positively influence environmental reduction of potassium losses by leaching during the period after biomass incorporation and supply potassium to the following crops (Deguchi et al. 2010).

8.5.2 The Influence of the Intercropping System on Soil Properties

The influence of intercropping on the physical properties of soil is not unambiguous and depends on various factors. For instance, intercropping with LM including inter-row vegetables may have a beneficial impact on soil structure and the water

Table 8.6 Indicators describing wet soil structure status in eggplant cultivation, depending on biomass of living mulch

Indicators describing wet soil structure status	Sowing term of living mulch 3 weeks before transplanting	On transplanting term	3 weeks after transplanting
Yield of biomass of living mulches (t ha^{-1})	12.23	5.91	5.52
Mean weighted diameter of aggregates (wet) (MWDg)	1.20	1.14	1.07
Water stability index (ΔMWD	2.30	2.26	2.08
Index of waterproof (Wod)	32.2	29.3	27.8

resistance of soil aggregate and other soil parameters. Wojciechowski et al. (2012) showed that a longer period of LM growth contributes to the production of a greater root system mass as well as that of aboveground parts of LM leading to a better effect on soil properties (Table 8.6). This research shows that the early date of LM seeding (3 weeks before planting eggplants) has a significant impact on forming the weighted mean diameter of the aggregate (MWDg), rate of water stability index (ΔMWD), and waterproof index (Wod). Later seeding of LM led to an improvement in the majority of examined physical features of soil; however, the differences were not statistically confirmed.

The beneficial impact of LM on soil in vegetable production results from:

- Soil surface covering with plant material, reduction of tillage, and protection treatments during growing seasons of plants cultivated with living mulches; beneficial influence on soil structure. LM cover the surface of wide inter-rows, protecting soil against degradation occurring as erosion.
- Living mulches protect soil surfaces between cultivated plants and also after their harvest from:
  - Surface flow
  - Direct activity of water drops
  - Blowing away of soil particles by wind

Living mulches improve the soil structure by increasing soil cohesion. This is caused not only by a direct decrease in the number of tractor and machine passes (lack of soil compaction) (Leary and De Frank 2000) but also its effect on the deep root system. It has a mechanical nature when during plant growth, roots are squeezed through soil particles, and after the decay of older roots, free pores remain and improve soil air exchange and water infiltration and flow. This beneficial, biological activity relies on the production of polysaccharides and gums by bacteria existing on roots of cultivated plants. These compounds combine to form soil aggregates and clump structures.

- Reduction of soil losses resulting from erosion has an impact on the decrease in the amount of nutrients leaching to deeper layers of the soil and groundwater (Tonitto et al. 2006, Kankanen and Eriksson 2007, Sturite et al. 2007).

- The loss of nutrients in ground-cover plant production is four times less (Leary and De Frank 2000). The leaching problem mainly concerns the nitrate form of nitrogen, which dissolves well in water. Release of $N\text{-}NO_3^-$ from plant biomass is lower in the case of its remaining on the soil surface, whereas more intense biomass decomposition and $N\text{-}NO_3^-$ release take place after biomass incorporation to the soil after tillage in autumn or winter seasons. More humid conditions in the soil profile and its stabilities lead to a higher occurrence and more diverse microorganism population. Leaving plants that were LM or living mulch biomass as a mulch for autumn-winter seasons has a positive effect on soil protection and the reduction of nutrient leaching. A 65–70% decrease in nitrogen losses can be achieved. The high value of LM as ground-cover plant has an important meaning for protection against soil erosion and nutrient leaching in crop production on rolling terrains on hillsides with slopes greater than 2% (Abdul-Baki et al. 2002a).

The influence of LM on soil moisture is ambiguous. Some authors have shown that LM helps to increase soil moisture and maintain it at a constant level, causing an increase the water capacity and filtration (Boyd et al. 2000, Abdul-Baki et al. 2002b). Living mulches such as vetch, red clover, and Kentucky bluegrass maintain high soil moisture levels during potato production. Other scientific research has not shown this effect of mulches on soil moisture. In eggplant production, the presence of LM does not change these parameters, i.e., soil moisture and its porosity (Wojciechowski et al. 2012).

- Interaction between LM and the temperature stabilization of surface soils is also beneficial. Living mulches protect soil against excessive overheating during periods of strong insolation or fast overcooling during cold seasons.
- During the growing season and thereafter, living mulches are an important source of organic matter. Living mulches may increase or at least maintain actual levels of organic matter in the soil (Sainju et al. 2002; Thomsen and Christensen 2004; Abdul-Baki et al. 2002b). During growth with cultivated plants, decaying roots and older parts of aboveground plants enrich the soil with organic matter. After harvesting, LM may be left on the field and fulfil the function of ground-cover plants or be ploughed as a green fertilizer. After biomass ploughing and mineralization, mineral components that were earlier stored up are released (Table 8.7).

Błażewicz-Woźniak and Mitura (2004), Adamczewska-Sowińska (2004), and Kołota and Adamczewska-Sowińska (2013) highlight the fact that LM is a valuable, secondary source of important macro- and micronutrients in the soil. It is particularly beneficial in the case of nitrogen, that is, a mobile element and easily lost from soil because of leaching to groundwater.

Incorporation to the soil of organic matter means that the substratum changes its physical, chemical, and biological properties. The intercropping system generally increases the diversity of the bacterial community in the rhizosphere and has an effect on the microbial biomass of nitrogen and ratio between the above parameters (Song et al. 2007). Heavy and compacted soil undergoes loosening, and light soil

Table 8.7 Fresh matter yield (t ha^{-1}) and the amount of nutrients incorporated into the soil by living mulches (kg ha^{-1}) (mean for various intercropping systems with vegetable production)

	Perennial ryegrass	White clover	Serradella	Winter rapeseed	Marigold	Average
Biomass yield	18.8	14.2	9.8	30.0	11.8	16.9
N	41.7	47.8	27.3	146.1	28.0	58.2
P	6.4	15.3	10.4	40.7	13.7	15.5
K	77.7	36.1	24.4	103.7	34.4	55.3
Mg	5.3	5.9	3.8	12.2	4.4	6.3
Ca	17.3	22.3	16.3	92.9	19.6	33.7

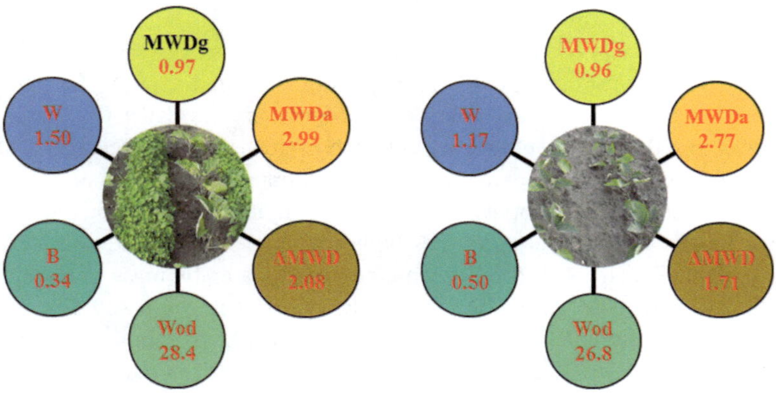

Fig. 8.16 Soil physical property indexes and soil aggregate stabilities for white clover living mulch and sole cropping system of eggplants. *MWDg* mean weighted diameter of aggregates (wet), *MWDa* mean weighted diameter of aggregates (dry), *ΔMWD* water stability index, *Wod* index of waterproof, *B* index of cloddiness of the soil, *W* index of soil structure

enriched with humus binds particles on aggregates, improves structures, and affects soil aeration and water holding capacity. Microorganism's increased content of humus in the soil causes its dark coloration which therefore favors faster heating.

Living mulches influence soil structure and improve soil physical indicators (Jędrszczyk and Poniedziałek 2009, Wojciechowski et al. 2012). Soil physical properties and aggregate stability are the basis of the different measurements and indexes. Soil samples were screened in different groups, and then the following measurements were taken: mean weighed diameters of aggregates – the wet method (MWDg) (in mm) – and mean weighed diameter of aggregates, the dry method (mm). The following indexes were also determined: water stability (ΔMWD), waterproof (Wod), cloddiness index (B), misting (S), and structure (W). Structure indicators depend on cover crops. White clover used as living mulch has a more favorable influence on soil structure than does perennial ryegrass (Wojciechowski et al. 2012). The authors paid special attention to the positive impact of LM on the water stability index and waterproof index (Fig. 8.16).

Wojciechowski et al. (2012) revealed a decrease in the cloddiness index of 0.16 and an increase in the structure index of 0.33 when eggplants were cultivated in a white clover mulch system. Fine soil particles are bound by humus or organic substances produced by fungi and bacteria during the decomposing organic matter process (e.g., glue) or by organic substances excreted from living mulch roots (polymers and sugars). In poor soil conditions, coarse clodds dominate as on degraded black soil.

8.5.3 Improvement in Soil Biological Activity

Soil organic carbon resources and tillage technics influence soil biological activities and soil microorganism dynamics (Cookson et al. 2008). Intercropping systems favor the development of various soil organisms and positively affect the biodiversity of the soil environment by different types of roots and diverse plant biomass. Finally, there is a positive influence on exudation processes in the rhizosphere (Bargaz et al. 2015). Polycultures significantly influence microorganism structure and microbial diversity compared to the monocropping system (Latati et al. 2014, Song et al. 2006). Hartwig and Ammon (2002) claim that, under living mulches, the biomass of earthworms is seven times higher than in soil cultivated in a conventional system. Pelosi et al. (2009) examined the greater biomass of earthworms living in soil under LM with white clover and bird's-foot trefoil and other species compared to the situation with traditional cultivation. This phenomenon is probably caused by the presence of a larger amount of vegetative organic mass incorporated into the soil, non-application of pesticides, and non-infringement of the biotope by ignoring or limiting mechanical soil cultivation and coverage of surface soil by plants. According to Schmidt et al. (2003), increased earthworm populations in the soil during wheat production in mulch with white clover result from the additional amount of organic matter delivered by the ground-cover plant. The presence of earthworms in the soil testifies to its fertileness and the lack of harmful contaminants. In research of wheat cultivation into white clover mulch, Sowiński (2004) showed that the number and mass of earthworms were significantly higher than the case in conventional systems (Figs. 8.17 and 8.18).

The great importance of earthworms results from the beneficial effect of earthworms on the structure and formation of the mineral-organic complex (coprolite) as well as the creation of soil macropores and the improvement of physical properties of soil (Photo 8.6).

Polycultures have a favorable impact on the development of soil microorganisms, as well as their growth after biomass incorporation. Soil organic carbon during the growing season of wheat intercropped with faba beans and maize is 47% or 37% higher compared to sole wheat cultivation (Song et al. 2007). The same tendency was also observed after incorporation of organic matter into the soil. The biological activity of organisms increases causing their decay, and thereby mineralization undergoes acceleration. On the process of microorganism biomass increasing with the same trends for species number (of bacteria, fungi and nematodes) up

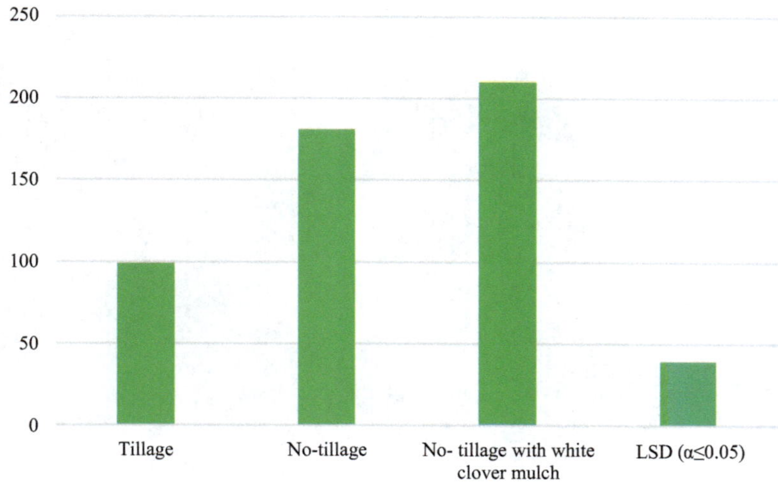

Fig. 8.17 Effect of tillage methods on number of earworms per m²

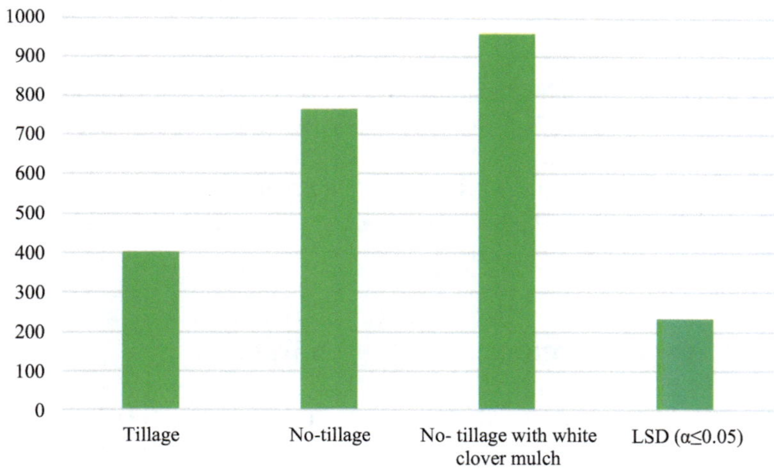

Fig. 8.18 Effect of tillage methods on weight of earworms (kg per ha)

too optimal proportion between species guarantee good soil productivities. Microorganism species composition in the rhizosphere depends on the cropping system and the plant species composition. The number of microorganisms such as collembola in the soil increases, causing the decay of organic matter. The formation of a large mass of roots by ground-cover plants such as rye or mixture clovers with grass is the main reason for increases in soil microbiological activity. It is claimed that no-tillage cultivation causes the number of arbuscular fungi to be higher in the soil and the colonies of roots of the majority of cultivated plants leading to an increased intensity of mycorrhiza. These arbuscular fungi mediate in the uptake and

Photo 8.6 Earthworm, coprolite, and soil macropores remained after relocating of annelids (Adamczewska-Sowińska K.)

supply of plants with mineral components. Deguchi et al. (2007) report that when the host of these fungi is white clover used as a living mulch, this causes the intensity of maize root colonization by mycelium and increased phosphorus assimilability, even at lower temperatures. Xu et al. (2008) point out that tomato roots are characterized by a much higher number of mycorrhizal fungi when cultivated cumulatively with Kentucky bluegrass.

8.6 Nitrogen Fixation and Nitrogen Transfer

Legumes (seeds or forage crops) play a crucial rule in sustainable agriculture and organic crop management. Legumes are a unique crop with a high protein content in biomass as well as in seeds with very high amino acid nutritional values, which could be used for food and animal feed. Legumes have a positive effect on cropping systems, increasing biodiversity and reducing negative impacts on the environment – decrease greenhouse gas emissions (Kirkegaard et al. 2008; Nemecek et al. 2008; Peoples et al. 2009). The significant role of legume crops in each agricultural system is a result of biological nitrogen fixation by bacterial (genus *Rhizobium*) symbiosis with legumes. Nitrogen fixation may reduce the application of artificial nitrogen fertilizers and decrease the negative impact of agricultural activities on the environment. Legumes are the only crops to fix atmospheric nitrogen by symbiosis with bacteria. Globally, each year 40–60 million tons of nitrogen (N) are produced by biological bacteria fixation (Graham and Vance 2003). Peoples et al. (2009) stated that 85 million tons of N are applied every year as artificial fertilizer.

Nitrogen is the most important compound for crops, less important only than light, temperature, and water (Thilakarathna et al. 2016). The amount of nitrogen derived from atmospheric N_2 ranges from a few kg (2 kg per ha for *Cicer arietinum* L.) to around 200 kg per ha per year (176 kg per ha for *Vicia faba* L.) (Büchi et al. 2015). Some of the biologically fixed N is also available for other crops as a subsequent effect derived from the mineralization and decay of legume residue biomass. Preissel et al. (2015) estimated this following effect of legume cultivation on cereal yields to be increasing in European conditions. For Southern Europe (Mediterranean conditions), cereal yields following legumes are from 0.2 to 1.0 t ha^{-1} higher than that for cereals cultivated following cereals. For temperate conditions, yield increases range from 0.5 to 1.6 t ha^{-1}.

The positive effect on following crop productivities is due to different combinations of factors (it is not only soil nitrogen that increases). Legumes stimulate the accumulation of soil organic matter from 0.5 to 1.0 t ha^{-1} due to tillage reduction and the incorporation of high amounts of biomass. The break-crop effect is the specific benefit of legumes to improve soil properties and increase organic matter content (Hernanz et al. 2009).

Residue biomass improves not only soil physical properties but also soil fertilities and reduces pest populations and disease infection. Legumes combined with tillage reduction have positive effects on water retention and harvested yields (Soldevilla-Martinez et al. 2013). Beneficial effect on the subsequent yield of crops even occurred on the third crops after legume cultivation (Evans et al. 2003).

Fixed N is not only beneficial for legumes cultivated on pure stands but also in polyculture, i.e., legume, or other species grown in mixtures, intercropping, and agroforestry (Thilakarathna et al. 2012; Chapagain and Riseman 2014, 2015; Nygren and Leblanc 2015). This nitrogen nutrition process is called a bidirectional system with legumes as one component. Directly during the vegetation season, nitrogen is transferred between two companion plants: one is called the "N donor" (i.e., exporter) and the other the "N receiver" (importer) of nitrogen (Yong et al. 2015). Nitrogen is transferred via intra- and interspecies interactions. Most often a higher N transfer occurs as interspecies interaction between legumes and non-legumes crops (Lesuffleur et al. 2013).

Thilakarathna et al. (2016) presented three nitrogen transfer pathways:

- Companion crops uptake nitrogen fixed by bacterial symbiosis with legumes after biomass decomposition and mineralization (Fustec et al. 2010). This transfer system is described as indirect. Receivers uptake nitrogen after mineralization.
- Realization by legume crops of soluble nitrogen to the soil solution and accumulation by non-legume crops – root nitrogen by exudation (Gylfadóttir et al. 2007; Paynel et al. 2008; Wichern et al. 2008).
- By mycorrhizae nitrogen exchange (Fig. 8.19).

The amount of transfer nitrogen could satisfy from 0% to 80% of non-legume crop demand for nitrogen. This amount depends on many biotic and abiotic factors:

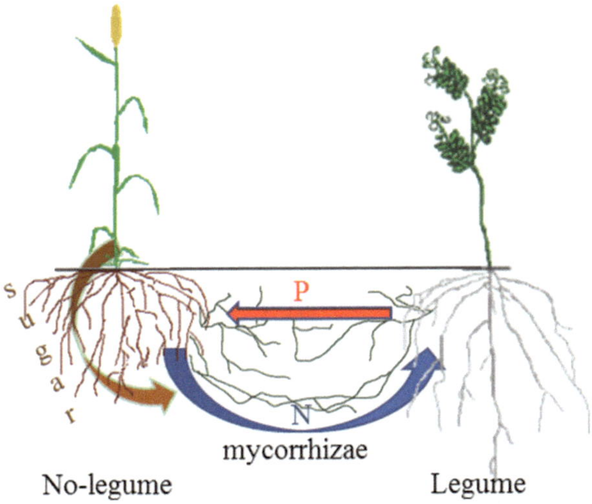

Fig. 8.19 Mycorrhizae compounds transfer phenomena

water stress, temperature, light, root contact and distance between roots, crop densities, growing stages, and defoliation stress (Moyer-Henry et al. 2006; He et al. 2009; Chalk et al. 2014). White clover quite often presents as the largest nitrogen donor crop for companion species (Pirhofer-Walzl et al. 2012; Rasmussen et al. 2012). For direct N transfer, root distance for legumes and non-legumes is the most important factor for the efficiency of exudation (Xiao et al. 2004; Daudin and Sierra 2008; Meng et al. 2015). Rasmussen et al. (2013) revealed that clover transferred to grass from 1.0% to 72% of nitrogen at the shortest distance between roots (from 0 to 5 cm). In the same conditions, reverse nitrogen transfer (from grass to clover) ranged from 0.3% to 17%. At longer root distance conditions between donor and receiver (5–15 cm), the amount of transferred nitrogen was much lower. The maximum confirmed distance for N transfer was different on the experimental site and for both donor species (grass and clover) amounted to 65 cm.

Fungi play an important role not only in the nitrogen transfer process, but also mycorrhizae fungi transfer carbon compounds (sugars and lipids) and phosphorus.

8.7 Conclusion

Present processes in agriculture have reduced the use of cultivation technologies that are not profitable from the management and economic point of view. Introducing an industrial character into the crop production technology of producing plant raw materials involves the same process assumptions as in industry, and this is one of the main trends in agriculture. This system is aimed at large-scale production, mostly monoculture and frequently with the inefficient use of pesticides and mineral fertilizers. This contributes to the production of food in systems inconsistent with the principles of agroecology, such as minimum soil distribution (cultivation), permanent soil cover, and crop rotation (Altieri and Nicholls 2012). In many regions of the world, but mostly in developed countries, the efficiency of human work is increasing, reducing inputs of the main types of resources, and large-scale production is currently characterized by the use of technology in agriculture. Today, about one billion people around the world are starving, and additional hundreds of millions suffer from dietary diseases such as obesity, various heart diseases, many types of cancer, or diabetes. Industrial agriculture has negative ecological and social impacts, and it is necessary to use appropriate safe technologies and produce healthy food for all of the world's population (Altieri and Nicholls 2012, Ikerd 2010).

However, we still need to search for a solution that can be applied, such as technologies that use natural mechanisms to regulate plant communities. An example of this is PS in which, as repeatedly stated, there is an improvement of soil quality and its physical, chemical, and biological qualities. Polycultures are difficult to implement mainly due to intraspecific and interspecific competition within plant communities but also due to difficulties in the mechanization of some elements of cultivation. The cultivation of polyculture is important in the maintenance of soil fertility, limiting soil degradation processes and protecting it by improving biodiversity.

8.8 Future Perspectives

On a global scale, present and future human generations may face many problems such as hunger, depletion of fossil fuel resources, environmental degradation, lost biodiversity, and the destruction of natural ecosystems. In the worldwide perspective among natural polycultures, biomes dominate on grassland, forest, and other ecosystems. An opposite trend has been observed in crop production with a reduction of diversity leading to a monoculture system. According to Dewar (2007), after reengineering the natural PS which will be adapted for food production may have a positive impact on the environment and production on a global scale. Furthermore, polyculture systems will become common farming practice and not only be used in small-scale farming. Dewar (2007) recommends this system for any scale of production, for small family farms as well as for large enterprises.

In the immediate future, the main impact on agriculture production and food security will come from weather variabilities and climate change processes. Trends for increasing temperatures connected with unregular rainfall patterns (rainfall occurrence timing and amount) will have a great influence (negative) on crop productivity (Porter et al. 2014). In some climate zones, the length and conditions of growing seasons will be modified (Sarr 2012), and yields will decrease from 6% to 24% (Waha et al. 2013). For current cropping systems, high yield variability from year to year will represent a pressing challenge for agriculture and food security for 9 billion people in 2050. Global agricultural production should be increased by 100–110% for the next 30 years (Tilman et al. 2011). New agricultural management strategies for cropping systems should be adopted. Agricultural systems (e.g., industrial farming) which destroy soil productivity and do not protect natural resources – soil, water, or air – will destroy our ability to produce food (Ikerd 2010).

Debaeke et al. (2017) proposed the creation of a model of climate-smart agriculture based on more diverse crop production combined with genetic crop improvement and the introduction of new varieties of crops resistant to environment stresses such as thermal shock, drought stress, waterlogging, low temperature, and also resistance to pests and diseases. A new breeding strategy is necessary and must be implemented for improving crops, making them easily adaptable to different conditions and stresses stemming from the adaptation of crops for earlier sowing dates at lower temperatures. In some cases, the development of winter varieties or creation of varieties better adapted for PS should be introduced into agricultural practice (Cutforth et al. 2007; Vadez et al. 2012).

The polyculture system (double cropping or relay cropping, main and cover crops) is a more effective method for increasing crop production and storage of carbon in the soil (Chataway et al. 2011; Rao et al. 2015). Crop diversities in rotations and polyculture cropping with a wide range of species and varieties also increase yield stability and reduce inter-annual and intra-season climatic variability.

Cropping systems based on biodiversity are an element of sustainable soil management with less (or no) use of chemical inputs: fertilizers and pesticides. More diverse cropping systems represent a new biological achievement for the creation of a future food production model. Tittonell (2015) said of this system that "agroecology is climate-smart." These systems, based on diversity principles, optimize the efficiency of resource use, resource recycling, synergies between crops, and factors based on natural regulation. The polyculture system is an imitation of the natural ecosystem. In such conditions, complementarity can be found between plants and the communities in almost every ecological niche.

Polyculture farming may be mostly implemented on those areas with a high potential for environmental degradation and biodiversity losses. This system, especially perennial PS, offers good protection for the soil year-round against erosion (water and wind) and other degradation factors (Dewar 2007). It should be assumed that sustainable crop management systems will gain in importance in the future in agriculture in general and not only in special conditions.

References

Abdul-Baki AA, Kotliński S, Kotlińska T (2002a) Vegetable production systems. Veg Crops Res Bull 57:11–21

Abdul-Baki AA, Teasdale JR, Goth RW, Haynes KG (2002b) Marketable yields of fresh-market tomatoes grown in plastic and hairy vetch mulches. Hort Sci 37(6):878–881

Adamczewska-Sowińska K (2004) Living mulches in tomato and pepper production and their residual effects on celeriac and carrot yields. Zesz Nauk AR we Wrocławiu 484, Rozprawy CCXIII (in Polish, abstract in English)

Adamczewska-Sowińska K, Kołota E (2002) Living mulches in field tomato production. Folia Hort 14(1):45–51

Adamczewska-Sowińska K, Kołota E, Winiarska S (2009) Living mulches in field cultivation of vegetables. Veg Crops Res Bull 70(1):19–29

Akanvou R, Kropff MJ, Bastiaans L, Becker M (2002) Evaluating the use of two contrasting legume species as relay intercrop in upland rice cropping systems. Field Crops Res 74:23–36

Altieri MA, Nicholls CI (2012) Agroecology scaling up for food sovereignty and resiliency. In: Lichtfouse E (eds) Sustain Agric Rev 11:1–29

Askegaard M, Eriksen J (2008) Residual effect and leaching of N and K in cropping systems with clover and ryegrass catch crops on a coarse sand. Agric Ecosyst Environ 123:99–108

Badri DV, Vivanco JM (2009) Regulation and function of root exudates. Plant Cell Environ 32:666–681

Bais HP, Weir TL, Perry LG, Gilroy S, Vivanco JM (2006) The role of root exudates in rhizosphere interactions with plants and other organisms. Annu Rev Plant Biol 57:233–266

Baldy C, Stigter CJ (1997) Agrometeorology of multiple cropping in warm climates. INRA, Paris

Balvanera P, Pfisterer AB, Buchmann N, He JS, Nakashizuka T, Raffaelli D, Schmid B (2006) Quantifying the evidence for biodiversity effects on ecosystem functioning and services. Ecol Lett 9:1146–1156

Bargaz A, Isaac ME, Jensen ES, Carlsson G (2015) Intercropping of faba bean with wheat under low water availability promotes faba bean nodulation and root growth in deeper soil layers, procedia. Environ Sci 29:111–112

Baumann DT, Kropff MJ, Bastiaans L (2000) Intercropping leeks to suppress weeds. Blackwell Sci. Ltd. Weed Res 40:359–374

Beets WC (1982) Multiple cropping and tropical farming system, grower. London, Britain, and West Views Press, Colorado, p 156

Belz RG (2007) Allelopathy in crop/weed interactions e an update. Pest Manag Sci 63:308–326

Ben-Hammouda M, Ghorbal H, Kremer R, Oueslati O (2001) Allelopathic effects of barley extracts on germination and seedlings growth of bread and durum wheats. Agronomie, EDP Sciences 21(1):65–71

Biedrzycki ML, Jilany TA, Dudley SA, Bais HP (2010) Root exudates mediate kin recognition in plants. Commun Integr Biol 3:28–35

Blanco-Canqui H, Lal R (2010) Cropping systems. In: Principles of soil conservation and management. Springer, Dordrecht, pp 165–192

Błażewicz–Woźniak M, Mitura R (2004) Wpływ uprawy konserwującej na zawartość składników mineralnych w glebie i korzeniach pietruszki. Rocz AR Poznań 356:3–11. (in polish)

Bonanomi G, Sicurezza MG, Caporaso S, Assunta E, Mazzoleni S (2006) Phytotoxicity dynamics of decaying plantmaterials. New Phytol 169:571–578

Boyd NS, Gordon R, Asiedu SK, Martin RC (2000) The effect of living mulches on tuber yield of potato (*Solanum tuberosum* L.). Biol Agric Hortic 18(3):203–220

Brainard DC, Bellinder RR (2004) Weed suppression in a broccoli-winter rye intercropping system. Weed Sci 52:281–290

Brooker RW (2006) Plant–plant interactions and environmental change. New Phytol 171:271–284

Brooker RW, Bennett AE, Cong W, Daniell TJ, George TS, Hallett PD, Hawes C, Iannetta PP, Jones HG, Karley AJ, Li L, McKenzie BM, Pakeman RJ, Paterson E, Schöb C, Shen J, Squire G, Watson CA, Zhang C, Zhang F, Zhang J, White PJ (2015) Improving intercropping: a synthesis of research in agronomy, plant physiology and ecology. New Phytol 206:107–117

Bruns HA (2012) Concepts in crop rotations. Agric Sci, Godwin Aflakpui (Ed.), InTechopen.com, available from: http://www.intechopen.com/books/agricultural-science/conceptsin-crop-rotation/ dated 03/05/2018

Büchi L, Gebhard CA, Liebisch F, Sinaj S, Ramseier H, Charles R (2015) Accumulation of biologically fixed nitrogen by legumes cultivated as cover crops in Switzerland. Plant Soil 393:163–175

Bullock DG (1992) Crop rotation. Crit Rev Plant Sci 11(4):309–326

Carof M, de Tourdonnet SP, Saulas P, Le Floch D, Roger-Estrade J (2007) Undersowing wheat with different living mulches in a no-till system. II. Competition for light and nitrogen. Agron Sustain Dev 27:357–365

Chalk PM, Peoples MB, Mcneill AM, Boddey RM, Unkovich MJ, Gardener MJ, Silva CF, Chen D (2014) Methodologies for estimating nitrogen transfer between legumes and companion species in agro-ecosystems: a review of 15N-enriched techniques. Soil Biol Biochem 73:10–21

Chapagain T, Riseman A (2014) Barley–pea intercropping: effects on land productivity, carbon and nitrogen transformations. Field Crop Res 166:18–25

Chapagain T, Riseman A (2015) Nitrogen and carbon transformations, water use efficiency and ecosystem productivity in monocultures and wheat-bean intercropping systems. Nutr Cycl Agroecosyst 101:107–121

Chataway RG, Cooper JE, Orr WN, Cowan RT (2011) The role of tillage, fertiliser and forage species in sustaining dairying based on crops in southern Queensland 2. Double-crop and summer sole-crop systems. Anim Prod Sci 51:904–919

Cookson WR, Murphy DV, Roper MM (2008) Characterizing the relationships between soil organic matter components and microbial function and composition along a tillage disturbance gradient. Soil Biol and Biochem 40:763–777

Corre-Hellou G, Fustec J, Crozat Y (2006) Interspecific competition for soil N and its interaction with N-2 fixation, leaf expansion and crop growth in pea-barley intercrops. Plant Soil 282:195–208

Cu STT, Hutson J, Schuller KA (2005) Mixed culture of wheat (*Triticum aestivum* L.) with white lupin (*Lupinus albus* L.) improves the growth and phosphorus nutrition of the wheat. Plant Soil 272:143–151

Cutforth HW, McGinn SM, McPhee KE, Miller PR (2007) Adaptation of pulse crops to the changing climate of the Northern Great Plains. Agron J 99:1684–1699

Czarnota MA, Paul RN, Weston LA, Duke SO (2003) Anatomy of sorgoleone secreting root hairs of Sorghum species. Int J Plant Sci 164(6):861–866

Damour G, Garnier E, Navas ML, Dorel M, Risède JM (2015) Using functional traits to assess the services provided by cover plants: a review of potentialities in banana cropping systems. Adv Agron 134:81–133

Daudin D, Sierra J (2008) Spatial and temporal variation of below-ground N transfer from a leguminous tree to an associated grass in an agroforestry system. Agric Ecosyst Environ 126:275–280

Debaeke P, Pellerin S, Scopel E (2017) Climate-smart cropping systems for temperate and tropical agriculture: mitigation, adaptation and trade-offs. Cah Agric 26 pp 12 www.cahiersagricultures.fr

Deguchi S, Shimazaki Y, Uozumi S, Tawaraya K, Kawamoto H, Tanaka O (2007) White clover living mulch increases the yield of silage corn via arbuscular mycorrhizal fungus colonization. Plant Soil 291:291–299

Deguchi S, Uozumi S, Touno E, Tawaraya K (2010) Potassium nutrient status of corn declined in white clover living mulch. Soil Sci Plant Nutr 56:848–852

Depuydt S (2014) Arguments for and against self and non-self root recognition in plants. Front Plant Sci 5:614

Dewar JA (2007) Perennial polyculture farming. Technical report, RAND Corporation occasional paper series. https://www.rand.org/content/dam/rand/pubs/occasional_papers/2007/RAND_OP179.pdf/ dated 30/06/2018

Doré T, Sene M, Pellissier F, Gallet C (2004) An agronomic view of allelopathic phenomena. Cah Agric 13:249–256

Evans J, Scott G, Lemerle D, Kaiser A, Orchard B, Murray GM, Armstrong EL (2003) Impact of legume 'break' crops on the yield and grain quality of wheat and relationship with soil mineral N and crop N content. Aust J Agric Res 54:777–788

Falik O, Reides P, Gersani M, Novoplansky A (2003) Self/non-self discrimination in roots. J Ecol 91:525–531

Farooq M, Jabran K, Cheema ZA, Wahidb A, Siddiquec K (2011) The role of allelopathy in agricultural pest management. Pest Manag Sci 67(5):493–506

Fustec J, Lesuffleur F, Mahieu S, Cliquet JB (2010) Nitrogen rhizodeposition of legumes. A review. Agron Sustain Dev 30:57–66

Garrity D, Dixon J, Boffa JM (2012) Understanding African farming systems: science and policy Implications. http://aciar.gov.au/aifsc/sites/default/files/images/understanding_african_farming_systems_report_for_aifsc_conference.pdf/ dated 03/05/2018

Ghafarbi SP, Hassannejad S, Lotfi R (2012) Allelopathic effects of wheat seed extracts on seed and seedling growth of eight selected weed species. Int J Agric Crop Sci 19:1452–1457

Głąb L, Sowiński J, Bough R, Dayan FE (2017) Allelopathic potential of Sorghum (*Sorghum bicolor* (L.) Moench) in weed control: a comprehensive review. Adv Agron 145:43–95

Gliessman SR (1986) Plant interactions in multiple cropping systens. In: Francis CA (ed) Mubiple crcpping systems. Macmillan Publishing Company, New York, pp 82–95

Graham PH, Vance CP (2003) Legumes: importance and constraints to greater use. Plant Physiol 131(3):872–877

Gylfadóttir T, Helgadóttir Á, Høgh-Jensen H (2007) Consequences of including adapted white clover in Northern European grassland: transfer and deposition of nitrogen. Plant Soil 297:93–104

Hart JP (2008) Evolving the three sisters: the changing histories of maize, bean, and squash in New York and the greater northeast. Current northeast Paleobotany II. New York state museum bulletin 512:87–99. http://www.nysm.nysed.gov/publications/bulletins/ dated 15/04/2018

Hartwig NL, Ammon HU (2002) Cover crops and living mulches. Weed Sci 50(6):688–699

He X, Xu M, Qiu GY, Zhou J (2009) Use of 15N stable isotope to quantify nitrogen transfer between mycorrhizal plants. J Plant Ecol 2:107–118

Hernanz JL, Sanchez-Giron V, Navarrete L (2009) Soil carbon sequestration and stratification in a cereal/leguminous crop rotation with three tillage systems in semiarid conditions. Agric Ecosyst Environ 133:114–122

Hiltbrunner J, Liedgens M (2008) Performance of winter wheat varieties in white clover living mulch. Biol Agric Hortic 26:85–101

Hiltbrunner J, Streit B, Liedgens M (2007) Are seeding densities an opportunity to increase grain yield of winter wheat in a living mulch of white clover? Field Crops Res 102:163–171

Hinsinger P, Betencourt E, Bernard L, Brauman A, Plassard C, Shen J, Tang X, Zhang F (2011) P for two, sharing a scarce resource: soil phosphorus acquisition in the rhizosphere of intercropped species. Plant Physiol 156:1078–1086

Hofmann RW, Lin W, Stilwell SA, Lucas RJ (2007) Comparison of drought resistance in strawberry clover and white clover. Proceed New Zealand Grassland Assoc 69:219–222

Ikerd J (2010) Industrialization of agriculture; consequences and challenges of sustainability. Nuffield Scholars Program 2010 Conference, Washington, DC, March 8, 2010. http://web.missouri.edu/ikerdj/papers/Nuffield%20-%20Industrial%20Agriculture.htm/ dated 30/06/2018

Jędrszczyk E, Poniedziałek M (2009) Influence of living mulches on selected soil properties and weed infestation in sweet corn cultivation. Zesz Probl Post Nauk Rol 539:265–272. (in Polish with English summary)

Jensen ES (1996) Barley uptake of N deposited in the rhizosphere of associated field pea. Soil Biol Biochem 28(2):159–168

Jensen ES, Peoples MB, Boddey RM, Gresshoff PM, Hauggaard-Nielsen H, Alves BJR, Morrison MJ (2011) Legumes for mitigation of climate change and the provision of feedstock for biofuels and biorefineries: a review. Agron Sustain Dev 32:329–364

Kankanen H, Eriksson C (2007) Effects of undersown crops on soil mineral N and grain yield of spring barley. Europ J Agron 27:25–34

Kirkegaard JA, Christen O, Krupinsky J, Layzell D (2008) Break crop benefits in temperate wheat production. Field Crops Res 107:185–195

Knörzer H, Graeff-Hönninger S, Guo B, Wang P, Claupein W (2009) The rediscovery of intercropping in China: a traditional cropping system for future Chinese agriculture – a review. In: Lichtfouse E (ed) Climate change, intercropping, pest control and beneficial microorganisms, sustainable agriculture reviews. Springer, Dordrecht, pp 13–44

Kołota E, Adamczewska-Sowińska K (2013) Living mulches in vegetable crops production: perspectives and limitations (a reviev). Acta Sci Pol Hortorum Cultus 12(6):127–142

Kumar V, Brainard DC, Bellinder RR (2009) Suppression of Powell amaranth (*Amaranthus powellii*) by buckwheat residues: role of allelopathy. Weed Sci 57(1):66–73

Latati M, Blavet D, Alkama N, Laoufi H, Drevon JJ, Gérard F, Pansu M, Ounane SM (2014) The intercropping cowpea-maize improves soil phosphorus availability and maize yields in an alkaline soil. Plant Soil 385:181–191

Latati M, Bargaz A, Belarbi B, Lazali M, Benlahrech S, Tellah S, Kaci G, Drevon JJ, Ounane SM (2016) The intercropping common bean with maize improves the rhizobial efficiency, resource use and grain yield under low phosphorus availability. Eur J Agron 72:80–90

Leary J, De Frank J (2000) Living mulches for organic farming system. Hort Technol 10(4):692–698

Lesuffleur F, Cliquet JB (2010) Characterization of root amino acid exudation in white clover (*Trifolium repens* L.). Plant Soil 333:191–201

Lesuffleur F, Paynel F, Bataillé MP, Le Deunff E, Cliquet JB (2007) Root amino acid exudation: measurement of high efflux rates of glycine and serine from six different plant species. Plant Soil 294:235–246

Lesuffleur F, Salon C, Jeudy C, Cliquet JB (2013) Use of a 15N2 labelling technique to estimate exudation by white clover and transfer to companion ryegrass of symbiotically fixed N. Plant Soil 369:187–197

Li W (2001) Agro-ecological farming systems in China. Man and the biosphere series, ed. by Jeffers JNR:26

Li L, Li SM, Sun JH, Zhou LL, Bao XG, Zhang HG, Zhang F (2007) Diversity enhances agricultural productivity via rhizosphere phosphorus facilitation on phosphorus deficient soils. PNAS 104(27):11192–11196

Li ZH, Wang Q, Ruan X, Pan CD, Jiang DA (2010) Phenolics and plant allelopathy. Molecules 15(12):8933–8952

Liang YL, Zhang CE, Guo DW (2002) Mulch types and their benefit in cropland ecosystems on the loess plateau in China. J Plant Nutr 25(5):945–955

Lithourgidis AS, Vlachostergios DN, Dordas CA, Damalas CA (2011) Dry matter yield, nitrogen content, and competition in pea-cereal intercropping systems. Eur J Agron 34:287–294

Liu TD, Song FB (2012) Maize photosynbthesis and microclimate within the canopies at grain-filling stage in response to narrow-wide row planting patterns. Photosynthetica 50:215–222

Lu Y, Watkins K, Teasdale JR, Abdul-Baki AA (2000) Cover crops in sustainable food production. Food Rev Int 16:121–157

Lucero DW, Grieu P, Guckert A (1999) Effects of water deficit and plant interaction on morphological growth parameters and yield of white clover (*Trifolium repens* L.) and ryegrass (*Lolium perenne* L.) mixtures. Eur J Agron 11:167–177

Mahapatra SC (2011) Study of grass-legume intercropping system in terms of competition indices and monetary advantage index under acid lateritic soil of India. AJEA 1(1):1–6

Meng L, Zhang A, Wang F, Han X, Wang D, Li S (2015) Arbuscular mycorrhizal fungi and rhizobium facilitate nitrogen uptake and transfer in soybean/maize intercropping system. Front Plant Sci 6:1–10

Moyer-Henry KA, Burton JW, Israel DW, Rufty TW (2006) Nitrogen transfer between plants: a ^{15}N natural abundance study with crop and weed species. Plant Soil 282:7–20

Nemecek T, von Richthofen J-S, Dubois G, Casta P, Charles R, Pahl H (2008) Environmental impacts of introducing grain legumes into European crop rotations. Eur J Agron 28:380–393

Nygren P, Leblanc HA (2015) Dinitrogen fixation by legume shade trees and direct transfer of fixed N to associated cacao in a tropical agroforestry system. Tree Physiol 35(2):134–147

Okigbo BN, Greenland DJ (1976) Intercropping Systems in Tropical Africa in multiple cropping. ASA Spec Publ 27:63–101

Orr CH, Leifert C, Cummings SP, Cooper JM (2012) Impacts of organic and conventional crop management on diversity and activity of free-living nitrogen fixing Bacteria and Total Bacteria are subsidiary to temporal effects. PLoS One 7(12):e52891

Paynel F, Lesuffleur F, Bigot J, Diquélou S, Cliquet JB (2008) A study of 15N transfer between legumes and grasses. Agron Sustain Dev 28:281–290

Pelosi C, Bertrand M, Roger-Estrade J (2009) Earthworm community in conventional, organic and direct seeding with living mulch cropping systems. Agron Sustain Dev 29:287–295

Peoples MB, Brockwell J, Herridge DF, Rochester IJ, Alves BJR, Urquiaga S, Boddey RM, Dakora FD, Bhattarai S, Maskey SL, Sampet C, Rerkasem B, Khan DF, Hauggaard-Nielsenm H, Jensen ES (2009) The contributions of nitrogen fixing crop legumes to the productivity of agricultural systems. Symbiosis 48:1–17

Pirhofer-Walzl K, Rasmussen J, Høgh-Jensen H, Eriksen J, Søegaard K, Rasmussen J (2012) Nitrogen transfer from forage legumes to nine neighbouring plants in a multi-species grassland. Plant Soil 350:71–84

Pleasant M (2006) The science behind the three sisters mound system: an agronomic assessment of an indigenous agricultural system in the Northeast. In: Staller JE, Tykot RH, Benz BF (eds) Histories of maize multidisciplinary approaches to the prehistory, linguistics, biogeography, domestication, and evolution of maize. Elsevier Academic Press, Amsterdam, p 672

Pleasant M (2016) Food yields and nutrient analyses of the three sisters: a Haudenosaunee cropping system. Ethnobiology Letters 7(1):87–98

Plucknett DL, Smith NJH (1986) Historical perspectives on multiple cropping. Multiple cropping systems (Francis C.A., ed.). Macmillan Publishing Company, pp 20–39

Porter JR, Xie L, Challinor V, Cochrane K, Howden SM, Iqbal MM, Lobell DB, Travasso MI (2014) Food security and food production systems. In: Climate change 2014: impacts, adaptation, and vulnerability. Cambridge University Press, Cambridge/New York, pp 485–533

Preissel S, Reckling M, Schläfke N, Zander P (2015) Magnitude and farm-economic value of grain legume pre-crop benefits in Europe: a review. Field Crop Res 175:64–79

Prithiviraj B, Paschke MW, Vivanco JM (2007) Root communication: the role of root exudates. Encycl Plant Crop Sci:1–4

Ramseier H. Legume screening for cover crops: weed suppression, biomass development and nitrogen fixation. https://www.hafl.bfh.ch/fileadmin/docs/Forschung_Dienstleistungen/Agrarwissenschaften/Pflanzen/Legume_screening_for_cover_crops.pdf/ dated 30/06/2018

Rangarajan A (2012) Crop rotation effects on soil fertility and plant nutrition. Chapter in. Production of Crop Rotation on Organic Farms: A Planning Manual was made possible with funding from Sustainable Agriculture Research and Education (SARE). https://www.sare.org/Learning-Center/Books/Crop-Rotation-on-Organic-Farms/Text-Version/Physical-and-Biological-Processes-In-Crop-Production/Crop-Rotation-Effects-on-Soil-Fertility-and-Plant-Nutrition/ dated 17/06/2018

Rao VN, Meinke H, Craufurd PQ, Parsons D, Kropff MJ, Anten NPR, Wani SP, Rego TJ (2015) Strategic double cropping on vertisols: a viable rainfed cropping option in the Indian SAT to increase productivity and reduce risk. Eur J Agron 62:26–37

Rasmussen J, Søegaard K, Pirhofer-Walzl K, Eriksen J (2012) N_2-fixation and residual N effect of four legume species and four companion grass species. Eur J Agron 36:66–74

Rasmussen J, Gylfadóttir T, Loges R, Eriksen J, Helgadóttir A (2013) Spatial and temporal variation in N transfer in grass-white clover mixtures at three Northern European field sites. Soil Biol Biochem 57:654–662

Sainju UM, Singh BP, Whitenhead WF (2002) Long term effects of tillage, cover crops and nitrogen fertilization on organic carbon and nitrogen concentrations in sandy loam soils in Georgia USA. Soil Til Res 63:167–179

Sarr B (2012) Present and future climate change in the semi-arid region of West Africa: a crucial input for practical adaptation in agriculture. Atmos Sci Lett 13:108–112

Schmidt O, Clements RO, Donaldson G (2003) Why do cereal-legume intercrops support large earthworm populations. Appl Soil Ecol 22(2):181–190

Semchenko M, Saar S, Lepik A (2014) Plant root exudates mediate neighbour recognition and trigger complex behavioral changes. New Phytol 204:631–637

Sobkowicz P, Podgórska-Lesiak M (2007) Experiments with crop mixtures: interactions, designs and interpretation. EJPAU 10(2). http://www.ejpau.media.pl/volume10/issue2/abs-22.html/

Soldevilla-Martinez M, Martin-Lammerding D, Tenorio JL, Walter I, Quemada M, Lizaso JI (2013) Simulating improved combinations tillage-rotation under dryland conditions. Span J Agric Res 11:820–832

Song YN, Zhang FS, Marschner P, Fan FL, Gao HM, Bao XG, Sun JH, Li L (2006) Effect of intercropping on crop yield and chemical and microbiological properties in rhizosphere of wheat (*Triticum aestivum* L.), maize (*Zea mays* L.), and faba bean (*Vicia faba* L.). Biol Fertil Soils 43:565–574

Song YN, Zhang FS, Marschner P, Fan FL, Gao HM, Bao XG, Sun JH, Li L (2007) Effect of intercropping on crop yield and chemical and microbiological properties in rhizosphere of wheat (*Triticum aestivum* L.), maize (*Zea mays* L.), and faba bean (*Vicia faba* L.). Biol Fertil Soils 43:565–574

Sowiński J (2004) The effects of tillage method and nitrogen rates on winter wheat harvested for silage and grain. Zesz Nauk AR. Rozprawy CCXVI 490

Sowiński J (2005) The influence of winter wheat and white clover bi-cropping system on white clover sward parameters. XX International Grassland Congress Dublin Ireland 26 June-1 July 2005, p 385

Sowiński J (2014) The effect of companion crops management on biological weed control in the seeding year of lucerne. Biol Agric Hort 30(2):97–108

Sowiński J, Wojciechowski W (2018) Nitrogen efficiency of winter wheat on different tillage methods for whole crops silage. Fresenius Environ Bull 27(1):230–235

Steenhoudt O, Vanderleyden J (2000) Azospirillum, a free-living nitrogen-fixing bacterium closely associated with grasses: genetic, biochemical and ecological aspects. FEMS Microbiol Rev 24:487–506

Sturite I, Henriksen TM, Breland TA (2007) Winter losses of nitrogen and phosphorus from Italian ryegrass, meadow fescue and white clover in a northern temperate climate. Agric Ecosyst Environ 120:280–290

Thilakarathna RMMS, Papadopoulos YA, Rodd AV (2012) Characterizing nitrogen transfer from red clover populations to companion bluegrass under field conditions. Can J Plant Sci 92:1163–1173

Thilakarathna MS, McElroy MS, Chapagain T, Papadopoulos YA, Raizada MN (2016) Belowground nitrogen transfer from legumes to non-legumes under managed herbaceous cropping systems. A review. Agron Sustain Dev 36:58. https://link.springer.com/article/10.1007%2Fs13593-016-0403-9/

Thomsen IK, Christensen BT (2004) Yields of wheat and soil carbon and nitrogen contents following long-term incorporation of barley straw and ryegrass catch crops. Soil Use Manag 20:432–438

Thorsted MD, Olesen JE, Weiner J (2006a) Mechanical control of clover improves nitrogen supply and growth of wheat in winter wheat/white clover intercropping. Eur J Agron 24:149–155

Thorsted MD, Olesen JE, Weiner J (2006b) Width of clover strips and wheat rows influence grain yield in winter wheat/white clover intercropping. Field Crops Res 95:280–290

Tilman D, Balzer C, Hill J, Beforta BL (2011) Global food demand and the sustainable intensification of agriculture. Proc Natl Acad Sci U S A 108(50):20260–20264

Tittonell P (2015) Agroecology is climate smart. In: climate smart agriculture 2015: global science conference 3, Montpellier (France), p 19

Tonitto C, David MB, Drinkwater LE (2006) Replacing bare fallows with cover crops in fertilizer-intensive cropping systems: a meta-analysis of crop yield and N dynamics. Agric Ecosyst Environ 112:58–72

Vadez V, Berger JD, Warkentin T, Asseng S, Ratnakumar P, Rao KPC, Gaur PM, Munier-Jolain N, Larmure A, Voisin AS, Sharma HC, Pande S, Sharma M, Krishnamurthy L, Zaman MA (2012) Adaptation of grain legumes to climate change: a review. Agron Sust Dev 32:31–44

Vance CP, Graham PH, Allan DL (2000) Biological nitrogen fixation: phosphorus – a critical future need? In: Pedrosa FO, Hungria M, Yates G, Newton WE (eds) Nitrogen fixation: from molecules to crop productivity, current plant science and biotechnology in agriculture. Springer, Dordrecht, pp 509–514

Vandermeer J (1989) The ecology of intercropping. Cambridge University Press, Cambridge, p 237

Vandermeer J, Van Noordwijk M, Anderson J, Ong C, Perfecto I (1998) Global change and multi-species ecosystems: concepts and issues. Agric Ecosyst Environ 67:1–22

Waha K, Müller C, Bondeau A, Dietrich JP, Kurukulasuriya P, Heinke J, Lotze-Campen H (2013) Adaptation to climate change through the choice of cropping system and sowing date in sub-Saharan Africa. Glob Environ Chang 23:130–143

Wichern F, Eberhardt E, Mayer J, Joergensen RG, Müller T (2008) Nitrogen rhizodeposition in agricultural crops: methods, estimates and future prospects. Soil Biol Biochem 40:30–48

Wojciechowski W, Adamczewska-Sowińska K, Krygier M (2012) Effect of living mulches on selected soil structure indicators in eggplant cultivation. Veg Crops Res Bull 77:49–59

Xiao YB, Li L, Zhang FS (2004) Effect of root contact on interspecific competition and N transfer between wheat and fabacean using direct and indirect N-15 techniques. Plant Soil 262:45–54

Xu Z, Ma G, Shah RP, Qin FF (2008) Japanese organic tomato intercropped with living turf-grass mulch. Cultivating the future based on science. Volume 1: Organic Crop Production. Proceedings of the Second Scientific Conference of the International Society of Organic Agriculture Research (ISOFAR). Modena, Italy, 18–20 June 2008, pp 619–623

Xu Z, Yu Z, Zhao J (2013) Theory and application for the promotion of wheat production in China: past, present and future. J Sci Food Agric 93:2339–2350

Yang F, Huang S, Gao RC, Liu WG, Yong TW, Wang XC, Wu XL, Yang WY (2014) Growth of soybean seedlings in relay strip intercropping systems in relation to light quantity and red far-red ratio. Field Crops Res 155:245–253

Yong T, Liu X, Yang F Song C, Wang X, Liu W, Su B, Zhou L, Yang W (2015) Characteristics of nitrogen uptake, use and transfer in a wheat-maize-soybean relay intercropping system. Plant Prod Sci 18:388–397

Zak DR, Holmes WE, White DC, Peacock AD, Tilman D (2003) Plant diversity, soil microbial communities and ecosystem function: are there any links? Ecology 84:2042–2050

Free Lipid Biomarkers in Anthropogenic Soils

9

Irena Atanassova, Harizanova Milena, and Martin Banov

Abstract

Free lipid biomarkers are sensitive indicators of the extent of climatic and anthropogenic disturbances in soils, in contrast to "bound lipids" that are tightly incorporated in the soil organic matrix and may be fixed for long periods of time. The studies reported in this chapter describe signature free lipid biomarkers in anthropogenic soils and their role as indicators of the degree of pedogenesis and degradation processes, e.g., technogenic pollution and water repellency. Soils separated from different horizons of a Technosol on which sewage sludge from paint and print industry was deposited indicated small quantitative changes in the major compound classes. Free lipids show similarities in the free lipid signature and predominant microbial sources, as indicated by the presence of even number and branched alkanes (C_{16}–C_{33}), short-chain (C_6–C_{18}) fatty acids, and C_{11}–C_{32} fatty alcohol distributions. No xenobiotics at measurable quantities were detected, except some metabolites.

Lipid compounds from a major coal mine area in Southeastern Europe, i.e., alkanes, fatty acids and fatty alcohols, as well as coal biomarkers (phyllocladane), were more abundant in the coal ash-amended water-repellent Technosols. The alkane distribution was monomodal, maximizing at C_{29}, while fatty acids maximized at C_{26}. These compounds indicate the ongoing pedogenesis in the long-time reclaimed spoils-turned soils. Statistical analysis reveals that total organic carbon (TOC) and the lipid fraction (the long-chain > C_{22} fatty acids, alcohols, and alkanes) were the drivers of soil water repellency in the studied water-

I. Atanassova (✉) · H. Milena
Department of Agrochemistry, Agroecology and Farming Systems, Institute of Soil Science, Agrotechnologies and Plant Protection, Sofia, Bulgaria
e-mail: i.d.atanassova@abv.bg

M. Banov
Department of Genesis, Geography and Soil Classification, Institute of Soil Science, Agrotechnologies and Plant Protection, Sofia, Bulgaria
e-mail: banovm@abv.bg

R. S. Meena (ed.), *Soil Health Restoration and Management*,
https://doi.org/10.1007/978-981-13-8570-4_9

repellent Technosols. In the non-water-repellent Technosols, lacking coal ash amendment, lower concentrations of the alkanes (< 2 μg/g) including the long-chain ones (C_{29-33}) and the presence of < C_{24} branched alkanes were recorded. The alkane distribution points at two origins for the alkanes, i.e., higher plants and microbial sources. The analysis of free lipid extracts allowed for the assessment of sensitive molecular indicators of biogenic terrigenous and coal origin and the degree of soil restoration following >40 years of post reclamation period.

The surface horizons of an Anthrosol amended with 3 t/ha biochar contained alkanes <C_{24} with prevailing even over odd predominance (EOP) of homologues and carbon preference index (CPI) of 10.6. The free lipid signature indicates anthropogenic sources and/or predominant microbial contribution to soil organic matter; however, thermally disrupted (from biochar) and microbial alkanes were difficult to discriminate.

Keywords
Free lipid biomarkers · Anthropogenic soil · Technosol · Sewage sludge · Coal · Biochar

Abbreviations

ACL	Average chain length
CEC	Cation exchange capacity
EC	Electrical conductivity
EOP	Even over odd predominance
FOC	Fulvic organic carbon
GC/MS	Gas chromatography-mass spectrometry
HOC	Humic organic carbon
LAS	Linear alkylbenzene sulfonates
MF	Match factor
MN	Mineral nitrogen
PAHs	Polycyclic aromatic hydrocarbons
PBDEs	Polybrominated diphenylethers
PCBs	Polychlorinated biphenyls
PCDDs	Polychlorinated dibenzo-*p*-dioxins
PCDFs	Polychlorinated dibenzofurans
PNEC	Predicted non-effect concentrations
POPs	Persistent organic pollutants
RF	Response factor
SOC	Soil organic carbon
SOM	Soil organic matter
TMAH	Tetramethyl ammonium hydroxide
TOC	Total organic carbon

WDPT	Water drop penetration time
WWTP	Wastewater treatment plant
CPI	Carbon preference index

9.1 Free Lipid Biomarkers and Their Environmental Significance

Studies of organic molecular markers and their biological precursors comprise a major field in contemporary organic geochemistry. Qualitative indicators, characterizing organic matter, are based on identifying homologous groups of organic compounds or individual compounds which carry important geochemical information. These compounds are known as organic molecular markers, biomarkers, chemofossils, etc. Their high informativeness is the reason for the growing interest in organic geochemical studies during the past >30 years.

The term "free lipids" refers to "labile lipids," which are differentiated from the terms "residual lipids" or "bound residues" operationally used by geochemists to distinguish between the compounds extracted by an organic solvent (or a mixture of solvents) which does not affect the structure of extracted compound or the structure of the sorbent unextractable pool of the compound and those extractants which affect the chemical structures mentioned above. The major methods for chemical degradation used in organic geochemistry are alkaline or acid hydrolysis, oxidation, and thermochemolysis with tetramethyl ammonium hydroxide (TMAH) applied for depolimerization of the polymolecular matrix of soil organic matter (SOM) (Kogel-Knabner 2000; 2002), while the resulting compounds are usually analyzed by gas chromatography-mass spectrometry (GC/MS). Formation of "bound residues" of hydrophobic organic pollutants in soils is also known as "sequestration," especially in situations of deposition of organic wastes from wastewater treatment plants (WWTP).

Lipid biomarkers are sensitive indicators of ecosystems and are dependent on specific sources of soil organic matter (Atanassova 2017; Meena and Meena 2017). Through their analysis, valuable information and more profound understanding of geochemical sources in a complex multifunctional ecosystem such as soil can be attained (Eganhouse 2004). Some of these anthropogenic markers are known as persistent organic pollutants (POPs). Their input in soil takes place through deliberate anthropogenic actions, e.g., in sludge from wastewater treatment and its deposition on land, application of organic fertilizers and ameliorants, in oil spill occasions or pipeline leakages, soil reclamation activities in coal mine regions, in plant protection (pesticides and herbicides), atmospheric fallout, etc. Although organic pollutants can be degraded, a full soil cleanup is rarely achieved, because of the formation of "bound residues" which influence POPs distribution, bioavailability, and toxicity to living organisms. Some molecular markers are biosynthesized, while others have accidental or deliberate source. Polycyclic aromatic hydrocarbons (PAHs) can have anthropogenic or biopedogenic origin (Atanassova and Brümmer 2004; Ashoka et al. 2017), while polychlorinated byphenyls have anthropogenic sources only. Lipid biomarkers are indicative of the extent of degradation of SOM and are diagnostic of the endogenic, exogenic, and soil-forming processes (Ambles et al. 1993; Bull et al. 2000). Carbon preference index (CPI) of n-alkanes is a widely used index suggested by Bray and Evans (1961) and reveals the level of catagenesis (maturity) of SOM and progressively decreases with SOM maturation, reaching the value of 1. In sapropelic organic carbon, *n*-alkanes with <20 C atoms in the chain, maximizing at C_{17}, predominate, while humic organic matter is characterized with high concentrations of *n*-alkanes in the range of n-$C_{23} \div n$-C_{35}, maximizing at C_{31} (Lichtfouse et al. 1998).

Soil organic matter (SOM) is composed mainly of plant residues, organic degradation products, microbial biomass, and humic matter (Zech et al. 1997). Content and composition of free lipids in SOM depend on the type of plant cover and the physicochemical soil properties (Dinel et al. 1990; van Bergen et al. 1997). It's considered that neutral lipids can be regarded as "molecular proxies," medium-term biomarkers in soil, or long-term biomarkers in aquatic sediments (Kuzyakov et al. 2014; Yadav et al. 2018b).

The free lipid fraction of SOM can be isolated directly from soils through nonpolar solvent (hexane, dichloromethane, iso-propanol, etc.) extraction and comprises 2–50 $g.kg^{-1}$of SOM (Stevenson 1994). Lipid compounds in soil originate from degradation products of plant biomass, shoot and root cuticular tissue, and microorganisms (Bull et al. 2000; Naafs et al. 2004; Otto et al. 2005; Wiesenberg et al. 2010a, b) and include fatty acids, sterols, terpenoids, long-chain homologous series of hydrocarbons, fats, waxes, resins, and more complex compounds (Stevenson 1994; Jandi et al. 2002; Jansen et al. 2006; Buragohain et al. 2017). Lipids are very important in retarding the mineralization of SOM (Ambles et al. 1993) and have diagnostic properties with respect to SOM and pedogenesis (Bull et al. 2000).

"Molecular fingerprint" or signature of lipid extracts, e.g., C-chain length; the domineering homologue; and the distribution of compound chains with even and odd number of C atoms are used as a measure of origin of lipids in soil. Predominance of odd over even number of C atoms in hydrocarbons and higher values of C_{odd}/C_{even} ratios (high CPI) and distribution in the C_{25}–C_{33} range is owing to plant-introduced epicuticular waxes (Kolattukudy et al. 1976), while *n*-alkane products of microbial activity are characterized by similar distribution of even and odd homologues and/or distribution in the lower carbon numbers (Stevenson 1994). Short-chain homologues with average chain length ($< C_{25}$) originate from multiple sources (Kuhn et al. 2010), microbial biomass (Kolattukudy et al. 1976), or degradation products from biomass burning (Eckmeier and Wiesenberg 2009; Atanassova and Teoharov 2010a, b; Atanassova and Doerr 2011; Kumar et al. 2017b; Ram and Meena 2014). Short-chain alkanes have been identified in soils irrigated with wastewater (Pascual et al.1997), and their occurrence has been associated with microbially produced lipids (Marseillea et al.1999; Parlanti et al.1994). Other indices such as "average chain length ACL" (Simoneit et al. 1991; Duan 2000) are also being used as indicators of source and evolution of soil organic matter (Cranwell 1973; Duan et al. 1998;

Pancost and Boot 2004) or paleoenvironment (Ohkouchi et al. 1997; Pancost and Boot 2004; Simoneit et al. 1991). Other marker ratios are CPI of fatty acids and fatty alcohols. For fatty alcohols, the predominant ratios of C_{even}/C_{odd} homologues in the long-chain >C_{20} range indicate a plant source (Jansen et al. 2006), while lower ratios and prevalence in the smaller C range indicate the increasing role of microbial source. CPI of fatty acids in the C_4–C_{26} range, as well as branched homologues, determines a bacterial contribution, while predominance in the C_{26}–C_{38} indicates origin from higher plants (Ambles et al. 1993; Boeschker and Middelburg 2002; Kolattukudy et al. 1976; Naafs et al. 2004; Schnitzer et al. 1986; Zelles 1997). The absence or presence of few functional groups contributes to the higher stability of neutral lipids and their accumulation in soil as compared with other compounds (Koegel-Knabner et al. 2005; Meena and Yadav 2015).

9.2 Free Lipid Biomarkers in Technosols from Industrial Sludge Deposition

Sewage sludge from municipal, industrial, or mixed wastewater treatment of soils has been considered to improve soil fertility because of enhancing organic matter and macronutrient contents. However, effective use or disposal of these organic wastes for agricultural production has to take account of both risks and benefits. European policy on the quality of wastewater discharged and the sewage sludge produced requires stringent measures on the disposal to land using either land spreading at dedicated sites or other sludge-to-landfill operations, e.g., at municipal landfills. One of the most abundant organic compounds of sewage sludge is various steroids, such as cholesterol, β-sitosterol, stigmasterol, campesterol, and 5β-stanols (Ibañez et al. 2000; Meena et al. 2014). Other organic pollutants such as polychlorinated biphenyls (PCBs), polychlorinated dibenzo-*p*-dioxins (PCDDs), polychlorinated dibenzofurans (PCDFs), polychlorinated biphenyls (PCBs), polybrominated diphenylethers (PBDEs), linear alkylbenzene sulfonates (LAS), polycyclic aromatic hydrocarbons (PAH), and phthalate esters were also found in higher concentrations in sewage sludge-treated soils (McLachlan et al. 1978; Folch et al. 1996; Matscheko et al. 2002; Patureau et al. 2007; Dadhich and Meena 2014).

On the other hand, wastewater from textile industries is of complex and variable nature. Dyes have been identified as the most problematic contaminants (Carliell et al. 1994). Azo dyes are most frequently used for dyeing textiles, and ~ 10% of those used during the dying processes do not attach to fibers, which leads to release into the sewage treatment systems and therefore to surface and groundwater and land (Clarke and Anliker 1980; Reisch 1996).

In the study on free lipids extracted from soil profiles under spread-to-land sewage sludge from Sofia wastewater treatment plant (WWTP), lipid sources showed predominant bacterial contribution as shown by the alkane, fatty acids, and *n*-alkanol patterns. Some aromatics and xenobiotics (phthalates) were detected in the lipid extracts of the clay fractions in depth (0–85 cm) of the soil profile, implying possible leaching. Réveillé et al. (2003) found that carbon is more extractable in matured sludge and the most prevailing compounds in the least matured sludge are fatty acids, whereas steroids dominate the extracts from the most matured one.

9.2.1 Sites, Materials, and Methods

9.2.1.1 Soil Samples and Study Site

The study site was chosen in the area of Elin Pelin paint and print textile factory of Miroglio (Bulgaria) where sludge from the wastewater treatment plant (WWTP) was deposited in 2006 and which announced shutdown in 2009. The sludge was of miscellaneous nature and dark colored and mainly consisted of waste products from the painting manufacture with minor contribution from municipal wastewater. The main parameters analyzed, i.e., heavy metals and organic pollutants (xenobiotics), were well below the PNEC (predicted non-effect concentrations; Langencamp and Part 2001). Below the sludge layer, there were technogenically affected horizons of 0–86 cm depth with artifacts. Three soil horizons Bu_2 and Bgk with technogenic influence and artifacts were studied, as well as the lowest gleyic G horizon (Table 9.1, Fig. 9.1), where changes in the free lipids were encountered. In the other horizons, lipid signatures and contents did not show changes in signature or contents.

9.2.1.2 Lipids Analysis

Soil samples (10 g in duplicate, sieved <2 mm, and ground <250 μm) from the surface horizons were extracted with 50 ml dichloromethane (DCM) in a Tecator Soxtec extractor at 100 °C for 2 h. The solution from the extraction process was evaporated to ~ 6 ml,40 μl 0,06 μg. Hence, μl^{-1}2-nonadecanone was added as internal standard, filtered using a Pasteur pipette packed with 2 cm silica gel (63–200 μm) and 0.5 cm Na_2SO_4(s) as a drying agent to remove very polar constituents. Then extracts were dried and derivatized to their corresponding trimethylsilyl (TMS) ethers and esters by adding 50 μL of BSTFA (N,O-bis(trimethylsilyl)trifluoroacetamide) containing 1% TMCS (trimethylchlorosilane).

9.2.1.3 Gas Chromatography-Mass Spectrometry (GC/MS)

Gas chromatograph Agilent 7890A with a 5975 C mass-selective detector (splitless injection mode) was used for analysis of the lipid extracts (Atanassova and Teoharov 2010a, b). Separation was performed on HP-5 ms capillary column (30 m × 0.25 mm

Table 9.1 Major properties of selected soil horizons of a Technosol from the vicinity of industrial WWTP Miroglio, Elin Pelin

		pH	CEC $pH_{8,2}$	ExCa	Ex Mg	EC mS/cm	TOC %
Soil horizons	Sludge layer 0–14	7.0	59.3	54.4	2.7	0.780	27.84
	Bu_2 53–93	8.5	22.8	22.0	0.8	0.269	1.42
	Bgk 123–143	7.6	24.6	22.1	2.5	0.322	0.56
	G 168–200	7.2	25.0	22.0	2.5	0.343	1.22

Fig. 9.1 Morphological composition of the soil profile at Miroglio Plant near Elin Pelin *(Anthropogenic-Technogenic Meadow Cinnamonic, bogged (Urbic Mollic Technosol Calcaric Siltic))*, WRBSR (2006)

I.D., film thickness, 0.25 μm), and He was used as a carrier gas. The GC program was the following: initial T 60 °C, hold 1 min, linear ramp 10 °C/min to 180 °C, ramped at 4°/min to 300 °C, hold 15 min, MS detection full scan, mass to charge ratio (m/z) 50–1000, cycle time 2.28 scans/s, and EI ionization 70 eV. Identification was based on comparison of the mass spectra of chromatographic peaks to NIST-MS library, authentic standards, GC retention times, and additional interpretation of mass fragmentation patterns. Compounds which could not be identified with a match factor > 80 have been termed as "suspected."

9.2.1.4 Quantification

For quantitative analysis, the GC/MS was calibrated with pure reference compounds, i.e., *n*-alkanes (C_8–C_{40}). The fragment ions used for quantification were *m/z* 57 for the n-alkanes, *m/z* 75 for the n-alcohols, and *m/z* 117 for the n-fatty acids. The peak areas for each component were compared to the peak area of the fragment ion (*m/z*58) of the 2-nonadecanone standard on assuming response factor RF = 1.

The objectives of the present study were to (i) characterize the molecular composition of the free lipid fraction of soil organic matter in a Technosol under industrial sewage sludge deposition and provide information on the origin, pedogenesis, and organic contamination and (ii) determine the changes in the relative distribution of major compound classes of organic geochemical markers along the depth of the soil profile.

9.2.2 Variation of Lipid Biomarkers and Pollutants in Selected Horizons of Industrial Sludge-Treated Soils

9.2.2.1 Sewage Sludge Layer

The mass spectra corresponding to the depicted peaks of the compounds were very similar. Common fragment ions were detected pointing at the similarity of the compound structures and metabolites formed. Some characteristic compounds for municipal sewage sludge were detected such as astaxanthin, which is a carotenoid, lipid-soluble pigment (match factor MF 40); 3-acetoxy 7,8-epoxylanostan-11-ol (detected in oil spills), MF 62 (Fig. 9.2); polycyclic aromatic hydrocarbon (PAH) derivative 17-(1,5-dimethylhexyl)-10,13-dimethyl-3-styryl-hexadecahydrocyclopenta [a]phenanthren-2-one (Fig. 9.3); 5.beta.-cholest-23-ene, (Z)-; and androstane (5.beta.)-. These compounds were not detected in the soil layers below the sludge.

9.2.2.2 Below-Sludge Soil Horizons

There were significant changes in the relative intensities of organic compound classes such as *n*-alkanes, fatty acids, and alkanols along the soil profile. Major qualitative changes in the lipid signatures were observed as well (Figs. 9.4, 9.5 and 9.6).

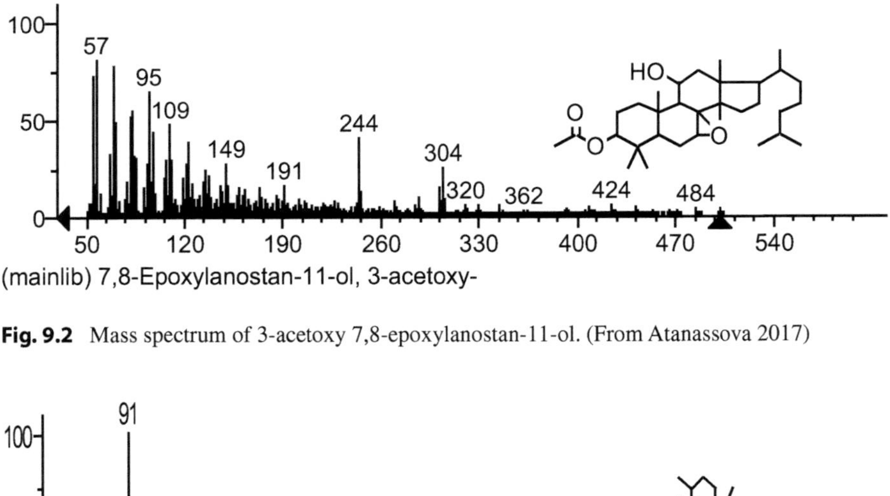

Fig. 9.2 Mass spectrum of 3-acetoxy 7,8-epoxylanostan-11-ol. (From Atanassova 2017)

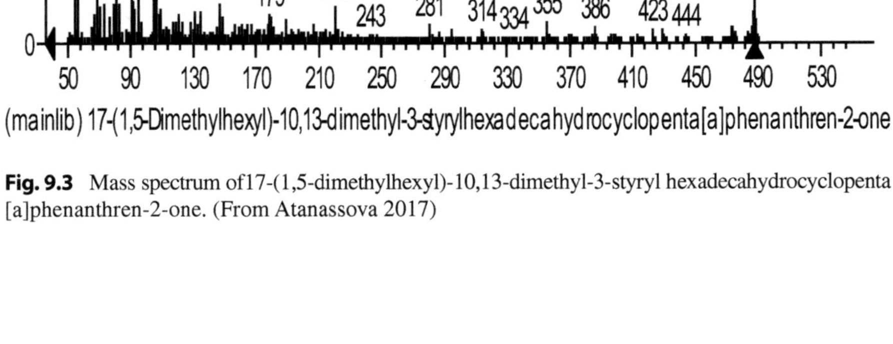

Fig. 9.3 Mass spectrum of17-(1,5-dimethylhexyl)-10,13-dimethyl-3-styryl hexadecahydrocyclopenta [a]phenanthren-2-one. (From Atanassova 2017)

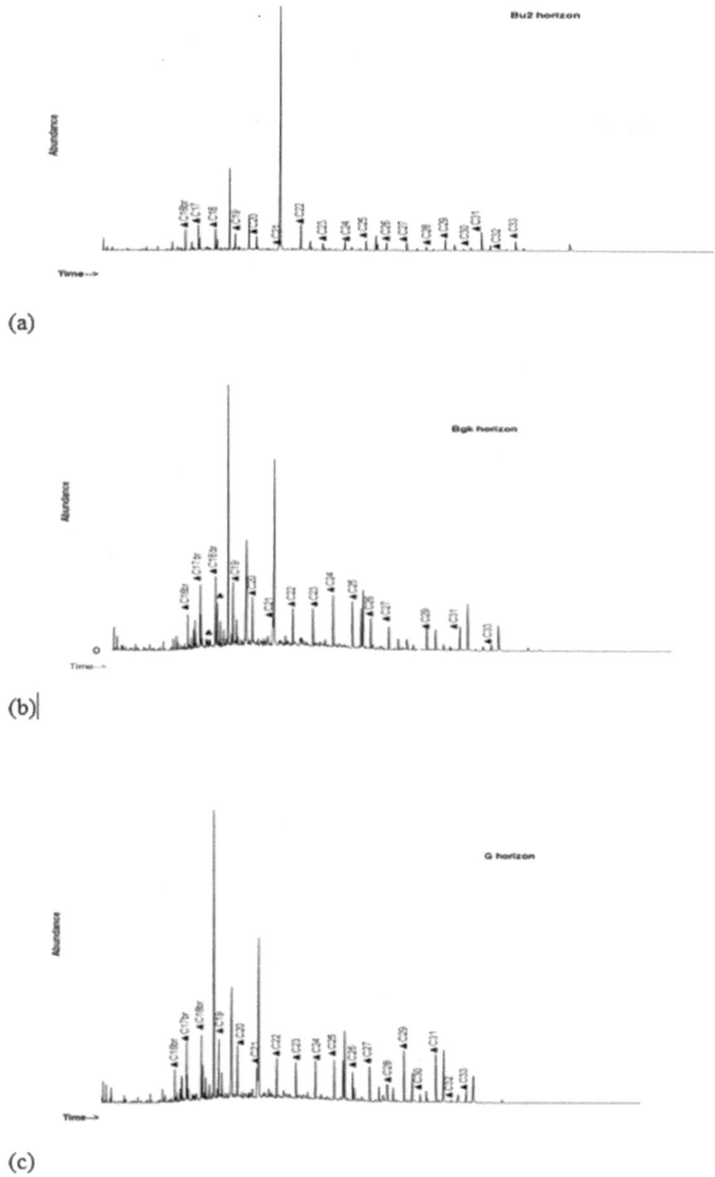

Fig. 9.4 (**a**) m/z 57 ion chromatogram showing the relative distribution of alkanes in Bu_2 53–93 cm; (**b**) m/z 57 ion chromatogram showing the relative distribution of alkanes in Bgk 123–143 cm;(**c**) m/z 57 ion chromatogram showing the relative distribution of alkanes in G 168–200 cm

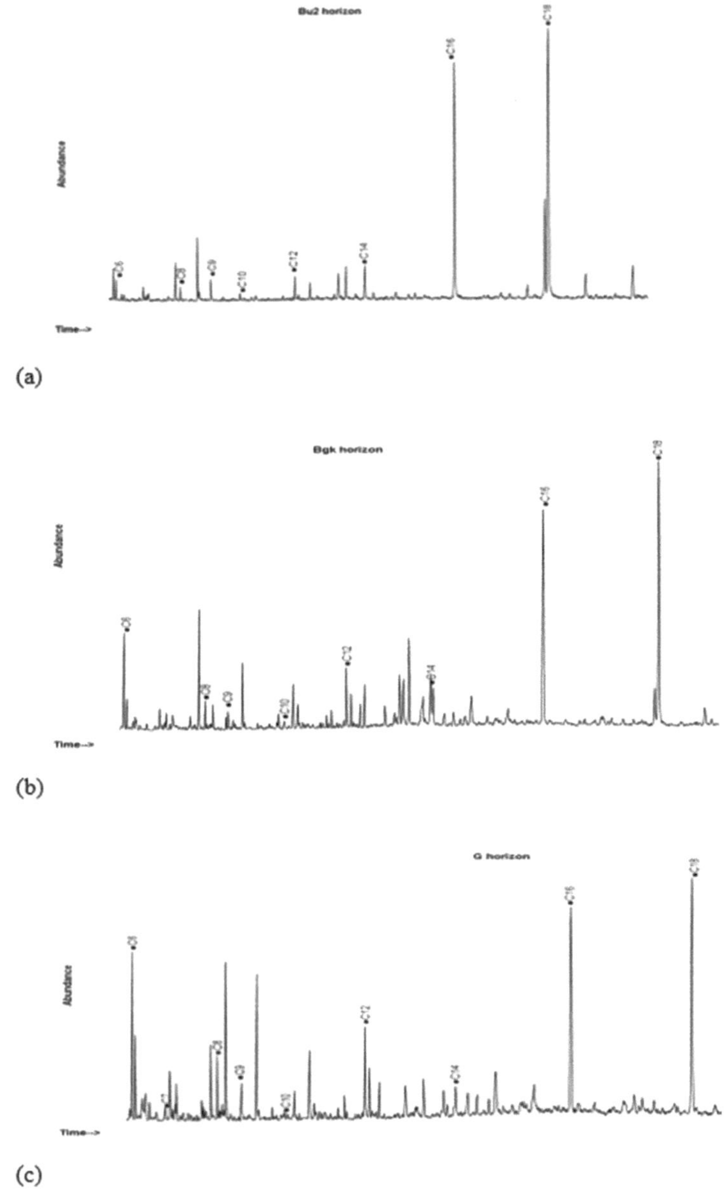

Fig. 9.5 (**a**) m/z 117 ion chromatogram showing the relative distribution of n-alkanoic acids in Bu_2 53–93 cm; (**b**) m/z 117 ion chromatogram showing the relative distribution of n-alkanoic acids in Bgk 123–143 cm; (**c**) m/z 117 ion chromatogram showing the relative distribution of n-alkanoic acids in G 168–200 cm

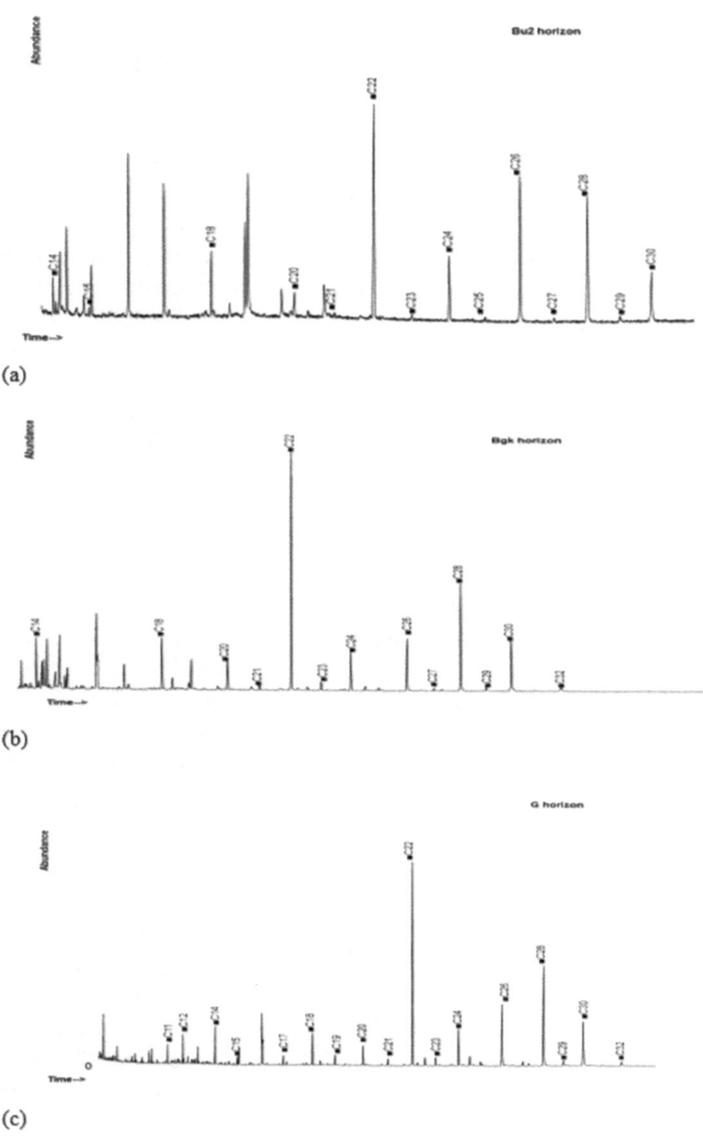

Fig. 9.6 (**a**) m/z 75 ion chromatogram showing the relative distribution of *n*-alkanols in Bu_2 53–93 cm; (**b**) m/z 75 ion chromatogram showing the relative distribution of *n*-alkanols in Bgk 123–143 cm; (**c**) m/z 75 ion chromatogram showing the relative distribution of *n*-alkanols in G 168–200 cm

9.2.2.3 Alkanes

It's widely accepted that n-alkanes with C chain length > 22 are considered as typical components of epicuticular waxes of higher plants (van Bergen et al. 1997). Upon analyzing the characteristic distribution pattern and carbon preference index (CPI), a measure of odd/even carbon preference of the alkanes (Tissot and Welte 1984), a clear distinction between anthropogenic and natural sources of hydrocarbons in the analyzed depths is possible. *n*-Alkanes are useful indicators in order to locate and quantify anthropogenic activities in the ecosystem. The *n*-alkanes show a C range from C_{16} to C_{33}. Also, peaks corresponding to complex mixtures of isomers of branched C_{16}, C_{17}, and C_{18} alkanes including *iso*- and *anteiso*- homologues were detected (Fig. 9.4a–c presents a relative distribution). These concentrations point at high microbial contribution to the *n*-alkanes CPI = 1.14 (Bu_2), 1.19 (Bgk), and 1.5 (G). A bimodal distribution was observed, with shorter chain C_{16}–C_{24} compounds dominating in Bgk (123–143 cm) and G (168–200 cm) horizon. When the alkane concentrations normalized to unit mass of soil organic carbon (SOC) were compared (Fig. 9.5), a predominance of short-chain alkanes (C < 23) was found in the soil from the lower-depth horizon Bgk, while in the lowest G horizon, higher-chain length (C_{29} and C_{31}) homologues were also extracted. This was also supported by calculation of the average chain length (ACL), ACL = $\sum(z_{n\,x}\ n)/\sum\ (z_n)$, with z_n = the relative amount of the n-alkane with n carbons, n = 16–33 (Eckmeier and Wiesenberg 2009).

ACL = 23.9, 22.6, and 23.8 for Bu_2, Bgk, and G horizons, respectively.

Short-chain compounds are known to be more easily biodegraded than long-chain homologues (Moucawi et al. 1981). The distribution of alkanes indicates a better preservation of microbial alkanes in the soil horizons analyzed. Since alkanes of C chain length > 22 are considered to originate from plant waxes (van Bergen et al. 1997), we assume dual origin, i.e., from plants and microorganisms. However, the presence of even long-chain *n*-alkanes with C > 22 in the soils from Bgk and G horizons and the weak dominance of the odd chain components also implies microbial origin (Ambles et al. 1989). The alkaline pH and the hydromorphic conditions of the bogged Technosol may cause intense biological activity through biotic and abiotic reduction of fatty acid precursors. On the one hand, the accumulation of long-chain (> C_{22}) alkanes can be derived from the soil surface through leaching from the top layers and, on the other, through metabolism of inherited alkanes and formation of new compounds of the same class. Anaerobic and redox processes may have occurred with different intensity at the different horizons of the soil profile affecting SOM and lipid composition. Different sorption in the SOM of the three soils is also possible, which will affect the alkane distribution in the free lipid fraction as expected (Fig. 9.4a–c and Fig. 9.7).

9.2.2.4 Fatty Acids

Alkanoic acids in the range C_6–C_{18} were detected in the three soils from Bu_2, Bgk, and G horizons. Fatty acid signature is characterized by even carbon number predominance and unimodal distribution maximizing at C_{16} palmitic acid (Fig. 9.5a–c). These acids are generally considered as bacterial markers (Lichtfouse et al. 1995;

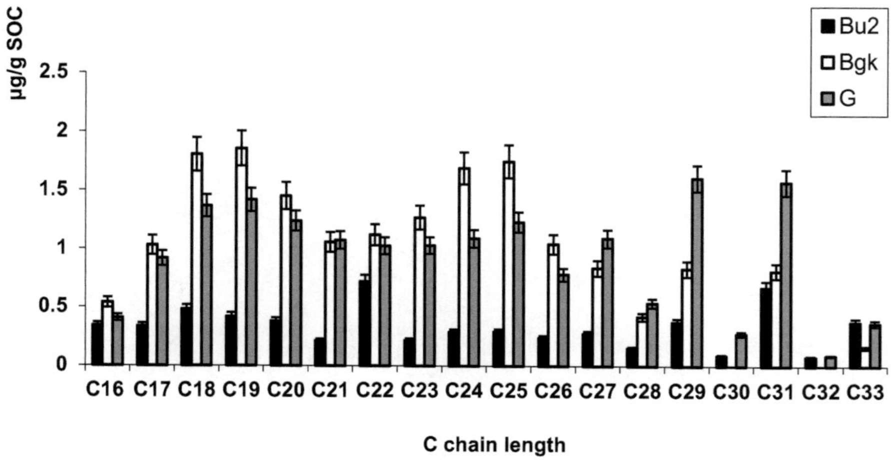

Fig. 9.7 Quantitative distribution of alkanes in three soil horizons of a Technogenic soil under industrial sludge deposition

van Bergen et al. 1997; Quénéa et al. 2004; Zelles 1999), and their high abundances point to contribution of bacterial acids in the free lipid fraction in the soils from the respective horizons. McElderry et al. (2005) found that short-chain fatty acids can be produced in anaerobic conditions by fermentative soil microbes, while long-chain acids (C > 22) have been detected in soil extracts originating from higher plant leaf waxes (van Bergen et al. 1997; Kolattukudy et al. 1976; Naafs et al. 2004; Datta et al. 2017a). The acid distributions and concentrations in the three horizons indicate that the origin of the extracted fatty acids is primarily microbes, because of the absence of the C > 18 homologues. Short-chain fatty acids (C_6–C_{10}) of potentially high phytotoxicity have been also isolated from a heathland soil (Jalal and Read 1983).

9.2.2.5 Alkanols

In the three soil depths, a series of even carbon-numbered alkanols C_{14}–C_{32} with a maximum at C_{22} and C_{26} (C_{28}) and C_{11}–C_{32} in the lowest G horizon following a bimodal distribution were observed (Figs. 9.6a–c and 9.8). These compounds are most likely derived from fungi, spore waxes, or hydrolysis products of esters through microbial activity (Naafs et al. 2004). Another pathway is through alkane oxidation (Ambles et al. 1989). Fatty acid reduction may also produce *n*-alkanols in the shorter C < 20 length range (Quénéa et al. 2006). Long-chain alkanols (C > 20) with even carbon numbers normally originate from higher plant leaf waxes (Rieley et al. 1991). It is considered that the identification of *n*-alkanols in the C_{12}–C_{30} range, with dominant <C_{22}, members together with odd C_{15}, $C_{17,}$ $C_{19,}$ and C_{21} alkanols may implicate bacterial input (Grasset and Ambles 1998; Naafs and van Bergen 2002; Meena et al. 2017c). We can speculate that the distribution observed in the studied soils indicates a contribution from higher plants as shown by the higher abundance of longer-chain alkanols (C > 20). However, the presence of *n*-alkanols with chain

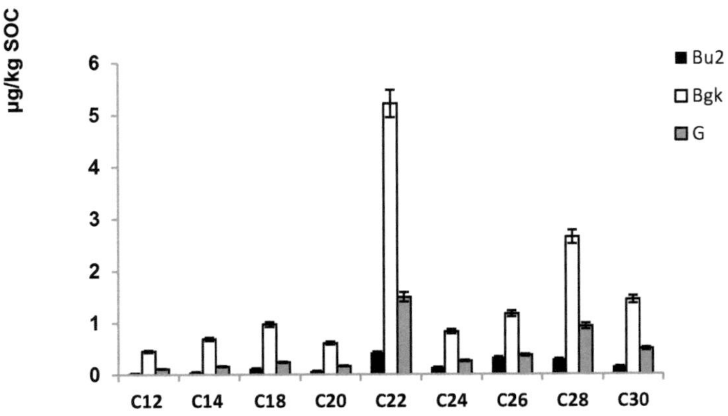

Fig. 9.8 Quantitative distribution of *n*-alkanols in three soil horizons of a Technogenic soil under industrial sludge deposition

length < C_{18} implicates sewage sludge origin for these alkanols as found by Mudge (2010) for alkanol profiles of WWTP influent samples. The C_{18} *n*-alkanol has also been detected in the root constituent suberin (Nierop et al. 2003) and in the lipids of some microbial spores (Ambles et al. 1989; Jaffé et al. 1996).

9.2.2.6 Other Compounds

Bu_2 and Bgk Horizons

In the technogenically affected horizons, Bu_2 and Bgk detected the following compounds: butanedioic acid (succinic acid) produced by microorganisms is a bidentate acid mobilizing heavy metals and organic carbon; exo-tricyclo[5.2.1.0(2.6)]decane C10H16 (component fuel studied for use in pulse-detonation engines) is a contaminant compound; 2-phenyl ethanol; 1-hydroxy-9-dodecyne belongs to the group of alkynes which are rare compounds in soil but occur in certain plants (Ichthyothere, Chrysanthemum, and other members of the Asteraceae family); 2-methyl-4-keto-2-hydroxy pentane (4-methyl-4-hydroxy-2-pentanone) is released to the environment as a result of its use as a solvent, additive, or synthetic intermediate for many materials but may be also of natural origin. It has been found to be part of the volatile compounds produced by *Stigmatella*, myxobacteria (Wang and Tao 2009). In soil, it is expected to exhibit very high mobility based upon the reported infinite solubility of the compound in water.

G Horizon

Isopropyl myristate also detected in the upper Bgk horizon is probably of plant origin. Myristic acid is a fatty acid common in plant sources such as nuts and palm seeds. Other compounds such as benzoic acid, 2,6-dihydroxy, and 2-phenyl ethanol are produced from phenolic hydrolysis and depolimerization of lignin (Kögel-Knabner et al. 1994). Contaminants (best match factor, MF, for

1,2-bisacenaphthylene) with characteristic fragment ions m/z 55, 73, 262, 328 b.p were detected in Bu_2 and Bgk horizons, but not in G horizon, and are suspected contaminants. In a study by Atanassova and Brümmer (2004), bioformation of some high-molecular PAHs such as dibenzo(a,h)anthracene and benzo(g,h,i)perylene was found in the lower soil layers of a Colluvisol profile from North Rhine-Westphalia with stagnant water and reducing conditions. These compounds were not detected in the Technosol profile in this study under the abovementioned experimental conditions of extraction, cleanup, and GC/MS analysis and identification.

The results obtained indicate that PAHs were either biodegraded following >6 years after sludge deposition, have formed "bound residues," and are adsorbed in non-bioavailable (non-extractable) form or represent <0.1% of total peak area of the analyzed components.

9.3 Free Lipid Biomarkers in Technosols from a Major Coal Mine Region in Southeastern Europe

Technogenic soils from mine areas are heterogeneous materials that often exhibit small-scale spatial variability of physical, chemical, and hydro-physical soil characteristics, e.g., water repellency, suggested to arise from the presence of irregularly distributed lignitic particles in overburden clay strata used for post-mining reclamation activities.

In former studies of Atanassova et al. (2017), the free lipid extracts of non-vegetated and pine-afforested spoils from brown coal production in Pernik region, Bulgaria, were analyzed, and lipid biomarkers and their environmental significance were assessed. Soil organic matter (SOM) of mine soils is characterized by the presence of alkanes, fatty acids, and fatty alcohols, as well as coal biomarkers. The alkane distribution in the reclaimed mine soils pointed at two alkane sources, i.e., terrigenous plants and ııre coal source. Fatty acid carbon preference indices ($CPI_{even/odd}$ = 10.4 and 8.1) reflect the predominant role of terrestrial vegetation in the formation of SOM. The higher concentrations of the long-chain alkanes, fatty acids, and alcohols are speculated as the reason for the severe and extreme water repellency observed with a natural soil in the vicinity of the mine (Vertisol) and the unreclaimed spoil. The presence of terrigenous steroid and triterpenoid markers in the reclaimed mine soils is an indication of the ongoing soil formation processes. The analysis of free lipids in SOM revealed sensitive molecular indicators of biogenic (terrigenous) and coal origin and the ongoing soil-forming processes after >20 years of spoil reclamation.

In a recent study of longtime coal ash-reclaimed Technosols (1970s) from Maritsa-Iztok lignite coal basin near Stara Zagora (Bulgaria), two plots (pine-vegetated and non-vegetated) were characterized by strong to extreme water repellency (water drop penetration time [WDPT] up to 14,400 s) and extreme acidity (pH 3–4) with finer texture at the pine-vegetated site.

These technogenic soils are heterogeneous materials exhibiting small-scale spatial variability of water repellency, suggested to arise from the presence of irregularly distributed black coal particles and lignite-coated sand and clay particles, thus affecting water and chemical elements transport (Gerke et al. 2001; Meena et al. 2018b; Sihag et al. 2015). However, water repellency is also a feature of undisturbed soils under various types of permanent vegetation. Organic compound classes thought to be associated with soil water repellency include alkanes, fatty acids, fatty alcohols, aldehydes, ketones, ω-hydroxy fatty acids, and α,ω-dicarboxylic acids characteristic of roots (Morley et al. 2005; Atanassova and Doerr 2010; Mao et al. 2015; Yadav et al. 2017c). The degree of water repellency was found to vary with moisture content and temperature (Dekker and Ritsema 1994).

In the Maritsa-Iztok, coal-producing region reclamation into forestry-dominated landscapes has started since 1970s. The conventional way consisted in depositing clay layers in landfills and reinstallment of humus-rich soil layer of the nearby natural soil (Vertisol) on top of the spoils. Another useful way of land reclamation has been to mix coal ash, a waste product from the thermal power plant with the geological overburden layers containing yellowish-green, grayish-green, and black clays overlaying the coal seams that have been excavated in the process of coal production. Tree planting (*Pinus nigra L.*) and ash treatment have been proved useful practices to increase the substrate pH and improve soil chemical and physical properties (Zheleva et al. 2004; Meena and Yadav 2014). The effects of excavated overburden mine wastes can lead to soil erosion, metal mobility, pollution, loss of biodiversity, etc. (Wong 2003; Sheoran et al. 2010; Kumar et al. 2018).

9.3.1 Sites, Materials, and Methods

The study was carried out at two plots separated by a distance of 30 m from each other at the village of Obruchishte, Maritsa-Iztok, coal mines. The experimental plots were located in an ~ 1 ha area, which was afforested with *Pinus nigra.* Site 1 is N 42.16, E 25.94, and site 2 is N 42.16, E 25.94 (Atanassova et al. 2018, Fig. 9.9).

(a) (b)

Fig. 9.9 Non-vegetated and pine-vegetated water-repellent Technosol (**a**) at Obruchishte and (**b**) water-repellent soil in soil core sampler

(a) (b)

Fig. 9.10 Non-humus-reclaimed Technosol (**a**) and humus-reclaimed Technosol (**b**)

A lack of plant density and water repellency was observed in plots of ~ 200 m^2 among a pine-vegetated area. Grids Δ 2 m, ~ 40 m^2 were constructed at a non-vegetated and another pine-vegetated site. Soil cores were sampled to a depth of 0–10 (0–5) and 10–20 cm. At both sites, the parent mine sediments contain irregularly distributed lignitic particles of sand grain size due to random mixing of black "greasy" clays with the yellowish-green and grayish-green clay strata overlaying the black clay strata back in the early 1970s when the spoils were created.

After sampling, soils were transferred to the laboratory, air-dried at room temperature, and sieved to <2 mm fraction. In the laboratory, the water drop penetration time (WDPT) was measured at controlled humidity and temperature (Doerr et al. 2002; Varma et al. 2017a).

Two other sites of non-water-repellent Technosols subject to humus and non-humus reclamation were analyzed near Mednikarovo village (N 42.11, E 26.04, and N 42.11, E 26.03, Atanassova et al., 2018, Fig. 9.10). At Mednikarovo site, the investigated soils were (i) humus layer-reclaimed soil of clay loam texture and (ii) non-humus-reclaimed soil of sandy loam texture. The surface horizon of ~ 40 cm depth is a translocated humus horizon of a natural Vertisol widely distributed around the mine territory. The sub-layers of ~2 m were composed of yellowish and greenish clays comprising the overburden sediments of the stratigraphic profile and possessing suitable physicochemical characteristics (pH ~7; Atanassova et al. 2018; Meena et al. 2016a; Dhakal et al. 2015). Both areas have been cultivated to alfalfa for years, but indigenous grass-forage species have grown everywhere.

Total organic carbon (TOC) was determined by oxidation with $K_2Cr_2O_7/H_2SO_4$ and fractionated into humic (HOC) and fulvic (FOC) organic carbon by 0.1 M$Na_4P_2O_7$ and 0.1 M NaOH (Kononova 1966). Cation exchange capacity (CEC) was determined as sum of titratable acidity (pH 8.2), exchangeable Ca, by saturation with K malate at pH 8.2 (Ganev and Arsova 1980). Electrical conductivity was recorded in soil water (1:5, ISO 11265:2002) and soil pH/Eh in soil/water slurry (1:2.5); texture was analyzed by the method of Kachinskii (1965). Soil water repellency was assessed by the water drop penetration time (WDPT) method (Doerr et al. 2002; Meena et al. 2018a; Yadav et al. 2017b). General soil properties are presented in Table 9.2.

Table 9.2 Main soil properties of the experimental soils

Site	Depth cm	WDPT[a](s)	pH		EC (mS/cm)		TOC %		$CEC_{8.2}$ (cmol. kg^{-1})		Clay %	
			Mean	± SD	Mean	± SD	Mean	± SD	Mean	± SD	Mean	± SD
Mednikarovo humus reclamation	0–10	1	6.9	0.01	0.04	0.01	1.50	0.02	56.3	0.2	38.6	5.21
	10–20	2	6.9	0.02	0.05	0.01	1.53	0.03	56.3	0.1	45.4	4.48
Mednikarovo, non-humus reclamation	0–10	2	7.2	0.01	0.04	0.005	1.04	0.02	29.6	0.2	24.4	1.25
	10–20	1	7.2	0.01	0.04	0.005	0.37	0.01	29.8	0.1	26.6	0.96
Obruchishte, pine-vegetated	0–5	14–9589 <u>1200</u>	4.6	0.20	0.19	0.14	5.09	0.42	41.27	3.89	28.0	7.21
	10–20	2–128 <u>10</u>	4.2	0.43	0.63	0.14	3.00	1.67	44.60	3.52	55.1	21.7
Obruchishte, non-vegetated	0–10	76–14,440 <u>1282</u>	3.2	0.11	1.72	0.76	6.43	3.98	67.77	6.16	17.8	3.86
	10–20	202–2470 <u>614</u>	3.1	0.10	1.61	0.69	6.11	0.96	69.90	8.70	17.0	2.37

From Atanassova et al. (2018), Simeonova et al. (2018)
[a]median of water drop penetration time (WDPT)

9.3.1.1 Lipids Analysis

The extractions were performed by sonication of the soils (4 g in duplicate) in 16 ml of acetone/hexane (1:1 vs/v) using a Julabo USR 3, 35 kHz, 200 W (Julabo LaborTechnik GMBH) for 10 min. The extraction was repeated twice with fresh solvent. The combined solvent extracts were dried with Na_2SO_4, vacuum-evaporated (Labconco CentriVap concentrator at 50 °C), and solvent exchanged into dichlorometane:hexane (1:1). Internal standard (2-nonadecanone) was added, vacuum dried, and then derivatized with BSTFA+ 1% TMCS (heating for 1 hour at 70 °C, Atanassova and Mills (2016), modified). After completion of derivatization, the derivatives were cooled, reconstituted to 1 ml with DCM, and analyzed by GC/MS. The analytical specifications of the GC/MS analysis and compound characterization were performed as in point **2.1**.

9.3.1.2 Statistics

Differences between compound contents in the samples were distinguished by Tukey's test at $p = 0.05$. Principle Component Analysis (PCA) and cluster analysis (CA) were carried out in order to study the simultaneous interaction of soil characteristics and properties with the lipid fraction (SPSS 22).

9.3.2 Lipid Biomarkers in Water-Repellent Coal Ash-Amended Technosols

The Technosols at both sites treated with coal ash were characterized by strong to extreme water repellency and extreme acidity (pH 3–4, Table 9.2). The GC chromatograms of the total lipid extracts indicate that the most abundant signature lipid compounds are diterpenoid coal biomarkers, such as phyllocladane and kaurene, and steroids, fatty acids, and fatty alcohols (>C_{22}, Fig. 9.11, Table 9.3). The higher concentrations of the long-chain fatty acids, fatty alcohols, steroidal compounds,

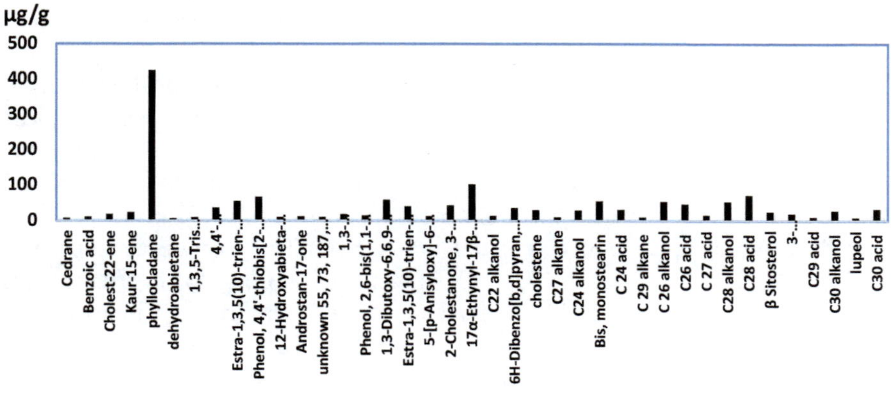

Fig. 9.11 Relative distribution of major compounds in free lipid extracts in a water-repellent-reclaimed Technosol (non-vegetated, WDPT = 14,440 s)

Table 9.3 Major compounds in lipid extracts of water-repellent Technosol

Cedrane; benzoic acid, 3-methoxy-4-hydroxy; cholest-22-ene, (5α); kaur-15-ene; phyllocladane; 7-isopropyl-1,1,4a-trimethyl-1,2,3,4,4a,9,10,10a-octahydrophenanthrene (dehydroabietane); 1,3,5-tris(hydroxy)benzene; 4,4′-diacetyldiphenylmethane; estra-1,3,5(10)-trien-17-one, 1-methyl-3; phenol,4,4′-thiobis[2-(1,1-dimethylethyl)-6-methyl-; 12-hydroxyabieta 1,8,11,13-tetraen-3-one; androstan-17-one; unknown m/z 55, 73, 187, 269, 285, 354; 1,3-benzenedicarboxylic acid, 4,5,6-tris(1,1-dimethylethyl)-, ester; phenol, 2,6-bis(1,1-dimethylethyl)-4-[(4-hydroxy-3,5-dimethylphenyl); 1,3-dibutoxy-6,6,9-trimethyl-6a,7,10,10a-tetrahydro-6H-benzo[c]chromene; estra-1,3,5(10)-trien-17-one,3-methoxy-2-[hydroxy]-; 5-[p-anisyloxy]-6-methoxy-8-[2,5-dimethylpyrr-1-yl]quinoline; 3-hydrloxypalmitic acid; 2-cholestanone, 3-phenyl-17α-ethynyl-17β-hydroxy-5β-estran-3-one; C_{22} alkanol; 6H-dibenzo[b,d]pyran, 6a,7,8,10a-tetrahydro-6,6,9-trimethyl-3-pentyl-1-[hydroxy]-, (6aR-trans); cholestene; C_{27} alkane; C_{24} alkanol; bis-monostearin; C_{24} acid

and diterpenoids in the more water-repellent samples are speculated as the reason for the severe and extreme water repellency observed (Fig. 9.12a–c). The concentration of long-chain (> C_{22}) homologous fatty acids, alkanols, and alkanes were significantly correlated with water drop penetration time tests (WDPT) (Fig. 9.13). Some contaminant compounds were present in high abundance, e.g., 4,4′-diacetyldiphenylmethane, also found in reclaimed soils from brown coal-producing region in Bulgaria (Atanassova et al. 2017; Verma et al. 2015c; Meena et al. 2015d). This contaminant is a product of the photocatalytic oxidation of coal. In addition, some PAH (6H-dibenzo[b,d]pyran and chromane) derivatives were detected (Fig. 9.11, Table 9.3).

The presence of dehydroabietane in the non-vegetated technogenic soils indicates the conifer vegetation origin of the coal-forming mires of Maritsa-Iztok coal basin. Pimaric and myristic acids detected in the pine-vegetated soils confirm the conifer vegetation dominance in the formation of soil organic matter (SOM). This is also confirmed by detection of hydroxy-benzoic and *iso-vanillic* acids in the non-vegetated site, indicating that conifers are the predominant coal progenitors in the peat bog of Maritsa-Iztok coal. Sesquiterpenoids in the soils were represented by α-cedrene at both sites. The abundance of steroids such as cholesterol and androstane derivatives indicates a source component from fauna, as these lands have been used for occasional grazing.

The PCA indicates that four principle components were identified with *eigenvalue > 1* accounting for 91.14% of the total variability (Tables 9.3 and 9.4). The first component is loaded by water drop penetration time (WDPT), CEC, % sand content (SAND), total organic carbon (TOC), humic (HOC) and fulvic organic carbon (FOC), mineral nitrogen (MN), electrical conductivity (EC), and % lipid compounds (homologous series of compounds with $C > 22$). This component contributed to 48.3% of the total variance. The 2nd component (18.5% of total variance) was loaded by the silt content (SILT) and total nitrogen (N), the 3rd by % moisture (M) and % sand content (SAND), and the 4th by % silt content (SILT) and % lipid compounds, which crossload 1st and 4th components. There was a significant positive correlation between WDPT and TOC ($R = 0.886{**}$), HOC ($R = 0.676{*}$), FOC ($R = 0.448{*}$), MN ($R = 0.651{*}$), and % lipid compounds (homologous series organic

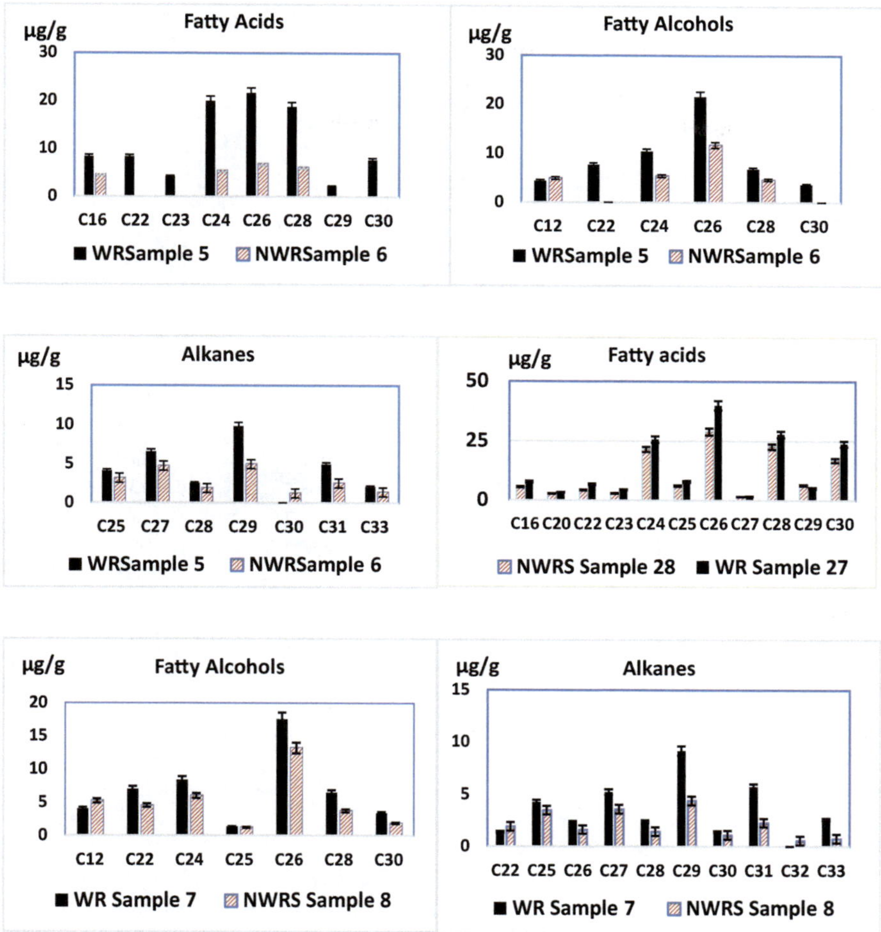

Fig. 9.12 Quantitative distribution of homologous series of alkanes, fatty acids, and fatty alcohols in water-repellent (WR) and non-water-repellent (NWRS) pine-vegetated soils amended with coal ash

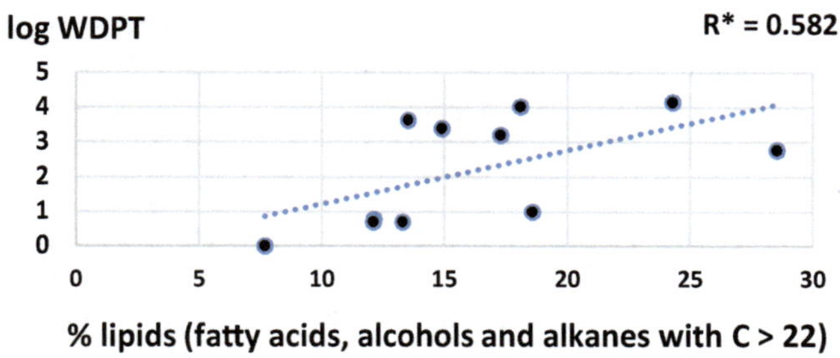

Fig. 9.13 Relationship between WDPT and % free lipids

Table 9.4 Components extracted with eigenvalue >1, % of variance, and cumulative variance

Component	Initial eigenvalues		
	Total	% of variance	Cumulative %
1	6.283	48.328	48.328
2	2.400	18.462	66.789
3	1.980	15.229	82.018
4	1.186	9.127	91.145

Table 9.5 Component matrix

	Component			
	1	2	3	4
WDPT	**.680**	.317	−.584	.166
CEC	**.864**	−.440	.178	−.086
SAND	**.727**	.075	**.639**	−.063
SILT	−.034	**.534**	.468	**.593**
CLAY	−.676	−.181	−.697	−.064
TOC	**.775**	.487	−.327	−.131
HOC	**.781**	.435	−.061	−.239
FOC	**.811**	−.039	.107	−.358
MN	**.934**	−.130	−.157	.024
EC	**.736**	−.622	.161	−.006
M	−.694	−.192	**.520**	−.176
N	−.123	**.931**	.205	−.146
Lip	**.553**	−.217	−.085	**.732**

compounds with C atoms >22, R = 0.582∗) and a negative correlation with the % of hygroscopic moisture (M) (R = −0.768∗∗). The cluster analysis supports the results from the PCA analysis (Fig. 9.14).

In order to elucidate possible sources, the following indexes were calculated: carbon preference index (CPI) = $\Sigma(C_{13}–C_{31})/\Sigma(C_{12}–C_{34})$ for the whole range of detected alkanes and average chain length (ACL) = 29. For the experimental soils, CPI varied from 2.19 to 10.06 and showed monomodal distribution with a maximum at C_{29}. Short-chain alkanes C<20 were not detected during the spring sampling period. As mentioned before, the predominant ratio of odd/even homologues is typical of *n*-alkanes in epicuticular waxes of plants (Bryselbout et al. 1998) and indicates a terrigenous plant source (Lichtfouse et al. 1997). The most abundant compound detected was phyllocladane 26.6 ± 6% > lipids (homologous groups of organic compounds C > 22) 24.9% ± 5% > steroids 18.3% ± 5% > organic pollutants (phenols, dibenzopyran, 4,4′-diacetyldiphenyl methane) 12.4% ± 6%. No relationship was established between WDPT and the concentration of phyllocladane or pollutants (polycyclic aromatic hydrocarbons [PAHs] and4,4′-diacetyldiphenylmethane) (Table 9.5).

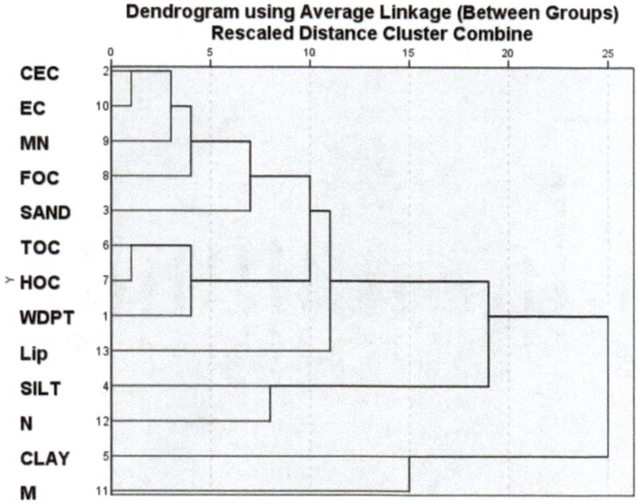

Fig. 9.14 Cluster analysis of 13 variables

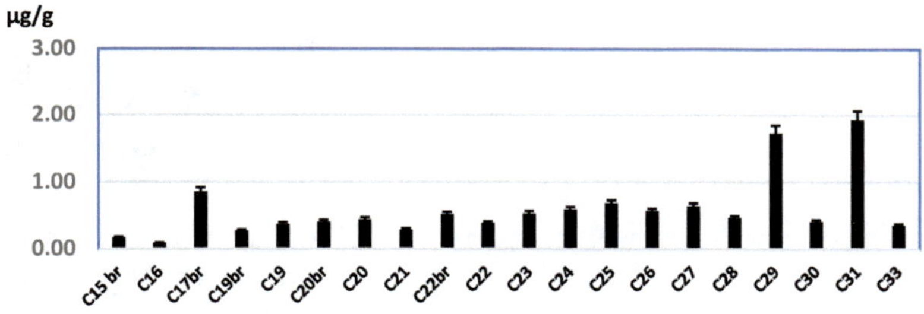

Fig. 9.15 Alkane distribution in humus-reclaimed Technosol

9.3.3 Lipid Biomarkers in Non-water-Repellent Technosols

9.3.3.1 Alkanes

The alkane signature in the humus-reclaimed Technosol from Mednikarovo region is represented by homologous members extending from C_{15} to C_{33} (Fig. 9.15) and exhibits a more complex distribution (trimodal) maximizing at $C1_{7br}$, C_{20}, C_{29}, or for simplicity bimodal, reflecting different sources. It is interesting to acknowledge the lower concentrations of the alkanes (< 2 µg/g) including the long-chain ones ($C_{29–33}$) and the presence of $<C_{24}$-branched alkanes in the non-water-repellent Technosols developed on overburden clays, however lacking black lignitic "greasy" clays, in comparison with the water-repellent Technosols (point **3.2**).

Long-chain alkanes have been implicated to cause water repellency in soils (Atanassova and Doerr 2010; Yadav et al. 2018a; Meena et al. 2015c). The origin of these compounds can be judged upon calculating the CPI. Alkanes can be classified in two main groups according to their sources. The first group comprises

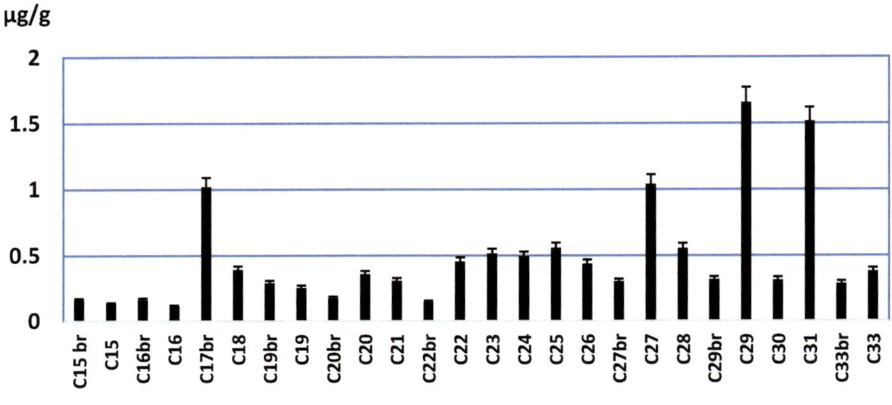

Fig. 9.16 Alkane distribution in a non-humus-reclaimed Technosol

Fig. 9.17 Fatty alcohol distribution in humus-reclaimed Technosol

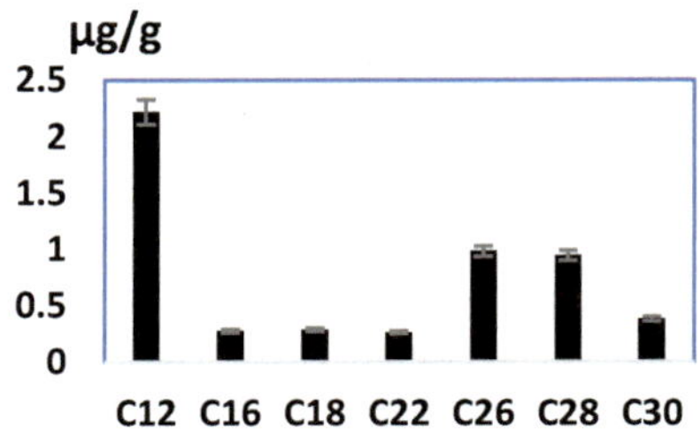

short-chain alkanes extending from C_{15} to C_{22} andCPIC_{15}–C_{22} = 0.718. Grimalt et al. (1986) also found that in coal seam samples from Hungary, even chain homologues surpass the odd ones. Wang et al. (2010) conclude that bacteria may produce even carbon-numbered n-alkanes below n-C_{22} and synthesize *n*-alkanes with no carbon predominance in the range C_{23}–C_{33}. Long-chain *n*-alkanes extending from C_{23} to C_{33} show an odd carbon-number predominance with not very high CPI values (2.9 for humus-reclaimed Technosol) and CPI =2.2 for the whole range of alkanes. It is speculated that a mixture of *n*-alkanes originating from higher plant waxes and bacteria can cause the lower CPI in the range $C_{23\text{-}33}$. Similar alkane distribution was found for the non-humus-reclaimed Technosol maximizing at C_{17br} and C_{29} (Fig. 9.16). CPI for the whole range is 2.0 and CPIC_{15}–C_{21} = 1.0 and CPIC_{22}–C_{33} = 2.5. These *n*-alkane distributions for the humus- and non-humus-reclaimed Technosols point at a dual origin for the alkanes, i.e., higher plants and microbial sources because of prevalence of short-chain branched and even alkanes.

9.3.3.2 Alkanols

Fatty alcohols (C_{22}–C_{32}) participate as wax coatings in terrestrial plants for prevention of desiccation, while aquatic organisms synthesize short-chain compounds. Bacteria normally produce fatty alcohols of odd carbon number or branched homologues (Mudge 2005). The abundance of C_{12} *n*-alkanol (Fig. 9.17) and the presence of low-molecular-weight (LMW) ($< C_{20}$) *n*-alkanols are linked to soil microbes and marine biota (Boreddy et al. 2018).

9.4 Free Lipid Biomarkers in a Long-Term Cultivated Alluvial Meadow Soil Treated with Biochar (Anthrosol)

Biochar is a promising ameliorant to soil influencing soil structure, texture, porosity, aggregate stability, mineral nutrition, carbon sequestration, etc. (Lechaman 2007; Atkinson et al. 2010; Montanarella and Lugato 2013; Dadhich et al. 2015; Kumar et al. 2017a). Biochar addition to soil is shown to improve soil functions, e.g., increase chemical and microbial stability, and bring about increased porosity and surface area, higher pH, and increased electrical conductivity (EC), cation exchange capacity, nutrient availability, and plant productivity (Atkinson et al. 2010; Meena et al. 2015b; Varma et al. 2017b). Experiments with biochars of various origin show that residence times varied between 4 and 29 years, and the type and quality of biochar promoted the abundance of various microorganisms, e.g., yeast-derived biochar-promoted fungi, while glucose-derived biochar was utilized by Gram-negative bacteria (Steinbeiss et al. 2009; Verma et al. 2015b; Meena et al. 2017b). However, Li et al. (2011) show that earthworms avoided soils containing 100 and 200 g/kg dry biochar at $p < 0.05$ due to insufficient moisture or the presence of polycyclic aromatic hydrocarbons (PAHs) formed during biochar production. In our recent studies (Petkova et al. 2018), biochar amendment of Alluvial meadow soil in field experiments with wheat and maize cultivated in crop rotation led to stimulation of soil microflora and increased CO_2 production. There are reports in literature of water-repellent biochar and water-repellent soil treated with biochar (Page-Dumroese et al. 2015). Concerning biochar stability, decomposition, and transformation in soil, it has been shown that neutral lipids showed higher stability as their content decreased ~ two times over 3.5 years (Kuzyakov et al. 2014; Yadav et al. 2017a).

9.4.1 Sites, Materials, and Methods of Analysis

The study was conducted in April 2017 at the experimental field of Nikola Poushkarov Institute of Soil Science, Agrotechnologies, and Plant Protection near the village of Tsalapitsa. Samples (0–10 cm) from a field experiment with wheat on Alluvial meadow soil amended with biochar produced by pyrolysis of oak residues applied at rate of 3 t/ha in 2017 were analyzed. Mineral fertilization was applied at rates N_{10} (urea), P_{12} (as triple superphosphate), and K_{10} (potassium sulfate). General soil characteristics were the following: total organic C % (0.8%–1%), total N (0.09–0.1%), and pH (6.0–6.3). Grids Δ 2 m and ~ 40 m^2 were constructed at four plots. Sample water repellency was measured by the water drop penetration time (WDPT) method (Doerr et al. 2002; Meena et al. 2016b; Datta et al. 2017b). The soil samples were weighed and subsequently equilibrated at the ambient air humidity before

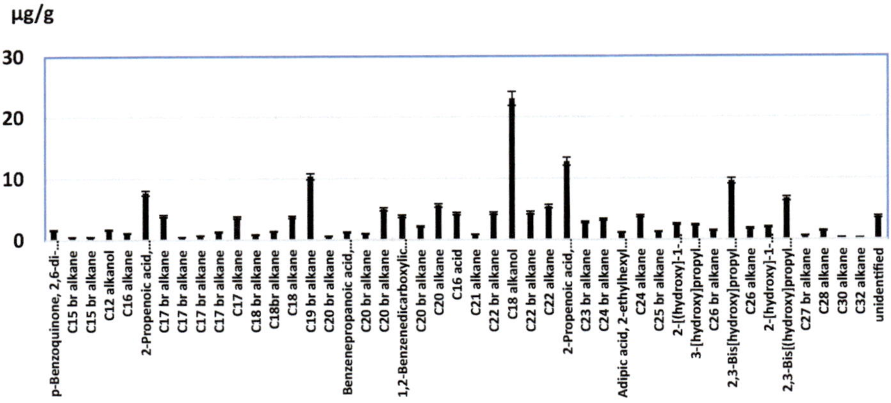

Fig. 9.18 Quantitative analysis of a pool of soil sample (0–10 cm) from a plot ~ 40 m^2, 3 t/ha added biochar, Alluvial meadow soil, and the compounds identified in the total lipid extracts

Table 9.6 Major compounds in biochar-treated Alluvial meadow soil

1 p-benzoquinone, 2,6-di-tert-butyl-; C_{15br} alkane; C_{12} alkanol; C_{16}alkane; benzofuran, 2,5-diacetyl-6-methoxy-; 2-propenoic acid, oxybis(methyl-2,1-ethanediyl) ester; C_{17br} alkane; C_{17} alkane; C_{18br} alkane; C_{18} alkane; phthalate; 7,9-di-tert-butyl-1-oxaspiro(4,5)deca-6,9-diene-2,8-dion; C_{20br} alkane; benzenepropanoic acid, 3,5-bis (1,1-dimethylethyl)-4-hydroxy-, methylester; 1,2-benzenedicarboxylic acid, butyl 2-methylpropyl ester; C_{20} alkane; C_{16}acid; trichloroacetic acid, hexadecylester; nonadecylpentafluoropropionate; octacosyl trifluoroacetate; C_{21} alkane; C_{22br} alkane; C_{18} alkanol; C_{22} bralkane; C_{22} alkane; C_{18} acid; 2-propenoic acid, pentadecylester; C_{23} bralkane; C_{24} bralkane; adipic acid, 2-ethylhexyl tetradecylester; C_{24} alkane; C_{25br} alkane; 2-[(hydroxy]-1; ([hydroxy]methyl)ethylpalmitate; 3-[hydroxy]propylpalmitate; C_{26br}alkane; 2,3-bis[hydroxy]propylpalmitate; C_{26} alkane; C_{27br} alkane; 2-[hydroxy]-1; ([hydroxy]methyl)ethyl stearate; 2,3-bis[(hydroxy]propyl stearate; C_{28} alkane; C_{30} alkane; C_{32} alkane

measuring water drop penetration time in the laboratory at recorded humidity and temperature. The WDPT measured at the laboratory for all soils treated with biochar including the control variant and the biochar used in the field trial showed that samples were hydrophilic (non-water repellent, WDPT <5 s).

Conditions for lipid extraction and analytical GC/MS analysis were as in points **3.1** and **2.1.**

9.4.2 Major Compounds and Biomarkers Distribution

An interesting trend in lipid signature was observed with the biochar-treated soil. Compounds distribution is presented in Fig. 9.18 and Table 9.6. The surface horizons of the wheat-grown soil contained alkanes (< C_{24}, Fig. 9.19) with prevailing even over odd predominance of homologues (EOP) and carbon preference index (CPI) 10.6, which was also found by Kuhn et al. (2010) for soils under forest and

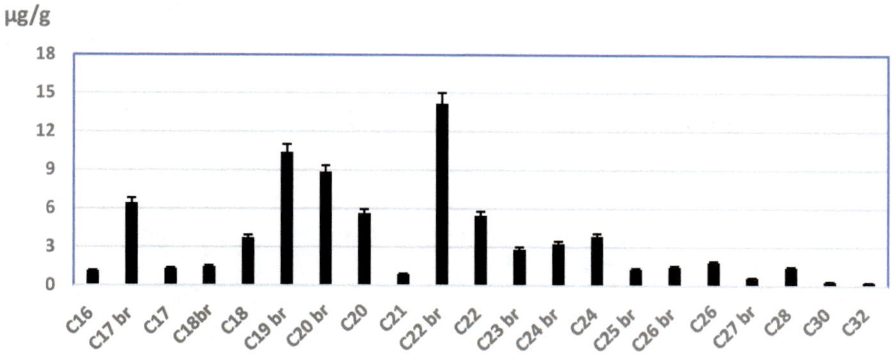

Fig. 9.19 Alkane distribution in lipid extracts of a biochar-treated alluvial meadow soil

grass cover. Preliminary studies have shown that burning soil organic matter may cause thermal disruption of long-chain alkanes (Atanassova et al. 2012a, b, Eckmeier and Wiesenberg 2009; Meena et al. 2015e). In the surface horizons of the Alluvial meadow soil, a source of short-chain alkanes might be the biochar or microbial biofilms found in arid soils (Dembitskiĭ et al. 2001; Verma et al. 2015a). These markers support findings for increased plant uptake of nitrogen from labeled biochar and increased soil respiration rates after biochar addition (Ameloot et al. 2013; Petkova et al. 2018; Meena et al. 2015a). Most abundant fatty acids were C_{14}, C_{16}, and C_{18}, while long-chain fatty acids and alcohols, which are implicated as higher plant sources to SOM and soil water repellency markers (Mao et al. 2015; Dhakal et al. 2016; Meena et al. 2017a) are lacking in this soil (Table 6). A contaminant, i.e., p-benzoquinone, 2,6-di-tert-butyl-, which has been detected in subsurface soils treated with sewage water (Barber et al. 1988) was present in the biochar-treated soils, implying that soils had been accidentally irrigated with treated wastewater.

The lipid signature indicates anthropogenic source and/or predominant microbial contribution to soil organic matter, i.e., thermally altered and microbial n-alkanes were difficult to discriminate, as also found by Knicker et al. (2013). The presence of esters of palmitic and stearic acids has been attributed to solubilization from micelle-like microparticles and colloidal dissolved organic carbon (DOC) (Atanassova et al. 2014).

9.5 Conclusion

Free lipid biomarkers, their contents, and their signatures in anthropogenic soils of various origin were analyzed, and their diagnostic properties assessed. The gas chromatograms of the free lipid extracts from soils separated from selected horizons of a Technosol on which sewage sludge from paint and print industry was deposited indicate some quantitative changes in the main homologous groups of alkanes and fatty alcohols. The process of gleyification in the lower horizon of the Technosol is characterized by hydromorphic conditions and limited removal of weathering products. Sources of lipid compounds are very similar in three characteristic soil horizons and point at predominant bacterial contribution because of the presence of even number and branched alkanes (C_{16}–C_{33}), short-chain (C_6–C_{18}) fatty acids, and C_{11}–C_{32} *n*-alkanols. The molecular composition of the free lipid pool has revealed no presence of xenobiotics at measurable quantities. However, some contaminants (metabolites) were detected in the lipid extracts of the soils in a free extractable form at depths (53–143 cm), which has implications for leaching down the soil profile and contamination of groundwaters in situations of technogenic influence and sludge deposition.

Biogenic terrigenous lipids (alkanes, fatty acids, and fatty alcohols), coal biomarkers (phyllocladane), as well as steroids are most abundant in a non-vegetated and pine-vegetated water-repellent Technosol from the area of the largest coal mine district in Southeastern Europe. These compounds indicate the ongoing soil formation processes in the long-time reclaimed spoils-turned soils. Statistical analysis confirms that total organic carbon (TOC) and the lipid fraction (the concentration of long-chain > C_{22}) fatty acids, alcohols, and alkanes were the drivers of soil water repellency in the studied Technosols. Free lipid signature in non-water-repellent Technosols lacking ash amendment, which have undergone long-term humus and non-humus reclamation, indicates dual origin for soil organic matter, i.e., higher plants, and microbial sources with no markers of water repellency, such as long-chain fatty acids, alkanes, and fatty alcohols.

Short-chain n-alkanes (C_{16}–C_{24}) with pronounced even over odd predominance (EOP) dominate the n-alkane assemblages in a wheat-grown biochar-amended Alluvial meadow soil. Previous studies have shown that pyrolysis-induced thermal disrupture of long-chain n-alkanes may produce short-chain homologues with an even over odd predominance. Presence or coexistence of polycyclic aromatic hydrocarbons derived from pyrolyzed biochar was not recorded; however, a marker (p-benzoquinone, 2,6-di-tert-butyl-) for wastewater-treated soil was found.

Free lipid biomarkers can serve as sensitive environmental indicators of the origin of soil organic matter, the degree of pedogenesis, anthropogenic disturbances, and degradation processes, e.g., soil water repellency.

Acknowledgments Studies reported in this chapter were supported by the National Science Fund (NSF), Ministry of Education and Science, project: DN 06/1 (2016–2019).

References

Ambles A, Jambu P, Ntsikoussalabongui B (1989) Evolution des lipides naturels d'un podzol forestier induite par l'apport d'engrais mineraux: hydrocarbures, cetones, alcohols. Sci du sol 27:201–214

Ambles A, Jambu P, Jacquesy J-C, Parlanti E, Secouet B (1993) Changes in the ketone portion of lipidic components during the decomposition of plant debris in a hydromorphic forest-podzol. Soil Sci 156:49–56

Ameloot N, Graber ER, Verheijen FG, De Neve S (2013) Interactions between biochar stability and soil organisms: review and research needs. Eur J Soil Sci 64(4):379–390

Ashoka P, Meena RS, Kumar S, Yadav GS, Layek J (2017) Green nanotechnology is a key for eco-friendly agriculture. J Clean Prod 142:4440–4441
Atanassova I (2017) Contemporary aspects of soil eco-chemistry and geochemistry in Bulgaria. Bulg J Soil Sci 1(1):33–59
Atanassova I, Brümmer G (2004) Polycyclic aromatic hydrocarbons of anthropogenic and biopedogenic origin in a colluviated hydromorphic soil of Western Europe. Geoderma 120:27–34
Atanassova I, Doerr S (2010) Organic compounds of different extractability in total solvent extracts from soils of contrasting water repellency. Eur J Soil Sci 61:298–231. https://doi.org/10.1111/j.1365-2389.2009.01224.x
Atanassova I, Doerr SH (2011) Changes in soil organic compound composition associated with heat-induced increases in soil water repellency. Eur J Soil Sci 62(4):516–532
Atanassova I, Mills G (2016) Biogenic and anthropogenic lipid markers in sediments from a marsh habitat associated with the LCP chemicals superfund site in Brunswick, Georgia, USA. Water Air Soil Poll 227:1–13
Atanassova I, Teoharov M (2010a) Nature and origin of lipids in clay fractions from a Fluvisol in a sewage sludge deposition field. Water Air Soil Pollut 208(1–4):295–304
Atanassova I, Teoharov M (2010b) Variation in lipid abundance and composition in a fire affected hillside from Lyulin mountain. Bulg Agric Sci Technol 2(3):153–159
Atanassova I, Doerr SH, Bryant R (2012a) Changes in organic compound composition in soil following heating to maximum soil water repellency under anoxic conditions. Environ Chem 9(4):369–378
Atanassova I, Velichkova N, Teoharov M (2012b) Heavy metal mobility in soils under the application of sewage sludge. Bulg J Agric Sci 18(3):396–402
Atanassova ID, Doerr SH, Mills GL (2014) Hot-water-soluble organic compounds related to hydrophobicity in sandy soils. In: Hartemink AE, McSweeney K (eds) Progress in soil science: soil carbon. Springer, Cham, pp 137–146
Atanassova I, Hristov B, Shishkov T, Doerr S (2017) Lipid biomarkers and their environmental significance in mine soils from Eastern Europe. Arch Agron Soil Sci 63(12):1697–1710
Atanassova I, Banov M, Shishkov T, Petkova Z, Hristov B, Ivanov P, Markov E, Kirilov I, Harizanova M (2018) Relationships between soil water repellency, physical and chemical soil properties in hydrophobic technogenic soils from the region of Maritsa-Iztok coal mine in Bulgaria. Bulg J Agric Sci 24(Suppl 2):10–17
Atkinson CJ, Fitzgerald JD, Hipps NA (2010) Potential mechanisms for achieving agricultural benefits from biochar application to temperate soils: a review. Plant Soil 337(1–2):1–18
Barber LB, Thurman EM, Schroeder MP, LeBlanc DR (1988) Long-term fate of organic micropollutants in sewage-contaminated groundwater. Environ Sci Technol 22(2):205–211
Boeschker HTS, Middelburg JJ (2002) Stable isotopes and biomarkers in microbial ecology. FEMS Microb Ecol 40:85–95
Boreddy SKR, Haque MM, Kawamura K, Fu P, Kim Y (2018) Homologous series of n-alkanes (C_{19}–C_{35}), fatty acids (C_{12}–C_{32}) and n-alcohols(C_8 –C_{30}) in atmospheric aerosols from central Alaska: molecular distributions, seasonality and source indices. Atmos Environ 184:87–97
Bray EE, Evans ED (1961) Distribution of n-paraffin as a clue to recognition of source beds. Geochim Cosmochim Acta 22:2–15
Bryselbout C, Henner P, Lichtfouse E (1998) Fossil fuel biomarkers in plant waxes as pollution parameters. Sci Total Environ 222:201–204
Bull ID, van Bergen PF, Nott CJ, Poulton PR, Evershed RP (2000) Organic geochemical studies of soils from the Rothamsted Classical Experiments-V. The fate of lipids in different long-term experiments. Org Geochem 31:389–408
Buragohain S, Sharma B, Nath JD, Gogaoi N, Meena RS, Lal R (2017) Impact of ten years of bio-fertilizer use on soil quality and rice yield on an inceptisol in Assam, India. Soil Res. https://doi.org/10.1071/SR17001
Carliell CM, Barclay SJ, Naidoo N, Buckley CA, Mulholland DA, Senior E (1994) Anaerobic decolorisation of reactive dyes in conventional sewage treatment processes. Water SA 20:341–344

Clarke A, Anliker R (1980) Organic dyes and pigments. In: Hutzinger O (ed) The handbook of environmental chemistry, vol 3. Part A. Anthropogenic compounds. Springer, Berlin/Heidelberg/New York, pp 181–215

Cranwell PA (1973) Chain-length distribution of n-alkanes from lake sediments in relation to post-glacial environmental change. Freshw Biol 3:259–265

Dadhich RK, Meena RS (2014) Performance of Indian mustard (*Brassica juncea* L.) in response to foliar spray of thiourea and thioglycollic acid under different irrigation levels. Indian J Ecol 41(2):376–378

Dadhich RK, Meena RS, Reager ML, Kansotia BC (2015) Response of bio-regulators to yield and quality of Indian mustard (*Brassica juncea* L. Czernj. and Cosson) under different irrigation environments. J App Nat Sci 7(1):52–57

Datta R, Baraniya D, Wang YF, Kelkar A, Moulick A, Meena RS, Yadav GS, Ceccherini MT, Formanek P (2017a) Multi-function role as nutrient and scavenger of free radical in soil. Sustain MDPI 9:402. https://doi.org/10.3390/su9081402

Datta R, Kelkar A, Baraniya D, Molaei A, Moulick A, Meena RS, Formanek P (2017b) Enzymatic degradation of lignin in soil: a review. Sustain MDPI 1163(9):1–18. https://doi.org/10.3390/su9071163

Dekker LW, Ritsema CJ (1994) How water moves in a water repellent sandy soil: 1. Potential and actual water repellency. Water Resour Res 30(9):2507–2517

Dembitskiĭ VM, Dor I, Shkrob I, Aki M (2001) Branched alkanes and other apolar compounds produced by the cyanobacterium Microcoleus vaginatus from the Negev desert. Bioorg Khim 27(2):130–140

Dhakal Y, Meena RS, De N, Verma SK, Singh A (2015) Growth, yield and nutrient content of mungbean (*Vigna radiata* L.) in response to INM in eastern Uttar Pradesh, India. Bangladesh J Bot 44(3):479–482

Dhakal Y, Meena RS, Kumar S (2016) Effect of INM on nodulation, yield, quality and available nutrient status in soil after harvest of green gram. Legum res 39(4):590–594

Dinel H, Schnitzer M, Mehuys GR (1990) Soil lipids: origin, nature, content, decomposition, and effect on soil physical properties. In: Bollag JM, Stotzky G (eds) Soil Biochem, vol 6. Marcel Dekker, New York, pp 397–429

Doerr SH, Dekker LW, Ritsema CJ, Shakesby RA, Bryant R (2002) Water repellency of soils. Soil Sci Soc Am J 66(2):401–405

Duan Y (2000) Organic geochemistry of recent marine sediment from Nansha Sea, China. Org Geochem 3:159–167

Duan Y, Song JM, Cui MZ, Luo BJ (1998) Organic geochemical studies of sinking particulate material in China sea area – I. Organic matter fluxes and distributional features of hydrocarbon compounds and fatty acids. Sci China (Ser D) 41:208–214

Eckmeier E, Wiesenberg GLB (2009) Short-chain n-alkanes (C_{16-20}) in ancient soil are useful molecular markers for prehistoric biomass burning. J Archaeol Sci 36:1590–1596

Eganhouse RP (2004) Molecular markers and their use in environmental organic geochemistry. The Geochemical Society Special Publications, Elsevier 9:143–158

Folch I, Vaquero MT, Comellas L, Broto-Puig F (1996) Extraction and clean-up methods for improvement of the chromatographic determination of polychlorinated biphenyls in sewage sludge-amended soils: elimination of lipids and sulphur. J Chromatogr A 719:121–130

Ganev S, Arsova A (1980) Methods of determining the strongly acidic and the slightly acidic cation exchange in soil. Soil Sci Agrochem 15:19–33 (in Bulgarian)

Gerke HH, Hangen E, Schaaf W, Hüttl RF (2001) Spatial variability of potential water repellency in a lignitic mine soil afforested with Pinus nigra. Geoderma 102:255–274

Grasset L, Ambles A (1998) Structure of humin and humic acid from an acid soil as revealed by phase transfer catalysed hydrolysis. Org Geochem 29:881–891

Grimalt J, Albaiges J, Alexander G, Hazai I (1986) Predominance of even carbon numbered normal-alkanes in coal seam samples of Nograd Basin (Hungary). Naturwissenschaften 73:729–731

http://ec.europa.eu/environment/archives/waste/sludge/pdf/organics_in_sludge.pdf

https://journal.agrojournal.org/page/en/archive.php?issue=87

Ibañez E, Borrós S, Comellas L (2000) Quantification of sterols, 5α- and 5β-stanolsin sewage sludge, manure and soils amended with these both potential fertilizers. Fresenius J Anal Chem 366:102–105

Jaffé R, Elismé T, Cabrera AC (1996) Organic Geochemistry of seasonally flooded rain forest soils: molecular composition and early diagenesis of lipid components. Org Geochem 25:9–17

Jalal MAF, Read DJ (1983) The organic acid composition of Calluna heathland soil with special reference to phyto- and fungitoxicity. Plant Soil 70:257–272

Jandi G, Schulten HR, Leinweber P (2002) Quantification of long-chain fatty acids in dissolved organic matter and soils. J Plant Nutr Soil Sci 165:133–139

Jansen B, Nierop KGJ, Hageman JA, Cleef AM, Verstraten JM (2006) The straight-chain lipid biomarker composition of plant species responsible for the dominant biomass production along two altitudinal transects in the Ecuadorian Andes. Org Geochem 37:1514–1536

Kachinskii NA (1965) Soil physics. Part I. High school Press, Moscow, pp 323 (in Russian)

Knicker H, Hilscher A, De la Rosa JM, González-Pérez JA, González-Vila FJ (2013) Modification of biomarkers in pyrogenic organic matter during the initial phase of charcoal biodegradation in soils. Geoderma 197:43–50

Koegel-Knabner I, von Lutzow M, Guggenberger G, Flessa H, Marschner B,Matzner E, Ekschmitt K(2005) Mechanisms and regulation of organic matter stabilisation in soils. Geoderma 128, 1–2

Kögel-Knabner I (2002) The macromolecular organic composition of plant and microbial residues as inputs to soil organic matter. Soil Biol Biochem 34:139–162

Kögel-Knabner I, De Leeuw JW, Tegelaar EW, Hatcher PG, Kerp H (1994) A lignin-like polymer in the cuticle of spruce needles: implications for the humification of spruce litter. Org Geochem 21:1219–1228

Kolattukudy PE, Croteau R, Buckner JS (1976) Chemistry and biochemistry of natural Waxes. Elsevier Science Publisher, Amsterdam

Kononova MM (1966) Soil Organic Matter: its nature, its role in soil formation and in soil fertility, 2nd edn. Pergamon Press, Oxford, p 544

Kuhn TK, Krull ES, Bowater A, Grice K, Gleixner G (2010) The occurrence of short chain n-alkanes with an even over odd predominance in higher plants and soils. Org Geochem 41:88–95

Kumar S, Meena RS, Pandey A, Seema (2017a) Soil acidity management and an economics response of lime and sulfur on sesame in an alley cropping system. Int J Curr Microb App Sci 6(3): 2566–2573

Kumar S, Meena RS, Yadav GS, Pandey A (2017b) Response of sesame (*Sesamum indicum* L.) to sulphur and lime application under soil acidity. Int J Plant Soil Sci 14(4):1–9

Kumar S, Meena RS, Bohra JS (2018) Interactive effect of sowing dates and nutrient sources on dry matter accumulation of Indian mustard (*Brassica juncea* L.). J Oilseed Brassica 9(1):72–76

Kuzyakov Y, Bogomolova I, Glaser B (2014) Biochar stability in soil: decomposition during eight years and transformation as assessed by compound-specific 14C analysis. Soil Biol Biochem 70:229–236

Kogel-Knabner I (2000) Analytical approaches for characterizing soil organic matter. Org Geochem 31:609–625

Langencamp H, Part P (2001) Organic contaminants in sewage sludge for agricultural use. European Commission Joint Research Centre, Institute for Environment and Sustainability, Soil and Waste Unit. Brussels, Belgium, pp 1–73

Lechaman J (2007) Bio-energy in the black. Front Ecol Environ 5(7):381–387

Li D, Hockaday WC, Masiello CA, Alvarez PJ (2011) Earthworm avoidance of biochar can be mitigated by wetting. Soil Biol Biochem 43(8):1732–1737

Lichtfouse E, Berthier G, Houot S, Barriuso E, Bergheaud V, Vallaeys T (1995) Stable carbon isotope evidence for the microbial origin of C_{14}-C_{18} n-alkanoic acids in soils. Org Geochem 23(9):849–852

Lichtfouse E, Bardoux G, Mariotti A, Balesdent J, Ballentine DC, Macko SA (1997) Molecular, ^{13}C, and ^{14}C evidence for the allochthonous and ancient origin of C_{16}-C_{18}*n*-alkanes in modern soils. Geochim Cosmochim Acta 61:1891–1898

Lichtfouse E, Wehrung P, Albrecht P (1998) Plant wax n-alkanes trapped in soil humin by noncovalent bonds. Naturwissenschaften 85(9):449–452
Mao J, Nierop KGJ, Rietker KM, Dekker SC (2015) Predicting soil water repellency using hydrophobic organic compounds and their vegetation origin. Soil 1(1):411–425. https://doi.org/10.5194/soil-1-411-2015
Marseillea F, DisnaraJR GB, Noackb Y (1999) n-Alkanes and free fatty acids in humus and A1 horizons of soils under beech, spruce and grass in the Massif-Central (Mont-Lozere), France. Eur J Soil Sci 50:433–441
Matscheko M, Tysklind M, de Wit C, Bergek S, Andersson R, Sellström U (2002) Application of sewage sludge to arable land–soil concentrations of polybrominated diphenyl ethers and polychorinateddibenzo-p-dioxins, dibenzofurans, and biphenyls, and their accumulation in earthworms. Environ Toxicol Chem 21:2515–2525
McElderry CF, Browning M, Amador JA (2005) Effect of short-chain fatty acids and soil atmosphere on Tylenchorhynchus spp. J Nematol 37:71–77
McLachlan MD, Horstmann M, Hinkel M (1978) Polychlorinated dibenzo-*p*-dioxins and dibenzofurans in sewage sludge: sources and fate following sludge application to land. Geochim Cosmochim Acta 42:1523–1532
Meena H, Meena RS (2017) Assessment of sowing environments and bio-regulators as adaptation choice for clusterbean productivity in response to current climatic scenario. Bangladesh J Bot 46(1):241–244
Meena RS, Yadav RS (2014) Phonological performance of groundnut varieties under sowing environments in hyper arid zone of Rajasthan, India. J Appl Nat Sci 6(2):344–348
Meena RS, Yadav RS (2015) Yield and profitability of groundnut *(Arachis hypogaea* L) as influenced by sowing dates and nutrient levels with different varieties. Legum Res 38(6):791–797
Meena RS, Yadav RS, Meena VS (2014) Response of groundnut (*Arachis hypogaea* L.) varieties to sowing dates and NP fertilizers under Western Dry Zone of India. Bangladesh J Bot 43(2):169–173
Meena RS, Dhakal Y, Bohra JS, Singh SP, Singh MK, Sanodiya P (2015a) Influence of bioinorganic combinations on yield, quality and economics of mungbean. Am J Exp Agric 8(3):159–166
Meena RS, Meena VS, Meena SK, Verma JP (2015b) The needs of healthy soils for a healthy world. J Clean Prod 102:560–561
Meena RS, Meena VS, Meena SK, Verma JP (2015c) Towards the plant stress mitigate the agricultural productivity: a book review. J Clean Prod 102:552–553
Meena RS, Yadav RS, Meena H, Kumar S, Meena YK, Singh A (2015d) Towards the current need to enhance legume productivity and soil sustainability worldwide: a book review. J Clean Prod 104:513–515
Meena RS, Yadav RS, Reager ML, De N, Meena VS, Verma JP, Verma SK, Kansotia BC (2015e) Temperature use efficiency and yield of groundnut varieties in response to sowing dates and fertility levels in Western Dry Zone of India. Am J Exp Agric 7(3):170–177
Meena H, Meena RS, Singh B, Kumar S (2016a) Response of bio-regulators to morphology and yield of clusterbean [*Cyamopsis tetragonoloba* (L.) Taub.] under different sowing environments. J App Nat Sci 8(2):715–718
Meena RS, Bohra JS, Singh SP, Meena VS, Verma JP, Verma SK, Shiiag SK (2016b) Towards the prime response of manure to enhance nutrient use efficiency and soil sustainability a current need: A book review. J Clean Prod 112:1258–1260
Meena RS, Gogaoi N, Kumar S (2017a) Alarming issues on agricultural crop production and environmental stresses. J Clean Prod 142:3357–3359
Meena RS, Kumar S, Pandey A (2017b) Response of sulfur and lime levels on productivity, nutrient content and uptake of sesame under guava (*Psidium guajava* L.) based agri-horti system in an acidic soil of eastern Uttar Pradesh, India. J Crop Weed 13(2):222–227
Meena RS, Meena PD, Yadav GS, Yadav SS (2017c) Phosphate solubilizing microorganisms, principles and application of microphos technology. J Clean Prod 145:157–158

Meena H, Meena RS, Lal R, Singh GS, Mitran T, Layek J, Patil SB, Kumar S, Verma T (2018a) Response of sowing dates and bio regulators on yield of clusterbean under current climate in alley cropping system in eastern U.P., Indian. Legum Res 41(4):563–571
Meena RS, Kumar V, Yadav GS, Mitran T (2018b) Response and interaction of *Bradyrhizobium japonicum* and *Arbuscular mycorrhizal* fungi in the soybean rhizosphere: A review. Plant Growth Regul 84:207–223
Montanarella L, Lugato E (2013) The application of biochar in the EU: challenges and opportunities. Agronomy 3(2):462–473
Morley CP, Mainwaring KA, Doerr SH, Douglas P, Llewellyn CT, Dekker LW (2005) Organic compounds at different depths in a sandy soil and their role in water repellency. Aust J Soil Res 43:239–249
Moucawi J, Fustec E, Jambu P, Jacquesy JC (1981) Decomposition of lipids in soils: free and esterified fatty acids, alcohols and ketones. Soil Biol Biochem 13:461–468
Mudge SM (2005) Fatty alcohols- a review of their natural synthesis and environmental distribution. Soap Deterg Assoc 132:1–141
Mudge SM (2010) Fatty alcohols in the terrestrial environment. American Cleaning Institute pp 70
Naafs DFW, van Bergen PF (2002) A qualitative study on the chemical composition of ester-bound moieties in an acidic andosolic forest soil. Org Geochem 33:189–199
Naafs DFW, van Bergen PF, de Jong MA, Oonincx A, de Leeuw JW (2004) Total lipid extracts from characteristic soil horizons in a podzol profile. Eur J Soil Sci 55:657–669
Nierop KGJ, Naafs DFW, Verstraten JM (2003) Occurrence and distribution of ester-bound lipids in Dutch coastal dune soils along a pH gradient. Org Geochem 34:719–729
Ohkouchi N, Kawamura K, Taira A (1997) Molecular paleoclimatology: reconstruction of climate variabilities in the late Quaternary. Org Geochem 27:173–183
Otto A, Shunthirasingham C, Simpson MJ (2005) A comparison of plant and microbial biomarkers in grassland soils from the Prairie Ecozone of Canada. Org Geochem 36:425–448
Page-Dumroese DS, Robichaud PR, Brown RE, Tirocke JM (2015) Water repellency of two forest soils after biochar addition. Trans ASABE 58(2):335–342
Pancost RD, Boot CS (2004) The palaeoclimatic utility of terrestrial biomarkers in marine sediments. Mar Chem 92:239–261
Parlanti E, HitaC JP, DinelH AA (1994) The internal double-bond insertion: a side reaction of aliphatic hydrocarbons degradation in soil. Soil Biol Biochem 26:1375–1378
Pascual JA, Garcia C, Hernandez T, Ayuso M (1997) Changes in the microbial activity of an arid soil amended with urban organic wastes. Biol Fertil Soils 24:429–434
Patureau D, Laforie M, Lichtfouse E, Caria G, Denaix L, Schmidt JE (2007) Fate of organic pollutants after sewage sludge spreading on agricultural soils: a 30-years field-scale recording. Water Practice Technol 2(1):wpt2007008
Petkova G, Nedyalkova K, Mikova A, Atanassova I (2018) Microbiological characteristics of biochar amended alluvial meadow soil. Bul J Agric Sci 24(Suppl 2):81–84
Quénéa K, Derenne S, Largeau C, Rumpel C, Mariotti A (2004) Variation in lipid relative abundance and composition among different particle size fractions of a forest soil. Org Geochem 35:1355–1370
Quénéa K, Largeau C, Derenne S, Spaccini R, Bardoux G, Mariotti A (2006) Molecular and isotopic study of lipids in particle size fractions of a sandy cultivated soil (Cestas cultivation sequence, southwest France): sources, degradation, and comparison with Cestas forest soil. Org Geochem 37:20–44
Ram K, Meena RS (2014) Evaluation of pearl millet and mungbean intercropping systems in Arid region of Rajasthan (India). Bangladesh J Bot 43(3):367–370
Reisch MS (1996) Asian textile dye makers are a growing power in changing market. Chem Eng News 15:10–12
Réveillé V, Mansuy L, Jardé É, Garnier-Sillam É (2003) Characterisation of sewage sludge-derived organic matter: lipids and humic acids. Org Geochem 34(4):615–627

Rieley G, Collier RJ, Jones DM, Eglinton G (1991) The biogeochemistry of Ellesmere Lake, U.K.-I. Source correlation of leaf wax inputs to the sedimentary lipid record. Org Geochem 17:901–912

Schnitzer M, Hindle CA, Meglic M (1986) Supercritical gas extraction of alkanes and alkanoic acids from soil and humic material. Soil Sci Soc Am J 50:913–919

Sheoran V, Sheoran AS, Poonia P (2010) Soil reclamation of abandoned mine land by revegetation: a review. Int J Soil Sed Water 3(2):13

Sihag SK, Singh MK, Meena RS, Naga S, Bahadur SR, Gaurav YRS (2015) Influences of spacing on growth and yield potential of dry direct seeded rice (*Oryza sativa* L) cultivars. The Ecoscan 9(1–2):517–519

Simeonova T, Benkova M, Nenova L, Atanassova I (2018) Chemical composition of soil solutions of technosols from a coal mine region in South-Eastern Europe. Bul J Soil Sci 3(1):4–12

Simoneit BRT, Sheng G, Chen X, Fu J, Zhang J, Xu Y (1991) Molecular marker study of extractable organic matter in aerosols from urban area of China. Atmos Environ 25A:2111–2129

Steinbeiss S, Gleixner G, Antonietti M (2009) Effect of biochar amendment on soil carbon balance and soil microbial activity. Soil Biol Biochem 41(6):1301–1310

Stevenson FJ (1994) Humus chemistry: genesis, composition, reactions, 2nd edn. Wiley, New York, p 496

Tissot B, Welte DH (1984) Petroleum formation and occurrence. Springer, New York, p 699

van Bergen PF, Bull ID, Poulton PR, Evershed RP (1997) Organic geochemical studies of soils from the Rothamsted classical experiments – I. Total lipid extracts, solvent insoluble residues and humic acids from Broadbalk Wilderness. Org Geochem 26:117–135

Varma D, Meena RS, Kumar S (2017a) Response of mungbean to fertility and lime levels under soil acidity in an alley cropping system in Vindhyan Region, India. Int J Chem Stud 5(2):384–389

Varma D, Meena RS, Kumar S, Kumar E (2017b) Response of mungbean to NPK and lime under the conditions of Vindhyan Region of Uttar Pradesh. Legum Res 40(3):542–545

Verma JP, Jaiswal DK, Meena VS, Meena RS (2015a) Current need of organic farming for enhancing sustainable agriculture. J Clean Prod 102:545–547

Verma JP, Meena VS, Kumar A, Meena RS (2015b) Issues and challenges about sustainable agriculture production for management of natural resources to sustain soil fertility and health: a book review. J Clean Prod 107:793–794

Verma SK, Singh SB, Prasad SK, Meena RN, Meena RS (2015c) Influence of irrigation regimes and weed management practices on water use and nutrient uptake in wheat (*Triticum aestivum* L. Emend. Fiori and Paol.). Bangladesh J Bot 44(3):437–442

Wang D-H, Tao W-Y (2009) Antitumor activity in vitro and volatile components of metabolites from myxobacteria Stigmatella WXNXJ-B. Afr J Microbiol Res 3(11):755–760

Wang Y, Fang X, Zhang T, Li Y, Wu Y, He D, Wang Y (2010) Predominance of even carbon-numbered n-alkanes from lacustrine sediments in Linxia Basin, NE Tibetan Plateau: implications for climate change. Appl Geochem 25(10):1478–1486

Wiesenberg GLB, Gocke M, Kuzyakov Y (2010a) Fast incorporation of root-derived lipids and fatty acids into soil – evidence from a short term multiple pulse labelling experiment. Org Geochem 41:1049–1055

Wiesenberg GLB, Dorodnikov M, Kuzyakov Y (2010b) Source determination of lipids in bulk soil and soil density fractions after four years of wheat cropping. Geoderma 156(3–4):267–277

Wong MH (2003) Ecological restoration of mine degraded soils, with emphasis on metal contaminated soils. Chemosphere 50(6):775–780

WRBSR (2006) A framework for international classification correlation and communication, World soil resources reports 103. FAO, Rome, p 132

Yadav GS, Babu S, Meena RS, Debnath C, Saha P, Debbaram C, Datta M (2017a) Effects of godawariphosgold and single supper phosphate on groundnut (*Arachis hypogaea*) productivity, phosphorus uptake, phosphorus use efficiency and economics. Indian J Agric Sci 87(9):1165–1169

Yadav GS, Lal R, Meena RS, Babu S, Das A, Bhomik SN, Datta M, Layak J, Saha P (2017b) Conservation tillage and nutrient management effects on productivity and soil carbon sequestration under double cropping of rice in North Eastern Region of India. Ecol Indic. http://www.sciencedirect.com/science/article/pii/S1470160X17305617

Yadav GS, Lal R, Meena RS, Datta M, Babu S, Das LJ, Saha P (2017c) Energy budgeting for designing sustainable and environmentally clean/safer cropping systems for rainfed rice fallow lands in India. J Clean Prod 158:29–37

Yadav GS, Das A, Lal R, Babu S, Meena RS, Saha P, Singh R, Datta M (2018a) Energy budget and carbon footprint in a no-till and mulch based rice–mustard cropping system. J Clean Prod 191:144–157

Yadav GS, Das A, Lal R, Babu S, Meena RS, Patil SB, Saha P, Datta M (2018b) Conservation tillage and mulching effects on the adaptive capacity of direct-seeded upland rice (*Oryza sativa* L.) to alleviate weed and moisture stresses in the North Eastern Himalayan Region of India. Arch Agron Soil Sci. https://doi.org/10.1080/03650340.2018.1423555

Zech W, SenesiN GG, Kaiser K, Lehmann J, Miano TM, Miltner A, Schroth G (1997) Factors controlling humification and mineralization of soil organic matter in the tropics. Geoderma 79:117–161

Zelles L (1997) Phospholipid fatty acid profiles in selected members of soil microbial communities. Chemosphere 35:275–294

Zelles L (1999) Fatty acids patterns of phospholipids and lipopolysaccharides in the characterisation of microbial communities in soil: a review. Biol Fertil Soils 29:11–129

Zheleva E, Bogdanov B, Tsolova M (2004) New eco-logical and technical problems of reclamation of disturbed terrains from Maritza-Iztok coal mines. Manag Sustain Dev, 1–2:323–328 (in Bulgarian)

Green Technologies for Restoration of Damaged Ecosystem

10

Shivani Garg and Rashmi Paliwal

Abstract

Different industrial, mining, agricultural and domestic activities produce a huge amount of wastes as by-products, which contaminate soil, surface water, and groundwater and cause ecological problems. Natural and traditional techniques are not very much sufficient to manage such type of pollutants/contaminants. The most affected area of environment is soil that indirectly affects biological interaction between plants and microorganisms. There is a need of highly eco-friendly approach to remove and manage such pollutants. Phytoremediation is a technique which remediates the contaminated site with and by the environmental phenomenon. Plants are the main tool of remediation in this technique. Phytoremediation includes the plant-mediated remediation of pollutants, like metal, organic, and hazardous wastes by its subclasses phytodegradation, phyto-volatilization, phytoextraction, etc. This chapter includes all the phytoremediation techniques used to treat different types of contaminant site. It is an environment-friendly green technique which focused on the combined use of more than one phytoremediation approach for the successful remediation of the polluted area under field conditions.

Keywords

Phytodegradation · Phytoextraction · Phytotechnology · Phytovolatilization · Soil degradation · Restoration

S. Garg (✉) · R. Paliwal
Institute of Environmental Studies, Kurukshetra University, Kurukshetra, Haryana, India
e-mail: shivanigarg397@gmail.com

R. S. Meena (ed.), *Soil Health Restoration and Management*,
https://doi.org/10.1007/978-981-13-8570-4_10

Abbreviations

Cd	Cadmium
EDTA	Ethylene diamine tetra-acetic acid
HMX	High velocity military explosive
PAH	Polycyclic aromatic hydrocarbons
PCB	Polychlorinated biphenyl
RDX	Research department explosive

10.1 Introduction

The natural environment is continuously polluted by toxic pollutant and/or contaminants day by day due to rapid industrialization and increasing human population. These contaminants cause serious health hazard due to their toxic effects and bioaccumulation properties throughout in food chain. In order to support the fast-growing human population, various developmental activities, for example, rapid industrial expansion, mineral mining, oil exploration, food processing, conventional crop management, intense agriculture practices, health care, vehicular expansion, etc., have dramatically added the heavy metal contaminants and other xenobiotic compounds (petroleum hydrocarbons (PHC) and polycyclic aromatic hydrocarbons (PAHs) into the ecosphere). In addition to this, roadways and automobiles also contribute substantial load of heavy metal to environment during traffic emissions (Onat et al. 2013; Yadav et al. 2018b). In many countries there are many contaminated sites that are affected with hazardous substances and causing significant harm to the human and environment. According to a recent estimation, there are approximately 250,000 contaminated sites in EEA member countries needing remediation (EEA 2007). In India, Gujarat, Maharashtra, and Andhra Pradesh have been reported as the most hazardous waste-contributing states (80%) (CPCB 2011).

Metals and other organic and inorganic contaminants are the most harmful environmental pollutants and very difficult to remediate in soils and sediments (Padmavathiamma and Li 2007; Jomova et al. 2011; Bernhoft 2012; Auger et al. 2013; Ashoka et al. 2017). Therefore, these days, restoration of sites damaged with such hazardous pollutants has become the most serious issue in the world. In the past, conventional methods for restoration of degraded sites have not only been proven as expensive, tedious, and inadequate but also generate secondary wastes. However, application of plant-mediated decontamination of polluted systems or sites, also known as phytoremediation, has many scientific and economic benefits as it is an elegant and low-cost approach. Phytoremediation uses plants to decontaminate soils as well as waters through various mechanisms such as extraction, sequestration, and detoxification. Application of phytoremediation techniques includes decontamination of sites affected by hazardous waste such as heavy metal where other techniques become very expensive and/or impractical. Phytoremediation is a collective term that is used for various applications, like phytodegradation, phytovolatilization, phytostimulation, phytostabilization, rhizoremediation, rhizofiltration, and phytoextraction. Application of various terrestrial and aquatic plants has also been reported to remediate the heavy metal-contaminated soil and aquatic systems, respectively (Sharma et al. 2015; Meena and Meena 2017).

Microbially assisted bioremediation strategies, such as phytoextraction or phytostabilization (Phieler et al. 2013; Meena et al. 2015d), may increase the beneficial aspects and can be viewed as potentially useful methods for application in remediation of low and heterogeneously contaminated soil (Giri et al. 2015; Pérez-Sirvent et al. 2017; Kumar et al. 2017b). These techniques are not only cost-effective but also aesthetically beneficial and acceptable for long-term applications. For phytoremediation process, more than 400 plant species of different families have been identified to withstand high metal concentration as well as fix them in their tissues (Guerinot and Salt 2001; Pal and Rai 2010; Yadav et al. 2018a). Heavy metal absorption and transformation by plant system is more likely to depend upon the solubility and complexity of heavy metal (Rungwa et al. 2013; Buragohain et al. 2017). However, plants have developed advanced mechanisms for extraction and accumulation of toxic metal species at high concentration. These mechanisms include immobilization of heavy metals in cell walls, preventing their contact with protoplasm, compartmentalization in vacuole and chelation in the cytoplasm (Pal and Rai 2010; Singh et al. 2015; Varma et al. 2017a).

Plants can also transform organic chemicals into less toxic forms. Root exudates and enzymes release stimulate the degradation of organic chemicals in rhizospheric zone, thereby building the soil organic carbon pool. Remediation of the technogenic barrens around the Pechenga nickel works on the Kola Peninsula resulted in the improvement of the soil properties, namely, in a decrease in acidity and enrichment with nutrients, which continued for several years (Koptsik et al. 2014; Meena and Yadav 2015). The landscape of mangrove wetlands on the east coast of India was restored by digging the canal side and planting species that tolerate a wide range of salinity such as *Aegiceras corniculatum, Bruguiera gymnorrhiza, Rhizophora apiculata, Rhizophora mucronata, and Xylocarpus moluccensis* along the slope; a total of 520 ha of degraded mangroves area were restored (Bhakta et al. 2016; Dadhich and Meena 2014). Vangronsveld et al. (2007) proposed an economically interesting strategy of phytoextraction for soil remediation. Selected food crops as rape seed, maize, and wheat are grown on contaminated soil, harvested and treated with different techniques, in order to obtain metal recovery and biogas production. The present chapter is designed to understand the different and advanced techniques of phytoremediation along with the traditional techniques for successful application of phytotechnology for degraded site reclamation.

10.2 Soil Degradation

Soil is unconsolidated rock with organic and inorganic material and considered as a fundamental source for the environment. Soil degradation is the process when soil loses its fertility and productivity capacity due to misuse or excessive use of soil as resource by humans. Degraded soil results in poor or no production and known as problem soil. Human activities including the intensification in agriculture practices significantly reduce the soil cover and accelerate the rate of soil degradation. Although the climatic and geological events also contribute to soil degradation, change in recent trends of human activities to fulfill the demand of increasing population has exceeded the natural erosion rates by several orders of magnitude. According to UNEP (1992), soil degradation is deterioration in soil productivity and land capability. The increasing population and the food gap put more pressure on land use. The adverse condition of which includes change in chemical structure of soil (alkalization, salinization, waterlogging, etc.) and continuous accumulation of toxic compounds in soil deteriorating the soil structure.

Soil degradation has become a serious problem in India. Land clearing and unsustainable development on forest land, deforestation, overgrazing, improper discarding of industrial effluents and wastes over lands, unmanaged landfills, surface mining, and industrial development are the other human activities responsible for different types of land degradation (Bhattacharyya et al. 2015; Meena et al. 2016a) (Fig. 10.1). Various factors that increase the process of soil degradation are discussed below.

10.2.1 Unsustainable Forest Resource Utilization

Unsustainable harvesting of forest products, fuelwood, and fodder extraction, extensive utilization of forest land for agriculture purpose, forest fires, overgrazing, etc., are all of the major degradation force for soil. Overgrazing and deforestation have resulted in development of more and more wasteland in different states of India (Bhattacharyya et al. 2015; Yadav et al. 2017c). Shifting cultivation/jhum cultivation/slash-and-burn agriculture is an agricultural practice predominant in

Fig. 10.1 Different types of threats to soil

northeastern parts of India. It involves conversion of forest area to agriculture land by clear cutting and burning of forest. The crops are cultivated till the soil nutrients are exhausted or the site is overtaken by weeds. Therefore, the process is mainly reported for the destruction and degradation of land in the region (Giri et al. 2015; Meena et al. 2015e).

10.2.2 Urbanization

Population growth, economic development, industrialization, etc. are the indirect causes of deterioration of soil quality. Processes of urban development progressively encroach the land that would otherwise be useful for agriculture, forestry, grassland, rangeland, pasture land, and wild vegetation growth. Ever-increasing human population and global food demand indicate that the production will need to increase by 40–70% by 2050 (World Resources Institute 2014; Karlen and Rice 2015; Kumar et al. 2018).

10.2.3 Mining

Mining and its subsequent activities cause a huge and long-lasting impact on land and ecosystem. It degrades the land to a significant extent that it took years to restore the associated ecosystem. Wastes generated from mineral/ore mining activities are generally inert solid materials and toxic in nature. The toxic substances present in the ore include heavy metals such as iron, mercury, arsenic, lead, zinc, cadmium, etc. (Giri et al. 2015; Meena et al. 2015b). Heavy metals leaching out of the stored waste piles contaminate the soil and associated water bodies in the environment. Mineral mining generates a huge quantity of waste or overburden and affect a large area of land severely (Table 10.1).

10.2.4 Unsustainable Agriculture

Unsustainable agricultural practices damage the land in several ways. The Green Revolution has brought about several technological developments leading to the production of high yield crop varieties and use of chemicals fertilizers and pesticides for

Table 10.1 Mineral production, waste generation, and land affected in 2005–2006

Mineral	Production (Mt)	Overburden/waste (Mt)	Estimated land affected (ha)
Coal	407	1493	10,175
Limestone	170	178	1704
Bauxite	12	8	123
Iron ore	154	144	1544
Others	9	19	–

Source: Bhattacharyya et al. (2015), Sahu and Dash (2011)

making the country self-sufficient. However, such developments intensify the land use and cover the land under irrigation, which further had negative impact on land and biodiversity (Bhattacharyya et al. 2015). Intensive agricultural activities lead to erode soil, bring waterlogging conditions to the soil, and make the soil sodic and saline. Some consequences of intensive agricultural activities are discussed as follows:

- Intensive cropping reduces the nutrients in soil.
- Excessive tilling practices degrade the soil quality in many ways, such as reduced soil organic matter, degrade the soil microbial activity and community structure, and change the soil physical property.
- Burning of crop residue causes loss of soil organic matter and other nutrients.
- Poor irrigation and improper management cause in development of waterlogging conditions in many areas.
- Poor irrigation and drainage management development of saline-sodic soils in many arid and semi-arid regions of India.
- Indiscriminate application of chemical fertilizers, pesticides, and sewage sludge or municipal wastes contaminates the soil with toxic substances that further affect the food chain and the ecosphere.

10.3 Techniques of Phytoremediation

Phytoremediation occurs by growing plants on polluted sites so that polluting components are absorbed by the root system of the plants and transported and translocated into various parts. Phytoremediation is dependent on various natural factors and a long-term process, i.e., plants take a long time to grow so that it is beneficial when it is used in combination with other processes (Garg 2017). Certain important factors require considering before choosing the plants for the remediation of contaminated sites; these include plant growth rate, adaptability and tolerance to edaphic conditions, root system, resistance to disease and pest, and the time required to achieve desired remediation results. The process also depends upon the associated microbial interaction to minimize the harmful effects of xenobiotics (Lee 2013; Rani et al. 2018). Heavy metals, such as arsenic (As), cadmium (Cd), lead (Pb), nickel (Ni), copper (Cu), mercury (Hg), chromium (Cr), etc., are well documented for causing serious health issues in environment (Kavamura and Esposito 2010; Li et al. 2013). Heavy metals in soil are also known to affect the soil microbial activities (Renella et al. 2006; Wyszkowska et al. 2009; Tripathi et al. 2014; Meena et al. 2018a; Datta et al. 2017a). Soil contaminated with heavy metals at high concentration can significantly reduce the plant growth. For example, Cd and Hg were reported to reduce the dry weight of root and shoot from 13.8% to 70.5% and seed yield by 40% (Ghani 2010). Various plants have been identified and reported to accumulate heavy metal (Table 10.2). Different techniques and mechanisms used by the plant for remediation of contaminated sites are discussed in Table 10.3. The process of remediation of each technique is different (Fig. 10.2), and medium of absorption is also varying from contaminants to contaminants.

Table 10.2 List of plant species used as tool for specific heavy metal

Metal accumulated	Plant species	References
Cr, As, Cd, Hg, Pb	*Eichornia crassipes*	Mhatre and Pankhurst (1997); Brooks (1998)
Cd, Zn	*Zea mays*	Felix (1997)
Cu, Co, Zn	*Thlaspi alpestre*	Brooks (1998)
Pb, Zn, Cu, Cr, Cd	*Populus nigra*	Wagner (1993)
Cu, Cd, Pb, Cr	*Taraxacum officinale*	Bini et al. (2000)
Cs	*Amaranthus retroflexus*	Fuhrmann et al. (2002)
Diesel fuel	Fern (*Azolla pinnata*)	Cohen et al. (2002)
Pb	Maize (*Zea mays* L.)	Hadi et al. (2010)
Pb, Cd, Cu	*Echinochloa crus-galli*	Kim et al. (2010)
Fe, Zn, Pb, Mn, Cu, Cd, Ni, and Cr	*S. cordifolia*	Anand et al. (2017)

Table 10.3 Different types of phytoremediation techniques and plant mechanisms

Techniques	Surface medium	Plant mechanism
Phytodegradation	Soils, groundwater within rhizosphere	Enhances the microbial degradation in rhizosphere
Phytovolatilization	Soils and groundwater	Plants evaporate or transpire selenium, mercury, and volatile hydrocarbon
Phytostabilization	Soils, groundwater, mine tailing	Root exudates cause metal to precipitate and become less available
Phytoextraction	Soils	Uptake and concentration of metal via direct uptake into the plant tissue with subsequent removal of the plants
Rhizofiltration	Surface water and water pumped	Uptake of metals into plant roots

Source: Vidali (2001)

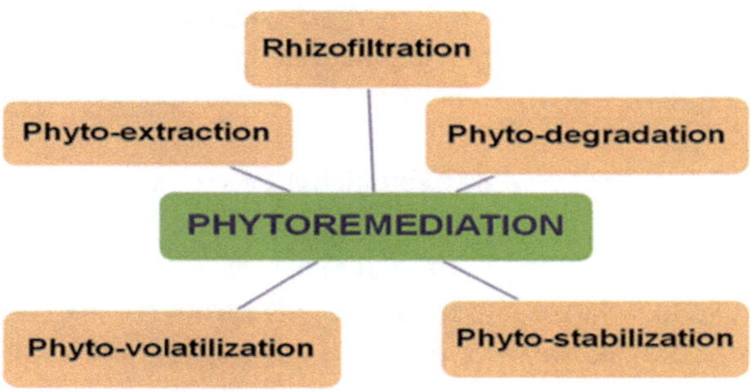

Fig. 10.2 Different types of phytoremediation techniques

10.3.1 Phytodegradation

Phytodegradation is also called phytotransformation. Phytodegradation is the ability of plants to take up and degrade the contaminants. Contaminants are degraded through internal enzymatic activity and metabolic processes. The metabolic processes for degradation are affected by the concentration and composition of contaminants, soil conditions, and plant species. Many different compounds can be removed from the environment by phytodegradation method such as solvents in groundwater, petroleum and aromatic compounds in soils, and volatile compounds in the air (Newman and Reynolds 2004). The biological processes occurred with contaminants within the plant itself and can be phytoextracted. Plants involve certain enzymes to catalyze the process of degradation, for example, oxygenases that have been identified in plants and are able to catalyze the degradation of aliphatic and aromatic hydrocarbons. Some of the plant enzymes can completely degrade many hydrocarbons that resulted into carbon dioxide and water as the end products (McCutcheon and Schnoor 2003; Meena and Yadav 2014). Phytodegradation of explosives, such as RDX (hexahydro-1,3,5-trinitro-1,3,5-triazine), GTN (Glycerol trinitrate), HMX (octahydro-1,3,5,7-tetranitro1,3,5,7-tetrazocine) and TNT (2,4,6-trinitrotoluene), etc., has always been a topic of interest for many workers as it is important to remove such contaminants from the affected soil and water that may otherwise accumulate in food chain.

Microorganisms associated with the plant roots have been reported to enhance the degradation of explosives (Van Aken 2009; Dhakal et al. 2015). Van Aken et al. (2004) reported an endophytic symbiotic bacteria *Methylobacterium populi* that lives within poplar, which can degrade RDX and HMX. Ekman et al. (2003) reported the possible involvement of detoxification enzymes such as cytochrome P450, glutathione S-transferase, an ABC transporter, and a probable nitroreductase in plant-mediated degradation of TNT. Diverse studies have also reported the phytodegradation of chlorinated hydrocarbons. Godsy et al. (2003) reported the anaerobic microbial degradation of trichloroethene (TCE) under the mature cottonwood plantation. Robson et al. (2003) assessed and reported four naturalized grasses (*Agropyron pectiniforme*, *Bromus inermis*, *Phleum pratense*, and *Poa pratensis*), three naturalized legumes (*Medicago sativa*, *Melilotus officinalis*, and *Trifolium repens*), two native forbs (*Artemisia frigida* and *Potentilla pensylvanica*), one native grass (*Bromus ciliatus*), and two native legumes (*Glycyrrhiza lepidota* and *Psoralea esculenta*) to exhibit phytoremediation potential on the basis of their in crude oil-contaminated soil.

10.3.2 Phytovolatilization

Phytovolatilization is the application of plant to remediate the contaminants that tend to volatize into the atmosphere from the aboveground plant parts such as leaf stomata and stems or from soil due to plant root activities (indirect phytovolatilization). Phytovolatilization occurs as the plants grow and take up the contaminants through their roots. Contaminants like mercury can pass through the plants to the leaves and volatilize into the atmosphere (Bolan et al. 2011; Hashmi et al. 2017; Meena et al. 2015c; Yadav et al. 2017b). Study on removal of highly toxic species of mercury, i.e., methyl mercury, was done through phytovolatilization by genetically modified tobacco plant. The toxic mercury species was converted to relatively safe and less toxic elemental species and phytovolatized by tobacco plant into the atmosphere (EPA 2009).

10.3.3 Phytostabilization

Toxic compounds or the pollutants are tending to migrate from one site to another through wind, water, soil dispersion and leaching, thereby enhance the chance of off-site contamination. Phytostabilization aims to hold or restrict the contaminants within the vadose zone or rhizosphere. This way, phytostabilization is a recommended technique to avoid the off-site contamination. Microbes associated with plant roots may further enhance the degradation of metals and organic toxicants due to restriction in mobility of the contaminants. Phytostabilization is best suited for enhancing the soil fertility and restoration of mining sites. Metal contaminants at waste sites are specifically phytostabilized as shown in Fig. 10.3. Plants either uptake or immobilize the pollutant, thereby reducing their mobility and transportation in the environment (Pulford and Watson 2003; Varma et al. 2017b; Meena et al. 2017c)

Plants naturally provide a barrier to the movement of contaminant by preventing erosion and leaching. Plants such as grasses, many herbaceous and wetland species with fibrous roots are generally used in the phytostabilization process to control bulk soil migration and/or prevent contaminant migration. Radziemska et al. (2017) evaluate the potential of *Festuca rubra* L. for phytostabilization of Cu-contaminated soil using halloysite as an immobilizing agent. The addition of halloysite significantly increased plant biomass, stabilizing the Cu in the roots, thereby reducing its toxicity to the aerial parts of the *F. rubra* L. (Radziemska et al. 2017; Dhakal et al. 2016). Therefore, phytostabilizing can be considered as a suitable approach to reducing the environmental risk. The process and potential actions of phytostabilization has been well documented by Bolan et al. (2011) (Fig. 10.4). Rhizosphere is the zone close to the plants' root system and highly influenced by plant roots and associated microbial assemblage activities. The microbial population and the plant growth are affected by soil characteristics such as pH, redox potential, nutrients availability, etc. (Fig. 10.4). However, some soil amendments such as nutrient supply and immobilization of contaminants can enhance the process of

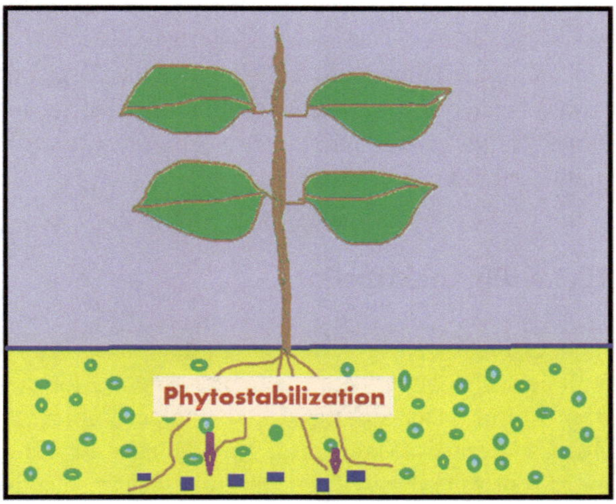

Fig. 10.3 Diagram showing process of phytostabilization

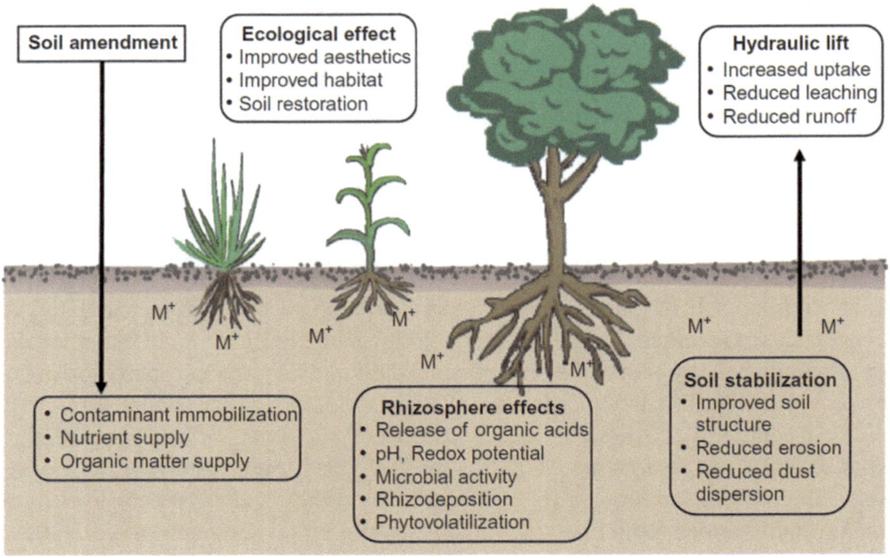

Fig. 10.4 Schematic diagram illustrating the potential action of phytostabilization on contaminants in soil. M^+ (Metalloid). (Source: Bolan et al. 2011)

phytostabilization (Adriano et al. 2004; Meena et al. 2016b). Microorganisms and plant exudates play an important role in this regard. Plant root exudates stimulate microbial activity and biochemical transformations mineralization of metals in the rhizosphere (Anderson et al. 1993; Paterson 2003; Bolan et al. 2013; Yadav et al. 2017a).

Phytotechnology mechanism to control the migration of surface water is also known as phytohydraulics, in which plant cover is used to control the infiltration control by rapid uptake of water by plants and remove through evapotranspiration processes and also by maximization of the evaporation from the soil. The plants' root system, densely developed near the water table act as an effective hydraulic pump and uptake water at very large rate followed by its removal through transpiration and evaporation from soil (Bolan et al. 2011; Datta et al. 2017b). Barton et al. (2005) studied the phytostabilization of a coal waste containing landfill site using loblolly (*Pinus taeda*) and Virginia (*Pinus virginiana*) and observed reduction in the runoff and drainage volume.

10.3.4 Phytoextraction

Phytoextraction also refers to phytosequestration, phytoabsorption, and phytoaccumulation. It is the process where plants uptake the contaminants by the root system from soil or water and translocate and accumulate them in plant biomass such as shoots and leaves (Mahar et al. 2016). Plants sequester certain contaminants present in the root zone through their physiological mechanism. Therefore, the constituent of the contaminants present in soil or water is first absorbed by the root system of plants and then transported through plant transport mechanisms to the aboveground tissues (Fig. 10.5). The plant may store the toxicant into the cell vacuoles after converting them into less toxic forms or may process the contaminants through phytodegradation mechanisms and phytovolatilize in the plant's transpiration stream.

Plants uptake bioavailable inorganic metals or elements such as As, Cd, Cu, Ni, Se, and Zn, which are often considered as environmental contaminants and also plant essential nutrients. Yang et al. (2017) reported that tobacco accumulates high Cd and produces large biomass after being cut and also showed enhanced removal efficiency of Cd from moderately contaminated acidic soils.

10.3.5 Rhizofiltration

Rhizofiltration is the process of removing recalcitrant compounds or xenobiotics from aqueous system through root system of plants. Plant roots absorb and concentrate the contaminants in roots or aboveground plant parts (Tomé et al. 2008; Meena et al. 2014). Rhizofiltration is effective in wetlands where all the contaminants in water are allowed to be absorbed by plants' root system. Contaminants such as lead, chromium (III), uranium, and arsenic (V) can easily be absorbed by the plants' roots. Hydroponically cultivated plants can rapidly remove heavy metals from water and concentrate them in the roots and shoots (Fig. 10.6). Roots of plants are capable of absorbing large quantities of lead and chromium from soil water or from water that is passed through the root zone of densely growing vegetation. Thayaparan et al. (2013) observed that the removal efficiency of a plant depends on the exposure period and reported that the maximum efficiency of *Azolla pinnata* to remove Pb (II) from aqueous solution through rhizofiltration is (1383 mg kg^{-1}) under the effect of nutrient concentration, initial metal concentration, and exposure period.

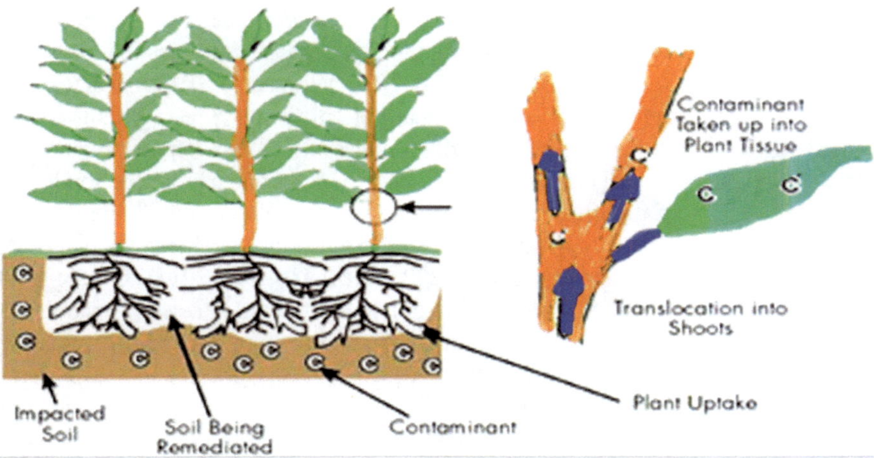

Fig. 10.5 Diagram showing process of phytoextraction. (ITRC 2001)

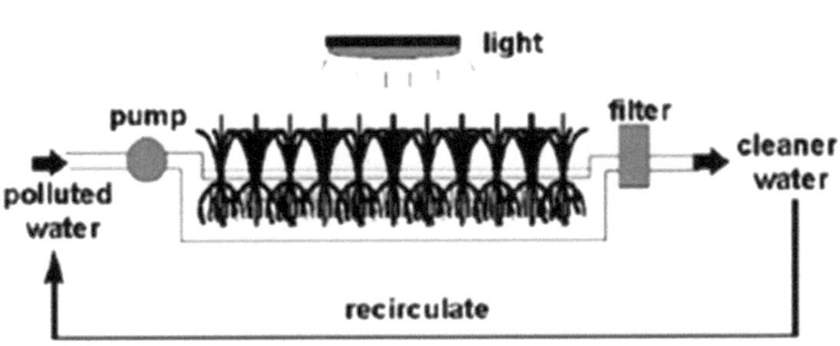

Fig. 10.6 Sketch of process of removal of contaminants from wastewater by hydroponically

Wetlands have always been a great success in treating metals for many years. The aerobic and anaerobic zones of wetlands comprising root systems and sediments facilitate the sorption and precipitation of toxic metals (MoEF 2011). After the treatment, the plants can be harvested and the contaminants like heavy metal can be recovered. Rhizofiltration involves various mechanisms to remove the contaminants from water, viz., surface absorption, intracellular uptake, deposition in vacuoles, and translocation to the shoot or precipitation by plant exudates. Therefore, like other phytoremediation techniques, rhizofiltration is also cost-effective and minimally disruptive environmentally (MoEF 2011).

10.3.6 Rhizoremediation

Rhizoremediation is an elegant plant–microbial-associated process of remediation and/or degradation of xenobiotic compounds. It involves the sequestration, immobilization, or retention of toxicants within the root zone and removal of contaminants from the soil/wastewater with the combined action of plants and associated microorganisms. Certain toxicants such as heavy metals may cause toxicity in plants. Microbes associated with the rhizosphere have been found to detoxify the metal phytotoxicity as they use various direct and indirect mechanisms to promote growth in plants (Kamnev and Lelie 2000; Gusain et al. 2017; Verma et al. 2015c). Rhizosphere support a diverse population of microbes that not only provide the nutrients to support plant growth but also able to degrade or convert the toxic compounds into less toxic compounds in soil. The plant root system penetrates and aerates the soil layers. Thereby, distributing the rhizobacteria through soil as well as solubilizing the toxicants in soil water. This way, the toxicants are biologically available to the plants and microbes (Kamaludeen and Ramasamy 2008; Dadhich et al. 2015). Gusain et al. (2017) reported the successful application of two Cd-resistant bacterial strains, viz., *Dietzia maris* and *Lysinibacillus* sp., and their consortium for rhizoremediation, which increased the growth and yield of wheat cultivars. *Pseudomonas putida* a potential root colonizer has been reported to possess the potent potential for rhizoremediation of pollutants and also to promote growth in plants and control the pests biologically (Beneduzi et al. 2012; Meena et al. 2015a). Lacalle et al. (2018) reported rhizoremediation of calcareous soils mixed with metals and diesel contamination with *B. napus* combined with an organic amendment. Toxic metals were removed from the contaminated soil when the plant roots absorb the inorganic ions and translocate them to above-ground plant parts.

Soil microflora of rhizosphere zone play a crucial role in xenobiotic transformation and metabolism. The plant roots and associated microbes produce certain rhizodeposits such as exudates, secretions, plant mucilages, mucigel, root lysates, siderophores, phytochelatins, amino acids and ACC deaminase into the rhizosphere, that further promote the microbial activities into the surrounding soil (Fig. 10.7). Consequence to which the chemical condition of the rhizosphere differs from bulk soil. These organic substances play a pivotal role to promote the plant interactions with their environment and consequently stimulate the microbial growth in rhizosphere, which in turn stimulate the degradation of soil contaminants by plants and provide a healthier soil environment for plant growth. Rhizobacteria enhanced the uptake of certain metal by plants by increasing the dissolution of metals (Table 10.4). Delorme et al. (2001) and Peng et al. (2005) reported that soil acidification increased the metal ion mobility in *T. caerulescens* and *Elsholtzia*, respectively. In field experiment conditions, it was observed that the pH in the rhizosphere soil of the Cu-accumulating plant species (*Elsholtzia*) was significantly lower than in the bulk soil when plants were grown in Cu and other metal-contaminated soil (Peng et al. 2005). Blake et al. (1993) reported the conversion of mobile and toxic Cr VI to non-toxic and immobile Cr III with the reduction of environmental mobility of other toxic ions (Hg, Pb, Cd), by *Pseudomonas maltophilia*.

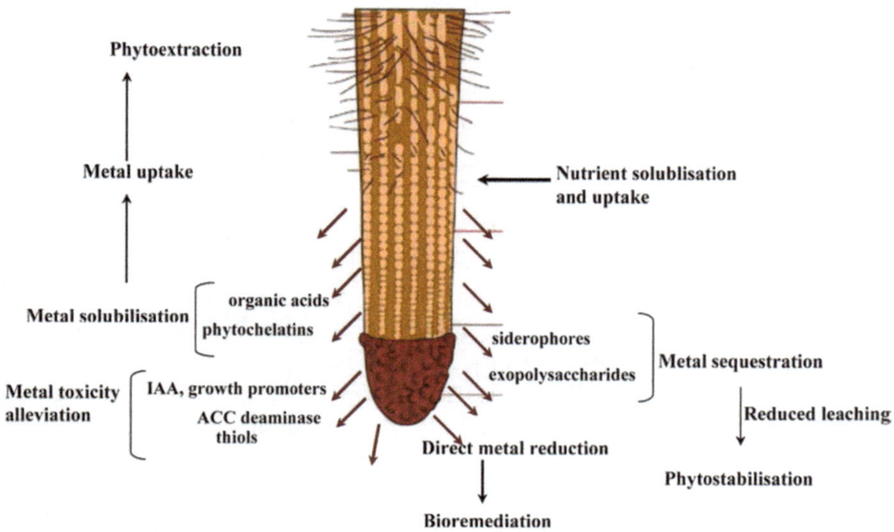

Fig. 10.7 Rhizodeposit production and their role in rhizoremediation. (Source: Kamaludeen and Ramasamy 2008)

Table 10.4 List of rhizobacteria reported to enhance the uptake of certain metal by plants

Bacteria	Plant	Heavy metals	References
Kluyvera ascorbata SUD165 *Kluyvera ascorbata* SUD165	Indian mustard, canola, tomato	Ni, Pb, Zn	Burd et al. (2000)
Pseudomonas sp.	Soybean, mungbean, wheat	Ni, Cd, Cr	Gupta et al. (2002)
Pseudomonas fluorescens	Soybean	Hg	Gupta et al. (2005)
Ochrobactrum intermedium	Sunflower	Cr (VI)	Faisal and Hasnain (2005)
Variovorax paradoxus, *Rhodococcus* sp., *Flavobacterium*	*Brassica juncea*	Cd	Belimov et al. (2005)
Bacillus subtilis SJ-101	*Brassica juncea*	Ni	Zaidi et al. (2006)
Xanthomonas sp. RJ3, *Azomonas* sp. RJ4, *Pseudomonas* sp. RJ10, *Bacillus* sp. RJ31	*Brassica napus*	Cd	Sheng and Xia (2006)
Pseudomonas sp., *Bacillus* sp.	Mustard	Cr (VI)	Rajkumar et al. (2010)
Ochrobactrum, *Bacillus cereus*	Mungbean	Cr (VI)	Faisal and Hasnain (2006)
Brevibacillus trifolium	Repens	Zn	Vivas et al. (2006)
Azotobacter chroococcum HKN-5, *Bacillus megaterium* HKP-1, *Bacillus mucilaginosus* HKK-1	*Brassica Juncea*	Pb, Zn	Wu et al. (2006)
Methylobacterium oryzae	*Lycopersicon*	Ni, Cd	Madhaiyan et al. (2007)
Pseudomonas aeruginosa KUCd1	Pumpkin, *Brassica juncea*	Cd	Sinha and Mukherjee (2008)
Pseudomonas aeruginosa MKRh3	*Vigna mungo*	Cd	Ganesan (2008)
Mesorhizobium sp. RC3	*Cicer arietinum*	Cr	Wani et al. (2009)
Serratia sp. SY5	*Zea mays*	Cd, Cu	Koo and Cho (2009)
Bacillus sp. PSB10	*Cicer arietinum*	Cr	Wani and Khan (2010)
Dietzia maris, *Lysinibacillus* sp.	Wheat	Cd	Gusain et al. (2017)

10.3.7 Riparian Buffers

Pesticides, herbicides, nutrients and other toxic pollutant from non-point sources like agricultural fields are the major threats for surface water bodies. People are now growing concerned over water quality issues due to increasing demand for safe water. Riparian buffers also known as riparian corridors/zones can be an effective control measure to avoid contamination of surface water and groundwater as well from non-point source (NPS) pollution. These are the vegetated areas growing with grasses, grass-like, forbs, shrubs, trees or other vegetation along the stream and protect adjacent water resources from NPS pollution. Riparian buffers control the erosion, filter the water, stabilize the bank and also provide habitat for aquatic and other wildlife. Shrubs, trees, and other vegetation absorbed the nutrients which come with the runoff from the agricultural and other operations that may otherwise

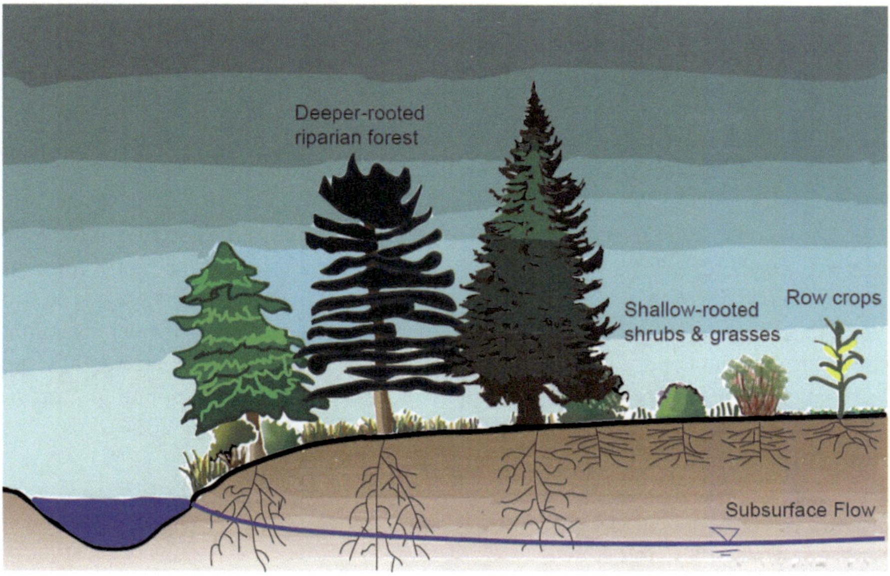

Fig. 10.8 Multispecies riparian buffers include trees, shrubs, and prairie grasses. (Source: Canning and Stillwell 2018)

be considered as the pollutant. Multispecies tree-shrub-grass riparian buffer systems are considered as efficient and cost-effective approaches to alleviate the agricultural pollution in heavily fertilized systems (Lin et al. 2004; Kumar et al. 2017a; Verma et al. 2015b). A multispecies riparian buffer consist of grasses, shrubs and tree where grasses clean and filter out contaminants from the surface runoff; the deep-rooted shrubs and trees uptake the nutrients and contaminants from groundwater levels (Fig. 10.8). The plants used in the riparian buffer follow the mechanism of phytohydraulics and change the hydrology of system. The root systems of riparian buffers' vegetation promote phytosequestration, rhizodegradation, phytoextraction, phytodegradation, and/or phytovolatilization, thus cleaning the contaminated site.

10.4 Some Case Studies of Phytoremediation

10.4.1 Mine Restoration (Salem County, New Jersey)

Acid mine restoration project, Quinton Township, Salem County, New Jersey, involved the plantation of swamp pink (*Helonias bullata*). Prior to restoration, the site was contaminated with acid-producing clays of mine, causing the pH of large soils areas less than 3.0. The restoration program involves the 59 acre mine area and the swamp pink plantation was removed and maintained during restoration by Rutgers University. The soil was treated with organic matter of mushroom compost and lime prior to replant. Native species and warmth season grasses were used in wetland parts and the upland areas, respectively (Bhakta et al. 2016).

10.4.2 Restoration of Sukinda Chromite Mines (SCM) Area of Orissa (India)

The site showed high levels of toxic hexavalent chromium. This was due to the flow of wastewater from chromium mines which exhibited a threat to the biotic community in the vicinity. A water hyacinth species *Eichhornia crassipes* was used to remediate the site. After the 15 days treatment of wastewater with water hyacinth, the chromium concentration in SCM water reduced. This was an experimental setup performed at laboratory scale to find out the efficiency of water hyacinth to remediate the site (Saha et al. 2017; Meena et al. 2017a; Ram and Meena 2014).

10.4.3 Remediation of Municipal Solid Waste (MSW) Using *Brassica juncea*

Municipal solid waste is huge problem to manage. To resolve this problem, a pot experiment conducted by growing hyper accumulator *Brassica juncea* variety 'Amulya' for a period of 3 months with various concentrations of MSW amends. The experiment was carried out to find out the applicability of phytoremediation of MSW for removal of heavy metals. The MSW used as a source of organic manure. The waste promoted the growth of plants and the study shown highly promising potential for removal of heavy metals such as Pb, Zn, Ni and Cu by phytoextraction. Among the four metals, the plants were shown better accumulation capacity over controls up to the last day of experiment (Namuduri et al. 2008; Sihag et al. 2015; Meena et al. 2017b).

10.5 Advantages of Phytoremediation

Phytoremediation is more economically viable, less disruptive to the environment and does not involve waiting for new plant communities to recolonize the site. It has the potential to treat sites polluted with more than one type of pollutant. It is more accepted by people as it is aesthetically pleasing. It avoids transport and excavation of polluted media, thus reducing the risk of spreading the contamination. The plant must be able to produce abundant biomass as plants store food in different parts (Anderson et al. 1998). The main advantage of phytostabilization is that it reduces the mobility and therefore the risk of contaminants without necessarily removing them from their source location (Suganthi et al. 2017; Verma et al. 2015a).

10.6 Disadvantages of Phytoremediation

It is dependent on favorable environmental conditions such as climate, geology, altitude, and temperature required by the plant for growth. Success is dependent on the plant's tolerance to the pollutant; contaminants absorbed by plant tissues are released back into the environment after death and decay. Contaminants collected in tissues when used as fuel pollute the indoor and outdoor environment. Contaminant solubility may increase the possibility of leaching. (Garbisu and Alkorta 2001; Meena et al. 2018b). Although current literature on phytoremediation is rather abundant, the physiological mechanisms that control and regulate the above processes are not yet elucidated, also because many of these are site-specific. The specificity depends first of all upon the kind of contaminant, its composition and possible transformation (evolution), and consequently, upon the vegetal species to be utilized for remediating a contaminated site. Further investigation, therefore, is needed in order to elucidate both the mechanisms involved and selection of plants, possibly applying genetic engineering to increase plant biomass. During phytoremediation process, the plant products can lead to bio-magnification.

10.7 Conclusion and Future Prospects

Phytoremediation is an important green technological method which uses plants of various genera for restoring ecosystem. Phytoremediation is known to be economically cost-effective and helpful to the environment that was once destroyed by mining activities, agriculture and logging industries and is applicable to any chemical pollutants in the environment. Contamination of soils with metal is a widespread problem around the globe and with different intensities in different regions because of the industrialization and mining of natural resources. Plants have a natural strength to detoxify, immobilize or eliminate contaminants during their growth by various biological processes. With the advancement in the field of genetic recombination technology, such transgenic plants can be instrumental in the phytoremediation approaches for remediate environment. Future studies should be focused on the phytoremediation approaches used in combination, with more than one technique for the successful remediation of the polluted area under field conditions and plants should be genetically engineered which grow rapidly. Remediation activities should always be economical and optimized, and the outcome should be balanced amid the benefits, risks, expenditure and feasibility. Some techniques should be employed for recovery of remediated contaminants so that they are recycled. The remediation measures of contaminated or restoring site should include regular monitoring of the soil acidity and nutrient status so that the necessary measures should be taken in order to prevent death of plants used. Microorganisms can easily adapt and grow in different conditions such as extreme environments with excessive salt, temperature, and oxygen and in anaerobic conditions in the presence of toxic compounds. Because of such quick adaptability, microorganisms can be used in site restoration with plants. If the phytoremediation techniques are worked with combination of microbes, the process remediates large contaminated sites.

References

Adriano DC, Wenzel WW, Vangronsveld J, Bolan NS (2004) Role of assisted natural remediation in environmental cleanup. Geoderma 122(2–4):121–142

Anand S, Kumar D, Bharti SK, Kumar N (2017) Phytotoxicity assessment of petrochemical industry effluent and phytoremediation potential of plants growing naturally at contamination site. Int J Green Herbal Chem 6(4):232–241

Anderson TA, Guthrie EA, Walton BT (1993) Bioremediation in the rhizosphere. Environ Sci Technol 27:2630–2636

Anderson CW, Brooks RR, Stewart RB, Simcock R (1998) Harvesting a crop of gold in plants. Nature 395(6702):553

Ashoka P, Meena RS, Kumar S, Yadav GS, Layek J (2017) Green nanotechnology is a key for eco-friendly agriculture. J Clean Prod 142:4440–4441

Auger C, Han S, Appanna VP, Thomas SC, Ulibarri G, Appanna VD (2013) Metabolic reengineering invoked by microbial systems to decontaminate aluminium: implications for bioremediation technologies. Biotechnol Adv 31(2):266–273

Barton C, Marx D, Adriano DC, Koo BJ, Newman L, Czapka S, Blake J (2005) Phytostabilization of a landfill containing coal combustion waste. Environ Geosci 12:251–265

Belimov AA, Hontzeas N, Safronova VI, Demchinskaya SV, Piluzza G, Bullitta S, Glick BR (2005) Cadmium-tolerant plant growth promoting rhizobacteria associated with the roots of Indian mustard (*Brassica juncea L. Czern.)*. Soil Biol Biochem 37:241–250

Beneduzi A, Ambrosini A, Passaglia LM (2012) Plant growth-promoting rhizobacteria (PGPR): their potential as antagonists and biocontrol agents. Genet Mol Biol 35:1044–1051

Bernhoft RA (2012) Mercury toxicity and treatment: a review of the literature. J Environ Public Health 2012:1–10

Bhakta JN, Jana BB, Lahiri S, Panigrahi A, Mandal SK, Rana S, Rokunuzzaman M et al (2016) Ecological restoration: an emerging eco- technology for sustainable environmental conservation. Int J Environ Tech Sci 2:26–30

Bhattacharyya R, Ghosh B, Mishra P, Mandal B, Rao C, Sarkar D, Das K, Anil K, Lalitha M, Hati K, Franzluebbers A (2015) Soil degradation in India: challenges and potential solutions. Sustainability 7(4):3528–3570

Bini C, Casaril S, Pavoni B (2000) Fertility gain and heavy metal accumulation in plants and soil, studied by means of a compost amended cultivation of *Taraxacum officinale*. Toxicol Environ Chem 77(3–4):131–142

Blake RC, Choate DM, Bardhan S, Revis N, Barton LL, Zocco TG (1993) Chemical transformation of toxic metals by a pseudomonas strain from a toxic waste site. Environ Toxicol Chem 12:1365–1376

Bolan NS, Park JH, Robinson B, Naidu R, Huh KY (2011) Phytostabilization: A green approach to contaminant containment. Adv Agron 112:145–204

Bolan NS, Choppala G, Kunhikrishnan A, Park JH, Naidu R (2013) Microbial transformation of trace elements in soils in relation to bioavailability and remediation. Rev Environ Contam Toxicol 225:1–56

Brooks RR (1998) Plants that hyperaccumulate heavy metals: their role in phytoremediation, microbiology, archaeology, mineral exploration, and phytomining. CAB International, Wallingford, p 380

Buragohain S, Sharma B, Nath JD, Gogaoi N, Meena RS, Lal R (2017) Impact of ten years of biofertilizer use on soil quality and rice yield on an inceptisol in Assam, India. Soil Res. https://doi.org/10.1071/SR17001

Burd GI, Dixon DG, Glick BR (2000) Plant growth promoting bacteria that decrease heavy metal toxicity in plants. Can J Microbiol 46:237–245

Canning JF, Stillwell AS (2018) Nutrient reduction in agricultural green infrastructure: an analysis of the Raccoon River Watershed. Water 10:749. https://doi.org/10.3390/w10060749

Central Pollution Control Board (CPCB) (2011) Hazardous metals and mineral pollution in India. A position paper. Indian National Science Academy, New Delhi

Cohen MF, Williams J, Yamasaki H (2002) Biodegradation of diesel fuel by an Azolla-derived bacterial consortium. J Environ Sci Health Part A 37(9):1593–1606

Dadhich RK, Meena RS (2014) Performance of Indian mustard (*Brassica juncea* L.) in response to foliar spray of thiourea and thioglycollic acid under different irrigation levels. Indian J Ecol 41(2):376–378

Dadhich RK, Meena RS, Reager ML, Kansotia BC (2015) Response of bio-regulators to yield and quality of Indian mustard (*Brassica juncea* L. Czernj. and Cosson) under different irrigation environments. J Appl Nat Sci 7(1):52–57

Datta R, Baraniya D, Wang YF, Kelkar A, Moulick A, Meena RS, Yadav GS, Ceccherini MT, Formanek P (2017a) Multi-function role as nutrient and scavenger off reeradical in soil. Sustain MDPI (9):402. https://doi.org/10.3390/su9081402

Datta R, Kelkar A, Baraniya D, Molaei A, Moulick A, Meena RS, Formanek P (2017b) Enzymatic degradation of lignin in soil: a review. Sustain MDPI 1163(9):1–18. https://doi.org/10.3390/su9071163

Delorme TA, Gagliardi JV, Angle JS, Chaney RL (2001) Influence of the zinc hyperaccumulator *Thlaspi caerulescens J.* and *C. Presl.* and the nonmetal accumulator *Trifolium pratense L.* on soil microbial populations. Can J Microbiol 47:773–776

Dhakal Y, Meena RS, De N, Verma SK, Singh A (2015) Growth, yield and nutrient content of mungbean (*Vigna radiata* L.) in response to INM in eastern Uttar Pradesh, India. Bangladesh J Bot 44(3):479–482

Dhakal Y, Meena RS, Kumar S (2016) Effect of INM on nodulation, yield, quality and available nutrient status in soil after harvest of green gram. Legum res 39(4):590–594

Ekman DR, Lorenz WW, Przybyla AE, Wolfe NL, Dean JF (2003) SAGE analysis of transcriptome responses in *Arabidopsis* roots exposed to 2,4,6-trinitrotoluene. Plant Physiol 133(3):1397–1406

EPA (2009) Technical/regulatory guidance – phytotechnology technical and regulatory guidance and decision trees, revised. http://www.itrcweb.org/Documents/PHYTO-3.pdf

European Environmental Agency (2007) Progress in management of contaminated sites (015). European Environment Agency, Copenhagen

Faisal M, Hasnain S (2005) Bacterial Cr (VI) reduction concurrently improves sunflower (*Helianthus annuus L.*) growth. Biotechnol Lett 27:943–947

Faisal M, Hasnain S (2006) Growth stimulatory effect of *Ochrobactrum intermedium* and *Bacillus cereus* on *Vigna radiata* plants. Lett Appl Microbiol 43:461–466

Felix H (1997) Field trials for in situ decontamination of heavy metal polluted soils using crops of metal accumulating plants. J Plant Nutr Soil Sci 160(4):525–529

Fuhrmann M, Lasat MM, Ebbs SD, Kochian LV, Cornish J (2002) Uptake of cesium-137 and strontium-90 from contaminated soil by three plant species; application to phytoremediation. J Environ Qual 31(3):904–909

Ganesan V (2008) Rhizoremediation of cadmium soil using a heavy metal resistant plant growth promoting *Rhizopseudomonad.* Curr Microbiol 56:403–407

Garbisu C, Alkorta I (2001) Phytoextraction: a cost-effective plant-based technology for the removal of metals from the environment. Bioresour Technol 77(3):229–236

Garg S (2017) Bioremediation of agricultural, municipal, and industrial wastes. In: Bhakta J (ed) Handbook of research on inventive bioremediation techniques. IGI Global, Hershey, pp 341–363

Ghani A (2010) Toxic effects of heavy metals on plant growth and metal accumulation in maize (*Zea mays L.*). Iranian J Toxicol 3(3):325–334

Giri K, Paliwal R, Suyal DC, Mishra G, Pandey S, Verma PK, Rai JPN (2015) Potential application of plant-microbe interaction for restoration of degraded ecosystems. In: Singh SO, Shriwastav K (eds) Handbook of research on uncovering new methods for ecosystem management through bioremediation. IGI Global, Hershey, pp 255–285

Godsy EM, Warren E, Paganelli VV (2003) The role of microbial reductive dechlorination of TCE at a phytoremediation site. Int J Phytoremediation 5:73–88

Guerinot ML, Salt DE (2001) Fortified foods and phytoremediation: two sides of the same coin. Plant Physiol 125:164–167

Gupta A, Meyer JM, Goel R (2002) Development of heavy metal resistant mutants of phosphate solubilizing *Pseudomonas* sp. NBRI4014 and their characterization. Curr Microbiol 45:323–332

Gupta A, Rai V, Bagdwal N, Goel R (2005) In situ characterization of mercury resistant growth promoting fluorescent *Pseudomonads*. Microbiol Res 160:385–388

Gusain P, Paliwal R, Singh V (2017) Rhizoremediation of cadmium-contaminated soil associated with hydroxamate siderophores isolated from Cd resistant plant growth–promoting *Dietzia maris* and *Lysinibacillus strains*. Int J Phytoremediation 19(3):290–299

Hadi F, Bano A, Fuller MP (2010) The improved phytoextraction of lead (Pb) and the growth of maize (*Zea mays L.*): the role of plant growth regulators (GA3 and IAA) and EDTA alone and in combinations. Chemosphere 80:457–462

Hashmi MZ, Kumar V, Varma A (eds) (2017) Xenobiotics in the soil environment: monitoring, toxicity and management. Springer, Cham

ITRC (2001) Phytotechnology technical and regulatory guidance document, interstate technology and regulatory cooperation, Costa Rica. Accessed at: www.itrcweb.org (12 October 2018)

Jomova K, Jenisova Z, Feszterova M, Baros S, Liska J, Hudecova D, Rhodes Valko M (2011) Arsenic: toxicity, oxidative stress and human disease. J Appl Toxicol 31:95–107

Kamaludeen SPB, Ramasamy K (2008) Rhizoremediation of metals: harnessing microbial communities. Indian J Microbiol 48(1):80–88

Kamnev AA, Lelie D (2000) Chemical and biological parameters as tools to evaluate and improve heavy metal phytoremediation. Biosci Rep 20:239–258

Karlen DL, Rice CW (2015) Soil degradation: will humankind ever learn? Sustainability 7:12490–12501

Kavamura VN, Esposito E (2010) Biotechnological strategies applied to the decontamination of soils polluted with heavy metals. Biotechnol Adv 28:61–69

Kim S, Lim H, Lee I (2010) Enhanced heavy metal phytoextraction by *Echinochloa crus-galli* using root exudates. J Biosci Bioeng 109:47–50

Koo SY, Cho KS (2009) Isolation and characterization of plant growth-promoting rhizobacterium, Serratia sp., SY5. J Microbiol Biotechnol 19:1431–1438

Koptsik GN, Koptsik SV, Smirnova IE (2014) Efficiency of remediation of technogenic barrens around the Pechenganikel works in the Kola subarctic. Eurasian J Soil Sci 47:519–528

Kumar S, Meena RS, Pandey A, Seema (2017a) Soil acidity management and an economics response of lime and sulfur on sesame in an alley cropping system. Int J Curr Microb Appl Sci 6(3):2566–2573

Kumar S, Meena RS, Yadav GS, Pandey A (2017b) Response of sesame (*Sesamum indicum* L.) to sulphur and lime application under soil acidity. Int J Plant Soil Sci 14(4):1–9

Kumar S, Meena RS, Bohra JS (2018) Interactive effect of sowing dates and nutrient sources on dry matter accumulation of Indian mustard (*Brassica juncea* L.). J Oilseed Brassica 9(1):72–76

Lacalle RG, Gómez-Sagasti MT, Artetxe U, Garbisu C, Becerril JM (2018) *Brassica napus* has a key role in the recovery of the health of soils contaminated with metals and diesel by rhizoremediation. Sci Total Environ 618:347–356

Lee JH (2013) An overview of phytoremediation as a potentially promising technology for environmental pollution control. Biotechnol Bioprocess Eng 18:431–439

Li X, Liu L, Wang Y, Luo G, Chen X, Yang X, Hall M, Guo R, Wang H, Cui J, He X (2013) Heavy metal contamination of urban soil in an old industrial city (Shenyang) in Northeast China. Geoderma 192:50–58

Lin CH, Lerch RN, Garrett HE, George MF (2004) Incorporating forage grasses in riparian buffers for bioremediation of atrazine, isoxaflutole and nitrate in Missouri. Agrofor Syst 63:91–99

Madhaiyan N, Poonguzhali S, Sa T (2007) Metal tolerating methylotropic bacteria reduces nickel and cadmium toxicity and promotes plant growth of tomato (*Lycopersicon esculentum L.*). Chemosphere 69:220–228

Mahar A, Wang P, Ali A, Awasthi MK, Lahori AH, Wang Q, Li R, Zhang Z (2016) Challenges and opportunities in the phytoremediation of heavy metals contaminated soils: a review. Ecotoxicol Environ Saf 1126:111–121

McCutcheon SC, Schnoor JL (2003) Overview of phytotransformation and control of wastes. In: Phytoremediation: transformation and control of contaminants, vol 19. Wiley-Interscience, Hoboken, p 358

Meena H, Meena RS (2017) Assessment of sowing environments and bio-regulators as adaptation choice for clusterbean productivity in response to current climatic scenario. Bangladesh J Bot 46(1):241–244

Meena RS, Yadav RS (2014) Phonological performance of groundnut varieties under sowing environments in hyper arid zone of Rajasthan, India. J Appl Nat Sci 6(2):344–348

Meena RS, Yadav RS (2015) Yield and profitability of groundnut *(Arachis hypogaea* L) as influenced by sowing dates and nutrient levels with different varieties. Legum Res 38(6):791–797

Meena RS, Yadav RS, Meena VS (2014) Response of groundnut (*Arachis hypogaea* L.) varieties to sowing dates and NP fertilizers under Western dry zone of India. Bangladesh J Bot 43(2):169–173

Meena RS, Dhakal Y, Bohra JS, Singh SP, Singh MK, Sanodiya P (2015a) Influence of bioinorganic combinations on yield, quality and economics of Mungbean. Am J Exp Agric 8(3):159–166

Meena RS, Meena VS, Meena SK, Verma JP (2015b) The needs of healthy soils for a healthy world. J Clean Prod 102:560–561

Meena RS, Meena VS, Meena SK, Verma JP (2015c) Towards the plant stress mitigate the agricultural productivity: a book review. J Clean Prod 102:552–553

Meena RS, Yadav RS, Meena H, Kumar S, Meena YK, Singh A (2015d) Towards the current need to enhance legume productivity and soil sustainability worldwide: a book review. J Clean Prod 104:513–515

Meena RS, Yadav RS, Reager ML, De N, Meena VS, Verma JP, Verma SK, Kansotia BC (2015e) Temperature use efficiency and yield of groundnut varieties in response to sowing dates and fertility levels in Western dry zone of India. Am J Exp Agric 7(3):170–177

Meena H, Meena RS, Singh B, Kumar S (2016a) Response of bio-regulators to morphology and yield of clusterbean [*Cyamopsis tetragonoloba* (L.) Taub.] under different sowing environments. J Appl Nat Sci 8(2):715–718

Meena RS, Bohra JS, Singh SP, Meena VS, Verma JP, Verma SK, Shiiag SK (2016b) Towards the prime response of manure to enhance nutrient use efficiency and soil sustainability a current need: a book review. J Clean Prod 112:1258–1260

Meena RS, Gogaoi N, Kumar S (2017a) Alarming issues on agricultural crop production and environmental stresses. J Clean Prod 142:3357–3359

Meena RS, Kumar S, Pandey A (2017b) Response of sulfur and lime levels on productivity, nutrient content and uptake of sesame under guava (*Psidium guajava* L.) based Agri-horti system in an acidic soil of eastern Uttar Pradesh, India. J Crop Weed 13(2):222–227

Meena RS, Meena PD, Yadav GS, Yadav SS (2017c) Phosphate solubilizing microorganisms, principles and application of microphos technology. J Clean Prod 145:157–158

Meena H, Meena RS, Lal R, Singh GS, Mitran T, Layek J, Patil SB, Kumar S, Verma T (2018a) Response of sowing dates and bio regulators on yield of clusterbean under current climate in alley cropping system in eastern U.P., Indian. Legum Res 41(4):563–571

Meena RS, Kumar V, Yadav GS, Mitran T (2018b) Response and interaction of *Bradyrhizobium japonicum* and *Arbuscular mycorrhizal* fungi in the soybean rhizosphere: A review. Plant Growth Regul 84:207–223

Mhatre GN, Pankhurst CE (1997) Bioindicators to detect contamination of soils with special reference to heavy metals. Report number. BIOSIS/98/05352

MoEF (2011) A state-of-the-art report on bioremediation, its applications to contaminated sites in India. Ministry of environment & forests government of India. http://www.moef.nic.in

Namuduri S, Kumar SK, Srksbl N, Balaram V, Rao TS (2008) Phytoremediation potential of *Brassica juncea* for municipal solid waste-a case study. Fourth international symposium on recent advances in environmental health research, poster session B

Newman LA, Reynolds CM (2004) Phytodegradation of organic compounds. Curr Opin Biotechnol 15:225–230

Onat B, Sahin UA, Akyuz T (2013) Elemental characterization of PM 2.5 and PM1 in dense traffic area in Istanbul, Turkey. Atmos Pollut Res 4:101–105

Padmavathiamma PK, Li LY (2007) Phytoremediation technology: hyper-accumulation metals in plants. Water Air Soil Pollut 184:105–126

Pal R, Rai JPN (2010) Phytochelatins: peptides involved in heavy metal detoxification. Biotechnol Appl Biochem 160:945–963

Paterson E (2003) Importance of rhizodeposition in the coupling of plant and microbial productivity. Eur J Soil Sci 54:741–750

Pérez-Sirvent C, Hernández-Pérez C, Martínez-Sánchez MJ, García-Lorenzo ML, Bech J (2017) Metal uptake by wetland plants: implications for phytoremediation and restoration. J Soils Sediments 17:1384–1393

Peng H-Y, Yang X-E, Jiang L-Y, He Z-L (2005) Copper phytoavailability and uptake by Elsh{o} ltzia splendens from contaminated soil as affected by soil amendments. J Environ Sci Health, Part A 40(4):839–856

Phieler R, Voit A, Kothe E (2013) In: Geobiotechnol I (ed) Microbially supported phytoremediation of heavy metal contaminated soils: strategies and applications. Springer, Berlin/ Heidelberg, pp 211–235

Pulford ID, Watson C (2003) Phytoremediation of heavy metal-contaminated land by trees-a review. Environ Int 29:529–540

Radziemska M, Vaverková MD, Baryła A (2017) Phytostabilization—management strategy for stabilizing trace elements in contaminated soils. Int J Environ Res Public Health 14:958

Rajkumar M, Ae N, Prasad MNV, Freitas H (2010) Potential of siderophore-producing bacteria for improving heavy metal phytoextraction. Trends Biotechnol 28:142–149

Ram K, Meena RS (2014) Evaluation of pearl millet and mungbean intercropping systems in arid region of Rajasthan (India). Bangladesh J Bot 43(3):367–370

Rani N, Sharma HR, Kaushik A, Sagar A (2018) Bioremediation of mined waste land. In: Handbook of environmental materials management. Springer, Cham, pp 1–25

Renella G, Egamberdiyeva D, Landi L, Mench M, Nannipieri P (2006) Soil microbial activity and hydrolase activity during decomposition of model root exudates released by a model root surface in cd-contaminated soils. Soil Biol Biochem 38:702–708

Robson DB, Knight JD, Farrell RE, Germida JJ (2003) Ability of cold tolerant plants to grow in hydrocarbon-contaminated soil. Int J Phytoremediation 5:105–124

Rungwa S, Arpa G, Sakulas H, Harakuwe A, Timi D (2013) Phytoremediation–an eco-friendly and sustainable method of heavy metal removal from closed mine environments in *Papua New Guinea*. Procedia Earth Planet 6:269–277

Saha P, Shinde O, Sarkar S (2017) Phytoremediation of industrial mines wastewater using water hyacinth. Int J Phytoremediation 19:87–96

Sahu HB, Dash S (2011) Land degradation due to mining in India and its mitigation measures. In: 2nd international conference on environmental science and technology IPCBEE, vol 6. IACSIT Press, Singapore, pp 1–5

Sharma S, Singh B, Manchanda VK (2015) Phytoremediation: role of terrestrial plants and aquatic macrophytes in the remediation of radionuclides and heavy metal contaminated soil and water. Environ Sci Pollut Res Int 22:946–962
Sheng XF, Xia JJ (2006) Improvement of rape (*Brassica napus*) plant growth and cadmium uptake by cadmium-resistant bacteria. Chemosphere 64:1036–1042
Sihag SK, Singh MK, Meena RS, Naga S, Bahadur SR, Gaurav YRS (2015) Influences of spacing on growth and yield potential of dry direct seeded rice (*Oryza sativa* L) cultivars. Ecoscan 9(1–2):517–519
Singh JS, Singh SP, Gupta SR (2015) Ecology, environment and resource conservation. S. Chand Publishing, New Delhi
Sinha S, Mukherjee SK (2008) Cadmium-induced siderophore production by a high Cd-resistant bacterial strain relieved Cd toxicity in plants through root colonization. Curr Microbiol 56:55–60
Suganthi M, Muthukrishnan P, Chinnusamy C, Gopi H (2017) Phytostabilisation for contaminant containment–review. http://krishikosh.egranth.ac.in/handle/1/5810030189
Thayaparan M, Iqbal SS, Chathuranga PKD, Iqbal MCM (2013) Rhizofiltration of Pb by *Azolla pinnata*. Int J Environ Sci 3:1811
Tomé FV, Rodríguez PB, Lozano JC (2008) Elimination of natural uranium and 226Ra from contaminated waters by rhizofiltration using *Helianthus annuus L.* Sci Total Environ 393:351–357
Tripathi S, Bhattacharyya P, Mohapatra R, Som A, Chowdhury D (2014) Influence of different fractions of heavy metals on microbial ecophysiological indicators and enzyme activities in century old municipal solid waste amended soil. Ecol Eng 70:25–34
UNEP (1992) World atlas of desertification. Publ E Arnold, London
Van Aken B (2009) Transgenic plants for enhanced phytoremediation of toxic explosives. Curr Opin Biotechnol 20:231–236
Van Aken B, Yoon JM, Schnoor JL (2004) Biodegradation of nitrosubstituted explosives 2,4,6-trinitrotoluene, hexahydro-1,3,5- trinitro-1,3,5-triazine, and octahydro-1,3,5,7-tetranitro-1,3,5-tetrazocine by a phytosymbiotic Methylobacterium sp. associated with poplar tissues (*Populus deltoides x nigra* DN34). Appl Environ Microbiol 70:508–517
Varma D, Meena RS, Kumar S (2017a) Response of mungbean to fertility and lime levels under soil acidity in an alley cropping system in Vindhyan region, India. Int J Chem Stu 5(2):384–389
Varma D, Meena RS, Kumar S, Kumar E (2017b) Response of mungbean to NPK and lime under the conditions of Vindhyan region of Uttar Pradesh. Legum Res 40(3):542–545
Vangronsveld J, Meers E, Dejonghe W, Geurds M, Diels L, Defoort B, Beeckman E, Smis J (2007) Phytoremediation for heavy metal contaminated soil and combined bio-energy production. In: Zu Y, Lepp N, Naidu R (eds) Biogeochemistry of trace elements: environmental protection, remediation and human health. Proc. IX ICOBTE, Beijing, pp 162–163
Verma JP, Jaiswal DK, Meena VS, Meena RS (2015a) Current need of organic farming for enhancing sustainable agriculture. J Clean Prod 102:545–547
Verma JP, Meena VS, Kumar A, Meena RS (2015b) Issues and challenges about sustainable agriculture production for management of natural resources to sustain soil fertility and health: a book review. J Clean Prod 107:793–794
Verma SK, Singh SB, Prasad SK, Meena RN, Meena RS (2015c) Influence of irrigation regimes and weed management practices on water use and nutrient uptake in wheat (*Triticum aestivum* L. Emend. Fiori and Paol.). Bangladesh J Bot 44(3):437–442
Vidali M (2001) Bioremediation: an overview. Pure Appl Chem 73:1163–1172
Vivas A, Biro B, Ruiz-Lozano JM, Barea JM, Azcon R (2006) Two bacterial strains isolated from a Zn-polluted soil enhance plant growth and mycorrhizal efficiency under Zn toxicity. Chemosphere 52:1523–1533
Wagner G (1993) Large-scale screening of heavy metal burdens in higher plants. In: Plants as biomonitors, pp 425–443
Wani PA, Khan MS (2010) *Bacillus* species enhance growth parameters of chickpea (*Cicer arietinum L.*) in chromium stressed soils. Food Chem Toxicol 48:3262–3267

Wani PA, Zaidi A, Khan MS (2009) Chromium reducing and plant growth promoting potential of Mesorhizobium species under chromium stress. Biorem J 13:121–129

World Resources Institute (2014) Creating a sustainable food future, Report 2013–2014: interim findings. World Resources Institute, Washington, DC

Wu CH, Wood TK, Mulchandani A, Chen W (2006) Engineering plant-microbe symbiosis for rhizoremediation of heavy metals. Appl Environ Microbiol 72:1129–1134

Wyszkowska J, Kucharski M, Kucharski J, Borowik A (2009) Activity of dehydrogenases, catalase and urease in copper polluted soil. J Elem 14:605–617

Yadav GS, Babu S, Meena RS, Debnath C, Saha P, Debbaram C, Datta M (2017a) Effects of godawariphosgold and single supper phosphate on groundnut (*Arachis hypogaea*) productivity, phosphorus uptake, phosphorus use efficiency and economics. Indian J Agric Sci 87(9):1165–1169

Yadav GS, Lal R, Meena RS, Babu S, Das A, Bhomik SN, Datta M, Layak J, Saha P (2017b) Conservation tillage and nutrient management effects on productivity and soil carbon sequestration under double cropping of rice in North Eastern region of India. Ecol Indic. http://www.sciencedirect.com/science/article/pii/S1470160X17305617

Yadav GS, Lal R, Meena RS, Datta M, Babu S, Das LJ, Saha P (2017c) Energy budgeting for designing sustainable and environmentally clean/safer cropping systems for rainfed rice fallow lands in India. J Clean Prod 158:29–37

Yadav GS, Das A, Lal R, Babu S, Meena RS, Saha P, Singh R, Datta M (2018a) Energy budget and carbon footprint in a no-till and mulch based rice–mustard cropping system. J Clean Prod 191:144–157

Yadav GS, Das A, Lal R, Babu S, Meena RS, Patil SB, Saha P, Datta M (2018b) Conservation tillage and mulching effects on the adaptive capacity of direct-seeded upland rice (*Oryza sativa* L.) to alleviate weed and moisture stresses in the North Eastern Himalayan region of India. Arch Agron Soil Sci. https://doi.org/10.1080/03650340.2018.1423555

Yang Y, Ge Y, Zeng H, Zhou X, Peng L, Zeng Q (2017) Phytoextraction of cadmium-contaminated soil and potential of regenerated tobacco biomass for recovery of cadmium. Sci Rep 7:7210

Zaidi S, Usmani S, Singh BR, Musarrat J (2006) Significance of Bacillus subtilis strain SJ-101 as a bioinoculant for concurrent plant growth promotion and nickel accumulation in Brassica juncea. Chemosphere 64(6):991–997